WITHDRAWN
UTSA Libraries

Historia

Transformations: Studies in the History of Science and Technology
Jed Z. Buchwald, general editor

Mordechai Feingold, editor, *Jesuit Science and the Republic of Letters*

Sungook Hong, *Wireless: From Marconi's Black-Box to the Audion*

Myles W. Jackson, *Spectrum of Belief: Joseph von Fraunhofer and the Craft of Precision Optics*

Mi Gyung Kim, *Affinity, That Elusive Dream: A Genealogy of the Chemical Revolution*

Janis Langins, *Conserving the Enlightenment: French Military Engineering from Vauban to the Revolution*

Wolfgang Lefèvre, editor, *Picturing Machines 1400–1700*

William R. Newman and Anthony Grafton, editors, *Secrets of Nature: Astrology and Alchemy in Early Modern Europe*

Gianna Pomata and Nancy G. Siraisi, editors, *Historia: Empiricism and Erudition in Early Modern Europe*

Alan J. Rocke, *Nationalizing Science: Adolphe Wurtz and the Battle for French Chemistry*

Historia
Empiricism and Erudition in Early Modern Europe

Gianna Pomata and Nancy G. Siraisi, editors

The MIT Press
Cambridge, Massachusetts
London, England

© 2005 Massachusetts Institute of Technology

All rights reserved. No part of this book may be reproduced in any form by any electronic or mechanical means (including photocopying, recording, or information storage and retrieval) without permission in writing from the publisher.

MIT Press books may be purchased at special quantity discounts for business or sales promotional use. For information, please email special_sales@mitpress.mit.edu or write to Special Sales Department, The MIT Press, 55 Hayward Street, Cambridge, MA 02142.

This book was set in Sabon by SNP Best-set Typesetter Ltd., Hong Kong, and was printed and bound in the United States of America.

Library of Congress Cataloging-in-Publication Data

Historia : empiricism and erudition in early modern Europe / Gianna Pomata and Nancy G. Siraisi, editors.
  p. cm — (Transformations)
 Includes bibliographical references and index.
 ISBN 0-262-16229-6 (alk. paper)
 1. Natural history—Europe—History. 2. Medicine—Europe—History.
 I. Pomata, Gianna. II. Siraisi, Nancy G. III. Transformations (M.I.T. Press)

QH135.H57 2005
508'.094—dc22

2005041560

Contents

Acknowledgments   vii

   Introduction   1
   Gianna Pomata and Nancy G. Siraisi

I   The Ascending Fortunes of *Historia*   39

1   The Identities of History in Early Modern Europe: Prelude to a Study of the *Artes Historicae*   41
   Anthony Grafton

2   Natural History, Ethics, and Physico-Theology   75
   Brian W. Ogilvie

3   *Praxis Historialis*: The Uses of *Historia* in Early Modern Medicine   105
   Gianna Pomata

4   White Crows, Graying Hair, and Eyelashes: Problems for Natural Historians in the Reception of Aristotelian Logic and Biology from Pomponazzi to Bacon   147
   Ian Maclean

5   Antiquarianism and Idolatry: The *Historia* of Religions in the Seventeenth Century   181
   Martin Mulsow

6   Between History and System   211
   Donald R. Kelley

II The Working Practices of Learned Empiricism   239

7 Conrad Gessner and the Historical Depth of Renaissance Natural History   241
   Laurent Pinon

8 *Historia* in Zwinger's *Theatrum humanae vitae*   269
   Ann Blair

9 Histories, Stories, *Exempla*, and Anecdotes: Michele Savonarola from Latin to Vernacular   297
   Chiara Crisciani

10 *Historiae*, Natural History, Roman Antiquity, and Some Roman Physicians   325
    Nancy G. Siraisi

11 Description Terminable and Interminable: Looking at the Past, Nature, and Peoples in Peiresc's Archive   355
    Peter N. Miller

Bibliography   399
Contributors   473
Index   475

Acknowledgments

The genesis of this volume was the workshop "*Historia*: Explorations in the History of Early Modern Empiricism" held at the Max Planck Institut für Wissenschaftsgeschichte in Berlin, 1–30 June 2003. We wish to express our gratitude to the Institute for its unusually generous sponsorship, which made the workshop and the subsequent volume possible and greatly facilitated our work as organizers and editors. The Institute's financial support and hospitality gave us the resources to convene an international group of scholars for a month—an invaluable asset when the goal is, as it was in our case, to bring together scholars from different disciplines, not to mention different languages and intellectual traditions. A month of scholarly cohabitation at the Institute allowed us to pursue the cross-disciplinary goals of our project in an intensively interactive way. Following the practice of working groups at the Max Planck Institute, papers were precirculated in draft to facilitate discussion during the first week of our meeting in Berlin, when each paper was debated in its own terms and with a view to general unifying themes. Contributors then enjoyed the Institute's gracious hospitality for two weeks devoted to revising the papers in the light of group discussion, and to completing new research as necessary, using the Institute's excellent research library service. Our month together ended with another week of renewed general discussion of the revised papers, with particular attention to the shared themes to be addressed by the editors in the introduction. The Institute's hospitality greatly facilitated our work in planning, preparing, and editing a collection of individually authored essays that are also, we hope, genuinely related to one another.

While we gratefully acknowledge the Institute's support, we particularly wish to thank its director, Lorraine Daston, for her intellectual contribution to our work. She was a driving force of the project from beginning to end. Hers was the idea of recruiting an international research group on the early modern *historia*, and in inviting us to organize the workshop she provided many useful suggestions and insights that were crucial for our original framing of the project. We would like to thank her, moreover, for her active participation in our Berlin workshop, which included an incisive summing-up of our final discussion.

Finally, we gratefully acknowledge the Institute's support in assigning us a research assistant, Fabian Krämer, who greatly helped with the editorial stage of our task. We thank him for his diligent work in identifying sources for the introduction and unifying the book's bibliography. We would also like to thank Elizabeth McCahill for help in preparing the manuscript for submission.

We are most grateful to Jed Buchwald, general editor of the Transformations series at the MIT Press, for his interest in our project and his helpful and expert advice for the completion of the volume. Thanks are also due to Matthew Abbate and Sara Meirowitz of the MIT Press for their careful editorial work on the manuscript, and to Janet Abbate for meticulous and thoughtful copyediting.

Last but not least, we would like to thank all the contributors to the volume for their enthusiastic sharing of our interest in *historia*, and their willingness to move beyond individual interests to join a group exploration of the intricacies of the early modern intellectual landscape. It was a great pleasure to share with such distinguished practitioners the challenges and the delights of *historia*'s time-honored craft.

Gianna Pomata and Nancy G. Siraisi

Historia

Introduction

Gianna Pomata and Nancy G. Siraisi

> Il y aurait un beau livre à écrire sur le mot d'*histoire* et ses variations de sens au cours des périodes successives... de l'histoire.
> —Lucien Febvre (1936) reviewing Karl Keuck, *Historia. Geschichte des Wortes und seiner Bedeutungen in der Antike und in den romanischen Sprachen*, 1934)[1]

Historia before Historicism

Writing about the prominence of natural history in the seventeenth century, such a deep connoisseur of the early modern "life sciences" as Jacques Roger noted that "the history of the word *historia* itself would deserve an accurate study."[2] The present book offers a contribution in that direction by bringing together several essays that investigate, in a collaborative manner, what *historia* meant in the early modern vocabulary of knowledge and the roles that *historia* played in early modern encyclopedism. The starting point of our inquiry is the ubiquity of *historia* in early modern learning—the fact that it featured prominently in a wide array of disciplines ranging from antiquarian studies and civil history to medicine and natural philosophy. This book is in fact the result of a workshop, sponsored by the Max Planck Institut für Wissenschaftsgeschichte in Berlin, where scholars working in these different fields (from the history of the medical and natural sciences to that of scholarship and historiography) convened with a common focus on *historia* as a key epistemic tool of early modern intellectual practices.[3]

Bringing together scholars from such disparate backgrounds was indeed indispensable to our project, because in striking contrast to the modern use of the term "history," the early modern *historia* straddled the distinction between human and natural subjects, embracing accounts of objects in the natural world as well as the record of human action. One may say, in fact, that from the early Renaissance to the eighteenth century, nature was fully part of the field of research called *historia*. Most studies of early modern intellectual life, however, have focused exclusively on *historia* as civil history, the sense closest to modern usage. A formidable body of scholarship has been devoted to humanist and Renaissance writings on *historia* as the record of the human past. Well known and much studied are the humanists' reevaluation of the rhetorical and moral uses of political and civil history and the new attention to the reading and writing of history signaled by the new genre of the *artes historicae*.[4] Other studies have traced such indications of the increasing significance attached to *historia* in the early modern period as the emergence of disciplinary histories, the new genre of confessional history, and the institution of chairs or lectureships of history in universities and academies.[5]

By contrast, the role *historia* played in early modern natural philosophy and medicine has been all but ignored, though there is plenty of evidence of the new significance that *historia*, in several variants and genres, also had in these fields. It is well known that a prominent feature of early modern science was the rapid and exponential growth of *historia naturalis*, but it would be more correct to talk of a proliferation of natural histories—in the plural—all with different philosophical pedigrees and correspondingly different notions of what *historia* was about. Aristotelian and Plinian models of natural history, to give the most obvious example, implied very different concepts of *historia*.[6] In medicine also—a discipline whose ancient connections with history were noted long ago by Arnaldo Momigliano[7]—*historia* enjoyed a special vogue. A veritable explosion of clinical and anatomical reports written in the *historia* format began in the late sixteenth century, to proliferate even further in the following century. Case histories (called *historiae* or *observationes*) and autopsy narratives multiplied, while in anatomy *historia* acquired the status of a specific, even technical term. When Fabricius of Acquapendente or William Harvey, for instance, wrote the results of their anatomical investigations, they regularly started with what they called a *historia*, meaning a thorough description of the structure of bodily parts, preliminary to an understanding of their function, or "use."[8] Next to the anatomists' *historia*, we also find in late sixteenth-century medical literature a new acceptation of *historia* as *casus*, the report of an individual case history, a usage circulated by Renaissance commentators on Hippocrates' *Epidemics*. Thus both in natural history and in medicine *historia* was an increasingly significant epistemic tool, covering different intellectual objectives but with an overall reference to the knowledge and description of particulars (including description from direct observation).

Was there any link between these new uses of *historia* by early modern physicians and naturalists and the Renaissance intensive discussion and reappraisal of *historia* as antiquarian knowledge? This question has been mostly evaded so far—not surprisingly, since until recently the history of historiography and the history of the natural sciences (including medicine) have mostly proceeded on distinct and separate paths.[9] The reasons for this separation are not difficult to surmise. A key factor has certainly been the increasing disjunction of classical studies and scientific training in the European educational system since the nineteenth century—a disjunction that reached momentous proportions in the challenge to the positivist view of the unity of science posited by the rise of historicism. At the turn of the nineteenth century and in the first half of the twentieth, the attempt to give an autonomous epistemological foundation to the historical sciences, undertaken by the founding fathers of neo-Kantian historicism—Windelband, Rickert, Dilthey—led to a sharp distinction between the "generalizing" (or nomothetic) natural sciences and the "individualizing" (or idiographic) historical sciences.[10] Whether the distinction between the two was defined as ontological (as in Dilthey) or logical and conceptual (as in Windelband and Rickert), the epistemology of the historical sciences was seen as fundamentally different from that of the natural sciences. It is not surprising, therefore, that the histories of the two groups of sciences were also perceived as largely unrelated—two autonomous, separate intellectual developments. Moreover, the conventional view of the origins of *Historismus*, as pioneered by Friedrich Meinecke, placed the rise of historical consciousness and method in the eighteenth century—a chronology totally unrelated to the scientific revolution.[11] Later studies of the development of historical method moved back the chronology, recognizing that already in the sixteenth century there was a significant move from the rhetorical or literary concept of *historia* to a new sensitivity to issues of method and source criticism, especially in the French *artes historicae*. But the significant connection drawn here was with jurisprudence—the spilling over into history writing of the critical methods of the jurists' *mos gallicus*—not with the natural sciences.[12] Though Arnaldo Momigliano long ago pointed out the similarities between the mental outlook of the early modern antiquarians and that of contemporary naturalists, his insight did not lead until recently to further studies in that direction, partly also, perhaps, because he himself saw antiquarianism as an intellectual tradition largely apart from historiography.[13] The underlying assumption that went unchallenged was that the humanist *historia* of the Renaissance was basically a literary genre, not an epistemic genre.

A significant breakthrough from this view was Arno Seifert's survey of early modern definitions of *historia* in a seminal book published more than a quarter of a century ago.[14] Combining historical semantics with traditional history of ideas, Seifert reconstructed in detail the *Begriffsgeschichte* of *historia* as a term of learned language, establishing beyond doubt that in the early modern period, in addition to the literary genre of *narratio rerum gestarum*, *historia* indicated primarily a *modus cognoscendi*—a cognitive category. Seifert argued that the epistemological status of *historia* underwent a momentous change in the transition from Scholasticism to humanism. In Scholastic philosophy, *historia* had meant incomplete knowledge—the Aristotelian *apodeixis tou oti* (or Scholastic *demonstratio quia*)—that simply offered a description of "how the thing is" without explaining why it is so. *Historia* in this sense was ancillary to philosophy, the humble prelude to the philosophical knowledge of causes. In contrast, the humanists rediscovered the ancient Greek (pre-Aristotelian) usage of the word as knowledge in general, not limited to *res gestae* but fully including nature—an intellectual turning point stressed by Seifert as the premise to *historia*'s triumph in Renaissance intellectual life. Reviewing the various meanings attributed to *cognitio historica* from the humanists to Kant, Seifert showed that starting with the Renaissance emphasis on *historia* as *vera narratio*, the term came to be associated with a cluster of concepts all variously related to descriptive, nondemonstrative knowledge (*cognitio quod est*), knowledge based on sense perception (*sensata cognitio*), and knowledge of particulars (*cognitio singularium*). This trend culminated in the

Baconian identification of *historia* and *experientia*, which gave the *historia* of nature an unprecedented significance as the foundation of a true natural philosophy. *Historia* kept this association with the semantic field of factual knowledge (*nuda facti notitia*) well into the philosophical language of the eighteenth century, as finally exemplified by Kant's definition of "historical knowledge" as "cognitio ex datis." Thus for about two hundred years, Seifert argued, *historia* came to signify in learned language the general field of "prescientific empirical knowledge" (*vorwissenschaftliche Empirie*), which had previously been unnamed because it had not constituted a significant epistemic category.[15] The main thesis of Seifert's book is that *historia* was the *Namengeberin*, the "name-giver" or godmother of early modern empiricism.

Seifert's work remains the most significant contribution to the history of *historia* as an epistemological category, and one to which our project is much indebted.[16] Our goal here, however, is to revisit the history of early modern *historia* with a primary emphasis not just, as in his book, on philosophical definitions, but rather on *historia*'s actual uses across disciplinary practices. The essays we present in this volume show that the uses of *historia* can throw light on the scientific and intellectual history of the early modern period in two ways. First, they reveal intellectual practices common to natural and human sciences at a time when what were later to be called the "two cultures" were still relatively undifferentiated. Second, because of *historia*'s scope as the descriptive knowledge of particulars, these essays offer a close look at the ways in which practices of information gathering, observation and description developed in early modern empiricism.

First and foremost, a closer look at *historia* brings into sharper focus the peculiar characteristics of the early modern system of the sciences. The versatility of the early modern *historia*, equally applicable to the domain of natural knowledge and to the study of human action, points to a salient feature of early modern encyclopedism: the lack of a clear-cut boundary between the study of nature and the study of culture. The early modern system of knowledge was a far cry from the sharp distinction of nomothetic versus idiographic disciplines envisaged by nineteenth-century historicism. If anything, much of early modern natural knowledge seems by modern standards intensively "idiographic," highly suspicious of generalization and primarily bent on capturing the protean world of particulars through strongly analytical and descriptive skills. The early modern *historia* seriously challenges our assumptions about nature and culture as separate fields of inquiry. It forces us to realize that, just as the early modern notion of nature was widely different from ours,[17] so did the early modern notion of *historia* vary in fundamental ways from what we now call "history" or "culture." Early modern *natura* and *historia* were not antithetical terms, nor were the boundaries between them drawn as we draw them now. Precisely because the contrast with nature is at the very core of our conception of history, it is often misleading to translate as "history" the early modern term *historia*, which often referred to natural objects. Though the expressions "natural history" and medical "case history" are obvious remnants in our current language of the early modern term, "history" tout court is all too often an inadequate if not erroneous rendition for a word that meant primarily a description with little, if any, reference to temporality. For these reasons, it is often difficult to translate accurately the early modern *historia* into the modern languages. Many authors in this collection choose to steer clear of anachronism by systematically using the Latin term, as we also do in this introductory essay and in the title of the book.

The early modern *historia* was a fundamental *trait-d'union* not just between conceptual domains but also between different scholarly activities. Indeed, it is especially by looking at actual intellectual practices that we can notice how questionable and grossly anachronistic are the disciplinary boundaries that are sometimes projected onto early modern culture. Thus a focus on the various forms of *historia* may bring us to a better understanding of intellectual connections that seem puzzling by present-day disciplinary divisions, such as the close proximity of antiquarian studies with medicine and natural philosophy—a phenomenon not yet fully described and accounted for, though often noticed in passing by historians of early modern intellectual life, such as Giuseppe Olmi in the case of Italy and Barbara Shapiro in that of England.[18] The connections among the uses of *historia* across disciplines were often personal as well as intellectual. It is not just that the scholars who engaged in antiquarian research or in the descriptive natural sciences shared the common culture of late humanism; in several cases, they were actually the same person. Learned physicians and naturalists not only incorporated *historiae* into their own disciplines but were often active contributors to the antiquarian and historical culture of their age. In fact, some early modern physicians openly advocated their polymathic competence in various areas of learning.[19] The very large part played by men with medical training in the development of natural history and of men with legal training in the development of civil history is well known, but there are also numerous examples of physicians who wrote civil history or engaged in antiquarian pursuits.[20] Furthermore, when patterns of authorship and personal interests are examined, the division between history and antiquarian studies of the material culture of the past becomes less clear than is usually assumed. The polymathic or polyhistorical character of early modern scholarship has been conventionally viewed under the negative angle of erudite (if not pedantic) eclecticism, but much innovative research into both nature and antiquities was carried on by scholars who were polyhistors in the sense of being the authors of *historiae* on various subjects.[21] A close look at the multifarious practices of *historia* shows that between the fifteenth and the seventeenth centuries, academic education still led into an intellectual world where some disciplinary boundaries were largely permeable, if indeed they existed at all. Though individual disciplines involved extensive specialized knowledge, many aspects of humanistic erudition, with its philological training, were common to learned discourse across disciplines, spanning both the study of nature and other more rhetorical or literary pursuits.

The essays in this volume bring out some of these connections by tracing the ascending fortunes of *historia* across a range of descriptive sciences that applied humanist scholarship to both nature and the human past. They show how closely intertwined were the practices of natural history, medicine, philology, and antiquarianism, how indeed the same scholar could nimbly move from one to another, or even combine them in the same piece of work. Most significantly, they highlight, through a rich set of examples, one of the most intriguing features of the early modern descriptive sciences, namely the interlocking of observational skills and philological learning—the coupling of empiricism and erudition. They point out striking parallels between ways of observing and ways of reading, close links between firsthand observation and book learning, thus bringing into focus the peculiar brand of scholarly or "learned" empiricism that characterized the practice of *historia* and the world of European learning before the hardening of the distinction between the humanities and the natural sciences.

The Ascending Fortunes of *Historia*

The cognitive valence of *historia* as inquiry into both natural phenomena and human deeds is as old as the word itself. Examinations of the meaning of the ancient Greek ἱστορία uniformly stress that the term applied not only to Herodotus's investigations on people and their histories but also to knowledge in general, including the knowledge of nature.[22] Ἱστορία seems in fact to have been part of the vocabulary of "proto-scientific" activity in the Greek culture of the late fifth and early fourth century BCE. Its use in one of the earliest Hippocratic treatises, *On Ancient Medicine*, in Euripides, and in Socrates' reference to his own early interests in natural things, as mentioned in Plato's *Phaedo*, suggest that in Herodotus's time and later, ἱστορία was intimately paired with the inquiry into nature (including the nature of man), as in the intellectual pursuit of the natural philosopher and the doctor. Indeed, a recent reexamination of Herodotus himself shows that his use of ἱστορία delved into questions of natural knowledge, and shows remarkable coincidence of interests with the early medical writers of the Hippocratic corpus.[23] The notion of ἱστορία as used by Herodotus is probably best categorized as belonging to a "predisciplinary world," or at least a world where the boundaries between what later became the separate disciplines of philosophy, medicine, ethnography and history were much less clear-cut than they have been taken to be in hindsight.

A decisive shift in the semantic history of the Greek ἱστορία was introduced by Aristotle, who narrowed the term down to indicate a particular mode or stage of knowledge: namely, the empirical inquiry into the how of things, preliminary to *episteme*, the philosophical knowledge of their why, or final cause.[24] This notion of *historia* was fated to have a long life in European intellectual history. But this new specific meaning of the word did not obliterate the common acceptation of ἱστορία as knowledge in general, which was sometimes still used by Aristotle himself.[25] Transplanted to the Roman world, *historia* kept in Latin the original Greek sense of "knowledge of many things," but it gradually came to denote primarily, in learned language, the literary genre of history writing, or more generally a "tale," either true or fictional, as in theatrical or pictorial representation.[26] In the Middle Ages, *historia* maintained this complex range of meanings, including the reference to firsthand or autoptical knowledge, as indicated by Isidore's *Etymologies*.[27] The notion that *historia* may refer not only to the narration of *res gestae* but also to the description of places and of natural objects, ranging from the description of the microcosm to that of the macrocosm, also survived in the Middle Ages.[28] Vincent of Beauvais's thirteenth-century encyclopedia encompassed *historia naturalis* (in which he included *historia sacra* from the Creation to the Last Day, cosmology, and natural history) and *historia temporalis* (human history).[29] But though the Greek application of *historia* to natural knowledge did not disappear, the main field of reference of the medieval *historia* seems to have been the recounting and commemorating of the human past, as indicated for example by two remarkable extensions of the term in this period, to pictorial and musical language. *Historia* as synonym for "picture" or "image" was already present in classical Latin, and it further develops in medieval Latin, where it is used to indicate historical or biblical scenes depicted in frescoes or woven in carpets and tapestry, as well as illustrations in books in general.[30] This pictorial meaning of *historia* would become a technical term in Renaissance art treatises (indicating a group of figures, often with allegorical meaning, as distinct from a portrait or individual figure).[31] The musical *historia*, on the other hand, developed in medieval liturgy, where it referred to a saint's *officium*, or more precisely the sung parts of the breviary (antiphons and responsories) of the divine office sung for the saint's day, which were based on episodes from the saint's legend. Thus the musical *historia* was "chanted memory," a sung, rhymed or rhythmic version of the saint's *Vita*. This form of *historia* also was fated to have a long life, developing in Protestant countries in the sixteenth and seventeenth centuries into a musical setting for a biblical story—a forerunner of the eighteenth-century oratorio.[32] But whether as chanted or woven memory, in the medium of image or in the medium of song, these new medieval extensions of *historia* related to the commemoration of the human past.

In the late Middle Ages, however, the Aristotelian notion of *historia* as nondemonstrative knowledge was revived and further developed in

Scholastic philosophy. As the apprehension of particulars, unable to reach the knowledge of universals and of final causes, *historia* was thus categorized as an inferior way of knowing, utterly distinct from, if not opposed to, *philosophia* or *scientia*.[33] This low status of *historia* in the Scholastic hierarchy of knowledge is neatly exemplified by the disdainful comment of the early fourteenth-century philosopher, physician, and astrologer Pietro d'Abano, according to whom "*historiae* . . . and anything in which particulars are considered" lack order, are "prolix and laborious," do not allow for demonstration, and hence cannot be classified as true *scientia*. Elsewhere, Pietro dismissed interest in *historiae* of multiple and diverse things as an entertainment for schoolboys.[34]

In striking contrast with the Scholastic attitude is the late Quattrocento humanists' glowing enthusiasm for *historia*—the *laus historiae*, or eulogy of *historia*, quickly becoming, as is well known, a sort of humanist genre, an informal prelude to the more codified attempt to establish the nature of *historia* and to spell out its rules in the sixteenth-century *artes historicae*.[35] Angelo Poliziano included *historia* "vel fabularis, vel ad fidem," either fabulous or trustworthy, in his *Panepistemon*, in which he purported to list exhaustively all kinds of knowledge, imitating in thoroughness—he declared—"the dissections of the physicians, which are called anatomies, and the inventories of notaries." Trustworthy *historiae* included "the *historia* of places, that is geography; of times, such as in chronicles; of nature, such as the *historia* of animals and plants; or of human deeds, such as annals and histories."[36] This was not a novelty, as we have seen from the cataloguing of *historia* in Vincent de Beauvais; but a strikingly new note can be heard in Poliziano's eulogy of *historia*, which adds to the conventional Ciceronian praise ("temporum testis, lux veritatis, vita memoriae, nuncia vetustatis") a new aggressive challenge to the conventional assumption of *historia*'s inferiority to philosophy. On the contrary, Poliziano argues, *philosophia* is insufficient by itself to deal with public and private affairs without "the variety and richness of *exempla* provided by *historia*." "If anyone would doubt this, let him open . . . the books of Aristotle himself, where he will not find a single general principle [*praeceptum*] or teaching [*documentum*] that does not derive from *historia* as if from a fountainhead, perennial spring and plenteous source."[37]

Even more striking in its novelty is Poliziano's next argument, namely that *historia* is indispensable for the philosophical knowledge of nature:

How will he deal with nature, he who is not versed in the *historia* of nature? Though some clever philosophers of our times may happen to deny this, who are not willing to approve of Pliny's singular work on this topic, for the reason namely that it is most elegantly written, nor to peruse Aristotle's *historia de animalibus*, because it has started to speak Latin thanks to Theodore Gaza's translation.[38]

Poliziano pours withering scorn on those philosophers of his time who refuse to acknowledge the significance of Plinian natural history and of the new humanist translation of Aristotle's *Historia animalium* by Theodore Gaza.[39] They are the undeserving heirs of antiquity, "like serfs, who made suddenly rich by the legacy of their master, yet are not capable of wearing the master's garments, nor of keeping the master's rule."

It is against this background of the humanist revaluation of *historia* that we can read several of the following essays, which reconstruct the new cognitive value and meaning attributed to *historia* in several areas of early modern intellectual life. In the opening chapter of the volume, Anthony Grafton revisits the humanist *artes historicae* from a new angle and with exciting new results. The conventional view of the *artes historicae*, particularly those of Italian origin, was that they remained mostly confined within the tradition of rhetorical history. Yet Grafton shows here that some *artes historicae* stressed the connection between the writing of civil history and some forms of empirical knowledge. For example, several theorizers of *ars historica* pointed out the importance of geography and chronology and the need to study maps and chronological tables, while also reminding their readers that useful knowledge about nature could be derived from reading history (as in the case of Thucydides on the Athenian plague). Furthermore, Grafton shows that the connections between antiquarianism and the tradition of the *artes historicae* were closer than has been usually recognized. Thus Francesco Patrizi, the author of one of the most significant *artes historicae* of the sixteenth century, included antiquarian forms of research within the scope of *historia*, while several other writers in the *ars historica* genre were themselves practicing antiquarians. In particular, Grafton uses changing Renaissance and early modern evaluations of the practice of introducing fictitious speeches into historical works as a test case for reexamining humanist writing on history in relation to the rise of new forms of historical criticism and of empiricism. He argues that the radical critique of rhetorical history that emerged in the late seventeenth century had antecedents—more modest, but clearly in place—in earlier humanist writings. Grafton shows, for instance, that the ancient Roman historian Quintus Curtius, who was lambasted for his fictitious speeches by Jean Le Clerc in his *Ars critica* in 1697, had already been criticized for the same issue by the humanist Angelo Decembrio in the 1440s in terms that clearly anticipate later skepticism about the credibility of rhetorical history. This suggests that in some fifteenth-century humanist circles the interest in *historia* was accompanied by a new sensitivity to the issue of the empirical validation of knowledge. Grafton's essay, like other recent historiography and Crisciani's chapter in this volume, suggests the need to revise older views of humanism as anti-empirical as regards both civil history and the descriptive natural sciences.[40] Rather, some of the intellectual skills of the humanists—including humanists who wrote in the vernacular—now appear to have been highly relevant to the development of a "factual sensibility."[41]

An important link between the humanist revaluation of *historia* and the development of Renaissance natural history is suggested here by Brian Ogilvie. Ogilvie describes the emergence and career of *historia* as term and concept in accounts of the natural world, arguing that the rise of natural history was a Renaissance phenomenon, unprecedented in spite of its ancient and medieval versions. He shows that the sixteenth- and seventeenth-century discipline of natural history displayed distinctive features that separated it both from its classical and medieval antecedents and from its later developments. A key factor in the growth of Renaissance natural history, he argues, was the surge of a positive attitude toward the study of nature among the humanists. More specifically, he points out that the distinctive identity of Renaissance natural history was shaped by the humanists' ethical conception of *historia*. When, in the mid-sixteenth century, humanistically educated naturalists began to adopt the term *historia* for descriptive accounts of nature, they also incorporated from civil history the idea that the *historia* of nature, as that of human deeds, should have an exemplary, moral-didactic function. Hence Renaissance natural history often had an ethical component, but was set off from earlier forms of moralizing about nature (for example, medieval bestiaries) by a new insistence on accurate description of natural particulars. Ogilvie also shows that this moralizing component would have an enduring significance in the seventeenth century, marking the relation of natural history to theology. As Aristotelian final causes and older ideas about general design in the cosmos gave way, the details of natural history became important arguments for design in the "physico-theology" of the later seventeenth century.

The spectacular rise of *historia* as a tool for the description of natural objects is paralleled by similar developments in early modern medicine, which are the subject of Gianna Pomata's essay. Linking changing medical epistemology to broader intellectual developments, she first examines the rise of the cognitive status of *historia* as reflected in expanding definitions of the term adopted by fifteenth- and sixteenth-century humanists. Foremost among the intellectual trends that fueled *historia*'s revaluation were shifts within Renaissance Aristotelianism, particularly as taught at the university of Padua, that led to a new appreciation of Aristotle's books on animals as a model for zoological and anatomical research. Within this modified Aristotelian framework, *historia* acquired new significance as a preliminary but indispensable step toward the knowledge of final causes. Meanwhile, the intensive humanist editing and retranslating of Galenic and Hippocratic sources brought new attention to texts such as the Hippocratic *Epidemics*, which gave large space to the empirical report of individual cases, as well as to the Galenic writings on the ancient medical sect of the Empirics, both of which were crucial for retrieving the ancient empiricist notion of *historia* as the report of direct (or indirect) observation. Against this background, Pomata traces the uses of *historia* in both anatomical and medical writings of the sixteenth and seventeenth centuries. Starting from the framework of Aristotelian biology, sixteenth-century anatomists came to view *historia*—the description of the part as directly observed in dissection—as the first of three stages of anatomical knowledge, the other two being the establishment of function and of ultimate purpose, or final cause. The emphasis was originally on the third stage, the knowledge of final cause, as the only one that qualified anatomy as a demonstrative *scientia*. But in the early seventeenth century, for innovative anatomists such as Harvey and Aselli, the *historia*, namely the narrative of anatomical observation and discovery, acquired a new prominence. In also developed a new emphasis on firsthand observation as a historically certifiable event (with details of time, place, and witnesses) and became highly valued whether or not it led to knowledge of final causes. In the same years, approximately 1580–1630, when the anatomical *historia* gained new significance, there was also a rise of the medical *historia*, namely the report of cases from medical practice. Accounts of the treatment of individual patients, usually termed *exempla* or *experimenta*, had been occasionally present in medieval texts on practical medicine, but they acquired unprecedented prominence in new late Renaissance medical genres such as the collections of case histories named *curationes*, *observationes*, or—as they were also increasingly called—*historiae*. The emergence of these new genres testifies to the spreading among physicians of the practice of consistent record keeping and of a new sense of the value of empirical knowledge acquired through professional practice. Pomata's essay shows that over this period the term *historia* acquired, in the medical context, a more decided empiricist connotation. This is true of both the anatomical and the medical *historia*, in spite of their different philosophical backgrounds: Galenist-Hippocratic in the case of the *historia medica* and Aristotelian in that of the *historia anatomica*.

Some of the ways in which the Aristotelian framework was modified in the sixteenth century to make room for a more empiricist attitude are reconstructed in this volume by Ian Maclean. Though not dealing directly with the concept of *historia*, Maclean's essay clarifies many of the intellectual premises that made the rise of *historia* possible. He argues that Aristotle presented sixteenth-century students of plants and animals with an exceptionally powerful model for attention to the diversity of life forms, but also an exceptionally intractable and—for some—thought-provoking set of problems regarding the epistemological status of biological knowledge. Most natural historians were trained in an Aristotelian mode of thinking and had to grapple with a number of its presuppositions. The logic they had learned told them that causal knowledge was to be prized above all else, that accidents were not informative about a thing's essence, and that species were defined by properties (*propria*). Aristotle's zoological works, on the other hand, allowed for less ambitious (but more easily attainable) cognitive goals, more informal classification, and description that took into account the variability of natural kinds and the difficulty of separating one genus from another. They even admitted hearsay evidence from unlettered but informed sources, such as fishermen and huntsmen. At the same time, they were strongly teleological. The natural historians who attempted to come to terms with all this had to find a way of uniting a descriptive approach to natural objects with a recognition of their essence defined through function. Ultimately, Maclean argues, this led to the fusing of the categories of *proprium* and accident, the acceptance of informal determination of species, and eventually the explicit bracketing out of final causes in the work of Bacon. The intellectual consequences of this process—that is, the increasing emphasis on sensory evidence, the switch from formal definition to description, and especially the acceptance of a lower threshold for biological knowledge—were precisely what made possible the validation of *historia* as descriptive knowledge and its increasing use in both natural history and anatomy, which are described in Ogilvie's and Pomata's essays.

Martin Mulsow's essay takes us into late sixteenth-century antiquarianism, where he traces the gradual development of a new form of *historia*, namely a history of religion characterized by antiquarian attention to the interpretation of texts and the gathering of evidence from material remnants of the past. He shows that sixteenth- and seventeenth-century scholars pursued the history of religions in several different contexts—among them antiquarian investigations of artifacts and customs, travel accounts of remote peoples, and philological investigation of ancient texts—and under different names (*fabula, superstitio, idolatria*). But though the term *historia* was seldom used (or, if it was, it was qualified as *historia ludicra*, a lightweight pursuit, since it dealt by definition with an untrue object, namely the fabulous beliefs of ancient paganism), older ways of writing about pre-Christian religion, including much allegorized or moralized mythography, gave way to methods of inquiry borrowed from antiquarianism and natural history. The new *historia* of religion, based on a combination of textual evidence, eyewitness descriptions, and study of material objects, called for the assessment of evidence and the recognition of the possibility of differences of interpretation. By the mid-seventeenth century, as antiquarians and philologists assembled more and more particulars about ancient religion from multiple and multilayered sources, the picture became more and more complex and ambiguous, allowing for radically differing hypotheses. Thus increasingly the way was opened for discussions of such topics as the origins of religion, natural religion, and the possibility that monotheism underlay some pagan cults—and, ultimately, for the idea that Christianity was just one religion among others, showing that even a *historia ludicra* could have far-reaching consequences in promoting critical thought.

Differing from the other contributors to the volume not only in perspective but also in genre, Donald Kelley offers a broad overview of the meanings of history rather than a case study. In the *longue durée* perspective he adopts, the early modern *historia*, embracing both nature and human action, is just a circumscribed episode in the millenarian history of history, which for him is quintessentially the narration of the human past. His essay provides a broad framework in which to understand the process through which history took over from *historia* in the late seventeenth century. In his view, the epistemological heyday of particulars in the sixteenth and early seventeenth centuries gave way to a search for system, method, philosophy of history, and universal history in the eighteenth century. He argues that larger ambitions for rational form and "system" underlay not only the "conjectural history" of the Enlightenment and the "historical science" (*Geschichtswissenschaft*) developed in the German universities, but also the genre of universal history as pursued from the time of Vico and Herder down to the present vogue of "global history." In the course of these eighteenth-century developments, the human and the natural sciences, once united by a shared set of philological and empirical practices, came to be perceived as divided, as erudition became increasingly irrelevant to the new knowledge of nature and natural knowledge conversely became extraneous to philological pursuits.

The essays in the second part of this volume allow us to take a closer look at European intellectual life before all this happened: at a time, that is, when a separation between "two cultures," scientific and humanist, was largely yet to come. The first part of this volume shows that the concept of *historia* connected fields of research ranging from civil history and antiquarianism (Grafton, Kelley, Mulsow) to natural history and medicine (Ogilvie, Maclean, Pomata). The essays in the book's second part examine this connection in fuller detail by looking at intellectual practices that were shared across disciplines, highlighting in particular the specific conjunction of empiricism and erudition that was a typical feature of the early modern *historia.*

The Working Practices of "Learned Empiricism"

There is no doubt that the empiricism of the early modern *historia*, in all of its varieties, must be qualified as erudite or textual in nature. Direct observation was preceded and accompanied by laborious compilation, based on the culling of information from earlier texts. In natural history and medicine, for instance, the palaeographical and philological study of manuscripts and the empirical investigations of plants, animals and diseases were often related aspects of the same activity. Empirical observation and philological reconstruction complemented one another, with the result that, as has been noted by Vivian Nutton, "in botany and, to some extent, in medicine, practical problems could easily be categorized as textual problems."[42] *Historia* was thus often *cognitio aliorum sensibus*[43]—the report of what had been noted by other people; the evidence of "experience" was apt in many instances to be drawn from learned texts as well as from personal observation. We are dealing here with a highly scholarly or learned variant of empiricism—an *empirisme érudit*, we may call it, echoing the well-known label of *libertinage érudit* that has been applied to some circles of early modern intellectual life.[44]

Empiricism and book learning have conventionally been seen as almost antithetical. Scholars have found it hard to reconcile the emphasis on direct observation, in Renaissance anatomy for instance, with the enormous baggage of philological skill and antiquarian learning that Vesalius and his peers brought to the dissecting table. This philological and antiquarian apparatus has been seen mostly as a handicap, an oppressively constraining theoretical filter that limited and distorted observation—and in some cases it undoubtedly did.[45] But there is also evidence to the contrary, evidence, namely, that the linguistic sophistication and tremendous familiarity with ancient texts that were the hallmark of humanist training could be harnessed to the cognitive goals of direct observation so as to complement or even enhance them. By focusing on the actual practices of *historia* through detailed case studies, the essays in the second part of this collection offer strong evidence that this could indeed be the case.

Laurent Pinon examines what *historia* meant in a masterpiece of the Renaissance genre of *historia animalium*, the zoological work of the Swiss physician and scholar Conrad Gessner. Unlike Aristotle's *Historia animalium*, Gessner's work was not a preliminary step to the philosophical knowledge of animals. His goal was practical, not explanatory or classificatory. He aimed at offering an inventory of whatever had been said about animals from the standpoint of practical utility (their uses for human needs, such as in cookery or medicine). But assembling all the information available in ancient sources meant, as Pinon points out, that Gessner's description of animals had a "historical depth": the animals observed in his day are systematically compared to those previously described by the ancients. Such "historical depth" is particularly important when Gessner deals with rare or hard-to-identify species, as in the case of some animals described by Caesar in *De bello Gallico*. In order to identify these animals, Gessner proceeded by comparing Caesar's evidence with the reports of other travelers and of his own correspondents. When natural observation could not be repeated at will, as in the case of monsters and rare animals, Pinon argues, the natural historian had to rely on historical and antiquarian knowledge. More generally, it was important to know the ancient animal lore so that an old species (one that already had a name) would not be renamed arbitrarily. If a modern species was not known to be recorded in an ancient reference work, the naturalist had to establish whether it was a new animal or merely one whose ancient description had not yet been discovered. Any claim to novelty in the identification of a species required therefore a thorough historical/philological investigation of the ancient texts for possible references to the supposedly unidentified species. In this respect, every Renaissance naturalist had to master ancient knowledge, and therefore needed the skills of the historian and the antiquarian. Pinon argues that such "historical depth"—namely the use of historical evidence in species identification—is a common feature of Renaissance natural history.[46] It can be found not only in Gessner but also, for instance, in Ippolito Salviani and Ulisse Aldrovandi, who listed all available bookish references as a preliminary to their own observations. Even Pierre Belon, much less of a philologist and much more of a traveler than Gessner, used antiquarian evidence (an ancient medal and piece of statuary he had seen in Rome) in his discussion of the hippopotamus, an animal he had not personally observed. The philological skills provided by a humanist training were an indispensable tool for establishing claims to new knowledge in natural history.

A parallel can be drawn here with claims of discovery in anatomical research, as we can see from the discussion of the anatomical *historia* in this volume.[47] When the early seventeenth-century anatomist Gaspare Aselli presented his discovery of the vessels that he called the "lacteals" (the lymphatics of the thorax), an integral part of the process of discovery as he understood it (and recounted it) was a *historia* that included not only the report of his observations but also a thorough investigation of all possible references to such vessels in ancient sources, as a premise to establishing that they indeed had not been previously identified. Only in this way—that is, through a philologically informed examination of the ancient medical texts—could one be sure of avoiding "discovering what has already been discovered," as Aselli said.[48] For an early modern anatomist, the ability to master the medical tradition by means of humanistic and philological learning was thus an indispensable skill in the path to discovery. Moreover, the inconsistencies, contradictions and gaps evidenced by the philological scrutiny of the ancient texts could raise research questions that could be elucidated through dissection. Fostering the mental agility required for moving back and forth between the book and the dissecting table, the interplay between the thesaurus of accumulated learning and new firsthand observation could be a powerful engine of intellectual change.

How did early modern naturalists and physicians acquire their astonishing mastery of ancient knowledge? Methods of compilation, as for instance in Gessner or Aldrovandi, were based on systematic excerpting from ancient and modern sources. In a variety of fields the *ars excerpendi*, the art of selecting relevant sections from texts in order to rearrange them and later use them in other contexts, was a fundamental aspect of training in early modern scholarship. The wider epistemological significance of the reading and writing habits fostered by the practice of excerpting are analyzed here by Ann Blair, in an essay that deals with a masterpiece of this genre, the widely read *Theatrum humanae vitae* by Theodor Zwinger. Zwinger, like Gessner, was a physician by training and one of the shapers of the Renaissance concept of *historia*.[49] Zwinger meant to provide his readers with a *naturalis historia humana*: a massive inventory of examples of human behavior excerpted from all possible authors and arranged not in chronological order but by commonplace headings. Zwinger's work is a testimony to the humanist belief in the cognitive value of particular historical *exempla* over general philosophical *praecepta*. His *Theatrum* is a storehouse where *exempla* are removed from their original source, associated with commonplace headings, and thus held ready for use by the reader in ways that are not predetermined. The traditional rhetorical use of the *exemplum* was to illustrate preexisting moral truths, but in Zwinger, in contrast, the role of the *exemplum* is open-ended. He refrained from assigning an ethical significance to each *exemplum*, leaving the moral conclusion to the reader. He also tried to suggest various possible ways of using an *exemplum* by repeating the same passage under several headings, thus stressing the independence of *exemplum* from context. He maximized the way in which each *exemplum* could be accessed and dictated no single use of a passage. Indeed, he conceived the *Theatrum* as a permanent work in progress where the reader could step in as an active collaborator, as shown by the fact that some headings were left empty in the first 1565 edition, inviting the readers to supplement them on their own.

Zwinger's *Theatrum* indicates the extent to which humanist reading encouraged the practice of detaching "nuggets of text" from their original context and rearranging them in ever-new combinations—a constant recycling of culture, as it were. Anton Francesco Doni, the Florentine typographer, polygraph, and author of a popular bibliographic repertory, *La libraria* (1550), described this combinatorial nature of reading and writing through the striking image of a "wheel of words":

We stand in front of a huge pile of books that contain a flood of words, and of those word-mixtures we make other mixtures, and so out of many books we make up one book. Whoever comes in, takes these and those facts [*fatti*] again and again, and by mixing words with other words he starts a new devious route [*anfanamento*] and a new work. Just so does this wheel of words go round and round, up and down, a thousand times an hour.[50]

The conventional image of the humanists' relationship to the ancient texts is epitomized by Petrarch collecting every fragment of Livy he could lay hands on, piecing together such precious relics with fervor akin to religious piety, in the hope of one day reassembling the entire text—the humanist as retriever and restorer of the lost integrity of works all too often known only as *membra disjecta*, dismembered parts.[51] But there is also plenty of evidence that this very same familiarity with the ancients could foster the opposite attitude—a propensity to dismember their works, ransacking them for detachable fragments that could be snatched and appropriated for new uses, in the effort to possess and mobilize more intensively the treasury of knowledge and experience inherited from antiquity. This attitude is amply indicated by the Renaissance vogue of compilations such as *florilegia*, anthologies, collections of *dicta et facta memorabilia*, *exempla*, compendia, and repertories of *sententiae* and commonplaces—the huge literature of which Zwinger's *Theatrum* is such an outstanding example. In the Renaissance these collections had often a distinctive practical bent, catering to the professional needs of readers from various levels of learning and social standing.[52] Their diffusion indicates that excerpting and recombining were deeply ingrained and widespread habits of reading and writing, habits that encouraged the fragmenting and decontextualizing of classical texts to serve the new purposes of new readers.[53]

One obvious function of this literature, as Ann Blair has argued elsewhere, was to provide readers with a way of coping with "information overload."[54] But one could also suggest that the decontextualizing integral to the practice of excerpting may have had an anti-dogmatic side effect, somehow eroding the mental habit of dependence on tradition and on the authority of *auctores*.[55] This may apply especially to natural history and medicine, where the practices of excerpting and anthologizing the ancient texts and rearranging them by commonplaces were particularly widespread.[56] Such practices of excerpting and compiling may seem at first sight far removed from the activities of observation in the natural sciences. But Lorraine Daston has suggested recently that commonplace books, as a device for collecting and storing information for future use, may have had a role in the development of the early modern notion of "fact" as a "nugget of experience" detached from doctrine and theory.

Because the habit of excerpting and quoting, also familiar from florilegia and collections of adages, tended to decontextualize the commonplace entry, there is a certain analogy with the fragmentation of experience into facts... There is also some evidence that the practice of keeping commonplace books led to the kind of combination and recombination of items that Bacon had foreseen for the facts of natural history—independent and even in defiance of the intentions of the original author. Although it is a leap from reading practices to observing practices, the engrained habits of excerpting, ordering and recombining the entries of commonplace books offer a suggestive parallel for at least the recording of facts about nature.[57]

As Blair also suggests in this volume, the collecting of "facts" in seventeenth-century England may well owe something to the humanist practices of excerpting, sorting out and compiling compendia that could serve for the storage and easy retrieval of information.

Gessner was a town physician; Zwinger had medical training. Medical men seem to have occupied a special place among the membership of the Renaissance Republic of Letters who practiced *historia* in many formats. In this book, the essays by Chiara Crisciani and Nancy Siraisi deal with physicians who moved easily from writing civil history to natural history or *historia medica*—indeed, in the examples analyzed by Siraisi, they used several kinds of *historia* within the same piece of work. In the case of Michele Savonarola, one of the most prominent humanist physicians of the Quattrocento, Crisciani argues that his writings on civil history and practical medicine were linked not only by the conceptual structure both genres required (moving from generalities to particulars and the extensive use of *exempla*) but also by the practical mandates of his role as a court physician, who was expected to be a counsellor not only on matters of sickness and health but on political conduct as well. Before moving to the Este court in Ferrara, Michele had been a successful practitioner of and lecturer on practical medicine in Padua. His move to Ferrara accentuated the practical bent of his work by giving him the occasion to write in the vernacular for a lay audience on the same medical subjects he had dealt with in his Paduan Latin treatises, written for medical students and practitioners. Comparing these two groups of texts, Crisciani finds that in the vernacular works Savonarola's "art of particularized description" is much enhanced by the frequent use of *casus* and *exempla*, reports of observations, anecdotes from personal memories, and "concrete metaphors" drawn from daily experience—all of them much more sparingly used in the Latin texts. The same can be said of Savonarola's historical work, where *exempla*, anecdotes and concrete metaphors are much more frequent in the vernacular than in the Latin version. In his historical work Michele sometimes calls his *exempla* or anecdotes *istorie*, but he does not do so in his medical writings. We are still, Crisciani suggests, in the prehistory of the medical *historia*. In its stead, the early language of Renaissance medical empiricism employed terms such as *casus* or *exemplum*. What we today would call "facts," Michele called *exempla*—a word that had implied since medieval times, in the language of preachers, a reference to "corporalia et palpabilia, et talia que per experientia norunt"—in a word, empirical knowledge.[58] In early medical humanism, Crisciani argues, the vocabulary of experience was still in the process of creation, though it was gaining momentum, especially in the vernacular.

Over a century later, for the Roman physicians studied here by Nancy Siraisi, *historia* had become, by contrast, a well-established medical term. In fact, Siraisi shows that these medical authors concerned themselves with *historia* in many forms: *historia* as *casus* (medical case history), *historia* as natural history, and *historia* as antiquarian knowledge based on the written record and the material remains of the past. Siraisi examines these various uses of *historia* in works by Andrea Bacci, Marsilio Cagnati, Alessandro Petroni, and Angelo Vittori, all practitioners in the papal city in the later sixteenth or early seventeenth century. She demonstrates the interlocking of medical and antiquarian interests, especially in the works of Cagnati and Bacci. Writing on the potability of the Tiber water—an issue of obvious practical relevance for public health—both Cagnati and Bacci combined natural historical, medical and antiquarian arguments. Writing on the flooding of the Tiber, Cagnati connected the river's natural history with what he called *historia epidemica* (a year by year summary of health conditions in Rome from 1568 to 1580). Bacci, in turn, combined the river's natural history with the history of ancient Roman practices in matters of water supply. He also reconstructed a history of the Tiber floods, using ancient historical sources such as Tacitus and Suetonius, antiquarian evidence such as inscriptions, and his own firsthand observation of the 1557 flooding. In both authors such versatile use of *historia* had a strong practical bent: their antiquarian and medical expertise was directed at improving the quality of the city's water supply.

Both Cagnati and Bacci stressed the practical usefulness of *historia* not only for the handling of public health issues but also for the cure of diseases, deferring like many other physicians of their times to the model case histories presented in the Hippocratic *Epidemics*. Of this group of Roman physicians, however, the only one who has left a collection of medical case histories in print is Vittori, perhaps because, unlike the others, he held no teaching position and was primarily a practitioner. Vittori declared he was trained in keeping records of the course of his patients' diseases by his teacher Petroni, and in turn he himself gave advice on the writing of case histories in a treatise addressed to a starting practitioner. There is evidence, however, that Cagnati also kept some record of his cases, as other physicians were starting to do in this period all over Europe.[59] Interestingly, he declares that he tested a generalization from a Hippocratic aphorism against a series of seven detailed *historiae* of his patients, and found that the generalization was not supported by their evidence—a very striking case of the use of medical *historia* as a probative, not simply illustrative, argument.

Siraisi's essay provides detailed evidence showing how the various practices of *historia* rested on the habit of integrating the analysis of texts with considerations of material evidence, whether antiquarian or based on direct observation. In the latter part of the sixteenth century, in fact, *historia* increasingly meant firsthand observation and could apply equally to natural or human subjects. In a striking autobiographical passage, the late sixteenth-century Dutch physician Pieter van Foreest, author of one of the first important set of *observationes*, or case histories of his patients, recounted how he had moved in his youth, for health reasons, from astronomical to medical observation:

I was one day engaged with my teacher Ophusius, the eminent mathematician, in examining those observations of the Harlem heavens, on which Johannes Regiomontanus wrote and to which additions were made by Johannes Schonerus of Karlstadt. In that occasion Ophusius boasted that he had made some observations of the inconsistency of the planet Mercury. Since that time, because I was of weak health and could not stand the harmful exposure to air when observing the stars at night, I decided, mindful of my profession, namely medicine, ... that I would make observations of the microcosm rather than of the heavens.[60]

For the practitioners of early modern "learned empiricism," observing the heavens and observing diseases were activities connected by similar habits of focused attention and by similar techniques of detailed, highly particularized serial description. A marvelous close-up of such techniques is offered here by Peter Miller's essay on Nicolas de Peiresc, the man whom no less than Arnaldo Momigliano called "that archetype of all antiquarians."[61] Miller takes us on a fascinating tour of the study of an early seventeenth-century antiquarian at work, providing extraordinarily rich evidence of the total ease with which the transition from the study of nature to the study of culture could be made daily by the same scholar. The working papers in Peiresc's archive (at various levels of elaboration, from raw notes to semi-polished essay) offer an unusual glimpse into early modern knowledge *statu nascenti*, in its raw state, allowing a closer look at actual cognitive practices than would be possible with more polished published materials. We can thus see that in Peiresc's work, notes on the movements of planets or the anatomy of animals stand next to notes on ancient customs and artifacts, in the same conceptual and material contiguity and profusion in which naturalia and artificialia stood next to each other in early modern museums and cabinets of curiosities, brought together by the same omnivorous passion for collecting and knowing. In Peiresc's archive we have something like the laboratory notes of a physicist, the notebooks of an anthropologist or a historian, the logjournal of an astronomer, the records of an anatomist, all mixed together in a cornucopia of sources that defies the compartmentalizations of disciplinary history.

Peiresc himself possessed to a high degree a quality we have identified as a standard feature of early modern "learned empiricism," a capacity for switching nimbly back and forth between book and direct observation. His friend and biographer Gassendi reports that Peiresc sent copies of Aselli's book on the discovery of the lacteals to his medical friends as soon as the book came out, and that furthermore

he determined to experiment in dog, sheep, ox and many other kinds of animals what had been written by Aselli on his singular discovery. I had meanwhile informed him that William Harvey, an English physician, had published a remarkable book on the transit of blood from the veins to the arteries, and from the arteries back to the veins, . . . whereupon he immediately decided to get the book and to explore the valves of the veins."[62]

It is clear, however, that in Peiresc's cognitive practices firsthand observation, autopsy in the Greek sense, took pride of place over compilation or book learning, and for Peiresc observation meant description. Miller shows that the fundamental link between the various "continents of Peiresc's world of learning"—astronomy, cartography, anatomy, antiquarian studies and the ethnography of rituals—was the cultivation of *ekphrasis*, the verbal description of the visual. We know of the close proximity of visual and verbal description from the work notes of seventeenth-century anatomists. The scattered and fragmentary records of Marcello Malpighi's anatomical observations on various animals, for instance, start typically with the verb *observavi* or *delineavi* ("I observed" or "I sketched").[63] Note taking was both verbal and visual; describing something in words and drawing it were part of the same practice of focused attention.[64] In Peiresc's case, as Miller notes, there are drawings interspersed throughout his papers, some in his own hand, some in that of artists he employed, and yet it is *ekphrasis*, the description by word, that dominates. This same art of description was applied to natural and human objects: whether Peiresc was observing two slugs caught in the act of "faire l'amour" (the expression he used), noting optical effects through self-experimentation and supplementing them with anatomical dissections of the eyes of various animals, mapping the sitting arrangements of the English Parliament or describing the funeral of a French cardinal, he described his "experiences" in notes as detailed, as comprehensive and as close to the actual act of observing as possible, often writing, like a reporter, while he was watching whatever was going on. He was a master of "thick description" *ante litteram*: his astronomical observations, Miller tells us, were "thicker" (including more details) than those of his contemporary Galileo. Such inclusiveness of details reflected a rejection or at least a bracketing out of inherited criteria of relevance, a resolute concentration on what Peiresc's friend Gassendi called "the *historia*, that is the phenomena themselves" with no intrusion of theoretical preoccupations.[65] Peiresc was trying to preserve information in all its richness for possible later uses, even if the significance of any given detail was hard to discern. In fact no detail could be too minute to be irrelevant: a facial hair of Charlemagne noted by Peiresc in the wax of the seal impressed on an early medieval charter could provide evidence that a coin depicting a beardless Charlemagne was of dubious authenticity.

With the same ease with which he switched from natural to human objects, Peiresc's mind was constantly moving from the present to the past. Miller's essay shows how he would search for clues about old customs and institutions not only from surviving material objects but also from present-day rituals. Peiresc's descriptions of the ceremonies and rites he witnessed reveal not only a quasi-ethnographic attention to spatial details but also a heightened "historical sensibility." He was constantly trying to dig up the fossilized remains of the past that were buried in ritual. The sequence of participants in a ceremonial procession, for instance, could provide a clue to the chronological order in which different abbeys had been established, and the present-day location of the public criers in Paris was an indication of the smaller circuit the city had occupied in former times. For Peiresc the past could be read backward from ritual, and the experience of the present had a built-in historical depth.

Writing about seventeenth-century experimental literature, Peter Dear has argued that one can notice a "historicization" of the concept of experience. The reports sent to the Royal Society, for instance, typically had a historical format: the observer reported the course of the event in the first person, presenting it as a historical event tied to a specific time and place. This implied, Dear noted, a basic shift in the notion of experience, which moved from indicating "a general statement about how things habitually behave" (as in the Aristotelian/Scholastic sense) to "statements describing specific events."[66] Peiresc's thick descriptions are a very strong example of this new sense of the locally and temporally situated character of experience in early modern scientific writing. A factor in the development of this new notion of scientific experience was probably the increased significance and diffusion of *historia* in scholarly culture. Indeed, the Baconian reform of natural history may be seen as the culmination of the reappraisal of *historia*'s epistemic value set in motion by the early humanists. What Boyle called "natural experimental history"— the detailed, fully circumstantial report of experiments—seems to have grown out of the rich humus provided by *historia*'s soil. It is no chance that a large number of experimental accounts by seventeenth-century natural philosophers bear titles borrowed from the subgenres of civil historiography—memoirs, historical account, general and particular history.[67] The new "natural experimental history" may be seen as the climax of the rapprochement between historical culture and naturalistic interests fostered by *historia*'s new uses and value.

On the other hand, Peiresc's perception of historical depth in social processes and rituals shows us another aspect of the "historicizing" of experience that was going on in much seventeenth-century scholarly culture.[68] It shows that the antiquarian's traditional penchant for synchronic description[69] was giving way to a focus on the diachronic arrangement of data. In other words, antiquarianism was getting closer to history. Peiresc's archive shows with absolute clarity that a factual sensibility and a historical sensibility developed together, though they were destined to be cultivated by two separate groups of sciences.

Covering about two centuries of *historia* practices (circa 1450–1650), these essays allow us to see two main shifts in the uses of the term over this period. In the early Renaissance, *historia* referred either to compilation from learned sources or to direct observation, and quite often to a combination of the two. By the late sixteenth century, in contrast, and increasingly so over the seventeenth, in both natural history and medicine, *historia* referred preponderantly to firsthand experience, though an experience informed by a thick web of references to scholarly learning.[70] Over this period, and especially in the early seventeenth century, the empirical connotation of *historia* seems thus to have prevailed over its erudite component. These essays also show that *historia* could apply equally to synchronic description (as was mostly the case in natural history and anatomy) and to diachronic description (as in medical case histories or civil history). But by the early seventeenth century, just as it was gaining a more pronounced empirical emphasis, the term was also acquiring a stronger temporal dimension. The conjunction of empirical and temporal connotations, however, was not destined to last long. Just as the naturalia and artificialia that were held together in seventeenth-century *Wunderkammern* would find separate homes in different museums by the late eighteenth century, so also the empirical and the temporal components of *historia* would part ways. As temporality moved to its core, the early modern *historia* would turn into the modern "history."

* * *

A growing and multifaceted literature has been devoted in recent years to early modern empiricism, spurred by a broadening of interest in the historiography of early modern science to include the cultures of natural history and the practices of early modern collectionism.[71] In particular, the studies on sixteenth- and seventeenth-century museums and cabinets of curiosity have challenged the conventional compartmentalization of the histories of antiquarianism and natural history by pointing out that the two activities were intimately interrelated. So the new surge of interest in natural history has been accompanied by a revival also of the history of antiquarianism,[72] which has brought into sharper focus the shared sets of mental tools and practices common to the early modern study of nature and of the past.

Meanwhile, recent studies have brought renewed attention to the development of empirical approaches to nature in the late Middle Ages and early Renaissance, as signaled by a new interest in the description of natural particulars.[73] More generally, these studies have highlighted the rise in intellectual status of empirical or descriptive knowledge that accompanied the enormous growth of information about both the natural and the ancient world between the fifteenth and the seventeenth century. Contrasting with (but also complementing) a high-profile tradition in the history of science—from Alexandre Koyré to Thomas Kuhn—that emphasized the role of theory and theoretical change in the scientific enterprise, these new studies have focused instead on the practices of observation and data gathering in early modern science, investigating the specific nature of early modern European empiricism both in its practices and its conceptual tools. Within this context, new light has been brought on the history of the early modern categories of scientific experience, particularly on the origins of the notion of "matters of fact."[74]

This book, we hope, will contribute to this ongoing historiographic effort by suggesting further research questions. While the role of natural history in the rise of early modern empiricism has by now been fully integrated into our picture of the scientific revolution, much remains to be done in other areas, first of all in one that, regrettably, could not be included in this collection, namely the law and legal culture. Recent studies suggest that legal discourses and procedures had a significant impact on the development of empirical methods of inquiry and on the conceptualization of the notion of "fact."[75] *Historia* was also a term of legal language, and an investigation of its usage in the legal context would probably contribute to a better understanding of both the early modern *historia* and early modern "learned empiricism."[76]

Much remains to be done also in an area, in contrast, that is fully included in this volume, namely medicine. On the sociology of *historia*, first of all. We have seen how significant physicians were as authors of *historia* in many forms—but not all physicians equally so. As some essays in this volume suggest,[77] a special role was played, in the development of the empiricist medical *historia*, by court physicians and town physicians, who felt more strongly the pressure to harness knowledge to the imperatives of success in practice. The link with practical goals of knowledge was strong also in the development of antiquarianism. This suggests that the rise of *historia* and related empiricism should be further studied in connection with the social processes and contexts that legitimated knowledge on the grounds of professional practice and connoisseurship, rather than academic training or title.[78]

More also remains to be investigated on the wider significance of medicine in early modern epistemology, although some valuable contributions have recently been made.[79] We know how important was the role of a renewed Aristotelianism in early modern natural history and anatomy, but we know much less about the influence of Galenic epistemology, inside and outside medicine, especially in connection with issues of empiricism.[80] Galen's writings on the ancient sect of the Empirics were intensively discussed in the early modern period, and were probably the primary source for the empiricist notion of *historia* as knowledge based on direct or indirect sensory evidence.[81] An investigation of the early modern discussion on ancient medical empiricism in relation to the new cognitive trends in and beyond medicine would be very useful.

Finally, more could be done on topics to which this volume contributes some preliminary groundwork. Further light could be brought on the early modern vocabulary of experience through a reconstruction of the changing use of its key words. Next to *historia* itself, related terms such as *exemplum, casus, res, experimentum, observatio*, and *phenomena* deserve further scrutiny, in order to bring into full view the whole range of early modern semantic and conceptual contexts in which the neologism "fact" developed.[82] We need also more studies on the sociology of early modern "learned empiricism"—on the social, professional and institutional contexts and networks that favored the interplay of textual and observational practices, promoting the connection between the arts of excerpting, compiling and rearranging information from texts and the arts of collecting, observing and describing all sorts of empirical data.

In conclusion, we may say that these essays confirm Seifert's view of *historia* as the "godmother" of early modern empiricism. We may also recall, in this context, that *historia*'s association with birth goes back to ancient Greek culture, as shown by the fascinating story of Historis told by Pausanias in his *Description of Greece*.[83] Historis, the daughter of the seer Teiresias, was assisting in childbirth the queen of Thebes, Alcmena. The labor was held up by the malevolent presence of the Pharmakoides, but the resourceful Historis tricked them by exclaiming, with a loud cry of joy, that Alcmena had been delivered. Whereupon the Pharmakoides decamped, deeming it useless to further oppose the birth, and Hercules was born. As these essays suggest, the birth of the New Science, the Baconian *temporis partus masculus*, was aided and sped forward by the surge of buoyant intellectual confidence, the casting aside of old limits, and the expansiveness in the quest of knowledge fostered by that mistress of the art of maieutics, the early modern *historia*.

Notes

Sources cited in boldface type appear in the Primary Sources section of the bibliography at the back of this volume.

1. Febvre 1936, 301–302.
2. Roger 1980, 264.
3. The contributors to this volume met at the Max Planck Institut für Wissenschaftsgeschichte in Berlin from 1 to 30 June 2003 for a workshop on "*Historia*: Explorations in the History of Early Modern Empiricism" organized by the editors and sponsored by the Institute. All the contributors have revised their essays in the light of intensive discussions of each paper during several weeks of meeting, so as to maximize the collaborative nature of the project. We highlight cross references between the essays in this volume by printing the authors' names in small capitals.

We would like to thank Lorraine Daston for providing gracious hospitality as Director of the Max Planck Institut für Wissenschaftsgeschichte and also, most warmly, for her active participation in our workshop. From beginning to end, work for this book has benefited from her intellectual input. Our original framing of the project was deeply indebted to her work on the history of the early modern notion of "fact," which has opened new perspectives also on the early modern *historia*. The finished volume, including this introduction, owes very much to her invaluable contribution to our Berlin discussion.

4. It is impossible to give a full bibliography of these studies. We just mention a few classic works such as Kelley 1970; Huppert 1970; Struever 1970; **Kessler** 1971; Bertelli 1973; Cochrane 1981; Barret-Kriegel 1988. Among more recent works see Zimmerman 1995; Bizzochi 1995; Kelley 1998; Garfagnini 2002. On

the *artes historicae* see Black, 1987, and the literature quoted in GRAFTON, in this volume, 51–58 and nn. 50, 60.

5. On disciplinary history see Kelley 1997 (with a useful section on histories of natural sciences). On confessional history see, among others, Scheible 1966; Cochrane 1981, 445–478; Neveu 1994; Ditchfield 1995; Zen 1994; Grafton 2001b, 21–38; Backus 2003, esp. 326–391. On the teaching of history see Sharpe 1982.

6. Both the Aristotelian and Plinian *historia* were descriptions of particulars, but while the Aristotelian *historia* was meant as a preliminary step to higher philosophical knowledge, the Plinian *historia* was a collection of *res*, pieces of useful information, without any direct link or claim to more ambitious epistemic goals. On this contrast see the essays by PINON and POMATA in this volume, 243–248 and 110–111.

7. Momigliano 1985.

8. See POMATA in this volume, 116–118 and the bibliography quoted there, nn. 63–65.

9. Among studies that have recently questioned the separation of the history of early modern science from the history of the humanities, see Shapiro 1979, 3–55; Shapiro 1991; Shapiro 1983, 119–162. See also Siraisi 2003 and Grafton 2001b, which points out interesting similarities between the Baconian program for renewing the study of nature and the earlier collective enterprise of the Magdeburg Centuriators for the reform of ecclesiastical history, suggesting that "the intellectual traffic between modern historians of science and of ecclesiastical learning should not all move in one direction" (38).

10. See Rickert 1902, trans. 1986, esp. 33–36, on the "logical opposition between nature and history." "The concept of what sets a limit to natural science, the concept of the unique and the individual, coincides with *the concept of the historical* in the most comprehensive sense of this term conceivable—in other words, the purely logical and formal sense. The result is a *logical* opposition between nature and history" (34). For a recent reflection on the dualism of the natural and the human sciences that also offers a comprehensive survey of the late twentieth-century debate between various neopositivist and neohistoricist views, see Habermas 1988.

11. Meinecke 1972 (originally published as *Die Entstehung des Historismus* in 1936).

12. On *mos gallicus* and historiography see the classic works by Kelley 1970 and Huppert 1970.

13. See Momigliano 1950; Momigliano 1958; Momigliano 1990, 54–79. For a reappraisal of Momigliano's views on the history of antiquarianism see Phillips 1996 and Crawford and Ligota 1995.

14. Seifert 1976.

15. Seifert 1976, 10.

16. Another useful, though more limited, study is Mandosio 1995. Though devoted to the semantic history of the German word *Historie*, Knape 1984 also

reviews the semantic history of the medieval and early modern Latin *historia*, including a useful section on the term's uses in natural history and medicine in the early modern period (433–439).

17. See Daston 1998.
18. Olmi 1992, 300–313; Shapiro 1983, 139–140, 152. See also Momigliano 1990, 58.
19. See for instance **Patin** 1684.
20. Among the many names that could be mentioned: Alessandro Benedetti, Symphorien Champier (see POMATA in this volume, 116, 123), Girolamo Mercuriale (see Siraisi 2003), Wolfgang Lazius (1514–1566), Gabriel Naudé (1600–1653), Charles Patin (1633–1693), Jacob Spon (1647–1685), and Giuseppe Lanzoni (1663–1730). See Siraisi 2000. Siraisi has in preparation a book-length study of the intersections of medicine and history in Renaissance culture that examines the contributions of physicians to civil history and antiquarianism, as well as aspects of historicism in medical literature.
21. For a brief introduction (with bibliography) on early modern polyhistory see Mulsow 2003. On the humanist background of the polyhistors see Grafton 1985. Among recent work on individual figures see Waquet 2000 (in particular, for its relevance to the present book, the essay by Bauer, 179–220) and Findlen 2004.
22. Keuck 1934, 6–8; Louis 1955; Mazzarino 1960, 1:132–133; Riondato 1961, 46–50.
23. Thomas 2000, 134–167, esp. 163–167. Thanks to Lorraine Daston for referring us to this book. On the empirical connotation of ἱστορία in Herodotus and the medical writers see Lateiner 1986, 1–20.
24. Riondato 1961, 65–78.
25. Louis 1955, 42.
26. Keuck 1934, 8–34.
27. Isidore 1962, 1.14.1.
28. On the concept of *historia* in the Middle Ages, from the fifth to the twelfth century, see Boehm 1965.
29. **Vincent of Beauvais** 1965, 1:14.
30. Keuck 1934, 80–96; Knape 1984, 84.
31. Knape 1984, 401–410. See also Grafton 1999; Baxandall 1988, 45–55.
32. On the musical *historia* see Knape 1984, 179–190, 430–433; Jonsson 1968. Arbusow (1951) has shown that this memory embedded in liturgy influenced medieval history writings.
33. Seifert 1977 (esp. 269–282 on the Scholastic conception of *historia*).
34. **Pietro d'Abano** 1482, 30.2 and 18.9.
35. Emblematic is Andrea **Alciati's** *encomion historiae* (1530) in the prefatory letter to Giacomo Bracelli, *Libri quinque de bello Hispaniensi*, where *historia* is

called "sola profecto certissima philosophia" (the only absolutely certain philosophy).

36. **Poliziano** 1971, *Panepistemon*, 1:471.

37. **Poliziano** 1971, *In Suetonium praefatio*, 1:501–502: "Philosophia quoque ipsa, neque de moribus agere, neque de domesticarum publicarumque administratione rerum, sine exemplorum multiplici copia, varietateque satis possit. Id si cuiquam incredibilius videatur, maximorum clarissimorumque de ea re philosophorum libros, atque in primis Aristotelis in manus sumat. In quibus nullum ferè aut praeceptum, aut documentum inveniet, quod non ex ipsa historia, tanquam e capite venaque perenni, et fonte aliquo uberrimo emanaverit." All translations from Latin are ours, unless otherwise stated. On Poliziano's revaluation of history against Aristotle's *Poetics* 1451b, see Godman 1998, 54–59. Godman attributes to Poliziano an "antiquarian approach" bent on emphasizing the particular over generalities, along the model of Hellenistic erudition (59).

Historia's usefulness for various goals of life was a leitmotif of humanist culture. See for instance Girolamo Cardano's comments on *historia* as providing a useful permanent record for three aspects of life: the discovery of things, such as in medicine, "because it is necessary for [the practice of] the art"; the ceremonies of religion, "so that they could be preserved"; and the lives and deeds of princes. *De libris propriis*, in **Cardano** 2004, 377–378. Our thanks to Ian Maclean for drawing our attention to this passage. On Cardano's medical humanism see Siraisi 1997.

38. **Poliziano** 1971, *In Suetonium praefatio*, 1:501–502: "Rerum verò naturam quo pacto tractabit is qui ipsam naturae historiam non calleat? Negent fortasse id argutuli quidam nostrae aetatis philosophi, ut qui nec Plinij singulare opus de eo negotio comprobent: idcirco videlicet quia est elegantissimè scriptum: nec Aristotelis historiam de animalibus evolvant quia latine iam et ipsa Theodoro Gaza interprete coepit loqui: ij mihi similes videntur servis, qui dominorum haereditate facti repentè divites, neque quo pacto induenda sit herilis vestis, neque omnino ipsius tenere ordinem norunt."

On the same argument as used by Lorenzo Valla see POMATA in this volume, 110.

On Poliziano's interest in natural history see Godman 1998, 101–106, 214.

39. On Poliziano's view of Gaza's translation see Godman 1998, 111, 114. On the significance of Gaza's translation see Perfetti 2000, 11–28.

40. On humanism and empiricism see especially Crisciani 2001 and CRISCIANI in this volume; also Ferrari 1996. For a reassessment of older views of humanist historiography, with specific reference to Leonardo Bruni, see Griffiths in **Bruni** 1987, 175–184. See also the reevaluation of Bruni as a historian, and of humanist historiography in general, in Cochrane 1981, 3–9.

41. A good example of the empiricist stance in early humanist antiquarianism is given by Curran and Grafton 1995. For a general reassessment of the conventional view of the opposition of "humanism" and "science" see Blair and Grafton 1992, esp. 539. See also Cochrane, 1976; Vasoli 1979. For a reevalua-

tion of the relation of late humanism to science between the late fifteenth and the early seventeenth century, see Grafton 1990b and 1991a.

42. Nutton 1993b, 26–27. On the central role of humanistic erudition in enhancing the power of learned tradition, which continued to inform much supposedly empirical knowledge, see Grafton with Shelford and Siraisi 1992; Blair 1997.

43. On this notion of *historia* see Seifert 1976, 139–149.

44. See the classical work by Pintard 1943 (rev. ed., 2000).

45. For a recent example of this view, which has a long pedigree in earlier histories of medicine, see Carlino 1994.

46. It is worth noting that the "historical depth" that Pinon identifies in Renaissance natural history would have an enduring significance in the history of taxonomy, precisely in relation to the problem of species identification. In an essay on the concept of *historia* in Linnaeus's natural history, Staffan Müller-Wille has noted that Linnaeus used *historia* to indicate the identification of a botanical species through a sequence of historical events: first naming by a botanist, first description and depiction, first finding in a certain geographic region, first introduction into a botanical garden and transmission from one garden to another. This series of events concerning the reproduction and circulation of a specimen among botanists provided for Linnaeus the empirical foundation for the concept of botanical species. See Müller-Wille 2001.

47. See POMATA in this volume, 118–121.

48. **Aselli** 1627, unpaged "Praefatio." We draw here on Pomata 2002.

49. On Zwinger's concept of *historia* and its reception in early modern philosophical lexicons see Seifert 1976, 79–88.

50. **Doni** 1972, 419, quoted by Ago 2004, 126. On the way in which a book could be constructed through the pasting together of other books (many of which were themselves anthologies and collections of *exempla*) see the essay by Cherchi (1997) on Tommaso Garzoni's *Piazza universale di tutte le professioni del mondo* (1585). On Garzoni see also Battafarano 1991.

51. For a reassessment of Petrarch's image as father of humanism see Witt 2000, chapter 6.

52. Renata Ago's study of the inventories of several seventeenth-century Roman libraries shows that compendia, anthologies, and *florilegia* were very common in book collections belonging to men of varying social status (noblemen, lawyers, craftsmen, artists). See Ago 2004.

53. The practice of anthologizing dates back to Hellenistic times and was common in Scholastic learning, but spread even further in the Renaissance. See Hamesse 1999; Blair 1996; Moss 1996; Zedelmaier in Waquet 2000, 75–92.

54. Blair 2003.

55. As suggested by Kessler 1999. Ian Maclean has argued that the anthologization of Galen led to many different, even conflicting versions of his doctrine: "it meant that new Galens could be construed by doctors" (Maclean 2002, 230).

56. On excerpting from Galen, see POMATA in this volume, 123–124, 135. On the literature of medical *loci communes* in the Renaissance and sixteenth century, a first guide to sources is provided by the heading *loci communes medicinales* in **Lipenius** 1679. Some useful general comments on the use of commonplaces in medical literature can be found in Maclean 2002; only a very brief allusion in Lechner 1962, 62.

57. Daston 2001b, 757–758.

58. Jacques de Vitry as quoted by CRISCIANI in this volume, 324, n. 97. Of the rich literature on the medieval *exemplum* see especially Lyons 1989; von Moos 1988; Auzzas, Baffetti, and Delcorno 2003.

59. See POMATA in this volume, 131–132.

60. **Foreest** 1584, *Praefatio ad lectorem*, 15–16: "Cum enim aliquando cum Ophusio mathematico insigni praeceptore meo observationes Harlemi in coelo perscrutarem, de quibus Johannes Regiomontanus scripsit, quas et Johannes Schonerus Carolostadius additionibus quoque auxit, atque Ophusius gloriaretur, se aliquot Mercurij planetae inconstantis observationes habere: ab eo quidem tempore cum debilis essem, nec iniurias äeris ferre possem, ut noctu stellas ita observarem, professionis meae, nempe medicinae memor... potius in microcosmo... observationes postea instituere decrevi, quam in ipso coelo." On van Foreest's *Observationes* see POMATA in this volume, 130–131, and literature there cited.

61. Momigliano 1990, 54.

62. **Gassendi** 1658, 5:300–301.

63. Malpighi's observation notes and drawings are held in Biblioteca Universitaria, Bologna, MS. 936. See for instance busta 1, K, cc. 21v, 26r, 31r–v, 39v, 44v, 49r; busta 2, A, B, C, D, E, F.

64. On the art of observation in early modern natural history see Marx 1974 and especially Daston 2004.

65. Gassendi, letter to Ludovicus Valesius (Louis de Valois), 27 June 1641, in **Gassendi** 1727, 6:98. "Vides proinde, quaecumque sint Scholarum pollicitationes, qua de scientia ego desperem; vides quam ausim prosequi, dum tois phainomenois, seu historia rerum, hoc est, ipsa effectorum tou Demiourgou, sum speculatione contentus" ("So you see, whatever be the promises of the Schools, what kind of scientific knowledge I despair of [ever attaining]; you see what kind of scientific knowledge I venture to pursue, by contenting myself with considering the phenomena, or the *historia* of things, i.e. the effects wrought by the Demiourgos").

On Gassendi's radically empiricist concept of *historia* see Baroncini 1992, 59–60, 124–125; and more in general Joy 1987, 195–226, which however provides only a somewhat unfocused discussion of what Gassendi meant by *historia* (223–249) and does not address the wide currency and significance of the word in early seventeenth-century natural philosophical literature.

66. Dear 1985, 152; see also Dear 1987; Dear 1995, 24–25, 85–92, 125. On the use of a historical format in experimental reports see also Shapin, 1984; Licoppe 1994.

67. Among Boyle's papers, for instance: "An Historical Account of a Degradation of Gold, Made by an Anti-elixir" (1678); "Memoirs for the Natural History of Human Blood" (1684); "Short Memoirs for the Natural Experimental History of Mineral Waters" (1685); "The General History of the Air" (1692). See **Boyle** 1772, 4:371–379, 595–759, 794–821; 5:609–750.

68. See the useful definition of "historicization" in Most 2001: "a specific mode of cognitive activity which defines a body of knowledge and in so doing determines that it is constituted in its essential meaning by its temporal structure. This cognitive activity has three moments or aspects which may be termed defamiliarization, recontextualization, and narrativization" (viii). On early modern historicist tendencies, see Sabine MacCormack's interesting case study of early modern Spanish historiography (MacCormack 2001).

69. On the antiquarian's penchant for synchronic description see Momigliano 1990, 61–62.

70. PINON notes in this volume (260) that at the end of the sixteenth century *historia* in the title of natural history books was reserved for the account of direct observation, while the listing and discussion of the literature would go under the name of "treatise." Similarly, in medicine the *historia* as the descriptive report of a case was increasingly separated from references to doctrine (see POMATA in this volume, 131).

71. The now extensive literature on early modern natural history, museums, and collecting suggests that these areas of knowledge and practice have been fully incorporated into the picture of fifteenth- to seventeenth-century scientific culture. Again, we can list here only a few selected items out of a growing body of work: Lugli 1983; Pomian 1987; Olmi 1992; Sparti 1992; Findlen 1994; Pinon 1995; Jardine, Secord and Spary 1996; Ferri 1997; Haskell et al. 1989; Freedberg and Baldini 1997; Daston and Park 1998; Olmi, Tongiorgi Tomasi, and Zanca, 2000; Raggio 2000; Freedberg 2002; Ogilvie (forthcoming). For additional references see the literature cited in the essays of OGILVIE and PINON in this volume.

72. The bibliography on Renaissance and early modern antiquarianism partly overlaps with the literature on natural history and collectionism quoted in the preceding note: see in particular Pomian 1987; Olmi 1992; Raggio 2000; as well as Solinas 1989; Osborne et al. 1997; Herklotz 1999. More in general, next to the fundamental works by Momigliano cited in note 13 above, see McKnight 1991; Cropper et al. 1992; Schnapp 1993; Barkan 1999; Miller 2000. See also *The Uses of Historical Evidence in Early Modern Europe*, special issue of *The Journal of the History of Ideas* 64.2 (2003).

73. See for example Jacquart 1990; Park 1999; and in general Grafton and Siraisi 1999.

74. See first of all the several studies by Daston (1991a, 1991b, 1994, 1996, 1997, and 2001b). See also Cerutti and Pomata 1999; Cerutti and Pomata 2001b; and the essays published in Cerutti and Pomata 2001a. For an approach limited to the English case see Shapiro 1994 and especially Shapiro 2000. Poovey 1998 is also limited to the English case (and generally less persuasive).

75. On the connection between the concept of fact and legal discourse, see Shapiro 2000, chapters 1 and 2, which is limited however to English early modern sources. For a different and broader approach that includes medieval Roman and canon law, see Cerutti and Pomata 2001b, 648–652, and especially Vallerani 2001. See also Bellomo 2000. For a highly contextualized research on the roots and developments of the concept of fact in early modern legal procedures see Cerutti 2003, 152–205 (especially 172–187).

76. On *historia* and early modern legal discourse see Hammerstein 1972. More specifically on forensic medicine see De Renzi 2001.

77. See the essays by Crisciani, Siraisi, and Pomata in this volume.

78. On legitimation by practice in the case of medical knowledge see Pomata 1996.

79. Most notably by Maclean 2002, 276–332 (see especially 337–338 on the relationship of medicine's structure of thought to the New Science of the seventeenth century).

80. On Galenic epistemology see Frede 1981 and 1985; Vegetti 1994. On Galenist epistemology in the early modern period see Wear 1995.

81. This has been noted (see for instance Burnyeat 1983, 2) but not yet fully investigated. On ancient medical empiricism the fundamental reference is still Deichgräber 1965 (first published 1930), but see also Frede 1987, Hankinson 1995, and Burnyeat and Frede 1997. Empiricism in early modern medicine seems to have been studied so far almost exclusively in connection with skepticism; see Pittion 1987. For a study of early modern skepticism based on literary and iconographic sources see Spolsky 2001.

82. On the history of the term *phenomena* see Baroncini 1992, 103–144. Relevant to the study of the early modern notion of experience, including scientific experience, are some of the essays in Veneziani 2002: see in particular Fattori 2002.

83. **Pausanias** 1935, IX.11.3, cf. *Der Neue Pauly*, s.v. Historis. The association of *historia* with birth and midwifery is documented also in Roman culture, as indicated by the cult of *Iuno historia* (on which see Renard 1953). On Historis's story see Bettini 1998, 47–51.

I
The Ascending Fortunes of *Historia*

1

The Identities of History in Early Modern Europe: Prelude to a Study of the *Artes Historicae*

Anthony Grafton

Quintus Curtius and the Gordian Knot of Tradition

In the years around 1700, a fragile imaginary mansion housed the citizens of the Republic of Letters. Though scattered from Edinburgh to Naples, they were connected by their passion for the central issues of the day: Newton's physics, Locke's politics, the chronology of ancient Egypt, and the mythology of ancient Greece. Touchy, alert, and fascinated by learned gossip of every sort, they constantly scanned the new review journals for every reference to their own work. Public arguments repeatedly flared up, and some of them seemed impossible to resolve—for example, the bitter debates between Jesuits on the one side and Jansenists and Benedictines on the other about the nature of historical evidence and scholarly method. Many of those who dwelled in the house of learning feared that it was in danger of going up in flames.[1] And no one tried more systematically to quench these fiery conflicts than Jacob Perizonius, professor of ancient history at Franeker and Leiden and a man who dedicated himself to putting out fires in the Republic of Letters—or at least in its philological and historical division. In detailed essays, couched in the serpentine Latin of late humanism and larded with quotations from sources in many languages, Perizonius did his best to show that a sensible historian could rescue the early histories of Egypt, Babylon, and Rome from the attacks of historical skeptics, without making dogmatic assertions of the reliability of ancient writers.[2] He dedicated himself to preserving what could be saved of *fides historica*—even as a new set of writers sharpened a new set of tools and prepared to mount a merciless attack on the scholarly and rhetorical traditions he held dearest.[3]

No one agitated Perizonius more than those self-appointed avatars of modernity, the captious critics who despised and dispraised the ancients. And no herald of the new traversed his central intellectual territories with more fanfare than Jean Le Clerc, journal editor and prolific writer on the themes of the time.[4] In 1697, Le Clerc issued what he described as a manual for a new kind of critical thinking and reading—the *Ars critica*, a massive introduction to philology and history. This substantial book offered its readers critical principles by which they could discern whether a particular book was genuine, whether a particular passage in a text was sound, and whether a particular historical source was credible. Le Clerc spoke a contemporary language when he claimed that he would teach the reader to test texts and traditions against the eternal principles of "right reason," insofar as these affected philology and hermeneutics. In practice, as when Le Clerc told the critic who had to choose between two readings to assume that authorial intent more probably lay in the *difficilior lectio*, he borrowed liberally from earlier humanists like Erasmus, whose works he edited in ten massive volumes.[5] But he tried to speak and write a period dialect of iconoclasm and innovation, even if he occasionally broke into more traditional forms of scholarly language. And he clearly took a special pleasure in using a classical, rather than a medieval or a modern, text as his central case in point—the object of his exemplary proof that, under critical scrutiny, any given work could reveal fatal flaws.

Part III of the *Ars critica* trained the harshly brilliant lamp of Le Clerc's critical principles on the Roman historian Quintus Curtius Rufus—a writer of the earlier imperial period who adapted Greek sources to tell the story of Alexander the Great.[6] His work, though incomplete, had won great popularity in the Renaissance, when illustrated versions of it in Italian made popular reading for princes.[7] Alfonso of Aragon, the connoisseur of history who staged "hours of the book" at his court in Naples, where humanists like Bartolomeo Facio and Lorenzo Valla debated passages in the text of Livy, read Curtius while ill and out of sorts, and recovered at once. He declared the work as effective and pleasant a remedy as anything in Hippocrates or Galen.[8] Numerous manuscripts and, after about 1470, many printed editions made the text accessible.[9] No less an authority than Erasmus declared Curtius ideal reading for those who wished "to maintain their rhetoric in a state of high polish." He even prepared an edition with marginal notes that called attention to some "novel turns of phrase" that could enrich the standard Latin lexicon.[10] The humanists who formulated influential protocols for reading ancient history in the later decades of the sixteenth century—Justus Lipsius and his allies—preferred Tacitus and Polybius to the historians that Alfonso and his contemporaries had loved most, especially Livy. Yet they shared their predecessors' love of Curtius. Lipsius spared no adjectives when he praised this "historian who is, in my opinion, as honorable and worthy of respect as any other. The felicity of his language and the charm of his way of telling stories are marvelous. He manages to be both concise and fluent, subtle and clear, precise and unpedantic. His judgments are accurate, his morals are shrewd, and his speeches show an indescribable eloquence." Scholars as distant from one another in space—if not in their tastes—as the Silesian humanist and poet Christopher Colerus and the first Camden praelector on ancient history at Oxford, Degory Wheare, agreed.[11] In particular, the speeches in Curtius compelled admiration as models of rhetoric well applied to history. Nicodemus Frischlin put Curtius first among the five authors from whom he drew an anthology of Latin speeches for the use of his students in Braunschweig.[12] In his lectures he analyzed in detail the ways in which the Roman historian made Darius narrate events, devise arguments, and create a feeling of loyalty and pathos among his soldiers. He made clear that he attributed these feats of rhetoric not to the Persian emperor, but the Roman historian—especially when he noted that Hannibal used one of the same arguments that Darius did "in book 21 of Livy."[13]

Le Clerc admitted that he too had long shared the traditional admiration for this master of classical rhetoric. At last, though, he had tested the work against two eternal touchstones: the particular rules of the art that Curtius professed, history, and the general rules of right reason, "which hold for all human beings, whatever nation they belong to and whatever age they may have lived in."[14] Le Clerc had also read the text several times—a method he recommended to those who wished to approach their sources in the truly critical frame of mind, without which reading seemed pointless: "It would be better for the ancients not to be read at all, than to be read as these gentlemen want them to be, that is, without critical judgment. For that sort of reading can only make us more foolish."[15] In the end, this close and repeated scrutiny revealed significant errors—errors that undermined Curtius's standing as a historian. Curtius had gone wrong in just about every way that a historian possibly could. True, Le Clerc admitted, he did not think it necessary, as others did, for a historian to have precise knowledge of every subject that might come up in the course of his account. But geography and chronology were the two eyes of history, at all times and in all places. And Curtius had mastered neither discipline. He thought that the Black Sea was directly connected to the Caspian, and he did not mention the years, or the seasons, during which the events in his account took place.[16] Reason demanded consistency, but Curtius's account swarmed with obvious errors and contradictions. When he described the scythed chariots of the Persians, he imagined that their blades projected through the spokes of their wheels, a manifest impossibility, rather than from their hubs.[17] Reason, finally, demanded independence from popular follies. Curtius supinely followed Greek writers when he portrayed the Persians and Indians as worshipping Greek divinities, rather than "barbarous" gods of their own with their own names and cults. From ancient texts and contemporary travel accounts, Le Clerc wove a compelling case against the *interpretatio Graeca* of foreign gods.[18]

Le Clerc traced most of Curtius's errors to a single source: the fact that he was a rhetorician rather than a historian. Good historians chose to follow the sources they thought most accurate. Bad ones like Curtius accepted the traditions that glorified their heroes and allowed them to spread their rhetorical wings, whether or not the facts were credible: "Those who have composed histories from ancient sources fall into two categories.... Some try to work out the truth, so far as that is possible, and examine everything diligently so that, when it is impossible to produce a certain account, they follow the more plausible narrative. Others take little interest in the truth, and choose instead to report the greatest possible marvels, since these are more susceptible of rhetorical adornment, and supply the matter for exercises in the high style."[19] Evidently, Curtius belonged in the second category.[20] That explained why he claimed that over one hundred thousand Persians and barely a hundred Greeks had died at the Battle of the Issus. "For this to have happened," Le Clerc commented with contemptuous clarity, "the Persians would have to have had wooden swords."[21] That explained, also, why Curtius never denounced Alexander's ambitious plans for world conquest, though they had caused terrible destruction, and the best ancient philosophers had made clear the immorality of such actions.[22]

Curtius vividly revealed the besetting weaknesses of the professional rhetorician trying to write history when he introduced formal rhetorical exercises into his narrative. He composed many speeches and put them in the mouths of his protagonists—so many, in fact, that it seemed amazing that his relatively short narrative could hold them all. In Le Clerc's view, a serious historian should never include speeches in his narrative, either in direct form or even in oblique summary.[23] Le Clerc knew, of course, that Curtius was following normal ancient practice when he did this, and that most modern critics approved when historians inserted speeches into their works, so long as they did their best to fit their words to the character of the speaker in question and used the speeches to explain policies and motivations. But this standard practice violated the historian's primary responsibility to tell the truth. Inventing a speech that the actor in question had not made was a lie, every bit as much as inventing an action that he had not carried out:

> I know what scholars have written, in an effort to show that historians may ascribe speeches to those whose deeds they describe—not the speeches they actually held, but the ones they could have, or indeed, in the judgment of the historian, should have held, so far as their content is concerned. Yet they admit that virtually none of those who make appearances as speakers could rival the historian's eloquence and wit. I also know that a good many of the ancients inserted this sort of speech into their histories. But, if we love the truth, we must admit that the elegance of the speeches that we read in their works deluded those who approve of their presence in histories. For there is no argument that could persuade a man who loves the truth that historians have the right, in this one regard, to tell lies. After all, it is hardly less of a lie to introduce a man saying what he did not say, because he could have said it, than to describe him as doing what he did not do, but could have done.[24]

Curtius's version of these practices, moreover, was especially ludicrous. He represented Greeks and barbarians, the illiterate and the educated, as speaking in the same rhetorically schooled, cultivated way, as if all of them belonged to a single happy Trapp family *avant la lettre*: "All the characters in Curtius declaim, and in a way that reflects the author's wit, not their own. Darius declaims, Alexander declaims, his solders declaim. Even the Scyths, completely ignorant of letters, make an appearance, duly singed by the rhetorical curling iron. This reminds of me of the family, all of whose members sang."[25] Traditionally, historians had defined their task as making their characters say the things appropriate to the situations in which they spoke. But doing this to a cast of characters as varied as Curtius's was ridiculous: "What more ridiculous invention could there be, than to make ignorant men or barbarians speak as eloquently as if they had spent many years studying rhetoric?"[26] Traditionally, historians had won credit and admiration by the skill with which they invented speeches. In fact, they had been fooling themselves and their readers. Worse, they habituated themselves to a kind of deceitfulness that infected other portions of their texts as well.[27]

The voice of modernity resounds, harsh and self-confident, through Le Clerc's denunciation of Curtius. Cutting himself off from a humanist tradition that had lasted for more than a millennium, he set out to show that history must no longer form a part of the ancient art of rhetoric—the art to which the greatest ancient authorities, above all Thucydides and Cicero, had assigned it, and to which most writers still attached it. The formal study of history, according to this tradition, was a matter of production rather than consumption: of defining the devices which enabled the historian to instruct, and at the same time to touch, the reader. Good history narrated past events in an accurate, prudent, and eloquent way. Readers studied it in the hope of understanding the political thinking of ancient leaders, as expressed in speeches, and of sharpening their understanding of moral precepts and their applications, as embodied in crisp, specific historical examples. They read in the same reverential, dedicated way in which young Romans had once gazed at the funeral masks of their ancestors while hearing descriptions of their deeds.[28]

Le Clerc, by contrast, saw the historian's task as centrally concerned with critical thinking and the intelligent weighing of evidence above all. He must examine his sources, take from them only what was demonstrably credible, and reproduce it in plain prose. He must introduce nothing of his own. Even if he had good reasons for ascribing particular plans or arguments to a given individual, he should simply lay these out in his normal prose, not in the form of a spurious oration.[29] The fact that Curtius had written under the Roman emperors did not confirm his classical standing: it underlined his obsolescence as a model, just as other errors proved his uselessness as a source. In the age of the New Philosophy, Le Clerc called for nothing less than a New History—a genre as rigorous, critical and devoid of traditional appeal as the Cartesian philosophy itself.

This radical attack on the humanist tradition naturally brought Perizonius hurrying into action. And no wonder: it represented the sharp end of a larger and more general attack on both the ancients and their students, an assault that Le Clerc mounted in more than one language and genre. The witty, miscellaneous *Parrhasiana* of 1699, supposedly pseudonymous but transparently the work of Le Clerc and written in accessible French, denounced humanist scholars as mere pedants who bit every back that was turned on them: "ils mordent tout le monde, ils se querellent entre eux pour des bagatelles."[30] In the same work, Le Clerc made clear that he wanted to see a new, nonclassical kind of history take shape. He denied that "one must be an orator in order to be a historian, as Cicero claimed."[31] Drawing a radical implication from this renunciation of rhetoric, he urged writers of history to consider adding full citations of sources to their works, even though the ancients had not done so. Indeed, he made the willingness to depart from ancient models a criterion of good sense and "right reason": "If the thing is bad in itself, the example of the ancients does not make it any better, and nothing must prevent us from improving on them. The Republic of Letters has finally become a country of reason and light, and not of authority and blind faith, as it was for so long. Numbers prove nothing, and cabals have no place here. No law, divine or human, forbids us to perfect the art of writing history, as men have tried to perfect the other arts and sciences."[32] Like earlier humanists, Le Clerc adopted a universal standard when he set out to criticize ancient texts. But where they had applied the methods of rhetoric in order to explain why certain devices worked and others did not, he applied the rule of right reason in order to show that ancient authority and its votaries had passed their sell-by date.

Perizonius saw his correspondent's public anatomy of an author as something far more threatening than an isolated critical aria. Perhaps, in fact, he took it as a personal attack—especially when Le Clerc sent him a copy of the *Ars critica*, accompanied by a letter in which he apologized for not having cited and praised Perizonius more often in the work, "since I had decided to abstain, so far as possible, from citing recent examples."³³ Certainly, Perizonius took the third part of the *Ars critica* as just the sort of captious, excessive criticism that had recently begun to flourish in all too many fields of scholarship. He replied to Le Clerc, briefly, in his 1702 edition of Aelian, arguing that no author could stand up to the sort of examination to which Le Clerc had subjected Curtius.³⁴ And in 1703 he published a lucid, careful essay, *Q. Curtius in integrum restitutus et vindicatus*, in which he did his best to show that Le Clerc's critique had missed its mark. A vigorous defender of tradition, Perizonius dismantled the pretensions of avant-garde critics as passionately as Le Clerc had denounced the pedantry of old-fashioned grammarians. The critics claimed to follow standards of taste and decorum far superior to those of the ancients. But they lacked both manners and judgment, which explained why some of them had even attacked one another physically, using books as weapons, at a sitting of the French Academy in 1683.³⁵ More seriously, many of them did not understand the language or the content of the classical texts they attacked: "the greatest error is committed by those who take pleasure in denouncing the ancients, in ignorance of their language, and also of events and histories, without knowledge of which they cannot be understood."³⁶ This kind of textual violence, Perizonius pointed out, formed a specialty of the French, who seemed bizarrely bent on ridding the world of the "elegance, wit, style, and skill" of ancient writers. And he set out to repel it on behalf of both the ancients and his fellow grammarians—who, he insisted, did not "reach senility still obsessed with grammatical trivialities," as Le Clerc maintained.

At the same time, Perizonius insisted that he himself was no uncritical admirer of the ancients.³⁷ From the start, he admitted that Curtius "had delighted in rhetorical descriptions of events, and perhaps more so than is appropriate to a serious historian."³⁸ Accordingly he defended his author against Le Clerc's accusations by invoking a standard radically different from the eternal "right reason" that Le Clerc had inscribed on his critical pennant. Perizonius, in fact, appealed his historian's conviction to the high court of history itself. Judgment, he pointed out, was always hard, even when one dealt with those who wrote in one's own language. "Nothing could be more ridiculous," Perizonius pointed out, "than to reject an author's judgment because he is following the received customs of his age and his people."³⁹ Le Clerc admitted that Curtius gave wrong geographical information. But he did so because he faithfully followed Greek writers of the time of Alexander. His errors derived from the cultural situation in which he wrote, not from a failure to meet eternal standards of which Curtius—like everyone else in his time—had necessarily been unaware.⁴⁰

Curtius's use of speeches could be defended on exactly the same grounds. True, Perizonius admitted that he accepted the traditional belief that a history should be stylistically unified, and to that extent his work lay within the main borders of the humanist tradition.⁴¹ But his central line of defense lay elsewhere. To condemn Curtius for inserting speeches into his text, when virtually every other ancient historian had done so, meant holding an ancient writer up to a modern standard. And this procedure, Perizonius insisted, made no sense, since every nation and every period had its own customs and standards: "They pass judgment on ancient matters from the standpoint of their own time and its customs. This is completely idiotic. Each people, and each period in the history of a given people, has its own customs."⁴² Here Perizonius seems as rigorously modern as Le Clerc, though in a different vein. Where Le Clerc found his standard for judging literary texts in the clear and distinct reason of Descartes, Perizonius found his in the historicism, the contextual reading and thinking more notoriously represented by Spinoza. Spinoza had used this tool to challenge the authority of Scripture. The Bible, he argued, had been written for a primitive people, not for sophisticated moderns, and thus could claim no authority in matters of history or metaphysics or natural philosophy. Perizonius used the same form of argument to save the coherence and interest of ancient texts. If these could be judged only by setting them into the contexts in which they had originally taken shape, then the severe modern critic who discovered their faults by applying a universal standard was making a category error. Perizonius agreed with Le Clerc that a modern historian should use only tested, credible sources—the sorts of monumental, numismatic, and archival evidence that defenders of *fides historica* held to be reliable. He also presumably agreed that a modern historian should not introduce fictional speeches or letters into his text. But he insisted that it made no sense even to ask why an ancient historian had done so—much less to condemn him. And in doing so he adumbrated the historicist rejoinder to the claims of universal reason that would prove so powerful in eighteenth- and nineteenth-century Germany.[43]

This debate, in other words, was not simply one skirmish in the Battle of the Ancients and the Moderns. Both Le Clerc and Perizonius argued for forms of critical history that lay outside the boundaries of the older rhetorical tradition, and not even Le Clerc abandoned the view that some of the ancients were genuinely classical—as he made clear when he insisted, against Perizonius, that ancient writers like Polybius and Dionysius of Halicarnassus could stand up even to the sort of critical acid bath in which Curtius had melted down.[44] Le Clerc may even have appreciated the terms in which Perizonius cast his defense more fully than he ever admitted. True, he fought back bitterly, accusing his adversary not only of misunderstanding his logic, but also of the last crime one would expect to see attributed to any polyhistor, laziness.[45] And he pulled every literary string he could on his own behalf.[46] Yet he too appreciated, as he noted in his inaugural lecture of 1712 on ancient history, that "the ancients must be read in a particular spirit, in order to learn what they thought and how they behaved, as if they had nothing to do with us—not what, in our opinion, they should have thought or done. Before we may judge them, we must ask how learned they were and how holy their customs were. Only when one knows these things properly, not before, will it be safe to judge them. If judgment precedes knowledge, we will think we have found in them whatever we like, not what is really there. And we will twist everything they said or did so that we may contemplate, in their history and books, an image conceived ahead of time in our minds."[47] Evidently Le Clerc too could deploy the rhetoric of tolerant historicism—at least when he wished to save the credit and authority of early Christian, rather than pagan, texts.

For all the divergence of their methods and results, both the firebrand and the fireman seem to be characteristically modern thinkers—figures that naturally belong in the vast fresco of *Radical Enlightenment* painted, with magnificent energy and panoramic vision, by Jonathan Israel in 2001 (as they did in that sly black-and-white masterpiece filmed in the 1930s by Paul Hazard, *La crise de la conscience européenne*).[48] Yet Le Clerc admitted, at least once, that he was not the first to wield some of the edged tools that he applied to dissecting Curtius. Gerhard Johannes

Vossius, seventeenth-century polyhistor and author of a treatise entitled *Ars historica*, had devoted two chapters of his work to the question of whether historians should include speeches in their work, and Le Clerc cited Vossius's work in one of his rare footnotes.[49] By doing so, moreover, he raised a question of central importance to the assessment of history in early modern Europe.

Vossius's work—as he himself made clear—was only one of the more recent entries in a vast bibliography of early modern works on the reading and writing of history. The genre of the *artes historicae* grew from deep roots in ancient and fifteenth-century thought, took a clear shape in the middle of the sixteenth century and assumed canonical form in 1579, when Johannes Wolf published his influential three-volume anthology, the *Artis historicae penus*.[50] Before describing Le Clerc or Perizonius as victors, proudly waving the scalp of a fallen humanist tradition, it seems necessary to examine the treatises on the *ars historica* that they knew. Certain questions arise at once. What critical approaches did the authors of these treatises devise and apply as they examined ancient historians? Did they adumbrate or formulate any of the shiny new methods and questions that gleamed like brand-new tungsten drill bits in the toolboxes of Le Clerc and Perizonius? Or did they remain—as Le Clerc and Perizonius evidently did not—within traditional boundaries, treating history—in the words of Felix Gilbert—as "a branch of rhetorics"?[51]

The *Artes Historicae*: Splendor and Misery of a Genre

The *ars historica* formed an organic part of the massive early modern effort to capture and use a world of particulars, and it did so in more than name only.[52] True, Jean Bodin argued, influentially, that one should distinguish *historia humana* from *historia naturalis* and *historia divina*. But as Bodin's astute critic Bartholomäus Keckermann pointed out, one could hardly maintain this separation rigorously, since human history unrolled within nature. In practice, Bodin himself agreed, since he treated climate as a determining factor in the development of each nation's *genius*.[53] Writers on the *ars historica* emphasized the resemblance between civil history and other forms of empirical knowledge—for example, by stressing that history had a strong visual component, best mastered by studying chronological tables, which revealed the course of history at a glimpse, and maps, which made it possible to know "the sites and distances of the kingdoms and places in which events are said to have taken place."[54] They highlighted the vital elements of knowledge about nature that could be derived only from reading history—for example, by noting the value of the detailed description of the plague given by Thucydides, and the explanations for it offered by Diodorus Siculus.[55] And they did their best, at times, to tie the creation of new forms of human history to the devising of equally original narratives in other fields. Campanella strongly emphasized the identity of all forms of *historia*, divine, natural, or human, as narratives that provided just the facts without offering causal explanations. He praised both Galileo's *Starry Messenger* and Baronio's history of the church as modern models of true history. With characteristic exuberance, he called both for someone to correct natural history by adding the many secrets of nature that had been discovered since Pliny, and for someone else to play the role of a "Baronius . . . mundi," weaving the traditions of all nations, the Chinese, the Japanese, the Tartars, and the inhabitants of the New World, into a single universal history of man.[56] The pathos and the power of *historia* were blazoned across these treatises.

Yet the authors of the *artes* took highly individual positions on many issues—positions not always predictable from their professions or their other interests. Francesco Patrizi, for example, enlarged the scope of history to include the new forms of research and writing widely practiced in his time by travel writers and antiquarians.[57] Some historians, he explained, "have not so much described events as customs, ways of life, and laws. . . . And there is another sort, those who, especially in our day, write in another way about the clothing of the Romans and the Greeks, the forms of armament they used, their ways of making camp, and their ships, their buildings, and other things of this sort, which are necessary for life. . . . And others, again, write in a certain novel way about the magistrates of the Romans and the Greeks, and others about the form of the Roman Republic, and those of the Athenians, the Lacedaemonians, the Carthaginians, and the Venetians. As you know, this is a most useful kind of writing."[58] Patrizi was himself a skilled antiquarian. He did pioneer work on the military affairs of the Greeks and the Romans long before Justus Lipsius made the subject fashionable, and it seems tempting to connect his novel view of history as a set of genres to his own historical practices.[59] But Patrizi's teacher, Francesco Robortello, also did elaborate antiquarian work, as did the Ferrarese scholar Alessandro Sardi, yet both wrote *artes historicae* that stayed firmly within the rails of the rhetorical tradition.[60] Reiner Reineck, historian and teacher of history at Wittenberg, Frankfurt an der Oder, and Helmstedt, produced a substantial *Methodus legendi cognoscendique historiam tam sacram quam profanam* in 1583. A member of the most sophisticated circle of antiquaries in the Holy Roman Empire, Reineck exchanged coins, medals, and site reports with expert collectors and learned travelers like Georg Fabricius, Joannes Sambucus, Johann Crato von Crafftheim, and Obertus Giphanius.[61] Yet these ventures into the study of material objects, which shaped Reineck's massive compilations on ancient chronology and genealogy, had little if any impact on his work as a theorist.

When the *artes historicae* took shape as a genre in the middle of the sixteenth century, moreover, individual treatises crystallized in a number of different contexts, each of which helped to shape them.[62] One matrix within which systematic discussion of history grew was literary. As early as the fifteenth century, humanists in Naples and elsewhere had begun to discuss the nature and value of history. Sometimes they remained within the relatively narrow confines of the rhetorical tradition, endlessly quoting Cicero's definition of history from *De oratore* 2.36 and looking in the texts for evidence that history really was "opus ... oratorium maxime" (*De legibus* 1.5). But from the start, stimuli from outside the historical tradition abounded, starting with Lucian's essay on the writing of history, which Guarino of Verona adapted in a formal letter on the subject.[63] One outsider spoke with special authority. Aristotle had argued in the *Poetics* that poetry offered profound and general truths, while history could tell only what a given person did or suffered. As early as the 1440s, Lorenzo Valla set out to counter this view in the prologue to his *Gesta Ferdinandi Regis Aragonum*. He admitted that certain great ancient philosophers had preferred the poet to the historian, "because they say that he comes closer to philosophy, since he deals with general points and offers teachings based on fictional examples that claim a universal validity."[64] But he sharply rejected this critique. After all, the whole purpose of history was "to teach us through examples."[65] Here Valla argued, contradicting Aristotle himself, that historians offered their readers profound general teachings. By the last decades of the fifteenth century, as Robert Black has shown, Pontano and others had begun to discuss the problems of historiography in a formal and sophisticated way. Their debates seem to have stimulated a number of practicing historians—Black cites Bartolomeo Scala, Giorgio Merula, Tristano Calco, and Francesco Guicciardini—to show a new energy at research and a new discrimination in the use of sources.[66]

In the middle of the sixteenth century, when modern Europe's first age of literary criticism blossomed, the *Poetics* continued to stimulate scholars to defend, and define, the art of history. Francesco Robortello, the first writer to produce a systematic treatise on history, also wrote a commentary on the *Poetics*.[67] Like Sperone Speroni, he tried to show that history could be treated as a formal "ars" in its own right. The Italian discussion took many forms. Francesco Patrizi argued in detail that no historian could gain access to a true account of the grounds for a major political decision—and that even if he did so, he could never hope to publish them.[68] Most others contented themselves with detailed reviews of the uses and forms of history. They highlighted its pedagogical function as a source of moral principles exemplified in action, which could teach the intelligent reader prudence, and its literary nature as a genre that could not only inform, but move, its readers. Over time, they made impressive efforts, as Eckhard Kessler above all has shown, to wield the traditional principles and rhetoric and the *loci classici* from Cicero and Lucian to define history as a genre and a domain with its own ends and rules.[69]

Outside Italy, crises in the evaluation of traditions played a larger role in giving rise to the new genre. Giovanni Nanni's (Annius of Viterbo's) 1498 publication of the lost (actually forged) *Antiquitates* of Berosus, Manetho, Metasthenes, and others forcefully challenged the preeminence of the classical histories. A good Dominican by training and a theologian by trade, Nanni turned his commentaries into systematic indictments of the individualistic, mendacious historians of Greece and Rome. He drew up the first formal criteria for the assessment of historians. Only priests, he argued, who wrote public annals based on archival documents, could preserve the past accurately. Berosus, Manetho, and the rest were priests, and their accounts rested on the proverbially accurate archives of their nations. Their accounts of ancient history—which identified Viterbo as the center of civilization, Janus as another name for Noah, and Isis and Osiris as the ancestors of the Borgias—were packed with appetizing details. In particular, they had a great deal to offer the heralds, panegyrists, and court historians of a parvenu world collectively bent, as all good parvenus are, on tracing its genealogy back to ancient roots.[70]

Other challengers flanked Nanni (and drew on his work). Northern historians, many of them inspired by him, ransacked the historical record to establish where the Franks and Anglo-Saxons had come from, and when. And the Reformation—which rested on the claim that the Catholic Church had departed, in head and members, from the doctrines and beliefs of the ancient church—posed the most fundamental challenge of all to the traditional histories of the church, which legitimated exactly the practices that the Reformers most disliked. The first modern collaborative historical enterprise, the Protestant church history institute established in Magdeburg by Matthias Flacius Illyricus, represented one effort to deal with this massive set of pressing intellectual and practical problems. Eventually, Catholic scholars like Sixtus of Siena, Antonio Possevino, and Cesare Baronio mounted large-scale replies to the Protestants—replies which, like the Protestant attack on tradition, rested on systematic efforts to collect, assess and use the entire source base of Christianity.[71] Baudouin came to the writing of his *Prolegomena* to the study of law and universal history fresh from examining the sources for the life of Constantine the Great and laying down rules of source criticism for the Magdeburg Centuriators. In his case and many others, the *ars historica* was as closely connected to the interest and practices of ecclesiastical history, an ancient and sophisticated form of historical writing, as to those of Roman lawyers.[72]

As individual challenges to the one Catholic Church turned into individual churches and the dialectic of proselytizing and repression gradually mutated into modern Europe's first religious wars, another set of traditions also called for assessment. Which laws, which constitution should a state adopt? Should modern lawyers assume that ancient Roman laws were still universally valid? Or should they note—as medieval lawyers concerned with modern issues like the law of citizenship had done before them—that modern circumstances required new legislation appropriate to the characters of new states and peoples? Legal codes and their contexts, like religious traditions, required sifting.[73]

Both in Spain and in France, where these problems dominated the discussion, writers on the art of history concentrated less on how one would create a single perfect narrative than on how one should assess the multiple conflicting narratives that actually confronted modern readers. Melchior Cano, François Baudouin, and Jean Bodin disagreed on many points—notably the value of the Annian forgeries, which Bodin cheerfully accepted, while the other two rejected them with contempt. Yet all of them took up Nanni's challenge and tried to state rules, as he had, for the evaluation and use of sources. Bodin, indeed, simply accepted and quoted Nanni's own precepts about the supreme value of priestly annals. Baudouin, by contrast, argued not only against Nanni, but for the sorts of oral, poetic traditions that preserved the history of both early Rome and the pre-Columbian New World. Cano, meanwhile, emphasized the individual historian's need for judgment and discretion as he tried to sort out proper rules for writing an honest, accurate and pious history of the church.[74] Many *artes historicae* written outside Italy, though not all, shared a distinctive concern for bibliographical control of the field that connects them to the world of *historia litteraria*—and that distinguishes them from less self-consciously critical efforts to amass facts and examples without regard for their original sources and contexts.[75]

Some historians of the *ars historica* have argued that all of these treatises show essential resemblances. Both Italian and northern scholars, they point out, dealt with the reading, as well as the writing, of history: Jacopo Aconcio was as conscious as Bodin or Baudouin of the need to read history systematically and critically, and like them he offered elaborate practical instructions to printers and readers. Aconcio, not Bodin, was the first to treat the study of history in material terms, as a problem in how to make useful notes—a practice that northern writers like Keckermann and Wheare would examine, in vastly greater detail, a century later.[76] More important, both northern and Italian scholars continued, throughout the sixteenth and seventeenth centuries, to see history as above all a form of rhetoric and a source of *exempla*—moral and prudential precepts worked out in the concrete form of speeches, trials, and battles. Thus George Nadel, Reinhart Koselleck, and Eckhard Kessler, while recognizing the individual differences between certain texts, cogently argue that the rise of the *ars historica* generally reinforced, rather than challenged, the rhetorical model of history. Nadel, moreover, unlike most other students of the genre, notes that it lasted until deep in the eighteenth century—whatever the challenges it received from innovators like Le Clerc and Perizonius. And Ulrich Muhlack has suggested that this outcome was only natural, since all humanists, however sophisticated their historical methods, applied them to attain the traditional ends of *historia magistra vitae* rather than to recreate an alien past *wie es eigentlich gewesen*.[77]

Others, notably the pioneering German and American scholars who revived the study of the *artes historicae* in the 1960s and their French successors, have highlighted the differences between the smooth, humanistic Italian treatises and the "crabbed," erudite northern ones. Both Baudouin and Bodin, after all, argued for a study of history that would be catholic, rather than confined to the traditional territory of learning, Greece and Rome. Both suggested to scholars that the histories of the New World, Asia, and Africa were as significant as those of Europe. Both emphasized the need to read and write history in a critical manner, with an eye always on the credibility of sources and the proper ways to combine and reconcile their testimony. Both assessed the value of the most prestigious ancient and modern historians, such as Polybius and Guicciardini, in part by their ability to draw on official documents and other reliable sources. Most important, both treated the *ars historica* as a hermeneutical discipline, a set of rules for critical readers, as well as (or, in Bodin's case, instead of) a set of canons for effective writers. To that extent, both began to treat history in a new way—as a comprehensive discipline that ranged across space and time, and as a critical discipline based on the distinction between primary and secondary sources.[78] Melchior Cano, as Albano Biondi has shown, did much the same, though he came from a very different theological and cultural world.[79] Unlike Bodin, moreover, he not only found his critical principles in sources other than Nanni's commentaries, but also used them to demolish the credibility of the spurious Berosus and Manetho. The Lutheran *artes historicae* of Chytraeus and others, for their part, show a concern for identifying the "scopus," or central point, of both ancient texts and the actors they described that links them to the new Protestant hermeneutics of Flacius Illyricus.[80]

The scholars who have emphasized these traits have put their collective finger on vital features of the sixteenth-century *artes*. Yet they rarely try to explain what became of these new forms of source criticism in the seventeenth and eighteenth centuries. The larger project of which this paper forms a first part will try to set out in detail the ways in which tradition and innovation fused and interacted in the *artes historicae*, Italian and northern. It will ultimately trace the larger contours of the *ars historica* into the eighteenth century, and it will work out the ways in which the *artes historicae* shaped, and the ways in which they reflected, the practices of contemporary readers and writers of history.

The Historian's Speeches: Rhetorical Decorum as a Hermeneutical Tool

One essay cannot, of course, come close to deciding the questions that remain open. But it can give us the chance to pull one thread from the variegated tapestry of the *artes historicae* and subject it to an examination, and the quarrel from which we began suggests the thread we should choose. As Le Clerc pointed out, Vossius, in his seventeenth-century *ars historica*, had already discussed the larger question of whether historians should compose speeches for their characters. Vossius, in turn, noted that a number of scholars had raised the issue before him.[81] Let us begin by asking what arguments about historical speeches feature in the *ars historica* tradition.

Lorenzo Valla laid down the main lines of one tradition long before the *ars historica* as such came into existence. His defense of history against Aristotle centered on the speeches that played a central role in making history instructive. Historians, Valla pointed out, did not record these word for word, but composed them, artistically, to teach the same sorts of general lessons that the poets embodied in the actions of their mythical heroes: "Does anyone actually believe that those admirable speeches that we find in histories are genuine, and not rather fitted, by a wise and eloquent writer, to the person, the time, and the situation, as their way of teaching us both eloquence and wisdom?"[82] Valla himself composed splendid examples of imaginary historical speeches to serve as cases in point. In his *Declamatiuncula* against the *Donation of Constantine*, he cited the anachronistic usage and medieval syntax of that document to show that it could not have been written in the fourth century. But in the same text, Valla also composed speeches in which the Senators, Constantine's sons, and the pope himself argued that the emperor should not give away the western empire. These were not, of course, historical documents. Rather, they represented Valla's versions of the speeches that these gentlemen should have delivered, at that time, in that place, in the light of their status and their circumstances. The rhetorical doctrine of decorum, which showed how to work out the ways of acting and speaking that were appropriate in a particular situation and to particular actors, provided Valla with his basic tool: the techniques that he used first to think himself into his protagonists' situation, and then to write what they should have said.[83] Decorum, in fact, was a technical and conceptual crossroads—the point where the protocols of rhetoric, which taught one how to compose speeches appropriate to a particular situation, met those of moral philosophy and political prudence. By inhabiting it, the historian could make his work as general as that of the poet. In the end, Valla argued, "so far as I can judge, the historians show more gravity, prudence and civil wisdom in their speeches, than any of the philosophers manage to in their precepts."[84]

Later humanists like the Neapolitan Giovanni Gioviano Pontano accepted Valla's view, though they generally ignored the philosophical stimulus which gave rise to it, and argued that the historian should introduce not only true speeches, but also compositions that had verisimilitude, as often as he possibly could. Pontano voiced the conventional wisdom about speeches and their role when he summed up: "Speeches greatly adorn a history—especially those cast in direct speech, where rulers are introduced speaking and acting in their own persons, so that one seems to be watching the event take place. But they must be fitted to their place and time, and decorum must be retained in every case."[85] Sebastian Fox Morcillo was only one of the numerous later writers who agreed that speeches could work splendidly so long as the historian observed decorum and made them appropriate to the actors in question: "but these speeches are to be made, when the context demands, in such a way as to maintain above all the decorum of the person speaking."[86] The speech served three central purposes: it forced the historian who composed it to think his way formally into the situation in which his actors had had to make and explain their choices; it enhanced the reader's prudence by enabling him to do the same; and it adorned the text with splendid examples of rhetoric in action.

At least one ancient historian—Justin, who in the third century abridged the Augustan world history of Pompeius Trogus—had already argued that historians should not compose speeches (38.3.1). But it was a modern, the ever-independent Paduan professor Francesco Patrizi, who put the methodological cat among the rhetorical pigeons. In his brilliant, ferocious work of 1560, *Della historia diece dialoghi*, he concentrated on epistemological and methodological issues. The first law of history, as Cicero had pointed out, was to tell the truth. But fear, prejudice, and the opacity of high politics in a world of courts made this impossible. And the traditional, rhetorical form of historiography also helped induce its practitioners to violate their own laws. For by including fictional speeches in their works, they introduced deliberate lies:

"Therefore," I added, "he will say neither more nor less than the truth." "That is true," they said. "And to tell the truth is to tell the facts as they are." "That is also true." "Therefore," I added, "he will not be an orator at all, nor will he compose speeches or anything of the sort." "O," said Cataneo, "how do these conflict with the truth?" So I answered: "Didn't you say that making speeches is the work of the orator?" "Yes, I did." "And doesn't the orator use words to make that which is less seem greater?" "Yes." "Well, that is how the work of the orator goes against the truth of the historian."

Even the defense that historians followed "the decorum of persons," Patrizi showed, could not hold. With manifest inconsistency, moderns represented ancient Romans, and Athenians represented Spartans, as speaking in the same way they did. But in doing so, they preserved stylistic decorum at the expense of the very historical and personal decorum they were supposed to observe: "There are many inventions, and these include the speeches. A clear proof of this is the well-known fact that the Romans never spoke as some recent historians have made them speak, and the Lacedaemonians never declaimed in the way in which a certain Athenian made them argue."[87] Decorum was not a tool for understanding political and military situations in a different time and place, but a Procrustean bed on which the historian maltreated wildly divergent individuals. A century and a half later, Le Clerc would wield exactly the same arguments against Curtius.

Again and again, authors of later *artes historicae* rebutted Patrizi. For the most part, they merely repeated the arguments that Pontano and others had advanced in favor of speeches before Patrizi wrote. Giovanni Antonio Viperano, Jesuit, historian, and bishop, knew school rhetoric well enough to admit that flavorless academic speeches could make a history slow reading.[88] But he insisted on the value of the intellectual and stylistic exercise required to compose historical speeches. Nothing could be harder, nothing more rewarding than learning to understand the "decorum" of one's characters and their situation. Clothing an actor's thoughts in one's own words did not amount to falsification. After all, one could rarely reconstruct them in detail; why not then follow probability, as one should when the course of events was obscure? "Some deny that a historian may insert any speech, if he has to make it up; for history can admit no fiction. But events work in one way, words in another. The former can be described as they took place. The latter cannot possibly be reported as they were pronounced. And if it is acceptable to follow probability when recounting obscure events, why is that not permissible for words as well?"[89] Five years later, the historian of Savoy and Genoa, Uberto Foglietta, directed even sharper rhetorical shafts against those who claimed that speeches destroyed a history's claim to truthfulness: "Finally, these [critics] attack speeches. These men, the sharpest intellects ever reported, deny that direct speeches are acceptable in history, though they do not attack indirectly reported speeches. By direct speeches, they mean orations by individual speakers; by indirect ones, those in which it is not the person himself who is brought on speaking, but when we narrate in our own words and person what someone said. They insist that direct addresses clearly violate the standard of truthfulness that ought to lie at the core of history."[90] Foglietta criticized those who insisted that speeches should only be given in their actual wording as "superstitious," since no one could recall even an everyday conversation word for word. Like Viperano, he admitted that historians who did not understand the difference between their world and calling and those of a statesman could spoil a history by composing long and pedantic speeches.[91] On the whole, however, he showed even more enthusiasm for historical oratory than his predecessor, and he devoted several pages to quoting and appreciating an exemplary speech from Livy. The Ferrarese scholar Alessandro Sardi, though less wordy than Foglietta, agreed on the main point. Careful attention to decorum could make speeches substantively truthful and rhetorically effective.[92] La Popelinière agreed no one could possibly recall speeches word for word. Orators, he argued, should not have the freedom to make up historical speeches as they liked, but "learned, experienced, and judicious" historians should be allowed to insert speeches into their work.[93]

Vossius, then, followed majority opinion against Patrizi and his few allies when he plumped for the tradition of speech writing and devoted two chapters of his work to describing in detail how each of the Greek and Roman historians had practiced it.[94] The main point he added was a practical one, the reflection of an experienced teacher of classical rhetoric who knew a great deal about ancient public life. Even if the historian could obtain stenographic transcripts of what his protagonists had said, Vossius argued, he would have to rewrite them in order to unify the style of his work. Otherwise its diction would be "hybrid and inconsistent."[95] For the most part, in other words, the waters of the *ars historica* closed over the great stone that Patrizi had hurled into them, and only the ripples that appeared as one authority after another briefly broke the surface and refuted his iconoclastic position revealed that he had ever written.

Yet one writer in the tradition responded to Patrizi far more edgily—and in a far more complex way—than his colleagues. Bodin had no love for speeches. When he rehearsed the ancient critiques of speeches that he turned up in Justin and Diodorus Siculus, he showed some sympathy for their authors. Even Thucydides had perhaps gone in excessively for oratorical composition. Livy certainly had: remove the speeches from his history, Bodin cracked, and only fragments would remain. That consideration, a reasonable one for once, had impelled Caligula to remove the works and images of Livy from all libraries.[96] Those who read histories only to pick out the speeches showed how superficially they understood the field: "Those who admire only the eloquence and fictional speeches in history are silly. In my view, it is impossible for someone who aims at giving pleasure when he writes to aim at the truth of things as well." True, Bodin derived his principles of source criticism from a forger.

Nonetheless, he applied them diligently enough that he had ceased to tolerate licensed rhetorical fantasies in his history.

Bodin also made clear that he did not approve of Patrizi's radical iconoclasm. Even the greatest historians had faults, but all historians provided indispensable knowledge, and the *ars historica* should provide rules not for devising some imaginary perfect history at a distant date in the future, but for reading the historians that already existed: "It is madness to hope for better historians than the ones we have. Even to wish for them seems a sort of crime. I do not see any point in the work of those who create for themselves an ideal of a perfect historian, of a kind that has never existed and never can exist, but ignore the ones that we actually read and reread."[97] To carry out this scrupulous but pragmatic examination of the sources, Bodin explained, one must abandon the prejudices natural to the time and place in which one lived. The great Hellenist Guillaume Budé had denounced Tacitus for writing against the Christians;[98] Bodin rejected this judgment: "Tacitus acted impiously, because he was not a Christian, but he was not impious when he wrote against us, since he was bound by the pagan superstition."[99] In fact, Tacitus would have committed an impiety if he had failed to defend his own religion—especially when he was conditioned to do so by the sight of Christians and Jews accused of abominable crimes and dragged off for punishment.[100] Modern scholars, in other words, needed to judge the texts and traditions they confronted in historical terms, the terms of their creators—not the anachronistic ones of their own, later times. Bodin could defend the historical tradition without claiming that it was perfect—just as Perizonius would, much later, in his reply to Le Clerc.

This single test case offers far too narrow a basis on which to judge the treatises *de arte historica,* but it is suggestive nonetheless. On the one hand, none of the authors of an *ars historica* anticipated the full, radical modernity of Le Clerc's work—his call for a kind of history that would emancipate itself in every way, substantive and stylistic alike, from the classical tradition. On the other hand, Patrizi and Bodin forged the sharpest tools that Le Clerc and Perizonius wielded when they went to work, in their different context and for their different ends, to criticize the ancients. Bodin advanced a tolerant historicism that seems to have much in common with that of Perizonius. And both men appreciated, much as Le Clerc and Perizonius did, the idea that the best historian was not a participant in events but a critical reader who appeared when the tumult and shouting had died and recreated what had happened from the sources, critically assessed.

Much remains to be learned. How far did scholars not only read, but apply, the teachings of the *artes historicae*? Le Clerc referred to these works explicitly, but Perizonius did not, and both referred with far more frequency to recent philological works, miscellanies and commentaries on classical texts. Did they encounter the new principles and practices of Patrizi and Bodin in these intermediate sources? Or did the philologists in whom they took such interest reformulate them independently? These and many other questions need answers, and the case of Curtius—a minor and derivative historian—cannot yield enough information to solve them. Yet we are not quite finished with his story and its implications.

Curtius in the Quattrocento: Back to the Future in Ferrara

Most writers on the *ars historica* assume one point without much question: that its authors devised the critical principles that they stated and applied, and that readers of history before the sixteenth century could not have anticipated their hermeneutical subtlety. Yet one further source in which Curtius comes up for critical discussion—the *De politia litteraria* of the Milanese humanist Angelo Decembrio—calls these assumptions into question. Decembrio described, in concrete and often strikingly accurate historical detail, the discussions on literary questions that had taken place in the court of Ferrara in the 1440s, when Guarino of Verona and his pupils dominated scholarly life in the city and the marquess himself, Leonello d'Este, played an active role in learned circles. A hybrid modeled on Aulus Gellius and Quintilian, a madly energetic combination of nostalgic literary dialogue and grindingly detailed textbook on orthography, the *Politia litteraria* offers everything from instructions on the proper use of diphthongs to lively accounts of philological and antiquarian discussions set in Ferrarese libraries and gardens.[101] It evokes both a particular culture of facts—chiefly ancient facts—and a particular modern ability to discriminate among them.[102]

Decembrio portrays Leonello and his friends as immersed in—almost obsessed with—ancient history and historians. They seem to have lost no opportunity to evoke the classical models whose actions great men should imitate. Leonello spent his winters in special quarters in the Este city palace, their walls adorned with images of Scipio and Hannibal: the former equipped, appropriately, with a horse and a servant, the latter with an elephant bearing a castle and an Ethiopian guiding it.[103] In fact, he took a special interest in the images of ancient heroes preserved in coins and statues. When a prosaic friend objected that he saw no intellectual merit in collecting works of art, as opposed to books, Leonello—at least in Decembrio's dialogue—joined the chorus of humanists that denounced the philistine. He took as much pleasure in seeing the faces of the ancient worthies, he explained, as in reading about them in texts "that are perceived only by the mind," just as his friend Giovanni Gualenghi felt reverence every time he looked at his picture of Jerome and the lion, and the young poet Tito Vespasiano Strozzi renewed his mourning for a dead girl every time he looked at her picture in a locket.[104] Leonello cultivated an acute visual sensibility. Though he loved and collected Flemish tapestries, he denounced the weavers who indiscriminately included in their works apocryphal stories about the Roman emperors.[105] No wonder that Leonello offered his support to Pisanello and Matteo de' Pasti, or that his court became the epicenter of the new classicizing taste for medals.[106] The Ferrarese humanists, in other words, envisioned the ancient world in three dimensions, if not in living color.

In the end, as one might expect, Leonello, his teacher and their friends preferred texts to all other relics of the ancient world, and they preferred historical texts to most others. Decembrio describes Guarino as praising the fine three-volume manuscripts of Livy, their title pages adorned with wreathed vine leaves and heroes on horseback, that were a specialty of the Florentine bookdealer Vespasiano da Bisticci—and that Leonello and his friends actually bought.[107] The dialogues in the *Politia* make clear, moreover, that these books were not just treasures to be contemplated. Leonello's courtiers scrutinized them, criticized their spelling, and even played games of *sortes Livianae* with them, passing the triple-decker Livy from hand to hand as they tried to find by lucky dips in the text exactly the passages that verified their beliefs and prejudices (Decembrio shows one of Leonello's courtiers, a great fan of Hannibal's, undergoing humiliation when he cites a passage that shows Hannibal acting in an undignified way when he is forced to leave Italy).[108] History, in Leonello's Ferrara as in Alfonso's Naples, meant the material texts of great ancient writers.

Unlike Alfonso, however, Leonello did not see Curtius's history of Alexander the Great as a sovereign remedy against ill health and that rundown feeling. When he fell ill, he worked through historian after historian. Curtius interested him because he described Alexander, himself very ill, paying no attention to warnings of poison and boldly gulping down a medicine prescribed by his doctor, Philip.[109] But the further Leonello read in Curtius, the more problematic he found the text.[110] It swarmed with contradictions. Why, after all, would Alexander drink a potion that he had been warned might be poisonous?[111] Why did Alexander find it impossible to get information about Darius's whereabouts? Curtius explained that no Persians ever deserted and that they kept the affairs of their rulers deeply secret. But he himself, a couple of books later, described a deserter informing Alexander about Darius's cavalry traps. Did the gods hide Darius, as if he were a Homeric hero having imaginary adventures?[112] And how could Curtius make Darius say, in his last speech, both that he had never contemplated running away, and that Alexander had twice put him to flight?[113]

Curtius also contradicted the clear testimony of nature and reason. He described the Indians' bows as so long and heavy that one could shoot them only by resting one end on the ground—an implausible technical error that would have made it impossible for bowmen to carry out their natural task of shooting quickly.[114] He claimed that the Indians' elephants terrified the Greeks' naturally timid horses—an inaccurate description that Leonello, like all the Estensi a connoisseur of horseflesh, dismissed with disdain.[115] Curtius even made fun of magic—not the ordinary, low sort well known in Leonello's world, which involved crowds of dead people leaving their tombs to dance by the light of the moon or groups of living people whose spirits left their bodies, but the learned form of magic by which, as everyone knew, the ancient Egyptians and Persians had successfully predicted future wars, pestilences, and changes in regime and religion.[116] Despite his grace and skill as a writer, Curtius posed special problems of credibility, at least to the reader whose sensibilities had been formed in this Ferrarese circle. Leonello or his friends, immersed in the classics and committed to using them, acutely sensitive to the visual and material world that the ancients had inhabited, found themselves falling down a rabbit hole, shocked and sickened by the speed of their fall and the strangeness of what they found at the bottom, when they went through passages that most fifteenth- and sixteenth-century readers traversed with pleasure and without incident.

Leonello did more than notice these inconsistencies. He also traced them back to his writer's character and formation. Curtius, he argued, was really a rhetorician rather than a historian. That explained why he praised his hero's self-restraint and moderation, even though he knew perfectly well that Alexander had enjoyed nothing more than a bout of sex with eunuchs and male prostitutes: "Therefore he claims that Alexander never indulged in extravagance, or indulged in passion in a way that might be seen as unnatural. Note how this conclusion contradicts what he wrote before."[117] That explained, too, why Curtius absurdly described the habits and rituals of a single "king of India," even though his own account made clear that, then as in Leonello's own time, India had had many rulers. Curtius, as he himself admitted, was less a writer (*scriptor*) than a copyist (*transscriptor*)—and at that he did not believe everything he copied. Yet he took over the implausible with the plausible, the vague with the precise, so long as it supported his flattering portrait of his hero.[118] Decembrio assures us that Leonello himself subjected the historian to this acid bath in historical and rhetorical criticism—one that ended, by his account, with the suggestive remark, unfortunately not pursued, that Herodotus too said many implausible things.[119]

Leonello—and Decembrio—did not anticipate all the methodological questions and suggestions to be found in the *artes historicae*, any more than those treatises adumbrated all the bold ideas of Le Clerc and Perizonius. Yet it seems clear that humanists began to pose new questions about history—questions about source criticism, about the problems inherent in rhetoric as the central discipline of historical writing, about the relation between natural and technical knowledge and historical texts, and about the general status of ancient writers—well before they began to draw up even the first, relatively traditional treatises on the subject. Leonello and his friends already saw the ancient historians less as inimitable, perfect accounts of events than as humans and historical sources, from whose accounts the real past must be reconstructed with the aid of philological and antiquarian learning.

A genial and helpful cicerone through the mansions of the Republic of Letters, Curtius cannot take us everywhere we need to go. But he has shown us much. By following him, we have learned to see past the deceptive smoothness of early modern rhetoric, to realize that novel ideas and practices took root and flourished in what scholars have often mistaken for a culture mired in tradition. He has led us from Leiden to Paris and from Paris to Ferrara, from the journals of the Republic of Letters to the disputations that took place in princely courts, and by doing so has revealed the existence of microenvironments where the pursuit of *historia humana* involved the creation of new practices and the cultivation of a new sensibility. And he has made clear the variety and richness of that small but distinctive sector of *historia* inhabited by those who wrote and read the *artes historicae*.

Notes

Sources cited in boldface type appear in the Primary Sources of the bibliography.

1. For some recent perspectives see Bots and Waquet 1994; Bots and Waquet 1997; Goldgar 1995; Miller 2000; Grafton 2001a; Malcolm 2002.

2. **Perizonius** 1685; **Perizonius** 1740a; **Perizonius** 1740b. See Erasmus 1962; Meijer 1971; and Borghero 1983.

3. Manuel 1959, 1963; Sartori 1982, 1985; Raskolnikova 1992; Grell 1983, 1993, 1995; Grell and Volpilhac-Auger 1994. For the wider context see also Borghero 1983; Völkel 1987; MILLER, MULSOW in this volume.

4. Barnes 1938; Pitassi 1987.

5. Bentley 1978.

6. For recent perspectives see Bosworth and Baynham 2000, especially Bosworth 2000 and Atkinson 2000. On the earlier popularity of the text, in diversely interpolated and adapted forms, see Cary 1956 and Ross 1988.

7. For a fascinating account of the way in which ancient historians were reconfigured to meet the tastes of courtly audiences in the Middle Ages and the Renaissance, see Dionisotti 1997 (for Curtius see esp. 540–541).

8. See François Baudouin's account in **Wolf** 1579, 1:706.

9. Winterbottom 1983.

10. Erasmus 1518 = **Allen et al.** 1906–58, no. 704, 3:129–131.

11. See **Wheare** 1684, 46, quoting the passage of Lipsius.

12. Frischlin 1588, 1–21.

13. Notes on Frischlin's lecture on Darius's speech in Curtius 1953 (4.14.9–26) appear in a copy of **Frischlin** 1588, which in turn forms part of a Sammelbändchen in the Herzog August Bibliothek, Wolfenbüttel (A: 108.3 Rhet. [3]). The quotation appears in **Frischlin** 1588, 4: "Annibal lib. 21 apud Livium eodem argumento utitur."

14. Le Clerc 1712a, Pars iii, 395–512, esp. 396.

15. Le Clerc 1712a, 402.

16. Le Clerc 1712a, 402–421, 457–475.

17. Le Clerc 1712a, 430–436.

18. Le Clerc 1712a, 448–457.

19. Le Clerc 1712a, 422. Le Clerc here rather resembles the contemporary theologians and natural historians who tried to extirpate marvels from other sectors of the encyclopedia in which they had traditionally played central roles.

20. Le Clerc 1712a, 423: "In posteriorum numero fuisse Q. Curtium res ipsa ostendit."

21. Le Clerc 1712a.

22. Le Clerc 1712a, 494–506.

23. Le Clerc 1712a, 488: "Ut nunc ad orationes veniamus, quas directas plurimas habet Curtius, ut vix totidem alibi occurrere in tam parvo volumine existimem; ante omnia, profiteri necesse habeo me esse in eorum sententia, qui in Historia gravi orationes omnes et directas et obliquas omittendas censent; nisi essent, aut earum sententia certissime sciri possit."

24. Le Clerc 1712a, 488.

25. Le Clerc 1712a, 490.

26. Le Clerc 1712a, 489.

27. Le Clerc 1712a, 490: "Sed et sensim, cum hisce figmentis, subit major fingendi licentia."

28. For eloquent explications of this exemplar regime, see Nadel 1964 and Landfester 1972.

29. Le Clerc 1712a, 488–489.

30. Le Clerc 1699, 1:249.

31. Le Clerc 1699, 175.

32. Le Clerc 1699, 145.

33. Le Clerc 1991, 256–257.

34. Aelian 1731, 678–679, 783–784. Le Clerc replied in 1702. See **Le Clerc** 1715, [*8v]–**2r, and Meijer 1971, 152–155.

35. Perizonius 1703, 28–30.

36. Perizonius 1703, 52.

37. **Perizonius** 1703, 37: "Sed nemini minus, quam mihi, opponi debebat nimia illa Antiquitatis admiratio, per quam fiat, ut vitia Auctorum Veterum neque ipse agnoscam, neque ab aliis commonstrari velim"; 191: "Nullus apud me tantae auctoritatis est Scriptor, si ab Sacris discessero, quem in iis, quae scripsit, nullum humanae imbecillitatis monumentum reliquisse putem, sive ille sit ex Antiquis, sive ex Recentibus."

38. **Perizonius** 1703, 3.

39. **Perizonius** 1703, 51.

40. **Perizonius** 1703, 160.

41. **Perizonius** 1703, 92: "quum historici munus sit, omnia in unum Historiae corpus redigere, et ita omnia uno etiam exprimere stylo, ne corpus illud evadat monstrosum aut inaequalitate sua deforme."

42. **Perizonius** 1703, 50.

43. Butterfield 1955; Antoni 1968; Antoni 1951; Marino 1975; Marino 1995; Bödeker et al. 1986; Blanke 1991; Leventhal 1994.

44. **Le Clerc** 1715, **2r.

45. **Le Clerc** 1712a, 247.

46. Barnes 1938, 151–152.

47. **Le Clerc** 1712b, 23–24.

48. Israel 2001; Hazard 1935.

49. **Le Clerc** 1712a, 488 and note. See also **Le Clerc** 1699, 1:182, with a characteristic barb in the tail.

50. Little work has been done on Johannes Wolf and his effort to produce a canon of *artes historicae*. These texts had attracted interest in Basel as early as 1551, when Oporinus reprinted Christoph Milieu's *De scribenda universitatis rerum historia libri quinque*, a highly original work that would reappear in Wolf's collection (and on which see Schmidt-Biggemann 1983 and Kelley 1999). But it seems likely, as Giorgio Spini suggested long ago (Spini 1948, 1970), that the phosphorescent letter of 20 November 1562 in which Jacopo Aconcio recommended Patrizi's newly published dialogues on history to Wolf played a vital role in making him see the interest of this new genre (**Aconcio** 1944, 350–355).

51. Gilbert 1965.

52. See the Introduction to this volume; POMATA in this volume; Seifert 1976.

53. **Keckermann** 1614, 2:1343; cf. Tooley 1953; Glacken 1967; Couzinet 1996.

54. See e.g. Chytraeus in **Wolf** 1579, 2:468–469.

55. **Franckenberger** 1586, 142–143.

56. **Campanella** 1954, 1244, 1252–1254; cf. Spini 1948; Spini 1970.

57. MILLER; MULSOW; POMATA, all in this volume.

58. Patrizi in **Wolf** 1579, 1:412.

59. **Patrizi** 1583; **Patrizi** 1594.

60. For Robortello see **Roberti** 1691–92, 1:593–686; for Sardi see **Sardi** 1577 [*8v].

61. **Reineck** 1583 documents these sophisticated circles in detail.

62. On the *ars historica* see in general the full bibliography of primary sources given in Witschi-Bernz 1972a. Contemporary lists include **Keckermann** 1614, II:1309–1388, with commentary, and **Draud** 1625, 1125–1126. The development of the genre has been traced by Bezold 1918; Moreau-Reibel 1933; Strauss 1936, chapter 6; Brown 1939; Spini 1948, 1970; Reynolds 1953; Pocock 1957; Klempt 1960; Franklin 1963; Kelley 1964, 1970, 1973, 1999; Nadel 1964; Cotroneo 1966, 1971; Huppert 1970; **Kessler** 1971, 1982; Landfester 1972; Witschi-Bernz 1972b; Cano 1973; Seifert 1976; Dubois 1977; Hassinger 1978; Schmidt-Biggemann 1983; Muhlack 1991; Grafton 1991b; Wickenden 1993; Couzinet 1996, 2000; Salmon 1997; Bellini 2002; Lyon 2003. Two anthologies, **Wolf** 1579 and **Kessler** 1971, contain the most influential *artes historicae* and serve as the chief foundation for the exposition that follows. Though [**Wolf**] 1576 lacks Wolf's name, it includes the same preface to the reader by Petrus Perna, the publisher, that appears in **Wolf** 1579. Evidently it is an earlier version of the collection.

63. For Guarino and Lucian see Regoliosi 1991, which provides a new text of the letter (28–37). For further evidence of diversity in Quattrocento discussions of history, see Regoliosi 1995a and Grafton 1999.

64. L. **Valla** 1973, 3.

65. L. **Valla** 1973, 5. On the complex problem of Valla's sources and for a full interpretation of his arguments, see Regoliosi 1994.

66. Black 1987.

67. Robortello in **Kessler** 1971 makes clear that his model for this anti-Aristotelian work is Aristotle's *Rhetoric*. Cf. **Robortello** 1968; **Robortello** 1975 for further evidence that this Padua-trained scholar, who had nothing to do with Ramus, shared Ramus's preoccupation with making humanistic studies into formal *artes*.

68. **Wolf** 1579, 2:457–465. For the larger context of these arguments see Vasoli 1989, 25–90.

69. **Kessler** 1971; Kessler 1982.

70. See Goez 1974; Stephens 1989; Grafton 1990a; Rowland 1998.

71. Backus 2003 surveys these developments; see also the comprehensive treatment in Cochrane 1981 and the detailed and insightful studies of ecclesiastical history as practice in Ditchfield 1995 and Hartmann 2001.

72. Lyon 2003 makes this case with great clarity.

73. Moreau-Reibel 1933; Franklin 1963; Kelley 1966, 1970; Huppert 1970; Couzinet 1996.

74. See Grafton 1991b; Ginzburg 2000; Biondi, introduction to **Cano** 1973; **Cano** 1776.

75. Zedelmaier 1992; Blair 1997; BLAIR in this volume.

76. Aconcio 1944, 305–313, reprinted in **Kessler** 1971; BLAIR in this volume.

77. See respectively Nadel 1964; Koselleck 1984; **Kessler** 1971, 7–47; Kessler 1982; Muhlack 1991.

78. See especially Brown 1939; Klempt 1960; Franklin 1963; Huppert 1970; Couzinet 1996.

79. See **Cano** 1973.

80. Chytraeus in **Wolf** 1579, 1:460.

81. See in general Wickenden 1993.

82. L. **Valla** 1973, 3.

83. Cf. especially Most 1984 and Kablitz 2001.

84. L. **Valla** 1973, 6.

85. Pontano, *Actius*, in **Wolf** 1579, 1:575–576.

86. Sebastian Fox Morcillo, *De historiae institutione liber*, in **Wolf** 1579, 1:794.

87. Patrizi, *Della historia diece dialoghi*, in **Kessler** 1971, 58v; **Wolf** 1579, 1:532–533.

88. Giovanni Antonio Viperano, *De scribenda historia liber*, in **Kessler** 1971, 40.

89. Viperano in **Kessler** 1971, 39–40.

90. Uberto Foglietta, *De ratione scribendae historiae*, in **Kessler** 1971, 31.

91. Foglietta in **Kessler** 1971, 31–33, 35: ". . . qui homines mihi non intelligere videntur quantum inter philosophum in scholis, atque in coetu doctorum et ingeniosorum virorum disputantem, praeceptaque vivendi tradentem; et civilem virum in concione, et in imperitorum turba verba facientem intersit; neque meminisse quantum utriusque munera inter se discrepent."

92. Alessandro Sardi, *De i precetti historici discorsi*, *Discorsi*, in **Kessler** 1971, 155–156. See also **Maccius** 1593, 117–129.

93. **La Popelinière** 1599, II, 75–81; see esp. 77: "Qui dit autre chose ment, non celuy qui la rapporte d'une autre façon."

94. **Vossius** 1699, cap. xx, 31–34; cap. xxi, 34–35.

95. **Vossius** 1699, 31: "Si vel historicus nactus sit orationes, prout eas notarii exceperunt, necesse tamen erit, ut caeterae historiae stylo accommodentur, ne dictio sit hybrida, ac dissimilis sui." Also interesting is the rich but inconclusive discussion in **Keckermann** 1614, 2:1322.

96. Bodin, *Methodus*, cap. iv, in **Wolf** 1579, 1:50: ". . . concionum tamen apparatum tacite reprehendit Diodorus. Eadem reprehensione utitur Trogus Pompeius (ut est apud Iustinum) adversus Livium et Sallustium, quod directas et obliquas conciones operi suo inserendo, historiae modum excesserint [38.3.11]. Nihil est enim, ut ait Cicero, in historia pura et illustri brevitate dulcius. Sin conciones de Livio detraxeris, exigua fragmenta restabunt. Quae causa Caligulam impulit, ut scripta Livii et imagines de bibliothecis omnibus prope amoveret."

97. Bodin in **Wolf** 1579, 1:49.

98. Bodin in **Wolf** 1579, 1:64: "Budaeus acerbe Tacitum scriptorem omnium sceleratissimum appellavit: quod nonnihil adversus Christianos scripsit."

99. Bodin in **Wolf** 1579, 1:64.

100. Bodin in **Wolf** 1579, 1:64: "ego vero impium iudicarem nisi quamcunque religionem veram iudicaret, non eam quoque tueri et contrarias evertere conaretur. Cum enim Christiani et Hebraei quasi venefici et omnibus sceleribus ac stupris infames ad supplicia quotidie raperentur, quis historicus a verborum contumeliis abstineret?"

101. For a general introduction see Grafton 1997. The text has now received a modern edition, **Decembrio** 2002. Though unsatisfactory in some respects, it has the merit of existing, and I will cite it in what follows.

102. CRISCIANI in this volume.

103. **Decembrio** 2002, 191 (2.14.1).

104. **Decembrio** 2002, 431–432 (6.68.20–22).

105. **Decembrio** 2002, 427–428 (6.68.10).

106. On this chapter of Decembrio see the edition and commentary in Baxandall 1963. For antiquarianism in Decembrio see further Curran and Grafton 1995.

107. **Decembrio** 2002, 459 (7.75.7).

108. **Decembrio** 2002, 230–231 (2.24.4).

109. **Decembrio** 2002, 411 (6.65.3): "Amicis igitur percunctantibus, quidnam potissimum per eam valitudinis intemperiem lectitasset, respondit historicos clariores diligentius tractavisse, Plinium maiorem, Livium, Salustium, commentarios Caesaris, Iustinum, sed Curtium quam antea studiosius. Cuius rei causa fuisset aegrotatio, ut de Alexandri modo disquireret: quam repente ob illius fluminis ingressum exanimatus, quanta mox fiducia, an temeritate magis, poculum sollicitudine plenum a medico suscepisset, utrum scriptoris de ipso rege an regis potius tam praesens, in dubio tamen eventu, sententia fuisset. Quanquam non in hac de Alexandro controversia solum, sed et aliis eius *historiae* locis Curtium sibi videri dicebat adeo contraria inter se protulisse, ut pene a naturali usu, ipsa etiam rerum veritate dissentire viderentur. Sperare aliquando ea demonstrandi tempus affuturum."

110. Petrarch had similar problems with Curtius, which he solved by portraying Alexander as "an unbalanced youth protected by Fortune, vain, wild, almost insane; and the resultant picture of him in the *De Viris Illustribus* was unique for the age in its single-minded, deliberate abuse" (Cary 1956, 266). For recent efforts to make literary and cultural sense of Curtius see e.g. Atkinson 2000 and Spencer 2002, and cf. more generally Carney 2000.

111. **Decembrio** 2002, 418–419.

112. **Decembrio** 2002, 419–420.

113. **Decembrio** 2002, 421–422.

114. **Decembrio** 2002, 422–423 (6.67.19): [Curtius says that Indians' bows were so large that they had to rest one end on the ground to shoot them (8.14.19,

8.9.28)] "Mirum est vel incredibile magis, quod ait historicus . . . Quod teli vel ballistae genus potissimum licet apud omnis gentes paululum forma vel qualitate sit differens, tamen ita fieri et excogitari solet ab hominibus, ut tractando feriendoque sit aptissimum." Decembrio 2002, 420 (6.67.11), on 8.14.23.

115. Decembrio 2002, 420 (6.67.11), on 8.14.23.

116. Decembrio 2002, 424 (6.67.23–24), on 7.4.8.

117. Decembrio 2002, 423–424 (6.67.22), on Curtius's praise of Alexander's self-restraint (10.5.32).

118. Decembrio 2002, 422 (6.67.16–17), on 8.9.23 ff—a splendid description of how the king of India is preceded by attendants bearing silver trays of incense, while he reclines in a golden litter set with pearls, listening to the singing of trained birds: "Neque enim vel Romanorum veterum, dum Curtius scriberet, seu Graecorum, si ad primos eius historiae scriptores intendimus, seu praesenti tempestate conveniunt, quae de Indorum rege memorantur. Quo satis perspicitur hunc Alexandrinae historiae transcriptorem potius, ut supra monstravi, quam scriptorem fuisse [note that the passage in question underlines Curtius's credulity: 9.1.34: Equidem plura transcribo quam credo; nam nec affirmare sustineo de quibus dubito, nec subducere quae accepi], qui, uti ab historico Graeco scriptum invenit, de Indorum aliquo rege illius tempestatis ita transcripserit. Sed quisnam tandem ille rex fuit? Nempe non ab hoc historico percipimus. Qui sic regis Indi mores vitamque describit, hominis appellationem non producens, ut tanquam unicus Indiae rex eiusdem semper vitae consuetudinisque describendus videatur ut Phoenix, ac quemadmodum nos unum summum pontificem veneramur, cuius seu Hispani seu Italici ad divinum cultum spectantes ritus cum omnium fere summorum pontificum, qui fuerint quique demum insequantur, religione conveniunt. Sed quid ambio? Nec Alexandri quidem temporibus unicus Indiae rex fuit, si eius acta perspexeris, sed complures reges reginaeque traduntur, necdum omnes ab Alexandro superati. Quod idem auctor paulo supra manifestat *Indiae regum* inquiens, non *regis*, luxuriam . . . [8.9.23: Regum tamen luxuria, quam ipsi magnificentiam appellant, super omnium gentium vitia]."

119. Decembrio 2002, 425 (6.67.25): "Nam omisso nunc Curtio, de quo satis hodierna disputatione memoravimus: quis non apud Herodotum multo plura deprehenderit, ac magis, quam in hoc auctore reprehendenda, hoc est minime credenda, quemadmodum historice referuntur. Quid ais, Guarine? At ille ita esse consentiens iandudum Leonelli subtilitatem iugemque memoriam, ut saepe alias, extollebat: adiecitque huius auctoris stilum inter caeteros interpretes perpolitum eminere ac eloquentiae cuiusdam separatae sic inter historicos, uti Plinii minoris esset in oratoribus."

2
Natural History, Ethics, and Physico-Theology

Brian W. Ogilvie

Introduction

Despite ancient models (Pliny, Aristotle, and Theophrastus), the modern discipline of "natural history" got its name only in the 1540s. The humanists who adopted the term *historia* for their investigations and descriptions of nature also incorporated with it the notion of exemplary history from the classical and medieval historiographical tradition. This alliance, which lasted for little more than a century, was responsible for two developments that brought natural history into close connection with ethics and theology. First, natural history reinvigorated a tradition of moral *exempla* drawn from nature. Bestiaries, lapidaries, and the like had all but disappeared from learned European culture by the sixteenth century, but the moral force of natural examples returned, for a time, in Renaissance natural history. Second, some naturalists and historians urged that natural history was an important step on the ladder to divine history: that is, to understanding the nature and attributes of God. This view of natural history as part of natural theology would help the latter meet the challenge posed in the seventeenth century by the "Epicurean" and Cartesian enemies of divine providence.

How Natural History Got Its Name

Renaissance natural history emerged out of the humanist movement,[1] but the study of nature was not originally part of the humanist program. In *De sui ipsius et multorum aliorum ignorantia*, Petrarch had dismissed knowledge of nature as useless to salvation. "What is the use—I beseech you—of knowing the nature of quadrupeds, fowls, fishes, and serpents and not knowing or even neglecting man's nature, the purpose for which we are born, and whence and whereto we travel?"[2] Fifteenth-century humanist pedagogues did not share Petrarch's contempt for the knowledge of natural things, but neither did they give it a prominent place in their ideal curricula. Pier Paolo Vergerio did note (ca. 1402–1403), "Knowledge of nature [*scientia de natura*] is especially suited to the human intellect; through it we know the principles and passions of things animate and inanimate, as well as the causes and effects of the motion and change of things in heaven and earth." But Vergerio thought of this *scientia* chiefly in terms of heavenly bodies and things "that have effects in the air and near the earth." He admitted that medicine was "a fine thing to know" but said little about it.[3] Leonardo Bruni did not mention natural history or natural philosophy in his letter to Battista Malatesta on study and letters, written in the 1420s.[4] Aeneas Sylvius Piccolomini (1450) urged boys to study geometry and astronomy, but he passed over other sciences in silence.[5]

Battista Guarino is an exception. In 1459, he wrote a treatise describing the educational practices and ideals of his father, Guarino of Verona. Guarino's curriculum was based on the traditional *studia humanitatis*: grammar, rhetoric, poetry, history, and ethics. But he also thought that boys should study books on their own, outside of the set curriculum. "When they begin to study on their own, they should be sure to examine those books that include a wide range of subjects, such as Gellius, Macrobius's *Saturnalia*, and Pliny's *Natural History*, which is no less varied than nature itself."[6] Guarino added Augustine's *City of God*, too, as a source of history and knowledge of pagan rites. Clearly natural history, for Guarino, was not a distinct discipline; students should read Pliny and other collectors of miscellanies to improve their stores of anecdotes and exempla.

Guarino had every reason to mention Pliny: his *Naturalis historia* was *the* book of natural history for medieval Europeans. But medieval manuscripts of Pliny were often partial, identified by subject or *incipit*, and by the thirteenth century, university masters paid it little attention.[7] Humanist scholars first reassembled Pliny's work into a coherent whole and then gave it to the press.[8] The editio princeps (Venice, 1469) bore the title *Libri naturalis historiae*; the term *naturalis historia* or variations on it was used in most incunabular editions, as well as those of the sixteenth century. There were occasional variations—the 1549 edition from Froben, in Basel, was published as *Historia mundi*.[9] But most late Renaissance readers encountered Pliny's work as a "natural history." Had humanists wished to use the name, it would have been readily to hand.

Pliny was not the only model for the *historia* of nature, though Greek examples were less evident to fifteenth- and early sixteenth-century writers. Aristotle's works on animals first saw print not as a *Historia animalium* but, instead, as *De natura animalium*.[10] On the other hand, Theophrastus's works on plants, the *History of Plants* and *Causes of Plants*, not only provided a model but also—even to the casual reader who merely encountered them in a library catalogue—implied a clear distinction between description and explanation. Unknown in the medieval Latin West, they were brought from Constantinople to Italy by Giovanni Aurispa in the early fifteenth century and translated into Latin by Theodore Gaza around 1450.[11] The editio princeps of the Greek came out only in 1497, but Gaza's translation, edited by Giorgio Merula, was first published in 1483.[12]

Historia, then, was available as a category for the study of animals and plants in the late fifteenth century, had humanists wished to use it (POMATA in this volume). But they did not. In the letter to Niccolò Leoniceno that opened the controversy over Pliny's reliability, Angelo Poliziano used "natural history" to refer to the title of Pliny's book, but he used the words "described" [*describit*] and "properties" [*proprietates*] to refer to Pliny's accounts of plants.[13] Leoniceno, meanwhile, titled his book "On the errors of Pliny and other physicians in medicine," and he referred to Pliny's "description" of plants and their "indicia," not to *historia*.[14] When it came to placing Pliny in a discipline, the choice for Leoniceno was between grammarians and orators on the one hand and philosophers and physicians on the other.[15] Leoniceno cited Theophrastus several times by name but never mentioned the titles of his books, and he referred to Aristotle's natural history simply as "Aristotle's books on animals, translated from Arabic to Latin."[16]

In his survey of discoveries and inventions, first published in 1499, Polydore Vergil drew on Pliny's *Natural History* extensively but did not consider "natural history" a discipline; for him, *historia* meant human history.[17] Like Leoniceno, he treated what would become the subject of natural history as a part of medicine.[18] Moreover, Vergil referred to his own work, *De inventoribus rerum*, as both *opus* and *lucubrationes*, not as a history of inventions or inventors.[19] The very breadth of Pliny's work perhaps discouraged imitation. Its thirty-seven books addressed astronomy, geography, the peoples of the earth, animals, plants, minerals, gems, medicine, and even the history of art. If this was "natural history," no wonder that Guarino thought the work no less varied than nature herself. There was certainly no discipline in Pliny's compilation, though the work is less credulous than scholars used to assume.[20]

The language used in the fifteenth and early sixteenth centuries to name and discuss works on natural history reflected a disciplinary and literary structure inherited from the Middle Ages. Bartholomaeus Anglicus's work *On the Properties of Things* was in print by the 1480s in Latin and vernacular translations.[21] Other works' titles used the word *natura*, following the models of Isidore of Seville and the bestiary tradition. Such works combined description, causal explanation, and a discussion of medical uses. The latter was also treated under the heading of "virtues" or "powers"; the 1484 *Herbarius* was subtitled "*de virtutibus herbarum*," and the 1510 edition of Macer began, "You who wish to know the different powers [*vires*] of plants."[22] Some books simply stated the subjects: Albertus Magnus's *De animalibus*, or a late fifteenth-century *Quadripartitum de quercu*.[23]

Albertus is instructive, for his work provides a clue about why the Aristotelian and Theophrastean notion of *historia* failed to take hold among late fifteenth-century medical humanists. The bulk of Albertus's work was a combined translation-commentary on Aristotle's works on animals: Albertus transcribed Michael Scot's translation from the Arabic and added his own glosses and digressions. The last six books, though, included descriptions of individual animals, in alphabetical order, largely taken from Thomas of Cantimpré's *De natura rerum*. Albertus admitted that it was not really philosophical to describe animals in that fashion [*hunc modum non proprium philosophiae*], but he wrote both for the learned and the ignorant, "and what is narrated individually [*ea quae particulariter narrantur*] better informs the rustic mind."[24]

Albertus used the verb "to narrate," but he did not call this addition a history, for it was not, in fact, *historia* in Aristotle's sense. In the *History of Animals* Aristotle did not collect individual animals' descriptions; rather, he considered their parts. Albertus distinguished both the *History of Animals* and the *Parts of Animals* from the other books on animals; the latter he called the "science that involves the animal's entire body," while the former was the "science of the parts and limbs of animals." This science, in turn, involved two parts: "their diversities in composition, operation, and generation; and their natural and proper causes."[25] Both were part of philosophy, unlike the particular narration that Albertus appended in the last five books of *De animalibus*.

Physicians like Leoniceno, who were concerned with properly identifying plants in order to use them as simple medicines, would have found the Aristotelian and Theophrastean approaches to *historia* useless. Instead, they turned to Dioscorides. Some natural philosophers occupied themselves with the knotty problems of properly determining species (see MACLEAN in this volume), but most physicians had other concerns. There was as yet no unified discipline devoted to the study of plants, animals, and minerals. Giorgio Valla's massive, encyclopedic work *De rebus expetendis et fugiendis* underscores this point. Valla's forty-nine books on the secular liberal arts discussed plants and animals extensively but under several distinct headings. Under "physiologia" and metaphysics, Valla discussed the soul (*anima*), generation, and growth, but he did not descend to individual species.[26] In his books on medicine, on the other hand, he enumerated dozens of species of stones, animals and their parts, and plants, but he discussed only their properties as simple medicines—and, as with vipers, scorpions, and even bees, the harmful effects of their venom.[27] Domesticated animals and cultivated plants found a place in the books on "Economy, or the administration of the home," which also included architecture.[28]

Valla's discussion of plants and animals is thus broken down into three distinct disciplinary categories: natural philosophy, medicine, and agriculture. Renaissance natural history developed out of the second of these disciplines. In the 1490s, the debate between Niccolò Leoniceno and Pandolfo Collenuccio over the medical errors of Pliny—many involving the identification of medicinal herbs—and Ermolao Barbaro's philological "castigations" of Pliny's encyclopedia raised two questions: first, did Pliny misunderstand the Greek sources for his names and descriptions, and second, how could the plants described by those sources (above all, Dioscorides) be identified with the plants that grew in fifteenth-century

Italy?[29] Both problems, but especially the second, could best be resolved by carefully studying the plants themselves. An additional complication was added when German students who had learned medical botany in Italy returned home and discovered that the plants that they had studied in Italy often differed significantly from those in the German countryside.[30] Classical descriptions sat alongside contemporary German illustrations in Otto Brunfels and Hans Weiditz's *Herbarum vivae eicones* (1532).[31] But the next significant herbal to be published, Hieronymus Bock's *Kreütterbuch* (1539), eschewed both illustrations and classical descriptions. Instead, Bock described plants himself, then ventured identifications with those described by the ancients.[32] Within a few years, what Bock did would be called "historia." Though its name echoed Aristotle and Theophrastus, the real model was, in fact, Dioscorides.

The earliest book on animals, plants, and minerals since Pliny to be called a "natural history" in fact predated Brunfels's and Bock's works. Nicolaus Marschalk's *Historia aquatilium* appeared in Rostock in 1517.[33] It was followed nine years later by Gonzalo Fernández de Oviedo's *Natural hystoria de las Indias* (also called the *Sumario*), published in Toledo.[34] But neither of these works, published at the edges of the sixteenth-century Republic of Letters, made a big splash. Oviedo would later become an important source for European naturalists, but only later in the century, after histories of nature had become a commonplace genre.[35]

"Histories" of nature entered the European consciousness more generally only in the 1540s. Conrad Gessner published his *Historia plantarum et vires* in 1541; this slim volume was followed the next year by Leonhart Fuchs's *De historia stirpium commentarii insignes* (1542).[36] The 1543 German edition of Fuchs underscored the newly emergent sense of *historia* by glossing the unfamiliar word: "*historia*, that is, name, form, place and time of growth, nature, strength, and effect."[37] Fuchs wrote that his "commentaries on the history of plants" were distinguished by his inclusion of "the whole history of every plant, with the superfluities cut out": that is, "their history, complete and reduced to order" [*historiam integram, & in ordinem digestam*], which neither the ancients nor the moderns had provided. Fuchs alternated between the singular *historia* and the plural *historiae*, an alternation that would characterize later Renaissance natural history in general: a *historia* was either an individual description or the work that contained such descriptions.[38]

For Fuchs, a *historia* in the first sense included not only the plant's names, physical description, times, and places, but also its medicinal virtues. His praise for Hieronymus Bock's precise descriptions was tempered with scorn for Bock's occasional misattribution of medicinal virtues to plants, the result of a misguided attempt to associate Dioscorides's descriptions with German plants.[39] But Gessner's distinction between *historia* and *vires* would eventually triumph, even if Gessner himself retained a more expansive notion of *historia* in his work on animals (PINON in this volume). By the early seventeenth century, Carolus Clusius's and Caspar Bauhin's *historiae* would include only names, descriptions, times, and places.[40]

Fuchs and Gessner realized that the Peripatetics had been interested in philosophical explanation—in causes—but they themselves were not. Fuchs emphasized *historia*, as he understood it, shoehorning the ancients into his category. In his catalogue of ancient writers on plants, he mentioned Theophrastus's "nine outstanding books of history" [*de historia*] but not *De causis plantarum*. He lumped Aristotle's books on animals together as "those remarkable books of their history."[41] Gessner, meanwhile, described his *Historia animalium* (1551–1558) as an "onomasticon or lexicon." Aristotle had written about animals philosophically, and those who wanted a continuous exposition should turn to his books.[42]

The flood of "histories" of animals and plants that followed on Gessner and Fuchs were descriptive. Writers who adopted a philosophical approach avoided the word: Andrea Cesalpino's work *De plantis* (1583) is an example. But there were few such works.[43] Whether written for a medical audience (as was the case through the 1550s) or for members of the new disciplinary community of naturalists, "histories" of plants and animals, above all the former, sprang from European presses. Rembert Dodoens's *Cruydeboeck* (1554) was translated into French as the *Histoire des plantes* (1557). Carolus Clusius and the Plantin press in Antwerp played an important role in disseminating the term; in addition to his *Historiae* of rare plants (1576, 1583, 1601), Clusius translated Dodoens into French and Garcia da Orta into Latin, using "history" in both titles. Dodoens's later Latin works on plants were also published as *Historiae*. Mathieu de l'Obel's first natural history was the *Stirpium adversaria nova* (1571), but when he started publishing with Plantin in 1576, he adopted the word "history." The word was used into the early eighteenth century; John Ray's *Historia piscium* (1686, published under the name of Ray's late friend Francis Willughby), *Historia plantarum* (1686–1704), and *Historia insectorum* (1710) are well-known examples.[44]

Such works were called *histories*. In book titles, the word was almost never qualified as "natural history." (An exception is Ferrante Imperato's *Dell'historia naturale* [1599], which owes its title to the wide range of natural phenomena discussed in its pages.)[45] When "natural history" does appear as a category, it is not in natural histories themselves but in classifications of the disciplines, from Jean Bodin's triplex division of history into human, natural, and divine (1566) through the seventeenth century (for example, in the works of Tommaso Campanella, for whom even Galileo's *Sidereus nuncius* was a natural history).[46] Such classifications recognized that not all histories were alike. Nonetheless, natural history shared, in the minds of Renaissance scholars, certain features with other forms of history. These included both ethical and theological dimensions. With its emphasis on careful reporting of true particulars about natural kinds, Renaissance natural history would transform medieval traditions in the ethical and theological use of creatures, producing two results: a didactic natural history that flourished in the sixteenth and seventeenth centuries but ultimately withered, and a revamped natural theology that would direct naturalists' attention to adaptation and function.

The Ethics of Natural History

For humanists, history was an ethical and pragmatic discipline. In Cicero's phrase, it was "magistra vitae" and "moral philosophy teaching by example."[47] Polydore Vergil emphasized this exemplary function of history in his account of the disciplines.[48] This moral function of *historia*, restricted initially to human history, came to be taken on by natural history as well; though anthropomorphizing *exempla* from the animal world had been the stock in trade of medieval bestiaries and preaching manuals, humanist naturalists insisted, instead, on the moral benefits of literal knowledge of the natural world. The process was gradual and can be traced from a number of angles.

In the most basic sense, humanists judged knowledge according to its pragmatic value—including ethics, which was, after all, the discipline of acting well (CRISCIANI in this volume). Giorgio Valla's humanist encyclopedia emphasized the moral value of all knowledge in its title: *Things to Seek and to Avoid*. Valla's implication that knowledge by itself was not only useless but even harmful was expressed by his son Gianpietro, who compared his father's encyclopedia, food for the mind, with dinner: just as those who eat too much must be purged, those who learn too much might suffer from mental indigestion.[49] Petrarch had earlier condemned seeking knowledge for its own sake. Nonetheless, many humanists in the early sixteenth century placed a high value on the knowledge of natural particulars. They could do so because they came to associate such knowledge with good learning and good character.

François Rabelais included knowledge of nature in the ideal humanist curriculum recommended by Gargantua to his son Pantagruel:

> I wish you to carefully devote yourself to the natural world. Let there be no sea, river, or brook whose fish you do not know. Nothing should be unknown to you—all the birds of the air, each and every tree and bush and shrub in the forests, every plant that grows from the earth, all the metals hidden deep in the abyss, all the gems of the Orient and the Middle East—nothing.[50]

Rabelais, who studied medicine with the naturalist Guillaume Rondelet (possibly the model for Dr. Rondibilis in the *Tiers livre*) at Montpellier, did not bother to make Gargantua explain the value of this knowledge; when he published *The Horrible and Ghastly Deeds and Feats of the Famous Pantagruel* in 1532, he could take it for granted.[51]

The connection between natural history and the good, indeed godly life had already been asserted by Desiderius Erasmus in the *Convivium religiosum* ("Godly feast"), one of his immensely popular Latin colloquies.[52] Erasmus had been educated in the orbit of the Brethren of the Common Life; he knew the *Imitatio Christi*, which had urged true believers to despise the created world, because it could only distract the imitator of Christ from the Creator Himself.[53] But Erasmus rejected this position, holding instead that the godly layman could and should know and appreciate the beauties of nature as a propaedutic to the spiritual life.

The godly feast is set in the garden of Eusebius, who has invited a group of friends from the city to his "suburban" estate. When they arrive, Eusebius gives them a tour of his gardens. A small flower garden greets them at the door, while a larger, more elegant garden is set inside of four walls. It is reserved for growing pleasant herbs, each type in its own area. The areas are labeled—not with the plants' names but with literary or proverbial allusions to them. Hence marjoram bears the label, "Keep away, pigs: I don't have a scent for you," referring to the belief that swine find the pleasant odor of the plant repulsive.[54] Erasmus underscored the importance of a *correct* knowledge of nature, and natural things, to esthetic enjoyment and spiritual contemplation. The plant labels were allusive, requiring that the reader be familiar with a range of herbal lore. Guests pointed out plants which were common in Italy but rare in northern Europe, for example aconite.[55] No olive trees grew in this garden, unlike those of medieval poetry.[56] All of its plants really grew there; and if its owner wished to regard rarer plants, he had them painted on its walls.

The godly feast was contrasted with other colloquia: for example, the profane and the poetical feasts.[57] By including the study of nature as part of a *godly* conversation, Erasmus stated his approval. Erasmus made some concessions to Christian morality: unlike Italian humanists' gardens, Eusebius's is decorated with images of Christ and the Evangelists, not with pagan herms and termini. His garden is a real one, however, not merely the metaphoric *hortus conclusus* of medieval theology, the closed garden of the Song of Songs, which was taken to represent both Mary and the Church. The plants in it are real, and Eusebius and his guests display proper knowledge of each kind. By knowing and delighting in nature, they prepared themselves to be better men and to approach more closely to God, the main subject of their conversation.

Conrad Gessner also emphasized the felicitous combination of visual pleasure gained by regarding nature and the intellectual pleasure of knowing what it was that one saw. In his description of the ascent of Mons Fractus, near Lucerne, he described the simple yet heady pleasures of the mountains. They would not, of course, appeal to the indolent or idiotic.

But give me a man of at least average mind and body, and liberally educated—not too much given to leisure, luxury, or bodily urges—and he should be a student and admirer of nature [*rerum naturae studiosum ac admiratorem*], so that from the contemplation and admiration of so many works of the Great Artificer, and such variety of nature in the mountains (as if everything were gathered in one great pile), a pleasure of the mind is conjoined with the harmonious pleasure of all the senses. Then, I ask you, what delight will you find in the bounds of nature which could be more honest, greater, and more perfect in every respect?[58]

Informed contemplation of nature could lead one to God. It could also inform one's moral conscience, as Gessner wrote in his *History of Animals*: "There want not instructions out of beasts, by imitation of whose examples, the lives and manners of men are to be framed to another and a better practice."[59] Gessner the naturalist was also a moralist, and he included moral exempla in the philological parts of his natural history.

That is not to say that Gessner confused the two, or that he did not recognize the distinction.[60] He clearly distinguished between accounts of things themselves and what he called "philology": the latter was a matter of words, not of things themselves.[61] But like human history, natural history could serve as the raw material for moral philosophy. That is, natural history could provide ethical *exempla*. In his philological sections, Gessner included *exempla* from history, proverbs, and emblems (PINON in this volume).

One of the most sustained such uses of *exempla* from nature is found in the works of Joachim Camerarius the younger (1534–1598). A naturalist who cultivated a medical garden in his native Nuremberg, corresponded with fellow naturalists throughout Europe, and prepared an epitome of Mattioli's well-known herbal, Camerarius was also an avid collector of emblems. He probably began before or during his journeys in Italy during the early 1560s.[62] Initially there was no unifying theme to these emblems and no commentary;[63] Camerarius, like many other sixteenth-century scholars—and emblematics, which required a solid knowledge of Latin, was preeminently a scholarly pursuit[64]—simply liked emblems.

When he decided to publish them, though, he gave them the form of a natural history: four *Centuriae symbolorum et emblematum*, published in Nuremberg between 1593 and 1604.[65] Camerarius began his first volume of symbols and emblems with a discussion of the reasons for their use. "Not only the writings of many philosophers but also the authors of Sacred Scripture," he noted, attest that it is an ancient custom and agreeable way of philosophizing and teaching to treat of things by images, similitudes, parables, hieroglyphs, and other such methods (which Plato, in his *Phœdrus,* grouped together under the name of *eikonologias*). Nor can anyone doubt that men's minds can be admonished and instructed in many ways by this compendious and ingenious doctrine.[66]

The reason for this, he claims, is that precepts taught in this fashion remain in the memory better, because we are simultaneously exposed to the admirable properties of nature and memorable histories, by which the omnipotence, good, wisdom, and providence of God are set out. In addition our judgment is informed and stabilized by comparing these histories [*historiae*] with each other.[67]

Camerarius's use of the word *historiae* underscores the place of his emblems in the exemplary tradition of the Renaissance. First he presented physical things, then he turned to moral sentences accommodated to them, interspersed with some notable histories.[68] (Camerarius used *historia* to refer both to examples and to narrations.) The most important purpose of the study of nature is to bring the mind to contemplation of God, for which purpose many of the holy Fathers, Prophets, and Apostles studied it, since "it clearly instructs us in the infinite omnipotence, benignity, and providence of God, joined with his infinite wisdom, and creates in us as it were a natural mirror of created things [*quasi naturale speculum rerum creatarum*]."[69] But there are other advantages to the philosophical study of nature: it is useful in instituting and conserving life, especially to doctors; and we can benefit from studying even the vilest creature, as Aristotle noted, "since there is nothing in nature in which there is not something miraculous placed." He discusses the utility of moral philosophy, and the use of historical examples in moral and political philosophy, concluding that "for all of these reasons, I judge it certain and more than certain to anyone, that I have sought nothing else in these lucubrations but that its readers may take something useful and pleasant from them."[70] Essentially, then, emblematics to Camerarius involved the pleasant presentation in a concise, easily memorable form of physical truths, moral exempla, and the occasional narrative tidbit.

Camerarius's work resembles, superficially, the medieval bestiary, and he alludes to the medieval tradition of *specula*, but the differences run deeper than the similarities.[71] Like his emblems, bestiaries drew moral lessons from the "natures" of animals (and occasionally plants, not to mention the prophet Amos!).[72] But the bestiary vanished in the early Renaissance. Theobald's verse *Physiologus* had been used frequently in medieval schools, but humanist pedagogues replaced it with classical texts.[73] Few bestiaries made the transition from manuscript to print.[74] The early printed works of natural history, such as the *Hortus sanitatis,* dealt with the natures and medicinal properties of animals, but not with their allegorical and moral meanings.[75] The *Physiologus,* so popular in various derived forms in the Middle Ages, reappeared in 1587, in Ponce de Leon's humanist edition of the Greek text, and regained a certain popularity, but this popularity was due to its presentation as a classical, rather than medieval, text.[76]

Moreover, neither bestiaries nor fables—in which animals and even plants behaved in crudely anthropomorphic fashion—displayed any concern for the truth of the *exempla* they drew from nature. Camerarius, however, insisted on verifying the *historiae* that he told before drawing morals from them. In order that the moral sense be true, the emblem had to deal with literal truth. (See CRISCIANI in this volume for an earlier parallel in the historical works of Michele Savonarola.) Camerarius could not have separated the two types of information with which he was primarily concerned in his emblems. Thus an emblem book dealing with animals and plants had to be written by someone competent to judge the truth of the stories told about them—a naturalist.

Indeed, Camerarius insisted on the difference between *exempla* drawn from nature and fables/or bestiaries. Many writers, he notes, have drawn moral examples from human actions, most notably Valerius Maximus among the ancients and Petrarch, Sabellicus, Fulgosus, Egnatius, Ravisius and others among more recent authors.

> But few have existed who have tried to apply *exempla* from natural things to human nature, unless we wish to place among them in some way the tellers of fables [*fabularum expositores*], who indeed should not be entirely excluded. Aelian alone, at least, offered something of this kind, but quite sparsely, and when he did offer one, he either briefly explained it, or omitted the use.[77]

He does not mention the compilers of bestiaries at all, unless he includes them under the "fabularum expositores." According to Camerarius, almost no one has written proper didactic natural history—that is, no one has founded it on natural history.

Aldrovandi included "emblemata" as one of the divisions of articles in his natural histories, and he frequently cited Camerarius in this section. He recognized, of course, that Camerarius's emblems were above all ethical: they were intended "for insinuating moral warnings to his readers" and notes under another, "By this example we are admonished."[78] But the moral is not separable from the animal's behavior, notes Aldrovandi, for Camerarius draws his moral from that behavior:

> Moreover, Camerarius, meditating on another property of the beaver, drew this animal gnawing on a tree trunk, with this motto: PERSEVERE. By which argument we are exhorted to constancy and perseverance in carrying out our labors, by the example of the beaver, who with such assiduous bites takes wood away from the tree until at last it falls.[79]

It is by following the true example of nature that we are to behave morally. Aldrovandi engages Camerarius seriously as an emblematist, even as his encyclopedic comprehension displays Camerarius's range as a natural historian.

Through Camerarius's work, moralizing natural history reached a wide audience. The complete work went through nine editions in the seventeenth century—eight of them in the second half.[80] Robert Burton referred to the first two centuries in his *Anatomy of Melancholy*.[81] Speaking of kind words, he notes that they "are cheerful & powerful of themselves, but much more from friends, as so many props, mutually sustaining each other like ivy and a wall, which Camerarius hath well illustrated in an Emblem."[82] The mighty and powerful are, from their high station, more likely to fall than the humble, "as a tree that is heavy laden with fruit breaks her own boughs, with their own greatness they ruin themselves: which Joachimus Camerarius hath elegantly expressed."[83] In these and other places Burton uses the moral sense of Camerarius's emblems to support his own views on melancholy and its related diseases.

Sir Thomas Browne also read and profited from Camerarius. Browne generally disliked emblems, but not because they were moralistic. Rather, they fostered vulgar errors by misrepresenting the truth. Of their stories, "some are strictly maintained for truths, as naturally making good their artificiall representations; others symbolically intended are literally received, and swallowed in the first sense, without all gust [taste] of the second."[84] Common folk, who cannot distinguish literal and figurative truth, can be easily misled by emblems, though the educated, Browne hopefully asserted, are not prone to such errors.[85] But Browne read Camerarius and cited him approvingly. Browne noted that many say and most believe that the lion fears the cock, but "how far they stand in feare of that animal, we may sufficiently understand, from what is delivered by Camerarius, [who writes that] in our time in the court of the Prince of Bavaria, one of the Lyons leaped downe into a neighbours yard, where nothing regarding the crowing or noise of the Cocks, hee eat them up with many other Hens."[86] Browne approved of Camerarius because he refused to follow vulgar error and showed some signs of discrimination, if not skepticism. Only the literal truth is capable of moral interpretation.

By the end of the seventeenth century, however, the didactic use of natural history for *ethics* had all but vanished. The 1702 Mainz edition of Camerarius's emblematic natural history specified that the emblems were "ethicopolitici," something that the author and his readers in the sixteenth and seventeenth centuries had taken for granted.[87] But eighteenth-century readers were less likely than their predecessors to read nature morally. The tradition of moralizing through fables continued— Aesop and La Fontaine remain popular to this day—but fables anthropomorphize animals; they do not attempt to "theromorphize" humans. And by the end of the seventeenth century, naturalists interested in the lessons to be drawn from natural history had a more immediate problem on their hands: proving, through a study of the Creation, that the Creator existed and cared for His creatures—that is, proving that God was the final cause of living things.

Natural History and Natural Theology

Though Camerarius's emblems taught moral lessons, his justification of emblematic natural history had begun with theology: the study of nature teaches God's "infinite omnipotence, benignity, and providence." This theme ran throughout Renaissance natural history from the beginning. Leonhart Fuchs, whose work helped spread the notion of the *history* of plants, promoted the study of plants as a means to a better comprehension of God. There are many reasons to study plants, he wrote, above all this, that the divine presence and benignity is illuminated by scarcely anything more than by the various forms and natures of plants, when we consider that they were all established for man's use. Whatever else, the diligent contemplation of herbs excites and confirms in our souls the knowledge [*opinionem*] that God cares for men, since he has worked so hard to aid and preserve them by adorning the earth with so many elegant and excellent plants with which they can ward off so many different diseases.[88]

Fuchs emphasized God's care in giving men plants to cure the illnesses with which they are afflicted. But God's gift was not only medical but, equally, esthetic. "What is more pleasant and delectable than regarding plants, which God has painted with so many and such varied colors, which he has crowned with so many elegant flowers whose colors no painter could ever adequately represent?"[89] For Fuchs, then, natural history served to prove both God's existence and his providence.

Such sentiments were commonplace in the engraved title pages of natural histories later in the century. Carolus Clusius's *Rariorum plantarum historia* (1601) admonished the reader, "God gave each plant its powers, and every herb teaches that God is present."[90] Thomas Johnson's 1633 revision of John Gerard's *Herbal* noted similarly, "Lest you forget the Author of the divine gift, every herb teaches that God is present," bolstering the claim by quoting Genesis 1.29: "Behold, I have given you every herb bearing seed, which is upon the face of all the earth."[91] God's providential care for his most important creation was proclaimed not only by the heavens but by the lowliest hyssop.

What took this conviction beyond the level of a commonplace was the precise method by which the observer proceeded from observation of the Creation to knowledge of the Creator. Conrad Gessner framed this process as an ascent: In the creation "doth Divinity descend, first to supernaturall things, and then to things naturall: and we must turne saile and ascend first by things naturall, before we can attaine at reach thinges supernaturall." Gessner leaves no doubt that this is the highest purpose of natural history: "For being thus affected and conversant, in beholding these neather and backer partes of God, confessing with thankesgiving that all these things doe proceede from his Divinity, we cannot stay but ascend uppe higher, to the worker himselfe, using all thinges in this life but as Prickes and Spurres, for occasion and admonitions to thinke uppon and reverence the prime Author."[92]

This ascent from the natural world to God was encouraged not only by naturalists but by the influential theorist Jean Bodin in his *Methodus ad facilem historiarum cognitionem* (1566). Bodin began his book by asking "Quid historia sit & quotuplex," but he answered the second question first:

> There are three kinds [*genera*] of history, that is, true narration: human, natural, and divine. The first pertains to man, the second to nature, the third to nature's parent. The first explains the actions of man living in society; the second deduces the causes placed in nature and their progression from the ultimate principle; the third contemplates the strength and power of almighty God and the immortal souls gathered in him. From these three forms of assent arise [*assentio triplex oritur*], probable, necessary, and religious, and the same number of virtues, i.e. prudence, knowledge [*scientia*], religion.[93]

Bodin's triple definition was taken up by Francis Bacon and Tommaso Campanella, though Bacon added a fourth category, literary history.[94] Rather than pursue the genealogy further, though, we should dwell on the implications of Bodin's approach.

Bodin's division was not merely horizontal. The bulk of the *Methodus* is devoted to the "historioscopic" task of explaining how to read and profit from human history. But in principle, the three genera of history could be arranged hierarchically. Bodin treated human history as uncertain, natural history as necessary, and divine as absolutely necessary. The reason is that God was a cause in himself, and nature had limited causes because it is always similar to itself. Human history, however, depends on human will, which is dissimilar to itself; from various causes proceed various effects.[95] However, human history is closer to ourselves; natural history is more remote but still accessible, while divine history is most important but also most removed from us. The three genera of history, taken in turn, provide an *ascensus ad Deum*. The reader of human history should be distracted from human turmoil and seek solitude; "thus despising the inconstancy and temerity of human affairs, he regards the most certain causes of nature; contemplating these, he feels such pleasure that, secure in the knowledge of such studies, he neglects kingly treasures." Even kings have voluntarily abdicated in favor of such studies. From these sciences, "whose subject matter concerns the senses," the "man well educated [*institutus*] by nature" will in turn ascend to those "that are perceived by the mind alone: that is, the strength and power of immortal minds, until he is rapt by swift wings, and seeking his first origin, he is wholly joined with God. In this lies the end of human actions, the ultimate rest, and the highest felicity."[96]

The certainty of the senses served as a warrant for the divine knowledge attained through their contemplation. The English divine Edward Topsell, whose natural history was largely translated from Conrad Gessner (as Topsell acknowledged), also emphasized the study of nature as a path to divine knowledge. The title page of his *Historie of Foure-Footed Beastes* proclaimed that it would reveal "the wonderfull worke of God in their Creation, Preservation, and Destruction." In dedicating his work to Richard Neile, Dean of Westminster, Topsell asserted that "no man ought rather to publish this unto the world, than a Divine or Preacher," for knowledge of the "creatures and works of God is divine." One of the Manichees' impieties was their belief that the creation was the work of an evil god; the pious Christian knows that what appears evil or ugly to us is, as part of God's design, beautiful and good. Moreover, that design can be perceived in the Creation; "every beast is a natural vision, which we ought to see and understand, for the more cleare apprehension of the invisible Maiesty of God." Finally, beasts also serve for moral "reproof and instruction," just as they did for Camerarius and other moralists.[97]

Like Bodin, Topsell emphasized the "necessity" of natural history, as against the merely probable knowledge of human history: "Now againe the necessity of this History [of beasts] is to be preferred before the Chronicles and records of al ages made by men, because the events & accidents of the time past, are peradventure such things as shall never againe come in use: but this sheweth that Chronicle which was made by God himselfe, every living beast being a word, every kind being a sentence, and al of them togither a large history, containing admirable knowledge and learning, which was, which is, which shall continue, (if not for ever) yet to the worlds end."[98]

And like Camerarius, Topsell insisted that his history of creatures be true, "for the marke of a good writer is to follow truth and not deceivable Fables." Because his sources were largely "Heathen writers," whose credulousness was well known, he insisted on repeating only what at least two or three, if not more, writers asserted. Nonetheless, he demanded not belief but assent from his readers: "For *Fides*, is *credere invisibilia*; but *consensus* is a cleaving or yielding to a relation untill the manifestation of another truth." He stood ready to retract any claim when it was proven to be "false and erroneous." Nonetheless, he insisted that his readers avoid taking a skeptical stance: "Although I doe not challenge [claim] a power of not erring, yet because I speake of the power of God, that is unlimitable, I will be bold to averre that for truth in the Booke of creatures (although first observed by Heathen men) which is not contrary to the booke of Scriptures."[99] God's omnipotence served to warrant the most outlandish claims about His creation, so long as they were supported by credible testimony—just as the same divine omnipotence and beneficence warranted the apparently incredible events described in Scripture.

Both Gessner and Bodin emphasized the *ascent* from sensible Nature to the insensible Creator, for otherwise, natural theology might lead not to proper reverence for God but to idolatry or even pantheism. Such an ascent was an integral part of earlier natural theology. But the sixteenth-century incorporation of natural history into natural theology transformed the latter in a striking way. Natural theology before the sixteenth century was succinct and emphasized celestial phenomena and general order. By the end of the seventeenth century, natural theology drew on detailed examples of design drawn from living creatures. The change was engendered by challenges from "materialists" and "atheists," but natural history was the midwife.

From Natural Theology to Physico-Theology

For Renaissance scholars, the paradigmatic presentation of natural theology was that of Cicero's *De natura deorum*. Balbus, the Stoic interlocutor in Cicero's dialogue, presented many arguments in favor of the existence of a God. Those most relevant to natural theology were based on the clear evidence of design. Balbus argued that design was so evident in the natural world that any reasonable observer would have to conclude there was a designer.[100] These arguments were taken up in the fourteenth century by Petrarch. In the same work in which he dismissed "knowing the nature of quadrupeds, fowls, fishes, and serpents," Petrarch employed Cicero's argument that the perfection of the heavens clearly established the existence of a benign Creator.[101] Petrarch's approbation of natural theology went hand in hand with his dismissal of what would become natural history. For him, natural theology was about the evident design of the world; the natural theologian did not need to descend into detailed knowledge of the natural world. (Theodore Gaza seemed to be responding to such arguments when he reproached "those who say that Aristotle said many things about the fly, the tiny bee, the worm, but few about God"; by studying the creation, one comes to "understand, admire, and worship God immortal."[102] But Gaza, who probably wrote these words in the 1470s, was in a distinct minority.)

The argument for design also received surprisingly little attention from Raymond of Sabunde, author of the *Liber creaturarum*, better known under the title *Theologia naturalis* first used for its incunabular editions. Raymond's work is best known, of course, not for itself but because it was the subject—and target—of Montaigne's longest essay, the "Apologie de Raimond Sebond," one of the most important early modern statements of skeptical philosophy.[103] But Raymond's work remained important. Despite a taint of Pelagianism—for Raymond implied that the doctrine of salvation could be learned from the Book of Nature, without the revelation contained in Scripture—his book continued to be reprinted through the middle of the seventeenth century.[104]

Raymond's approach to the question of design was straightforward, though in a different way from Petrarch's and Cicero's. He invoked the *scala naturae*, the great chain of being.[105] The distinction between inanimate matter, plants, animals, and man sufficed, for Raymond, to show that the world was the product of a prudent designer. Raymond's concern was not so much with design or divine prudence as with the harmony between the Book of Creatures and the Scriptures, and the ways in which the student of nature, inspired by the Holy Spirit, could read in the pages of the former the truths of the latter.[106]

Despite Raymond's popularity in the early sixteenth century, Balbus's Stoic arguments remained the starting point of natural theology down to the early seventeenth century. They were repeated by Leonard Lessius, whose work *De providentia numinis* (1613) helped define the parameters of later natural theology.[107] Lessius added to Cicero's examples only the recent telescopic discoveries by Galileo and others, which revealed in more detail the order and harmony of the heavens. Indeed, what is most striking about the arguments of natural theology (as opposed to their contexts) from antiquity through the sixteenth century is how little they changed. Whether urged against Epicureans, infidels, or nominalists, or developed as intellectual exercises to strengthen the faith, natural theology returned to the same arguments from universal consent and evident design in the grand plan of nature. If anything, theologians wanted their natural theology in moderate doses; too much attention to the creation might lead to neglect of the Creator or even to pantheism.[108]

Later seventeenth-century natural theology was different.[109] In physico-theology, its most fully developed form, it emphasized not the immediately perceptible regularities of the heavens and the *scala naturae* but, instead, the intricate contrivances of living organisms. John Ray's *Wisdom of God Manifested in the Works of the Creation* (first edition 1691) adduced the human hand and eye as the best proofs of design in the world. Ray's work drew upon a century and a half of natural history to prove three points: the existence of God, "his Infinite Power and Wisdom," and the reverence that human beings owe him.[110] His argument is sustained by the welter of detail from natural history that he offered.

Ray was the first great naturalist to write at length on natural theology. That undoubtedly shaped his work, and through him, the natural theology of the eighteenth and nineteenth centuries. But other factors were equally important. The seventeenth century was an age not only of natural theology but, more broadly, of "secular theology," in Amos Funkenstein's formulation.[111] This secular theology was based not on revelation but on reason; even though its proponents identified themselves as Christians, Christology played no role in their arguments about the nature and attributes of God.[112] John Toland wrote that "Religion is calculated for reasonable Creatures."[113] Raymond of Sabunde might have agreed up to a point, but he would have rejected Toland's claim that religion was "calculated"; whatever the Congregation of the Index thought when it banned his *Theologia naturalis*, Raymond had insisted that inspiration with the Holy Spirit was the starting point, not the ending point, for reading the book of creatures.

Ray and his contemporaries could make no such assumption, and though a fideist like Montaigne might have thought their insistence on the reasonableness of religion to be pathetic, they had reasons. Lessius, Ray, John Wilkins, Toland, Tillotson, and other seventeenth-century natural theologians took up the Stoics' strategies against the Stoics' enemies: skeptics who denied that human beings could, with reason, come to any positive knowledge of God's being and attributes, and especially "atheists" (i.e., Epicureans and deists) who denied his existence and providence.[114] Lessius had fought the "atheists" not only with the Stoics' strategies but with their weapons. What Ray and other naturalists brought to the fray were weapons far more powerful than those available to Cicero's Balbus, Petrarch, and Lessius. And these weapons were drawn precisely from the store of knowledge that Petrarch had despised: the nature of animals and plants. Without the conviction that humans could ascend from the knowledge of vile and lowly creatures to the Creator himself, natural theology—indeed, the entire project of secular theology—could not have faced the threats that arose in the seventeenth century.

Ray's *Wisdom of God* and other natural theologies of the later seventeenth century leave no doubt about the source of these threats: the "pretend" theism of Descartes, and, lurking in the background, the "atheism" or pantheism of Spinoza. Balbus's arguments had sufficed to demonstrate the *existence* of God, but they offered little to support divine providence. Cartesianism, at least as it was imagined by its enemies, was insidious in that it appeared to concede the former but, by denying the doctrine of final causes, in fact denied the latter.[115] By reducing everything to efficient causation, Descartes had removed final causes so far from nature that they may as well not have existed—a point driven home by his posthumous work *Le monde, ou Traité de la lumière*, which began by insisting that God made the world in six days but that it would have been just the same had he merely let the primal chaos sort itself out.[116]

In sum, the natural theology of the later seventeenth century was located in a very different argumentative context from that of the Renaissance, a context shaped by the end of teleology as a necessary category of explanation (MACLEAN in this volume). It is true, as Schneewind argues, that "the collapse of Aristotelian teleological thinking" in the seventeenth century was offset because "Christian teleology was available to replace Aristotle's."[117] But a world of difference lay between the two. In Aristotelian natural philosophy, final causes were located in the essence of a natural thing; that was precisely the difference between nature and craft.[118] In a natural thing, formal, final, and efficient causes were united. The formal, final, and efficient causes of artificial things, on the other hand, were located in the craftsman. By eliminating intrinsic final causes, the mechanical and corpuscular philosophies of the seventeenth century made nature into a work of art—or, to take the metaphor that was increasingly popular in early modern Europe, into a kind of machine.[119]

Most seventeenth-century thinkers replaced intrinsic teleology with extrinsic teleology: their world-machine was the work of God. It is certainly thus incorrect to say that seventeenth-century natural philosophy lacked a teleological component.[120] But it *is* correct to say that the corpuscular natural philosophies of the early seventeenth century were seen by many as *implying* that teleology was superfluous. Gassendi, for instance, insisted that his atoms were changed only through efficient causation; they had no nature or essence to be the seat of an intrinsic teleology. Instead, final causation was imposed from without, by God's design.[121] But how could this design be known? Where were physico-theologians when God laid the foundations of the earth?

Here, physico-theologians could turn to a paradoxical ally: Galen. Though a harsh opponent of Christianity, Galen was a thoroughgoing teleologist; moreover, he attributed final causes not only to things in themselves but also to the demiurge who created the world.[122] He taught physico-theologians that God's design could be understood only by detailed investigation of the natural world. As Robert Boyle put it, only a careful investigation into efficient causes provides real knowledge of the mechanism from which final causes (and hence divine purpose) may be deduced. Only detailed examination of the contrivances by which the eye (for example) functions, wrote Boyle, can convey a real understanding of the skill and wisdom with which God has arranged the parts.[123] Knowing the efficient causes was a prerequisite for inferring at least some of God's final causes. Hence physico-theology depended on the kind of detailed knowledge of the natural world that Petrarch had scorned in order to achieve, laboriously, the conclusions to which Petrarch had blithely leapt. Renaissance anatomy and natural history provided the framework on which physico-theology could take shape.

In short, against the *Deus otiosus et absconditus* of the Cartesians, natural theology had to offer retail, not wholesale proofs of divine providence. John Wilkins realized this: his 1649 contribution to natural theology emphasized "the beauty of providence" even in its "rugged passages," while his *Principles and Duties of Natural Religion* drew on Galen's biological teleology for its proofs of divine providence.[124] Naturalists were ideally situated to offer support for this argument, for the variety of nature and the function of animal and plant organs were so evident that it seemed absurd to deny them. They would continue to offer examples of natural functions as evidence of divine providence until two nineteenth-century challenges arose: romantic idealism and Darwinian evolution.

Conclusion

By the middle of the eighteenth century, "natural history" was no longer seen by the Republic of Letters as a part of history. The term "natural history," rare in the sixteenth and seventeenth centuries, came to supplant "history" in the titles of works on animals and plants—for instance, René-Antoine Fourchauld de Réaumur's *Mémoires pour servir à l'histoire naturelle des insectes* (1734–1742), and, of course, Buffon's *Histoire naturelle, générelle et particulière* (1749–1804). This trend had begun in the middle of the seventeenth century. John Johnston's natural histories, published in 1657, mark the shift.[125] By the eighteenth century, though, *natural* history was dominant. The *Encyclopédie* justified it explicitly, mentioning "natural history" in the article *Histoire* (published 1765) only to note that it was "improperly called history," being instead "an essential part of physics."[126]

The moralizing natural history of Conrad Gessner, Joachim Camerarius, and Ulisse Aldrovandi had not survived the seventeenth-century transformation of ethics, except in the debased form (as they would have seen it) of anthropomorphizing fables, a tradition which owed nothing to them. Natural theology, meanwhile, continued to be closely allied with natural history, whether in the *Spectacle de la nature* (1732–1735) of the Abbé Pluche or the natural theology of William Paley (1802) and the Bridgewater Treatises in England (1833–1840).

Neither of these developments would have occurred had not history and natural history been allied in the middle of the sixteenth century. By the seventeenth century, this alliance had been broken. Indeed, the alliance between the two—the classification of natural history as a species of the genus "history," with all it entailed for Bodin and others—was the product of a brief convergence of circumstances. Yet that convergence had momentous effects for the relation between natural history, ethics, and natural theology. For a brief period, "natural history" was not equivocal but, rather, multivocal.

Notes

Sources cited in boldface type appear in the Primary Sources of the bibliography.

1. Reeds 1976, 1991; see Ogilvie (forthcoming 2006).
2. **Petrarca** 1948, 58–59.
3. **Vergerio** 2002, 54.
4. **Bruni** 2002.
5. **Piccolomini** 2002, 250–252.
6. **Guarino** 2002, 5.
7. Borst 1995, 251–292.
8. Nauert 1979.
9. **Pliny the Elder** 1549.
10. **Aristotle** 1498.
11. Schmitt 1971, 246.
12. **Theophrastus** 1483; Schmitt 1971, 246.
13. **Leoniceno** 1492, sig. a1r-v.
14. **Leoniceno** 1492, sig. a2v, a3v, and passim.
15. **Leoniceno** 1492, sig. a3r.
16. **Leoniceno** 1492, sig. b1v.
17. **Vergil** 2002, 1.12, 106–112.
18. **Vergil** 2002, 1.21, 162–168.
19. **Vergil** 2002, 2–22.
20. On Pliny, see Bodson 1986; Beagon 1992; French and Greenaway 1986.
21. **Bartholomaeus Anglicus** 1485, 1488.
22. **Macer** 1510; **Herbarius** 1484.

23. Universitätsbibliothek, Basel, MS. K III 42, Conrad Weigand, *Quadripartitum de quercu* (fifteenth century).

24. **Albertus Magnus** 1495, lib. 22, tract. 1, cap. 1, fol. 212r. **Albertus Magnus** 1916–1921, 1349, reads "ea quae particulariter de particularibus narrantur, rusticam melius instruant contionem."

25. **Albertus Magnus** 1495, lib. 1, tract. 1, cap. 1, fol. 1r; **Albertus Magnus** 1916–1921, 2.

26. G. Valla 1501, books 20–23, 21.1–101.

27. G. Valla 1501, 24.24–105, 25.1–7, 26.24–26, 29.60–63.

28. G. Valla 1501, books 42–44; domestic plants and animals are discussed, in alphabetical order, in 42.25–131; book 26 takes up common barnyard animals from poultry to cattle.

29. Reeds 1976; Palmer 1985.

30. **Cordus** 1534; see Dilg 1969, 1971.

31. **Brunfels** 1532.

32. **Bock** 1539; Hoppe 1969.

33. **Marschalk** 1517–1520. My thanks to Laurent Pinon for this reference.

34. **Fernández de Oviedo y Valdes** 1526. On Oviedo, see Gerbi 1985.

35. Lowood 1995, 303–306, 308.

36. **Gessner** 1541; **Fuchs** 1542.

37. **Fuchs** 1543.

38. **Fuchs** 1542, sig. α6r, trans. **Meyer et al.** 1999, 1:210.

39. **Fuchs** 1542, sig. α5v–6r. The translation in **Meyer et al.** 1999, 1:209 misconstrues "sua ex Dioscoride nomina" as "his own names instead of Dioscorides'," which renders the passage incomprehensible.

40. Ogilvie (forthcoming 2006), chapter 4.

41. **Fuchs** 1542, sig. α4r, α2r.

42. **Gessner** 1551, sig. β2r.

43. For an example, see Bäumer 1990.

44. **Clusius** 1576, 1583, 1601; **Dodoens** 1583, 1568, 1557; **L'Obel** 1576; **Willughby** 1686; **Ray** 1686–1704, 1710.

45. **Imperato** 1599.

46. **Bodin** 1566, 9; **Bodin** 1572, 11; **Campanella** 1954, 1238–1252.

47. Landfester 1972.

48. **Vergil** 2002, 1.12, 106–113.

49. G. **Valla** 1501, 1.1; Gianpietro's dedication to Giovanni Jacopo Trivulzio, sig. [square]vii^v.

50. **Rabelais** 1990, 158.

51. **Rabelais** 1532.

52. **Erasmus** 1703, vol.1.

53. **Thomas à Kempis** 1941, 4, 154.

54. Cf. the emblem on this subject by **Camerarius** 1595, fol. 103, "Non tibi spiro."

55. **Erasmus** 1703, 1: 675.

56. Curtius 1953, 184, 94–97.

57. Subjects of two other Erasmian colloquies: the "Convivium profanum" and the "Convivium poeticum"—not to mention the "Convivium fabulosum."

58. **Gessner** 1556, 50.

59. Gessner in **Topsell** 1607, sig. ¶2r, translated from **Gessner** 1551, sig. α3r.

60. Cf. Ashworth 1996, 1990.

61. **Gessner** 1551, sig. α6r, β1v, β2v–3r.

62. Based on his testimony in the first volume and his frequent use of Italian authors; see also Harms 1985, 77.

63. On the lack of theme, see Harms 1985, 77.

64. Though there was a burgeoning vernacular emblem literature, especially in the seventeenth century, primarily in German and Dutch; see Praz 1964; Henkel and Schone 1967.

65. The engraved title page of the first volume gives 1590 as the date, but the dedicatory epistle is dated 1593. There is a modern reprint: **Camerarius** 1986.

66. **Camerarius** 1590, sig. A2r-v .

67. **Camerarius** 1590, sig. A2v.

68. **Camerarius** 1595, fol. 2r-v.

69. **Camerarius** 1595, fol. 2v.

70. **Camerarius** 1595, fol. 3r–4r.

71. The distinction between medieval exemplary literature and emblem books—a genre begun by Protestants and fellow travelers—is underscored by Diehl 1986.

72. On the bestiary tradition, see Clark and McMunn 1989; Stannard 1978; McCulloch 1960. George and Yapp 1991 argue, unconvincingly, that bestiaries were intended as scientific works, an anachronistic category. For the prophet Amos, who gets his own chapter in the *Physiologus*, the ur-bestiary, see *Physiologus* 1979.

73. Grendler 1989, 113, 17–22.

74. The ones that did were usually printed in small towns for popular, rather than learned, consumption. See, for example, *Libellus de natura animalium* 1958, which was probably published in 1508.

75. In *The Noble Lyfe,* an English translation of excerpts on animals from the 1485 *Hortus sanitatis,* there are no allegories and very few morals, the camel's hatred of incest being a notable exception. The work has been reprinted in facsimile: ***Hortus sanitatis*** 1954.

76. The slim volume presents the text in parallel Greek and Latin with commentary. It was attributed to Epiphanius, bishop of Cyprus. I referred to a reprint of the work: *Physiologus* 1588.

77. **Camerarius** 1596, sig. a3r.

78. **Aldrovandi** 1637, 208, 73.

79. **Aldrovandi** 1637, 286.

80. National Union Catalogue Pre-1956 Imprints, "Camerarius."

81. **Heusser** 1987, 299.

82. **Burton** 1893, 2:124.

83. **Burton** 1893, 2:170.

84. **Browne** 1981, 1:55–56.

85. **Browne** 1981, 1:54.

86. **Browne** 1981, 1:280–281. Camerarius's text differs slightly from Browne's version; Browne's version, in fact, is identical (except for punctuation and capitalization) to Aldrovandi's quotation of Camerarius, making it seem likely that Browne either did not read Camerarius or, at the very least, did not copy down the quotation directly out of his emblem book.

87. **Camerarius** 1702.

88. **Fuchs** 1542, sig. (2v.

89. **Fuchs** 1542, sig. (2v.

90. **Clusius** 1601, t.p.

91. **Gerard** 1633, t.p.

92. Gessner in **Topsell** 1607, sig. ¶3v–¶4r, translated from **Gessner** 1551, sig. α4r-v.

93. **Bodin** 1566, 9; **Bodin** 1572, 11. In practice, Bodin downplayed the distinctions; for instance, he held that climate determined much of human history.

94. F. **Bacon** 1973, 69.

95. **Bodin** 1566, 11–12; **Bodin** 1572, 14–15.

96. **Bodin** 1566, 26–27; **Bodin** 1572, 33–34.

97. **Topsell** 1607, sig. A3v–A5r.

98. **Topsell** 1607, sig. A5r-v.

99. **Topsell** 1607, sig. A5v–A6r.

100. **Cicero** 1933, 2.2, p. 124.

101. **Petrarca** 1948, 58, 80–83.

102. **Perfetti** 2000, 16–17.

103. **Popkin** 1979.

104. The following editions are listed in OCLC's WorldCat database, consulted May 13, 2003: Strasbourg 1496, Strasbourg 1501, Nuremberg 1502, Lyon (?) 1507, Paris 1509, Lyon 1526, Lyon 1540, Lyon 1541, Venice 1581, Frankfurt

1635, Lyon 1648, Amsterdam 1661 (the last before the nineteenth century). French translations were published in 1551, 1565, 1569, 1581, 1603, 1611, and 1641; all but the first two were Montaigne's translation. Since OCLC's database is drawn mostly from North American libraries, I have probably missed several editions.

105. **Raimundus Sibiuda** 1648, tit. 1–3, pp. 1–10; see Lovejoy 1964.

106. **Raimundus Sibiuda** 1540, prolog., pp. 1–4. The prologue was omitted from most post-Tridentine editions.

107. Buckley 1987, 48–53.

108. **Goodman** 1622, 6–7. See also MULSOW in this volume.

109. Noted by Glacken 1967, 392–393 (though he stresses the continuities) and, in the specific context of design arguments, by Barrow and Tipler 1986, 55. The caesura between medieval natural theology and the physico-theology of the late seventeenth and eighteenth centuries is emphasized by Stebbins 1980, 62–64. Thanks to Martin Mulsow for bringing Stebbins's work to my attention.

110. **Ray** 1691, sig. A7r–A8r.

111. Funkenstein 1986.

112. Noted by Buckley 1987, 65–67. Pascal thought that this was the principal weakness of natural theology, for "nous ne connaissons Dieu que par Jésus-Christ." **Pascal** 2003, 185–187.

113. **Toland** 1696, xv.

114. Before the eighteenth century, "atheist" was a term of abuse; the first self-described European atheists appeared during the Enlightenment. Febvre 1968; Buckley 1987.

115. **Ray** 1691, 20–40.

116. **Descartes** 1910.

117. Schneewind 1990, 44.

118. Aristotle *Physics* 2.1, 192b10–35.

119. See, e.g., Mayr 1986.

120. See Osler 2001.

121. Osler 2001, 159; see Joy 1987.

122. Hankinson 1989.

123. Osler 2001, 163–164.

124. **Wilkins** 1649; **Wilkins** 1675, 80–81.

125. **Jonstonus** 1657c, 1657d, 1657a, 1657b.

126. *Encyclopédie* 1751–1780, 8:220–221.

3

Praxis Historialis: The Uses of *Historia* in Early Modern Medicine

Gianna Pomata

"The word *historia* had for the Greeks a much wider meaning than it generally has today," Giorgio Valla noticed in his humanist encyclopaedia of the late Quattrocento.[1] Over half a century later, in his dialogues *Della historia* (1560), Francesco Patrizi portrayed himself in the grip of philosophical *furor*, obsessively asking everybody: what does *historia* mean? Patrizi's quest was prompted by the enormously expanded use of the term in his times, what he calls "*historia*'s infinite way of being": "I see that its ways are infinite, and infinitely varying amongst themselves. It seems impossible to me that they all may be covered by a single definition."[2]

It is in this period of dramatic change in *historia*'s meaning, at the heyday of the Renaissance, that we find the term used for the first time as a medical title. Giorgio Valla's friend and correspondent, the humanist physician Alessandro Benedetti, named his anatomical treatise *Historia corporis humani sive Anatomice* (written 1493–1496, published 1502).[3] Benedetti's text inaugurated the genre of *historia anatomica* and a trend that would peak a century later in such classics of anatomy as André Du Laurens's *Historia anatomica humani corporis* (1599) and Caspar Bauhin's *Anatomica corporis virilis et muliebris historia* (1609).[4]

Among humanist physicians of the early sixteenth century we also find the term *historia* referring to what will be later named *historia medica* or *medicinalis*—the embryo of what we now call the case history. The earliest example I have been able to find is *Galeni historiales campi* (1532) by Symphorien Champier, a French physician with many contacts with Italy, an admirer and correspondent of Alessandro Benedetti. Champier started the vogue for excerpting *historiae* from ancient medical authors, a genre that would also flourish later on, as testified by Giovan Battista Selvatico's *Galeni historiae medicinales* (1605) and Zacutus Lusitanus's *De medicorum principuum historia* (1636).

Mostly absent from the vocabulary of Scholastic medicine,[5] *historia* gained a new and prominent role in the language of humanist doctors. Over the course of the sixteenth century, starting from a marginal position, *historia* would become a key term of medical writing. The vogue for anatomical and clinical texts written in the *historia* format gained momentum in the second half of the Cinquecento and proliferated even further in the following century. The repertories of seventeenth-century medical writings have copious listings under such headings as *historia anatomica* and *historia medica*.[6] In this paper, I will argue that the development of these new medical genres was part of a wider reappraisal of *historia*'s epistemological value, and that it should be understood against the backdrop of the momentous shifts in the term's usage in general philosophical parlance. Consequently I will first survey the meanings of *historia* in the language of Renaissance scholars, and then examine the separate developments of *historia anatomica* and *historia medica* from the early sixteenth century to the first half of the seventeenth. In both cases, we can notice an increasing centrality of *historia*, connected with its use for recording, communicating and validating observation. In all these varieties, the medical *historia* carried a strong empiricist connotation. It meant *sensata cognitio*, knowledge based on sense perception, as well as the report thereof. The medical *historia*, as *historia* in general, held a prominent place in the early modern vocabulary of experience, and its spectacular rise as a form of medical writing is part of the wider history of the varieties of early modern scholarly empiricism.

Historia's "Infinite Ways of Being"

The treatises on *ars historica*, a new and distinctive Renaissance genre, provide ample documentation of the sixteenth-century debate on *historia*, culminating in the publication of the *Artis historicae penus* (1579).[7] So far, however, these texts have been studied mostly with an overriding concern for *historia* as civil history, the sense closest to modern usage. As a consequence, little attention has been given to *historia* as a tool for recording and communicating the observation of nature. Yet the *histo-*

ria of nature and the *historia* of *res gestae* were closely intertwined in the sixteenth century, particularly so in the case of medicine. In the preamble to his 1545 anatomical treatise, the humanist physician Charles Estienne drew a close parallel between the author who writes anatomical *historia* and the one who records the memory of human deeds. "Just as the historian should definitely not divulge fabricated fables, wherein one can perceive many ornaments, many fine forms of words and sentences, but very little truth; so he who writes the *historia* of the human body should take care, among other things, to know from direct experience the things he will describe; he should also make sure not to mention anything false, and have no fear of saying anything that is true."[8]

In his exhaustive survey of early modern definitions of *cognitio historica*, Arno Seifert lists so many notions of *historia* that one feels inclined to share Francesco Patrizi's dizziness at the protean semantic multiplicity of the term.[9] At the risk of oversimplification, however, I will suggest that in sixteenth-century philosophical parlance we can identify four basic notions of *historia*, each distinguishable although at times overlapping. There was first of all the meaning of *historia* inherited from medieval Scholastic philosophy, as knowledge that offered a description of something without explaining it. Scholastic philosophy had developed an elaborate distinction between two forms of demonstration, *demonstratio propter quid* and *demonstratio quia* (corresponding to Aristotle's *apodeixis tou dioti* and *apodeixis tou oti*). Only the *demonstratio propter quid* yielded true knowledge of causes, whereas *demonstratio quia* was limited to a description of "how the thing is," not why it is so. *Historia* came to be identified with the inferior *demonstratio quia* as knowledge of effects, lacking an understanding of final causes. In this specific sense, *historia* meant incomplete knowledge—an inquiry that does not pursue its subject in a final, conclusive way. So for instance Thomas Aquinas, referring to Aristotle's *De anima*, says: "It is called a *historia* because it deals with the soul without coming in this treatise to the final inquiry into all the things that are pertinent to the soul. And this precisely is the manner [*ratio*] of history."[10] This reductive notion of *historia* can be found also among the authors of medieval chronicles. For instance Gaetano Fiamma, a fourteenth-century Dominican and chronicler of Milan, interrupts the narration in one of his histories to introduce a

Scholastic disputation on some controversial issues, noting: "In doing this, we follow the way of the philosopher rather than that of the historian [*hystoriographus*], because I investigate the causes of what in another chronicle was simply described."[11] *Historia* in this sense was external to philosophy, a humble, ancillary enterprise contrasted with the lofty knowledge of causes.

This identification of *historia* with "a simple exposition of the thing without its cause" (*simplex rei enarratio sine causa*) is commonplace among the Aristotelian commentators of the Renaissance. Lodovico Boccadiferro, in his lectures on *De anima*, argues that there is a contradiction in the fact that Aristotle used *historia* to describe his inquiry in *De anima* (a truly philosophical work) and in *Historia animalium* (not a philosophical work). Only in *Historia animalium* is the term used in its proper sense, as *simplex rei enarratio sine causa*.[12] On the other hand, the fact that Aristotle had called *historia* not only the books on animals but the doctrine of the soul, the very core of his natural philosophy, was a stumbling block to the Scholastic view of *historia* as incomplete knowledge. The Scholastic philosophers tried to sort this out in several ways without finding a satisfactory solution. The humanists solved the issue philologically by going back to the ancient Aristotelian commentators, Philoponus and Simplicius, who had argued that Aristotle had used the word in this case in its generic meaning as knowledge. Accordingly, in new Latin versions of *De anima*, like that by Johannes Argyropulos, the Greek word *historia* was translated as *scientia*.[13]

Thus, in contrast with the restricted, Scholastic notion of *historia*, the humanists rediscovered the ancient Greek (pre-Aristotelian) usage of the word as knowledge in general—an intellectual turning point stressed by Seifert as the premise to *historia*'s triumph in Renaissance intellectual life.[14] This is what Giorgio Valla was thinking of when he noticed that the ancient notion of *historia* had been much wider than the present, meaning usually "quicquid vere graviterque explicatur" (anything that is expounded in a true and serious way).[15] Valla stressed especially how the ancient and the modern notion of *historia* differed in the range of subject matter. Today, he noticed, we usually call *historia* "the exposition of *res gestae*, the memory of the old order of cities and empires, the fame of illustrious men." For the ancients, in contrast, the subject matter of *historia* was much more multifarious: "So they filled up *historia* with various and manifold examples, . . . as Pliny did for the variety of nature with immense knowledge."

The Plinian model was probably a primary factor in this broadening of the notion of *historia*. Marino Becichemo, a humanist close to Benedetti's circles, turned a lecture on Pliny in 1503 into a disquisition on *historia*'s widening horizons. He also, like Valla, defines it as "quicquid vere et graviter explicatur" and strives to give the broadest acceptation possible of the term. *Historia*, he points out, does not refer only to contemporary history (as in the classical Latin opposition of *historia* and *annales*);[16] in fact, *historia*'s office, which is "to tell the truth," can cover a wide spectrum of subjects: besides "res gestae," it includes "the knowledge of places, hence geography; of times, hence chronicle; and of nature, hence the very same natural history of plants and animals."[17] For Becichemo, the close proximity, indeed the substantial unity, of *historia naturae* and *historia rerum gestarum* is such that he proceeds to praise Pliny by evaluating his work in the light of Lucian's precepts on history writing, which were based on Thucydides' model.

As evidence that a *historia* is "anything which is expounded in a true and serious way," Becichemo quotes Aristotle's *De anima*, but also the fact that Aristotle "entitled *historia animalium* some of his celebrated works on animals, as says Theodorus translator."[18] This view would become commonplace in mid-sixteenth-century discussion on *historia*. Francesco Patrizi would brush aside the narrow definition of *historia* as "things done by men " by pointing out that "Aristotle . . . has written the *historia* of animals, Theophrastus of plants, and Pliny of all nature."[19] On the same evidence plus Aristotle's *De anima*, Sperone Speroni would also conclude that "every form of writing that narrates or teaches anything whatsoever about the universe . . . may be called *historia*."[20] As late as the early seventeenth century, this broad understanding of *historia* as *expositio cuiuscumque rei* ("a description of anything") is recognized and mentioned even by those authors who want to limit it to the record of human *res gestae*.[21] There is no doubt that the sixteenth-century notion of *historia* was a far cry from the sharp distinction between the human and the natural sciences envisaged by nineteenth-century historicism.

In Renaissance epistemology the significant distinction seems to be not so much between *historia naturae* and *historia rerum gestarum* as between two different concepts of natural history, one derived from Pliny, the other from Aristotle's *historia animalium*.[22] In contrast to the Aristotelian view, the Plinian notion of *historia* did not imply the hierarchical ranking of philosophy or poetry over *historia*. The questioning of *historia*'s inferiority was also associated with Neoplatonist positions; as Cotroneo has pointed out, a thread of Neoplatonism runs through the defense of *historia* from Lorenzo Valla to Francesco Patrizi.[23] Lorenzo Valla, for instance, rejects the view of those "very great and very ancient" authors who disparaged *historia* by arguing that the poet's rank is higher than the historian's, because the poet, like the philosopher, deals with universals.[24] An argument in *historia*'s defense, Valla notes, is its contribution to the knowledge of nature: "If it is not irksome to tell the truth, from *historia* was drawn much knowledge of nature, which others then turned into general principles [*praecepta*]."[25] In even stronger terms the Platonist commentator Sebastián Fox Morcillo in *De historiae institutione* (1557) claims that "all sciences are *historiae* and can be properly called so" on the basis of philological evidence (in Plato's *Phaedo*, *historia* refers to the knowledge of physical things) and also, interestingly, on the evidence of medicine, which "reviews the parts of the human body, its diseases and their causes and signs, the kinds of remedies and all the rest, *quasi longam aliquam historiam*."[26]

On the other hand, in the late Renaissance Aristotle's *Historia animalium* was the vehicle of a more narrow notion of *historia*, in direct continuity with its Scholastic definition as inferior knowledge. A work of comparatively little interest to medieval readers, *Historia animalium* attracted a great deal of attention from the humanists, as shown by the new translation of both *Historia animalium* and Theophrastus's *Historia plantarum* by Theodore Gaza, first published in 1476. In the preface, Gaza recalls the Scholastic definition of *historia* as "an exposition of the thing as it is, which the schools of philosophers of our time use to call the *quia*." *Historia animalium* is properly called so, in contrast with *Parts of Animals* and *Generation of Animals*, which deal instead with the final and efficient causes. Gaza adds that the *historia*, the exploration of the thing as it is, or according to the *quia*, precedes the investigation of causes.[27] This view was developed by one of the most influential Aristotelians of the sixteenth century, Jacopo Zabarella: "The books on the *historia* of animals are a preparation to all the other books on animals."[28] For Zabarella, true philosophical knowledge is not historical but *scientalis*: it implies the discovery of causal first principles and the deduction from them of the particulars of observation. Compared to this, *historia* is a lightweight and underdeveloped mode of treatment. But it is, nonetheless, the preparatory stage to natural philosophical knowledge.

There is an interesting shift here: on the one hand, as in Scholastic philosophy, *historia* as *cognitio effectuum* is opposed to philosophy as *cognitio ex causis*; for instance, Zabarella says that of everything one can give either a historical (merely descriptive) or a philosophical (causal) account. But on the other hand, *historia* is also recognized as a preliminary stage of the inquiry that leads to the discovery of causes, and thus as a constitutive, foundational part of philosophical knowledge. The shift from Scholastic Aristotelianism to Renaissance Aristotelianism implied a shift from *historia* as knowledge without causes to *historia* as knowledge preparatory to the investigation of causes. This shift seems to have been an important component of the "Aristotelian empiricism" of the sixteenth century, as we shall see below when dealing with Fabricius's anatomy.[29]

But the main vehicle of an empiricist notion of *historia* in the sixteenth century seems to have been the medical rather than the natural philosophical tradition. Here we meet a third basic meaning of *historia*, besides the Scholastic "knowledge without causes" and the humanist "knowledge tout court": *historia* as *sensata cognitio*—knowledge based on sense perception or observation. Interestingly, in the early modern philosophical lexicons the paternity of this meaning of *historia* is attributed to Galen. For instance Goclenius, in his *Lexicon philosophicum*, says that for Galen, "*historia* is meant in the sense of observation, that is, a knowledge proceeding from one's own direct experience or collected from the trustworthy reports of what [was] experienced by others in the course of time."[30] In fact this meaning of *historia* as observation, either direct or indirect, derives from Galen's discussion of the ancient medical sect of the Empiricists. The ideas of this sect were known to medieval and early modern scholars mostly indirectly, through doxographic works. Galen in particular had reported the ideas of the Empiricists in two of his books, *De experientia medica* and *Subfiguratio empirica* (a third text, *De optima secta*, which also talks of the Empiricists, was attributed to Galen in the Renaissance but is not genuine according to modern scholarship).[31] Together with Celsus's *De medicina*,[32] these Galenic works were the main vehicle for the transmission of the ancient Empiricists' views to medieval and early modern culture. What matters, of course, is not so much Galen's own opinion of the Empiricists (as it happens, he was critical of them, disagreeing with their exclusive reliance on direct experience and insisting that experience should be guided by reason).[33] What is relevant here is that his enormously influential work ensured the survival of the Empiricists' views through medieval and early modern medicine. *Historia* in the ancient Empiricist sense meant, if we can trust Galen as doxographer, "the report of those things that have been seen or are as if they had been seen" ("nuntiatio eorum que visa sunt aut sicut visa"). In other words, it is the narration of what one has either seen oneself (also called in Greek *autopsia*) or read in books as having been seen by other people (*historia*). *Historia* in this second sense should be examined critically: only those histories should be accepted by the Empiricist physician on which there is agreement (*concordantia*) of many authors. Galen seems to assent to this principle, adding that "those things of sense experience on which there is agreement among men, are trustworthy for the purposes of life."[34] *Historia* here is direct observation plus a critically examined tradition of observational reports.[35]

There is finally one fourth meaning of *historia*, used by late Renaissance commentators on Hippocrates' *Epidemics* to refer to *casus*, the course of disease in an individual, along the model of *Epidemics I* and *III*. Commentaries on *Epidemics*, a fast growing genre in the second half of the sixteenth century,[36] are one of the places were the term *historia* acquired new currency in medicine as applying to the description of a single case over time. This was a novelty: previously, *exemplum* had been the word most often used in medical literature to indicate the individual case, or what we would roughly call a case history. In medicine, as in natural history, *historia* had usually stood instead for an orderly collection of examples.[37]

Surveying the notions of *historia* in the sixteenth century, two things stand out. First, it is clear that the term's meaning was in tremendous flux; second, that its trajectory was definitely upward. In the late Renaissance hierarchy of knowledge forms, *historia* was certainly in the ascendant. As pointed out by Seifert, the starting point of this ascent was the humanist rediscovery of the pre-Aristotelian meaning of the word as knowledge in general, which, together with the Plinian paradigm, seems to have played a significant role in the humanists' rejection of Scholastic epistemology. But even the Aristotelian notion of *historia* was fated to gain a new, much improved epistemological status. At the turn of the century the Aristotelian "description of the thing as it is" was the standard definition of *historia* in the philosophical lexicons.[38] But what had been a badge of inferiority in Aristotelian eyes was turned into an advantage in the Baconian program of a renewed natural history as "the granary and storehouse" of "matters of fact," the indispensable foundation on which to build a true knowledge of nature, free of "opinions, doctrines and speculations."[39] The decisive step in this radical reappraisal of *historia*'s epistemological value seems to have been its association with a rigorous induction—a step that distanced the term from the loose Plinian acceptation as a collection of examples. This new cognitive notion of *historia* became current among the followers of the Baconian program. Thus Francis Glisson, for instance, defined *historia* as "plena enumeratio experimentorum." The goal of this "part of the art of discovery," he wrote, is not "the mere narration, or the knowledge by *experimenta* of the thing narrated, nor the bare knowledge of life's variety.... Its first goal instead is to be a representation of all the nature investigated in an unknown subject, so that the intellect, thanks to sufficient induction, may rise to the abstract nature of the thing, and to new and infallible axioms.... Therefore, in order to discover the nature of hidden things, one has to collect the full recapitulation [*enumeratio*], i.e. the *historia* of all *experimenta* referring to the nature investigated. This collection or *historia* is the first part of the art of inventing the arts."[40]

Some of these new developments had been prepared by the humanist discovery that *historia* could not be confined to the Procrustean bed of its Scholastic Aristotelian definition. Francesco Patrizi was certainly aware of the cognitive implications of his philosophical game when he astonished his interlocutors by broadening indefinitely the borders of *historia*. He questioned, first of all, the commonplace distinction between historian and philosopher: "Eminent literati think that the historian's craft is to narrate just and exclusively the effects, and that searching for the cause of any thing whatsoever is nowadays the philosopher's office." But, Patrizi objects in a daring move, "a cause by its true nature, though being a cause of another fact, is in itself also a fact. And as such it falls within the historian's narration." *Pace* Aristotle, causes also belong in the historian's domain: the model here is *historia rerum gestarum*, where it is clear that "whether the historian be dealing with actions having to do with war, peace, or conspiracies, it is proper that he always tell by which cause the actor was moved to do a certain thing."[41] But *historia* does not deal only with "le cose degli huomini," it also covers "le cose di natura et sopra natura"; not only the past but also the present and the future; it is not only a written text but can be also represented in pictures.[42]

One senses, in the Renaissance excitement over *historia*'s endless versatility and variability, mirroring the infinite complexity of nature and life, something like the Victorians' enthusiasm for another genre, the novel, that also started from humble origins to reach dazzling heights. I am thinking of George Eliot's loving words about the novel: "There is no species of art which is so free from rigid requirements. Like crystalline masses, it may take any form, and yet be beautiful."[43] The upward mobility of *historia* in the epistemological framework of the late Renaissance is clearly evidenced in tracing the fortunes of *historia anatomica* and *historia medica*, as we shall do now.

Historia Anatomica

The exordium of *historia anatomica* confirms the close link between *historia* and anti-Scholastic attitudes. Alessandro Benedetti's presentation of his anatomical treatise as a *historia* from its very title was definitely meant as a thrust against Scholastic medicine. Using a metaphor that combined his double identity as physician and philologist, Benedetti stated that by means of his "libelli de historia membrorum hominis" he intended "to heal the thousand wounds" inflicted on the body of medicine by the replacement of the original Greek terms with barbaric, Arabic words.[44] His anatomical *historia* certainly differs from contemporary Scholastic anatomical summae such as Gabriele Zerbi's *Liber anathomiae* (1502). Whereas Zerbi's book is a thick and heavy *coacervum* of all the authorities, Benedetti's text is compact and crisp (he himself calls it a *breviarium*) and stresses open-ended inquiry and the report of directly observed cases, mostly on rare conditions, rather than a synthesis of received knowledge.⁴⁵

Benedetti mixes his own observations with anecdotes drawn from tradition, following the humanist practice of excerpting and using in ever-changing contexts *facta memorabilia* from the classical authors—those morsels of past experience that Ann Blair has called "factoids."⁴⁶ In the case of medicine, this humanist practice had a time-honored precedent in the tradition of medieval *practica* textbooks, which often reported anecdotes from a large repertory of *exempla*.⁴⁷ Benedetti was emphatically a practitioner: besides the *Anatomice*, he wrote a collection of medical aphorisms and a compilation on diseases and their remedies, along the "from head to toe" model of fifteenth-century *practica* textbooks.⁴⁸ His anatomy also served the needs of therapy, and for him therapeutics required registering and memorizing past experience. He followed Arnald of Villanova's precept: "Experimenta et certificata stilo brevi, et aphoristico scribere."⁴⁹ So in his practical texts as well as in *Anatomice* he reported successfully experimented cures (*experimenta*). His work was praised for "the novelty of its *experimenta*" by Jacopo Antiquario in 1493.⁵⁰

Experimentum means here—as it meant for Arnald of Villanova in the fourteenth century—"unius rei sine ratione cognitio" (the knowledge of one thing without rational justification), to use the definition given by Giorgio Valla. "As when for instance," Valla goes on, "somebody learns merely that a certain thing is helpful, and so he uses it though ignoring the reason why the thing is useful."⁵¹ This definition fitted perfectly the medical handing down of empirically found remedies (in fact collections of recipes were often called *experimenta*).⁵² Besides *experimentum*, the experience of individual cases is called by Benedetti *observatio* or most frequently *exemplum*. Benedetti's *exemplum* is different from Aristotelian experience, which was based on ordinary, repetitive events; it usually concerns extraordinary events, described historically with date, place, and witnesses.⁵³ For Benedetti the *exemplum* often captures experience *statu nascenti*, as it is perceived by the working anatomist or observer. Pliny's natural history was a model of this use of the *exemplum* for the collation of knowledge not yet capable of being systematized.⁵⁴ Benedetti was extremely interested in Pliny, to whom he devoted a philological work of emendation: the first by a physician, with particular attention to materia medica.[55] Though he also drew on Aristotle's *Historia animalium*, which he used as a fundamental source of Greek anatomical terminology,[56] his use of the *exemplum* suggests that his notion of *historia* is the broad and relaxed Plinian one, unencumbered by any reservations linked to an Aristotelian epistemology. On the contrary, he shared the humanists' lofty view of *historia*, based on the prestige of civil historiography. Like many other physicians of his and later times,[57] Benedetti moved easily from being the historian of the body to being the historian of *res gestae*. In 1496, exactly the same year in which he was completing the *Anatomice*, he wrote a history of Charles VIII's Italian campaign, *Diaria de bello Carolino*, based on his direct involvement in the war as military doctor to the Venetian army.[58] Appropriately, Johannes Cesarius, who published the Cologne edition of Benedetti's *Anatomice* in 1527, recommended his works not only to physicians but to all "those who want to train themselves in the reading of histories."[59]

Humanist anatomists like Benedetti, Estienne, Vesalius, and Falloppia spoke of *historia* in an anatomical context without any apparent awareness of the reductive Aristotelian meaning of the term.[60] In the second half of the sixteenth century, however, *historia anatomica* seems to conflate the Galenic meaning of *historia* as ocular inspection (the Greek *autopsia*) with the Aristotelian acceptation as description without causes. So for instance du Laurens, in his influential textbook, explained that there are two parts to anatomy: a less noble but more certain part, *historia*, gained by inspection, whether of vivisected animals, dissected men, or pictures in books; and a nobler but less certain part, *doctrina*, also called *scientalis*, which is concerned with causes and contains universal *theoremata* from which are built demonstrations.[61] This particular shift of the term's use can be clearly seen in the "philosophical anatomy" taught by Hieronymus Fabricius at Padua from 1565 to 1613, and later carried on, with sensational results, by his pupil, William Harvey. Padua was, of course, the hotbed of a renewed and aggressive Aristotelianism. Fabricius's program of anatomical research was peculiarly new in taking as a paradigm Aristotle's animal books, which traditionally had not been the object of systematic lecturing in the natural philosophical and medical university curriculum.[62] Unconventionally, Fabricius took the sequence of *Historia animalium*, *Parts of Animals*, and *Generation of Animals* as a research model that could be applied to discover new things about the animal and human frame.⁶³

In all the works he published after thirty years of practicing dissection, Fabricius followed a fixed format. To report anatomical research, he believed, one ought to follow a series of stages. First one should give the *historia* of the part in question (a full account from observation and dissection of everything noteworthy about the object). Secondly, one should inquire into the *actio* of the part; thirdly, identify its *usus* (or purpose, or final cause: the general reason why the part exists and why it is as it is); and lastly, explain the particulars of observation by deduction from the final causes. Only by going through all these stages can anatomical inquiry be truly philosophical: that is, able to reach statements that are universally true explanatory accounts of the causes of things. Most anatomists, Fabricius argued—among them even Vesalius—stop at the first stage, the *historia*, without pursuing the inquiry into the action and use of the parts. Fabricius's anatomical *historia* was thus an operational rendition of the natural philosophical category of *historia* as preliminary to the inquiry into cause. It should be noted that *historia* in this sense is not a repository of particulars conceived as independent of theory, but an account only of those particulars of observation that were found to be relevant in the light of the final cause. The *differentiae* (the specificities of some animals) included in the *historia* were only those that could be deduced from the general "use" of the part. Fabricius's *historia* is not a report of Baconian "matters of fact": it is, however, a very interesting example of early modern Aristotelian empiricism in a peculiarly Paduan version.⁶⁴

It is worthwhile now to examine the meaning that the anatomical *historia* came to have for Fabricius's most famous pupil, William Harvey. Harvey adopted Fabricius's program of a "philosophical anatomy": in his *Anatomical Lectures*, he taught, following his teacher, that anatomy had to contain three parts: a *historia*, or description of the part; an account of the action; and a demonstration of the use, or purpose.⁶⁵ But in *De motu cordis* Harvey does not reach the crowning stage of the inquiry: he does indeed establish the action of the heart, but does not provide an account of its "use" or final cause. Compared with Fabricius's works, *De motu cordis* is only a *historia* plus an account of the action; it lacks the crucial element of "use." The history of the reception of the circulation hypothesis shows that this deficiency in final cause was a major stumbling block to the acceptance of Harvey's view. A very common objection was that one could not accept circulation because Harvey had not explained its "use." Harvey answered this criticism by means of a principle of limited explanation: "One should first inquire into the Quod sit before dealing with the Propter Quid."[66] What he meant was not simply that the production of *historiae* comes before the search for causes (this was obvious), but that one may also stop there and yet achieve a significant result. The discovery of the circulation seems to have shaken Harvey's belief in *propter quid* knowledge as the fundamental goal of anatomy. As argued by Roger French: "Harvey's natural philosophy now included the notion that reliable *historiae*, mostly experimentally produced, were of value on their own and constituted knowledge, without an intellectual link to a final cause."[67]

A striking parallel to Harvey's reevaluation of the role of *historia* in anatomical knowledge is provided by the work of Gaspare Aselli, whose discovery of a new kind of vessels, which he called *venae lacteae* (the lacteals, or lymphatics of the thorax) was hailed by late seventeenth-century anatomists as one of the most important innovations of their times.[68] Aselli shared Fabricius's ambitious model of anatomy as *scientia*: demonstrative knowledge that, with the help of dissection, reaches a full and complete knowledge of every body part.[69] Anatomy so defined, he says, has three components: (a) *historia*, to use Aristotle's term: the description of the part; (b) *actio, seu functio*: the account of its action; (c) *usus*: the demonstration of its purpose.[70] But compared with Fabricius, Aselli vastly expands the place of *historia* in his account. His text on the discovery of the lacteals is structured in four parts (it was indeed originally delivered as four lectures at the University of Pavia in 1625): the *historia* takes three of these parts, while the last one covers both *actio* and *usus*.

Aselli first expands the *historia* by giving a very detailed "history of the discovery" (*inventionis historia*) "reported in a trustworthy way [*cum fide*]" in a graphic narrative.

> As regards the *historia*, this is how the thing went. On July 23 of that year (1622) I had taken a dog in good condition and well fed, to dissect it alive at the request of some friends, who very much wished to see the recurrent nerves. When I had

finished showing the nerves, it seemed a good idea to observe also the movements of the diaphragm, in the same dog and by the same operation. While I am trying to do this, and for that purpose I open the abdomen and pull down with my hand the intestines and the stomach stuck together into one mass, I suddenly see a great number of slender cords [*funiculos*], so to speak, extremely thin and white, scattered all over the whole mesentery and intestine, from almost innumerable starting points. Taking these at first sight to be nerves, I did not pause. But presently I realized I was wrong, since I noticed that the nerves of the intestines were distinct from these cords, and quite different from them, besides running apart from them. Therefore, struck by the novelty of the thing, I stood silent for some time, while there came into my mind the verbous and venomous disputes among anatomists concerning the meseraic veins and their function. And perhaps fortune helped, since a few days before I had looked into a little book by Giovanni Costeo, written about this very matter.[71] So gathering my wits together for the sake of experimenting, with a very sharp lancet I pricked one of the largest of these cords. I had barely touched it when I see a white liquid, like milk or cream, immediately gush out. Seeing this, I could not restrain my delight, and turning to those who were standing by, especially to Alessandro Tadino and Senatore Settala, both members of the supreme College of the order of Physicians and, at the moment of my writing, Head Officers of Public Health, I say with Archimedes: *Eureka*, and at the same time I invite them to the delightful spectacle [*jucundum spectaculum*] of such an unusual thing."[72]

What we can notice immediately in this narrative of the discovery is the richness of what we would call historical detail: the indication, for instance, not only of the year but even of the day and month of the event—a precision we rarely find in earlier anatomical texts. This wealth of particulars is clearly meant to enhance the credibility of the story, as is the list of the "very reliable witnesses" who were present at the dissection.

But Aselli's *historia* includes much more than just the historical account of the discovery. He is very keen on stressing the cognitive implications of the *historia*, which he spells out in great detail. *Historia*, he says, covers the foreknowledge (*praenotio*) of the thing.[73] "In every thing that is an object of knowledge it is necessary to grasp beforehand [*praecipere*] and to anticipate [*anticipare*] what the thing is (*oti esì*, or whether it exists, *ei esì*) and what it signifies . . . : of which the first refers to the existence [*ten uparxein*], the latter to the notion conveyed by the name [*nominis notio*]. . . . From the name's notion we are often led to an acquaintance with the thing itself."[74] Naming the new object is therefore a very important step in the cognitive process, and Aselli spends considerable mental work in selecting the right name for the new vessels.

To this task he summons all his philological knowledge of ancient and modern anatomical terminology. He finally decides to call the new vessels *venae lacteae*, because the word *lactes*, he notes, had been used by both ancient and modern medical authors in a variety of meanings, all however loosely associated with the mesentery—the part where the newly discovered vessels are located. Moreover, he notes, the common sense association of the word *lactes* with milk (*lac*) fits these vessels, since they carry a fluid very similar to milk. So Aselli picks a term that appeals to common sense but, more importantly, has a respectable scholarly pedigree: it belongs to the specialized tradition of medical language. As such, the term establishes the new thing not only in commonsense perception but also, more importantly, as an object of scholarly knowledge. We could say that by naming it in this way, Aselli turns the new thing he discovered into an epistemic object.

So much for the name. The existence of the vessels, he says, is sufficiently established by the reliability of the *historia* (*historiae fides*); and by *historia* he means here the account of the discovery plus the *historia* in its usual anatomical sense, namely the description of "what was clearly seen repeatedly [*evidentia consuetudinis*] over several dissections."[75] He then gives the *historia* in this latter sense, a detailed description of the new vessels' site, size, etc., referring throughout to the plates that accompany the text.[76] These plates are also a conspicuous part of the *historia*, and in fact Aselli refers to them only in the first part of his text. As visual *historia*, the plates represent a composite image of what he has seen over several vivisections (they are used to give the audience a vicarious experience not just of one but of many dissections) and serve to establish a clear foreknowledge of the thing in the audience's minds. The image, far from what we might think today, is not supposed to prove anything: it is simply an aid to perception and memory. The image does not demonstrate anything because it is simply another aspect, in a different medium, of the intellectual category called *historia*, namely the description, based on sense evidence, of *to oti*, how the thing is. Aselli concludes the *historia* by saying: "All of this was said not to prove anything but to put the thing in the palm of our hand, so to speak." The *historia* is ostensive, not demonstrative, knowledge. He is careful to point out that the detailed description of the milky veins, including the plates, is provisional knowledge (*quantum mihi cognoscere datum fuit hactenus*). And yet he also notes incidentally—as if to justify the attention and space he devoted to the *historia*—that "*to oti* is the first step and the beginning of knowledge: or rather, according to the most eminent philosophers, all of it virtually [*potestate*], or at least half, if not more than half of it."[77]

Aselli's text shows a remarkable widening of the meaning and use of *historia* in anatomical knowledge, Here *historia* includes, as we have seen, the historical narrative of the discovery, the philologically informed search for the appropriate name for the object of inquiry, and the object's meticulous description through textual and visual medium. Aselli's text also makes clear that *historia* was the most convenient medium for the introduction of novelty. Occupying the lowest rank in the hierarchy of anatomical knowledge, being the least "scientific" of its three parts, the least bound by the rules of rigorous demonstration, the *historia* could accommodate provisional knowledge, or even knowledge that conflicted with the received observational and theoretical frameworks. It was the area of argument most permeable to novelty, where the constraining force of tradition was least compelling. This is the reason why the *historia* takes up so much space in Aselli's account of his discovery. It is also the reason why the *historia*, both textual and visual, became more and more important in the anatomical texts of this period.

The *historia* also proved to be the most long-lived part of Aselli's work. In the *Bibliotheca anatomica* of Le Clerc and Manget (1685), a massive anthology of the anatomical discoveries of the moderns, the only part of Aselli's text that was excerpted and republished was the *historia*.[78] This was certainly because by then further research on the lymphatics had shown that Aselli had been wrong in the *usus* he claimed for the lacteals. But it was also because in the second half of the seventeenth century there was clearly a reading public that was interested in anatomical *historiae* for their own sake. It is symptomatic, for instance, that even Harvey's *Exercitationes de generatione animalium*, in which he had tried very hard to reconcile *historia* as observational report with an Aristotelian natural philosophical framework,[79] was avidly read precisely for the *particularis historia* of what went on in the uterus of does and hinds after intercourse.[80] In 1674 Harvey's book was republished in Amsterdam with all the philosophical parts excised and only the observational *historiae* left. The editor, the Dutch physician Justus Schrader, was an assertive advocate of the Baconian program and was convinced that there are two ways of pursuing knowledge in either medicine or natural philosophy: by way of systems or by way of *historiae*, that is *observationes* and *experimenta*—and that the second way was definitely the best. He decided therefore to "excerpt all those things that Harvey noted as seen by himself . . . and to leave out all his discourses and reasonings, be they ever so beautiful."[81] There was by then more interest in *historiae* as the report of *to oti* than in demonstrations *propter quid*—an increasing sense that for the purposes of anatomical knowledge, *historiae* or observational reports might even be the cream of the argument.

This rising epistemological significance of *historia* was a trend that went beyond anatomy. It was even more pronounced, as we shall see, in those genres that were more directly related to medical practice.

Historia Medica

Historia medica, like *historia anatomica*, also seems to have its origins in medical humanism, but its pedigree is certainly more ancient than that. Narrative accounts of the treatment of single patients were inserted, as *exempla* or *casus*, in the medieval textbooks of the genre *practica* from Archimatthaeus in the twelfth century to Michele Savonarola in the fifteenth.[82] But in medieval medicine *casus* and *exempla* are to be found in the folds of the text, so to speak: they do not emerge as a genre on their own. In the late Middle Ages, however, one can see the development of a genre apparently aimed at presenting and discussing an individual case: the *consilium*, a form of writing that would have a long life in medical literature, from the thirteenth to the seventeenth century. In what is still, as far as I know, the only monograph on the history of the case history, Pedro Laín Entralgo has stressed the role of the *consilium* in bringing the individual patient to the foreground of medical attention. In his view, the many collections of *consilia* from the fourteenth and fifteenth centuries show that, in spite of the Scholastic goal of "plura singularia ad universalem reducere" (turning many particulars into a universal), the attention to individual cases became more and more prominent in medical literature.[83] In their thorough study of the medieval *consilia*, however, Jole Agrimi and Chiara Crisciani have shown that the *consilium* dealt typically with a disease, not with a sick person.[84] At times biographical details about the patient were given, at times they were not, nor was the temporal order of symptoms always recorded. The narrative format was used only sporadically, and the description of symptoms was heavily interlarded with references to the authorities. Though the *consilium* starts from a *casus*, its goal is not describing the individual case per se but redefining it according to doctrine as a "universal made concrete."[85] Crisciani has shown that the very structure of the *consilium* as a codified genre was partly responsible for this cognitive bent: doctors who dealt with cases in abstract doctrinal terms in their *consilia* would describe the very same cases much more empirically, with a chronicle-like attention to temporality and detail, when writing about them in letters meant to convey politically sensitive information about the diseases (and thus the life chances) of high-ranking people in the early Renaissance courts.[86]

An early sixteenth-century example of the emergence of *historia* within the genre of medical *practica* is *Galeni historiales campi*, a selection of *historiae* excerpted from Galen with commentary, published in 1532 by the humanist doctor Symphorien Champier.[87] A Platonist, feminist, and satirist of Scholastic medicine, Champier also wrote histories of Savoy, Lorraine, and of the antiquities of the city of Lyon.[88] He seems to have been uncomfortably aware of the novelty of his *Historiales campi*. He writes diffidently about collecting in a small book such comments on Galen's *historiae* as he had been writing over a long time and left "scattered in the hands of his students all over France, or gathering dust in the dark corners of his library." He admits to some trepidation in publishing such "membra dissecta," especially at the thought of those of his contemporaries "who severely examine all little books, and with puckered brow and rhinoceros-like nose sniff at the productions of other people."[89]

What parts of Galen does Champier excerpt? He selects eighty-six extracts, each titled *historia*, each prefaced by the indication of the Galenic work it comes from (mostly *De locis affectis* and *De ingenio sanitatis*)[90] and followed by Champier's own comment. The topic of each *historia* is not an individual case but a disease or a problem of medical practice (for instance, "which diet is appropriate for a bilious complexion"). The extract from Galen, on the other hand, is usually an account based on direct experience—cures he performed, observations from dissections, extraordinary events he witnessed. If *historia*, predictably enough, seems to refer here to direct experience, it does not refer to individual cases. One *historia*, in fact, often includes several cases, called *exempla* (when it deals with an individual case, the excerpt is titled *exemplum*, not *historia*). In his comment, Champier usually discusses in general terms the disease of which Galen reported one or several instances. Only occasionally does he reconsider the case presented by Galen, and very rarely does he report a case of his own. When he does, it is the novelty and rarity of the experience that he wants to stress.[91] Generally, however, Champier does not use Galen's *historia* as a peg on which to hang his own experience. He is more interested in using it to establish general rules about treatment. So most of the excerpts are accounts of Galen's successful cures (*curationes*), presented as exemplary guides to practice.

Laín Entralgo has argued that, compared with the Hippocratic corpus, Galen's works contain very few accounts that can be properly called case histories.[92] Sixteenth-century medical readers would strongly disagree with him. In the *Theatrum Galeni*, a massive compilation of *loci communes* from Galen's work compiled in the 1540s and '50s by the humanist physician, botanist and antiquarian Luigi Mondella, the entry *historia* figures prominently, referring the reader to 174 excerpts from Galen; almost every one is a *casus*, an account of an individual cure (indeed, the headings *casus* and *historiae* are cross-listed).[93] No less than Johann Crato von Crafftheim—arguably one of the most influential physicians of the sixteenth century, an indefatigable medical networker and, like Mondella, a protagonist of the new medical humanist genre of *epistolae medicinales*—wrote in his *Epistola* "on the right way of reading Galen": "In reading Galen one meets many *historiae* and accounts of cures [*curationum recitationes*], which if taken as examples of rational method, considered together with the maxims of the art and collected in the right order, will bring not only much light and strength to Hippocrates' doctrine, but also uncommon aid and assistance to treatment."[94] For sixteenth-century physicians like Champier, Mondella, or Crato, Galen's *historiae* were collectables of great intellectual and practical value precisely because they were *curationum recitationes*, accounts of successful treatment. For Laín Entralgo the model of the case history is the Hippocratic noninterventionist account, as in *Epidemics I* and *III*—a description of the course of disease with very little, if any, mention of therapy.⁹⁵ Consequently, he traces the roots of the Renaissance "prehistory of the case history" to the Hippocratic revival of the second half of the century. In particular, he sees the commentaries to *Epidemics* as the primary impulse behind the late Renaissance new interest in the description of individual cases and the main force in the loosening of medical observation from the tight grip of doctrinal concerns that had severely limited its range in the *Consilia* literature.⁹⁶

There is much evidence, however, suggesting that the Renaissance embryonic case history was fashioned after a Galenic, not a Hippocratic model. Vivian Nutton and Nancy Siraisi have both stressed that the typical pattern of case narratives in Renaissance medicine is the Galenic *curatio*, the report of a cure with a strong component of self-congratulation, if not downright self-advertisement. Nutton has argued that it is hard to trace any direct influence of *Epidemics* upon most of the stories about individual patients told by sixteenth-century physicians and surgeons, which are typically, like Galen's, reports of successful cures.⁹⁷ Analyzing Cardano's *Curationes et praedictiones admirandae* (1562), Nancy Siraisi has shown that, although *Epidemics* is Cardano's averred model, one can find only sporadic traces in his text of the impersonal and noninterventionist stance of the Hippocratic physician. In marked contrast to the histories in *Epidemics*, Cardano's accounts of his practice were rich in autobiographical detail and strongly emphasized success. In fact, she argues, Cardano imitated Galenic case analysis and Galenic self-presentation, combining the case history with self-advertising autobiography.⁹⁸

I think there is no doubt that Nutton and Siraisi are right, and that the late Renaissance medical genre of the *curationes* was fashioned after a Galenic model. The question is, why? The collections of *curationes* that started to come out in the middle decades of the sixteenth century were certainly a new genre. For the first time, accounts of successful cures were presented no longer semi-hidden in the doctrinal framework of a text, as in the fifteenth-century *practica* textbooks, but prominently displayed as freestanding on their own, loosely organized by numerical order: for instance, in groups of hundreds, as in the *Centuriae curationum* that Amatus Lusitanus published in seven installments between 1549 and 1570.⁹⁹ Though undoubtedly a new genre, the *curationes* share some traits with an old genre, the medieval *experimenta*, and in several ways seem to be derived from it. Medieval collections of *experimenta* recorded remedies that had proved successful but whose efficacy could not be justified on doctrinal grounds. The format for storing and transmitting this empirical knowledge was the recipe (a list of ingredients with more or less detailed instructions), usually prefaced simply by the name of the disease to which it should be applied, but sometimes instead by a brief narrative of the case in which it had given good results.[100] The fourteenth-century *experimenta* of Arnald of Villanova, for instance, present both formats: the simple recipe and the recipe cum case narrative.[101]

Like the *experimenta*, the Renaissance *curationes* are presented as therapies legitimized mainly by efficacy rather than by doctrine. Like the *experimenta* and the *consilia*, moreover, they contain recipes, but with an important shift: the case narrative is now much more prominent—indeed, it is the main focus of attention.[102] We see this shift very clearly in Amatus's *curationes*: in each of them the recipe is inserted in a detailed narrative account of an individual case. The amount of historical detail is striking, especially considering its dearth in the traditional *consilium*. Amatus invariably tells the place of the cure, the name of the patient plus several details about his social rank and lifestyle, and sometimes (like Galen) even the amount of the fee he received for the successful treatment. No doubt his *curationes* were an advertisement of his skills, along Galen's model. But this self-promoting attitude was probably due not just to Galen's influence, but also to the particular conditions of medical practice in the sixteenth-century medical marketplace.

In fact, Amatus's *curationes* are strongly reminiscent of the "cure testimonials" we find among the records of sixteenth-century medical authorities such as the Italian colleges and protomedicati. These testimonials were presented by folk healers when applying for a permit to practice. Often undersigned by the patients, the testimonials listed successful cures and the composition of the remedies used. These documents indicate the coexistence of two sources of legitimization of medical practice: one by authority, from above (the license granted by the representatives of medical learning), and one by efficacy, from below (the patients' testimonials).[103] Though essential for folk healers, reliance on patients' testimonials was by no means limited to them. Learned physicians also had to compete daily in a medical marketplace where efficacy was key to success (even more so when a shadow might be cast on their learning, as in the case of Amatus, by the stigma of a Jewish identity). In fact, we find patients' testimonials appended to the little book published by an unlicensed healer, Costantino Saccardini, as well as to the autobiography of a learned physician such as Girolamo Cardano.[104] In both cases, the testimonials served to advertise the practitioner's good reputation and to anoint his practice with the chrism of efficacy.

Legitimization by efficacy was obviously more vital for lower-status practitioners, such as the surgeons; so it is not surprising that it is in their books, as Nutton has pointed out, that we find many instances of Renaissance *curationes*.[105] It was also essential for those practitioners who introduced new forms of treatment that were not justifiable by reference to the authorities. A case in point is the Paracelsian doctor Martin Ruland the Elder, who, like Amatus, published his aptly named *Curationes empiricae et historicae in certis locis et notis hominibus optime riteque probatae et expertae* in *centuriae* installments over the course of eighteen years (1578–1596).[106] He pointedly mentions, in his preface, the legitimizing role he attributes to his patients: "Over many years now of practicing medicine and curing the sick at home and abroad I've been taking daily note of the most remarkable and dependable cures, ... which the sick themselves, restored to health, ... urge me to record in writing."[107] In Ruland's *curationes* the name and place of each patient is carefully indicated not only in the text but even in the index. This is in fact the only early modern medical work I know where the index lists the names of the patients next to the diseases treated and the remedies employed. Never had the sick person's individuality been given so much space in a medical text! Such accuracy is obviously due to the fact that Ruland thought his patients had to be identifiable if they were to be probatory witnesses for the efficacy of his cures. A competitive medical marketplace put pressure on the practitioners to assert the legitimacy of their mode of treatment in terms of effectiveness, not just of prestige derived from learning. This possibly explains why the self-touting Galen (and not the self-effacing Hippocratic physician) was the model of the *Curationes*.

On the other hand, there is no denying the significance of the Hippocratic model in reorienting medical observation in this period. The

Hippocratic case histories seem in fact to have deeply influenced another medical genre that starts to emerge in the late sixteenth century—the *observationes*. The increasing interest in *Epidemics*[108] was associated with a break from Scholastic tradition, especially from its method of emphasizing the general features of a disease according to its description by the authors—a method that made it difficult to give full attention to the individual patient. Thus for instance Giovanni Battista da Monte, who in the years around 1540 radically altered the way of teaching medicine at Padua by taking the students to the patient's bedside in the hospital of San Francesco, recalled with disapproval the Scholastic way of examining (or should one say, ignoring) the sick: "When I was a young man, learning practice as an apprentice to the famous physicians of the time..., one had a confused way of holding consultations. First, [the physicians] would declare the temperament of the patient, before they had even pointed out the manifest signs, and so were trying to demonstrate the known through the unknown, which is a preposterous order. And when they had considered the temperament, they would proceed to name the disease. After which they would list its symptoms as described by Avicenna, whether or not they were actually present in the patient. And sometimes they would declare twenty signs, but when I approached the patient I could barely find two of them."[109] Late in his life da Monte lectured extensively on *Epidemics I*.[110] He took two Hippocratic cases (those of Philiscus and Silenus, the first two cases in *Epidemics I*) and discussed each symptom in great detail, in the chronological sequence of the original. He seems to have meant to provide the students in this way with a model of meticulous observation of an individual case, and he stressed the usefulness, both theoretical and practical, of such an approach: "Habebitis in singulo casu et theoricam et practicam."[111] He also gave practical advice to his students on what to note in a case: "You should ask either the sick [person] or those who attend him what are his habits and customs, what trade he exercises, whether he has anything specific to his nature, such as an aversion to cheese or wine, etc. Once you have learnt these things, you will make a catalogue of them all, proceeding in order. First you will place all those things that are apparent externally, and thus you will construct a simple *historia*."[112]

Da Monte's *Consultationes*, collected and published posthumously by his students, indicate the transition from the old genre of the *consilium* to the new observational approach. The first volume is made up of entirely conventional *consilia*, mostly dealing in fact with patients that da Monte had not seen at all and on whose condition he had been informed by letter. But the second volume, edited by Johann Crato, includes detailed reconstructions of cases based on direct observation, including the particulars of the treatment adopted. Crato also included in this volume twenty-two of da Monte's *curationes febrium*, which are also detailed accounts of directly observed cases and include the guidelines on how to write a case history mentioned above.[113]

A conscious effort to fashion his observations after the Hippocratic model was made by François Valleriola, a successful practitioner who was called to a chair of medicine in Turin by duke Emanuele Filiberto in 1572. Publishing his *Observationes medicinales* in 1573 after forty years of practice, Valleriola wrote: "[Hippocrates] wrote on tablets all that he saw occurring in the sick person, and narrated the complete *historia* of the disease and what happened to the sick each day, each hour, each moment, giving specifically the name of each person . . . as shown in the books of *Epidemics*. In the same way, following Hippocrates' custom, I reworked for general use the things I wrote down, taking into considerations only those diseases that appeared to me most dangerous and of dubious treatment. I have told in the first place the *historia morbi*, naming each patient individually as done by Hippocrates. Thereafter follows an ample *explicatio* of each observation, . . . wherein I recount in detail in which way and by which cause the patient fell into this or that disease."[114] Valleriola is aware of the difference between the Hippocratic model and the *curationes* genre. "I place after each *historia morbi* its own *curationes*. We know that Hippocrates in *Epidemics* omitted these, contenting himself with the narration of how the thing went. His goal indeed was only to relate what had happened, without discussing the remedies or even mentioning them. We however, lest something lacked in our work, added also the great number of remedies we used." In spite of Valleriola's programmatic intention, the *curationes* paradigm prevails over the Hippocratic model. The very title of each observation often draws attention to the "miraculous" cure it contains and—even more revealing of the close proximity of Valleriola's *observatio* to the conventional *curatio*—recipes of the remedies used are included in the *explicatio* of each case.

We should notice that Valleriola often calls the case *historia*. The *observatio* and the *historia morbi* are treated as synonyms. Indeed, it is especially with reference to the Hippocratic model that *historia* acquired a new meaning as the narration of an individual case. Amatus Lusitanus, for instance, had already referred to the cases in *Epidemics* as *historiae*.[115] In contrast, we may remember that Champier always called the individual case *exemplum*, and *historia* maintained for him its plural connotation as a collection of *exempla*. The influence of *Epidemics* seems to have brought about an important shift in *historia*'s meaning—from an emphasis on plurality to an emphasis on singularity. There is a strong awareness in the late sixteenth-century commentators on *Epidemics* that the Hippocratic model implied a fundamental restructuring of medical attention to focus on the individual case. Francisco Vallés stated in his important commentary on *Epidemics*, "The medical art is the art of healing, and treatment deals only with individual men, like Callias or Socrates. But there is no way of bringing Callias and Socrates into schools and books unless by means of *historiae*." Vallés is referring here, with clear polemic intent (and possibly a hint of medical skepticism) to a well-known passage in *Metaphysics* I.1, where Aristotle had disparaged as mere "matter of experience" the knowledge that some remedies had proved beneficial to specific individuals, such as Callias and Socrates.[116]

The effort of "bringing Callias and Socrates into schools and books" was gaining momentum in the 1580s. Two important collections of *observationes* were published in these years, Dodoens's *Medicinalium observationum exempla rara* (1581) and van Foreest's *Observationes et curationes medicinales* (1584). In the case of van Foreest, this was the first installment of an enterprise that would reach up to 32 volumes. A practitioner par excellence, van Foreest dedicated the first volume of his *Observationes* to the municipality of Delft, which he had faithfully served as a town doctor for 27 years.[117] He declared that in keeping such detailed records of his cases he followed the lead of Hippocrates, but also of Galen (indeed he quotes approvingly Champier's collection of Galen's *historiae*) and of Amatus Lusitanus.[118] As in Amatus, his *Observationes* give the name of each patient, the place and sometimes the date of the cure, as well as the recipe of the remedies used. The title of each *observatio* highlights the disease (e.g., "De febre ephemera") but also, at times, the individual patient and the most striking feature of the case ("De puella a febre putrida per fluxum sanguinis narium curata"). Again as in Amatus, the brief narrative account of the case is followed by more lengthy and learned *scholia*.

There is scattered evidence that in the second half of the sixteenth century the habit of keeping detailed records of cases (especially the most unusual ones), so strikingly evidenced by van Foreest's *Observationes*, was spreading among European physicians. Valleriola, for instance, published only sixty *observationes* but claimed he had collected six hundred in his daily practice that he hoped to publish at some later date. The *curatio* and *observatio* were clearly the new genres that tended to replace the *consilium*, although the latter did not disappear. Diomedes Cornarius, who became court physician in Maximilian II's Vienna, published in 1599 a collection of *consilia*, which are actually records of his *curationes* in chronological sequence from 1566 to 1597.[119] There seems to be a strong emphasis on chronology here: on the margin of each case the exact date is given, with day, month, year, and the course of disease and the treatment are described day by day (as in the Hippocratic *Epidemics*). Though they go under the conventional name of *consilia*, Cornarius's cases are striking evidence that new wine was been poured in the old bottles. Both the early modern *curatio* and *observatio* as the medieval *consilium* focused on an individual case, but the difference between them is clear: whereas in the *consilium* diagnostic and therapeutic reasoning was tightly anchored to a doctrinal framework, *curatio* and *observatio* give pride of place instead to the precise and detailed description of the single case, abstaining from interpreting it in the light of doctrine (or, if that is done, it is done in a separate part of the text, under the rubric of *scholion*).[120] What is of interest now is precisely the particularity and singularity of each case. The focus on singularity explains the selective attention for whatever is rare, exceptional, and unheard-of that is a constant feature of the late sixteenth-century collections of medical *observationes* and *historiae*.[121] A case seems noteworthy precisely when its singularity is so extreme that it challenges classification: the rare and odd is the epitome of individuality. An instance of the unforeseeable, the odd case proves the complexity and mysteriousness of nature, which defies Scholastic systems.[122] This interest in the rare is often signaled in the titles of these works, as for instance in Marcello Donati, *De medica historia mirabili* (1586) and in another work of Diomedes Cornarius, *Historiae admirandae rarae* (1595).[123] While Donati's *historiae* are accounts and discussions of his own cases (collated with similar cases as reported by other authors), Cornarius's *historiae* are an anthology of rare and indeed near-miraculous cases that he culled from the medical literature and from the earlier manuscript records of the Vienna Medical College, to which he belonged. Collecting documentation on medical rarities could thus involve what we would now call historical research on primary sources. In another work, titled *Observationes medicinales*, Cornarius described the rare cases he had observed in his own practice, prefacing each case with similar observations by other authors in chronological order (as if to establish a chronology of observation for each rare condition). Something similar had been already done by Dodoens in his 1581 *Observationes*. Dodoens declared that he was going to abstain from reporting what was already observed by others. His criterion for including a case was rarity and novelty.[124] It was the claim to novelty, perhaps, that led him to always specify the date when his observations were made—a new feature compared with the *curationes* model, where usually the place but not the date was given. In Dodoens's and Cornarius's *observationes* we can notice a shift beyond the *curationes* model. Most significantly, not all the cases are successful cures. For instance, in an *observatio* titled "De latente ac difficili angina," which describes an epidemic of acute sore throat in the summer of 1565, Dodoens reports the case of a butcher who died of the disease in the space of a day from the onset of the symptoms, describing what he saw in his lungs after opening the cadaver. The report of unsuccessful cures shows that the *observationes* literature was moving away from the *curationes* and was being structured more closely on the model of *Epidemics*.

Interestingly, Dodoens published together with his own *observationes* extracts from previous authors who, like him, had recorded "rara et abdita" (rare cases and hidden causes), thus suggesting a sort of genealogy of the new *observationes* genre. Next to excerpts of medical mirabilia from Velasco de Tarenta's early fifteenth-century *Practica* and from Alessandro Benedetti's works, he republished in its entirety Antonio Benivieni's *De abditis ac mirandis morborum et sanationum causis* (1507).[125] At the beginning of the sixteenth century, when it had first been published, Benivieni's text, with its brief accounts of extraordinary cases, had been an idiosyncratic experiment in medical observation and writing, definitely outside of all existing genres. At the end of the century, it had found a place in what Dodoens saw as a tradition of medical attention devoted to what was peculiar and puzzling in the manifestations of disease.

A much fuller genealogy of the new genre is spelled out in the introduction to Johann Schenck's *Paratereseon*, or *Observationes medicae, rarae, novae, admirabiles et monstrosae* (1584–1597), the text that signals the coming of age of the *observationes* as a primary form of medical writing. There is indeed a striking difference between Amatus's and Ruland's pocket-sized *curationes* (clearly a vademecum for the practitioner) and Schenck's splendid quarto volume, where he collected "rare cases, hidden causes, unexpected and incredible events, memorable cures and monstrous bodily conformations." The volume brings together, around the common focus on the new, the rare, and the anomalous, both anatomical and clinical records—the anatomists' and the practitioners' *observationes*. His declared goal is to "collect in one volume those new and wondrous things that the most celebrated physicians observed not so much by means of doctrine as by means of *experimentum*,... and that were transmitted to us piecemeal in a scattered, incidental and straggly way."[126] Schenck's volume is a masterpiece of medical excerpting and rearranging: the *observationes*, culled from hundreds of ancient and modern authors, are organized systematically according to the conventional head-to-toe order of textbooks of practical medicine. It offers, as Theodor Zwinger said in the prefatory letter to Schenck published in the volume, "historiarum et curationum microcosmi exempla, per locos communes digesta." It can thus be considered another example of the "method of commonplaces," whose significance in early modern natural philosophy has been pointed out by Ann Blair.[127]

In the preface, Schenck lists the authors that he found most useful in his enterprise, thus suggesting the broad lines of a genealogy of the observation of rare cases. He acknowledges Champier's *Historiales campi* as a predecessor in the art of extracting *observationes* from Galen, though he says Champier did not do the job very well, because several works of Galen were not available in translation in his times and especially because "he followed no order in putting the *historiae* together."[128]

Among the moderns, the pioneer in the observation of rare cases was Benivieni, followed by Cardano and Fernel (from whose works he declares to have taken over sixty *observationes*). Others include Amatus Lusitanus, Valleriola, Dodoens, Marcello Donati, van Foreest, and Ruland—the authors who, we have seen, were crucial in shaping the new genres of *curationes and observationes*.

Schenck stresses that his enterprise was made possible by a wide network of correspondents (he lists seventy-one) who helped either by sending him the references to rare observations they had come across in their reading or by allowing him to read and excerpt from their own unpublished *observationes*. Thus for instance Joachim Camerarius sent him his "*sylva* of observations, containing over fifty *historiae*"; Jean Bauhin sent him his private journals, containing *observationes* and *curationes* "written for his own private use." Here is more evidence that the habit of keeping records of what was observed in daily practice was spreading among European physicians, and that there was considerable interest in the circulation of such materials. Records of rare cases seem to have been in these years a prized item of collection and circulation, as were specimens of rare plants and animals; and their diffusion followed probably the same route through the same epistolary networks.[129] New publications in this field were eagerly awaited. Thus, for instance, Schenck notes having learnt from a correspondent that the Roman physician Alessandro Petroni "will publish in a few years the *historia* of the diseases that he cured." His work seems to be very promising, Schenck remarks, because "he adapts the method of healing to individual circumstances, and lists what happened each day to this or that patient, what remedies he administered, whether the treatment was beneficial, what he himself predicted would happen and what actually happened in his patients, citing by name the Romans in his care."[130]

That there was indeed a readership interested in such accounts of medical observation is indicated by the significant number of *observationes* and *historiae medicae* published in the first half of the seventeenth century.[131] These collections were clearly a primary medium for the circulation of information in the "res publica medica" of the early modern period.[132] By the first half of the seventeenth century medical *observationes* and *historiae* were interchangeable terms,[133] both indicating a condensed report of firsthand observation and meant as a cumulative contribution to a Europe-wide network of information exchange among scholars. The diffusion of periodicals published by the learned academies and medical associations in the second half of the seventeenth century seems to have been a development of the networking set in motion by the *observationes*. Remarkably, the reports sent to journals such as the *Ephemerides medico-physicae Academiae naturae curiosorum*, the *Acta medica et philosophica Hafniensia*, the *Acta eruditorum*, the *Journal des savants*, and the *Philosophical Transactions* often adopted the *observatio* format (as well as its penchant for the bizarre)—a sign of the genre's triumph as one of the paradigmatic forms of scholarly communication of the seventeenth century.[134]

The interest in excerpting medical *historiae* from the ancients was still alive at the beginning of the seventeenth century. So Giovan Battista Selvatico published a new selection from Galen, arguing that contemporary medical authors were busy following the Hippocratic model in *Epidemics*, while Galen's useful *historiae* were "left dormant" under layers of less useful doctrine.[135] Zacutus Lusitanus's *De medicorum principuum historia* (1636–1642) collected 200 *historiae* excerpted from Galen, plus 188 taken from other ancient medical authors, systematically organized for the purposes of consultation by the medical practitioner for what Zacutus himself defined as a *praxis historialis*.[136] It should be noted, however, that there was also in this period a growing sense of the superiority of the moderns over the ancients and, more precisely, the idea that the modern medical *historia* should go beyond the old models. Thus Gerolamo Perlini published a "method for succinctly writing medical *historiae*," which was meant as an improvement over the traditional form. "We well know that Galen and even more so Hippocrates wrote various medical *historiae*, and that the same was done by the ancient sect of the Empiricists; but both touched this subject lightly and guided by other concerns, noting only few of the symptoms—the principal and surprising—hardly sufficient for once more reconsidering and reexamining the case."[137] Perlini argues that one should keep distinct the physiological, pathological and therapeutic aspect of each *casus* and, most importantly, that a medical *historia* should be written so as to enable the reader to participate anew in an imaginary consultation over the case. The medical *historia* is here the means of sharing observation so that it can be used more intensively within a community of practitioners.[138]

From *experimentum* to *curatio* to *observatio/historia*, medical observation had come a long way. Whereas the *medieval experimentum* had been legitimized only by the chrism of efficacy, the new genres stemming from medical practice had gained, over the course of the sixteenth century, intellectual legitimacy and prestige. Like the anatomists' *historia*, the medical *historia* was also prized by sixteenth- and seventeenth-century practitioners for the cognitive possibilities it offered. Dodoens for instance (but it was by no means an original statement) praised the cognitive value of *exempla*, which for him was both didactic-mnemonic and heuristic. "Thanks to *exempla*, the content of previously received general knowledge [*praeceptiones*] is more easily recalled or committed to memory . . . Also, thanks to *exempla* the way is opened to new knowledge."[139] Indeed, the sixteenth-century medical literature clearly shows that the *historia* format was primarily used for registering the unexpected and the unaccountable, to open a chink in the armor of established expectations through which the perception of novelty could steal in.

What epistemology was behind the increasing value and currency of the medical *historia*? We have seen that the *historia anatomica* developed within the framework of an Aristotelian empiricism modified by a strong dose of belief in the value of *to oti* or descriptive knowledge. The *historia medica*, on the other hand, was fostered by what we could call a Hippocratic-Galenic "rational empiricism": a reading of Hippocrates and Galen that stressed the balanced cooperation of reason and experience in the production of medical knowledge. Here also, however, the belief in the empirical component of the cognitive process grew stronger and stronger over the sixteenth century. In fact, when defending the legitimacy and usefulness of the observation of individual cases, the authors of *curationes* and *observationes* often referred to arguments taken from Galen, but deriving actually from the ancient Empiricists.[140] The empirical connotation that the word *observatio* still conveys to us is not misleading. In the early modern medical language *observatio* was indeed "vox empyricorum" (a term used by the Empirics), as it is defined in the medical lexicons[141]—the Latin rendition of the Greek *parátéresis* used by the ancient Empiricists (and *parátéresis*, let us remember, is the first word in the title of Schenck's *Observationes*). This does not mean that the early modern authors of *observationes* saw themselves as members of the "empirical sect." Even Ruland, for instance, is at great pains to stress that his *curationes empiricae* are no "nuda experimenta." "I call these curationes empirical," he says, "not because they are made only of *experimenta*, as professed by the Empiricists' sect, but because they combine rational discipline with experience: indeed, in every *curatio* I inquire and thoroughly search for the diseases' causes and signs." A similar disclaimer was made by van Foreest. Professing himself a "rational physician," he stressed that his work, though called *observationes*, was not "mere observation, or *historia* of things observed, in the Empiricists' fashion."[142] And yet, in spite of such disclaimers, what we found in the *observationes* literature looks very much indeed like the ancient empiricist notion of *autopsia* plus *historia*: firsthand observation plus accounts of similar cases by other observers, often quoted word by word with scholarly reference to chapter, paragraph, and sometimes page number. For all their different philosophical backgrounds, both the anatomists' *historia* and the practitioners' *historia* shared an emphasis on the value of firsthand observation, and they both contributed to the early modern reappraisal of the cognitive value and dignity of experience.

Notes

I have discussed parts of this paper in several venues: the Centro Internazionale di Storia delle Università e della Scienza (Bologna), the Institut Louis-Jeantet d'Histoire de la Médecine (Geneva), the EHESS and the Centre A. Koyré (Paris), and the Department of History and Philosophy of Science of the University of Cambridge. I have very much profited from the comments of several participants in these discussions; I would like to thank in particular Pietro Redondi, Gabriele Baroncini, Paola Govoni, Andrea Carlino, Simona Cerutti, Dominique Pestre, Andrew Cunningham, and Ole Grell. I am also deeply indebted to all the participants in the Berlin workshop for insightful readings and constructive criticism.

Sources cited in boldface type appear in the Primary Sources of the bibliography.

1. **G. Valla** 1501, t. II, lib. XL, cap. XXXVII: "De scribenda historia, quidquid in ea posteris tradenda praecipuum sit." On Valla's *De expetendis et fugiendis rebus* as an encyclopedia of Greek and Latin sciences see Branca 1981.

2. **Patrizi** 1560, 7v. On Patrizi's *Della historia* see Otto 1979, 134–173.

3. **Benedetti** 1998. On the dates of the work composition see Ferrari 1996, 79, 87 n. 89, 172.

4. **Du Laurens** 1599; **Bauhin** 1609; see also, for instance, **Valverde** 1556, **Landi** 1605.

5. In the late Middle Ages, the term *historia* is occasionally used in translations of Galenic works with reference to anatomical description (see for instance the texts quoted in Sudhoff 1910, 361–363), but the term does not seem to have had much currency or significance in Scholastic medicine.

6. See the headings "Anatomicae historiae" and "Historia medica" in **Lipenius** 1679, **Linden** 1651, and **Mercklin** 1686.

7. **Wolf** 1579. On the extensive literature on the *artes historicae* see GRAFTON in this volume, 51–58, 71 n. 62.

8. **Estienne** 1545,*iir.

9. Seifert 1976.

10. **Thomas Aquinas** 1587, 1.

11. "In hoc, philosophi magis quam hystoriographi modum ymitamus, quia probo per causas quod simpliciter in alia cronica est conscriptum" (quot. by Arnaldi 1998, 135).

12. **Boccadiferro** 1566, 18v.

13. On all this see Seifert 1976, 44–52.

14. Seifert 1976, 40 ff. See also the Introduction to this volume, 4, 9–10. For several examples of the pre-Aristotelian use of ἱστορία to denote knowledge in general (referring for instance both to Herodotus's and Anaxagoras's work) see Mazzarino 1960, 1:132–133. A deep connoisseur of ancient Greek historiography, Mazzarino sharply criticized Eduard Schwartz's view that ἱστορία and philosophy were "polar opposites in Greek thought. This might be accepted for the Hellenistic period, but definitely and absolutely not for the pre-Aristotelian period" (132–133).

15. **G. Valla** 1501, t. II, lib. XL, cap. XXXVII.

16. On which see Aulus Gellius, *Noctes Atticae*, 5.18; Isidore of Seville, *Etymologiae*, 1.41.1. In Roman historiography, *historia* referred to an account of events witnessed by the author, whereas histories of bygone times went under the name of *annales*. See also Keuck 1934, 12–15.

17. **Becichemo** 1504.

18. **Becichemo** 1504. The reference is to Theodore Gaza's translation of Aristotle's *Historia animalium*, printed in 1476 (see Perfetti 2000, 13–14).

19. **Patrizi** 1560, 2v.

20. Speroni, *Dialogo dell'Historia* (1587), in **Speroni** 1740, 2:227.

21. Thus **Viperano**, *De scribenda historia* (1569): "Voco historiam non expositionem cuiuscumque rei; quam verbi notionem Plinius secutus, librum suum inscripsit, Historiam naturalem, et Historia animalium Aristoteles, et Plantarum Theophrastus; sed narrationem rerum gestarum Historiam voco" (**Wolf** 1579, 1:842). For similar statements see for instance **La Popelinière** 1599, 23–25, **Ducci** 1604, 2; **Dandini** 1610, 322; **Mascardi** 1859 (originally published 1636), 54–55.

22. See for instance Ermolao Barbaro's distinction, in a letter of 1480, between "particular and sensory reasons" in *historia naturae*, of which Pliny is the model,

and "universal and speculative reasons," of which the model is Aristotle (**Barbaro** 1943, 1:3, 10). Cf. Giovanni Pozzi, "Introduzione," in **Barbaro** 1993, 1:clii.

23. Cotroneo 1971, 187.

24. The original Aristotelian opposition was in fact between history and poetry: cf. Aristotle *Poetics* 1451b1–7.

25. L. Valla 1962, 2:5, "Si vera fateri non piget, ex historia fluxit plurima rerum naturalium cognitio, quam postea alii in praecepta redegerunt." For a similar claim by Poliziano see the Introduction to this volume, 11, 34 n. 38.

26. **Fox Morcillo** (1557), in Wolf 1579, 1:830.

27. *Theodori Gazae Thessalonicensis praefatio in libros de animalibus Aristotelis* in **Aristotle** 1513 (unpaged). Seifert (1976, 46–47, 66–68) suggests that such a strong opposition of *historia* to philosophical knowledge may well be an overstatement of Aristotle's original view. For a detailed examination of the meaning of ἱστορία in Aristotle see Riondato 1961, esp. 45–81.

28. **Zabarella** 1586, 108r.

29. For an interesting reappraisal of Charles Schmitt's well-known thesis on Zabarella's empiricism (Schmitt 1969) see Baroncini 1992, chap. 2, esp. 51–57. On sixteenth-century trends in the reading of Aristotle's biological works see MACLEAN in this volume.

30. Goclenius 1613, 626. On *historia* as "ocularis et sensata cognitio" see also Seifert 1976, 79–88.

31. See **Galen** 1985 for English translations of *De experientia medica* and *Subfiguratio empirica*. See Deichgräber 1965 for the doxographic evidence on the Empiricists and Galen's *Subfiguratio empirica* in the 1341 Latin version of Niccolò da Reggio. The editio princeps of *Subfiguratio* was the 1502 Venetian edition of Galen. *De experientia medica* was known to Renaissance readers only as a fragment titled *Sermo adversus empiricos medicos*, translated by A. Gadaldino and published in the Giunta edition of Galen in 1550 (see Perilli 2004, 44–45, 198–200). The full text was recovered in Arabic translation in 1931 and translated into English in 1944 (see **Galen** 1944). For the text of *De optima secta* see **Galen** 1821–33, 1:106–223; on its dubious authenticity see the update in Perilli 2004, 201–209. On the ancient Empiricists see Frede 1987, 243–260, and Perilli 2004.

32. See Pittion 1987, 107. Celsus's *De medicina* was first printed in 1478.

33. Frede 1981, however, argues that Galen shows "a considerable amount of sympathy for empiricism." Both *De experientia medica* and *Subfiguratio empirica* "on balance turn out to be defences of empiricism against certain standard rationalist criticisms of empiricism" (71).

34. "Quaecumque vero concordantiae fiunt de sensibilibus rebus in hominibus, fideles secundum vitam sunt" (*Subfiguratio empirica* in Deichgräber 1965, 67–68).

35. Cf. Deichgräber 1965, 298, and Seifert 1976, 148. Also Robin 1944, 181–196.

36. See Nutton 1989a.

140 Gianna Pomata

37. This view of *historia* would persist in the Baconian vocabulary of experience, where, as Daston has noted, "histories" were meant to be comprehensive surveys of particulars in a specified domain. "The inclusive 'histories' contrasted with the isolated 'observation,' which made no pretence to complete coverage of a topic" (Daston 2001b, 752).

38. For instance **Goclenius** 1613, 626–627.

39. F. Bacon (1620) in F. Bacon 1857–58, 4:*255,* 262.

40. British Library, Sloane Collection, MSS of Francis Glisson, MS 3315 n. 381: "De historia, sive de plena enumeratione experimentorum." "Finis sive scopus huius partis artis inventivae est, non mera narratio, aut cognitio experimentalis rei narratae, nec nuda cognitio varietatis existentiae . . . scopus autem eius primus est ut sit repraesentatio quaedam totius naturae quaesitae in materia non nota, ut inde intellectus, ex inductione sufficiente, ad naturam abstractam eiusdem, et axiomata nova et infallibilia ascendat . . . quare ad inveniendas rerum absconditarum naturas, plena enumeratio sive hystoria omnium experimentorum quae spectant ad naturam quaesitam colligenda est. Quae collectio sive hystoria primam partem huius artis de inventione artium constituit."

On the changing meaning of *experimenta* see below in this essay. One should note that the meaning of *historia* and that of *experimentum* changed together.

41. **Patrizi** 1560, 41v–42r.

42. **Patrizi** 1560, 14r: "Non havete voi signori Vinitiani, nella sede del vostro maggior consiglio, dipinta la historia di Alessandro III e di Barbarossa?" In fact pictures were sometimes called *historie* in early modern Italian (for a seventeenth-century example see Raggio 2000, 34). On *historia* as pictorial representation see the Introduction to this volume, 9, 33 nn. 30–31, and bibliography there cited.

43. George Eliot, "Silly Novels by Lady Novelists" (1856), in **Eliot** 1963, 324.

44. Ferrari 1996, 107 n. 9. The medical metaphor of "healing the text" had already been used by Guarino Veronese and Coluccio Salutati (Ferrari 1996, 206).

45. Ferrari 1996, 143–144; 341 n. 102 on *Anatomice* as a *breviarium.*

46. Blair 1992, 545.

47. See Crisciani in this volume, 308–311.

48. In this he differed from other humanist physicians who were not interested in practical medicine or in anatomical research, such as Leoniceno. On the contrast between Benedetti and Leoniceno see Ferrari 1996, 315–316. On medieval *practicae* see Agrimi and Crisciani 1988, chapters VI and VII.

49. "To write down remedies certified by experience in a concise and aphoristic style": Arnald of Villanova quoted in Agrimi and Cristiani 1988, 198. Ferrari 1996, 317 n. 34, 277–278, 249.

50. Antiquario's letter to Giorgio Valla quoted in Ferrari 1996, 279–280. On Antiquario see *Dizionario biografico degli italiani,* 3:470–472.

51. G. Valla, 1501, lib. I, cap. III "De vario cognitionis modo." On the history of the term *experimentum* in medieval medicine see Agrimi and Crisciani 1990, 9–49, and the essays collected in Veneziani 2002, in particular Hamesse, Spinosa, Fattori.

52. Agrimi and Crisciani 1990, 39–47.

53. Ferrari 1996, 329 n. 69. See for a contrast the conventional Aristotelian notion of experience as it was used by a contemporary of Benedetti, Alessandro Achillini, anatomist and reader of medicine in Bologna (Baroncini 1992, 13–22).

54. Ferrari 1996, 329–330 n. 69. On the use of the *exemplum* in Pliny see Gazich 1986, 143–169; Zorzetti 1980, 33–65.

55. **Benedetti** 1507 (see Ferrari 1996, 175–241).

56. Ferrari 1996, 109, 112, 119.

57. See SIRAISI in this volume.

58. **Benedetti** 1967.

59. Quoted in Ferrari 1996, 301.

60. The reference to Estienne is in note 8 above. In the *Fabrica*, Vesalius uses *historia* in the sense of orderly and perspicuous description (for instance **Vesalius** 1555, 1:103). In Falloppia's *Observationes anatomicae* (**Falloppia** 1606), *historia* is used in various ways: as in Vesalius to indicate anatomical description ("venarum historia," "arteriarum historia," 1:83) or to refer to Vesalius's work as "integerrima humani corporis historia" (1:37); but also to denote a brief case or *exemplum* from his practice (2:377).

61. **Du Laurens** 1602, 23 (cf. also 27, 38–39).

62. See Perfetti 2000, 3–4.

63. Cunningham 1985; Cunningham 1997, chapter 6.

64. On the Padua version of Aristotelian empiricism as exemplified by Zabarella see, besides Feyerabend's 1970 pioneering essay, Schmitt 1969 and Baroncini 1992, 39–62. On Baconian "facts" as conceptually different from Aristotelian experience see Daston 1991a.

65. French 1994, 66.

66. Harvey's *Exercitatio anatomica de circulatione sanguinis* (1649), quoted by French 1994, 277.

67. French 1994, 256.

68. In 1687 Marcello Malpighi, defending the anatomical research of his times from the attacks of traditionalists like his colleague Sbaraglia, listed Aselli's discovery next to the hypothesis of the circulation of the blood (**Malpighi** 1967, 550–551). See also **Le Clerc** and **Manget** 1699, 2:667–668.

69. **Aselli** 1627, 34. Aselli studied in Pavia, not in Padua, but a link with the Paduan anatomical model was possibly provided by his teacher in Pavia, Giambattista Carcano, who had been himself a pupil of Falloppia in Padua (on Aselli's life see *Dizionario biografico degli italiani*, 4:389–391).

70. **Aselli** 1627, 34.

71. The reference is to **Costeo** 1565.

72. Aselli 1627, 19–20.

73. He identifies this *praenotio* with what Aristotle in *Posterior Analytics* (1.1 and 1.3) called *prognosis* or *prolepsis*, and Cicero in *De natura deorum* (book 1) "a sketch of the thing taken by the mind in advance."

74. Aselli 1627, 22.

75. Aselli 1627, 27.

76. Aselli's plates are well known for being the earliest color-printed medical illustrations. See Premuda 1993, 228, and Roberts and Tomlinson 1992, 521–522. On the growth of illustrations in anatomical textbooks over the sixteenth century see Carlino 1999b, 5–45.

77. Aselli 1627, 43, 22.

78. **Le Clerc** and **Manget** 1699, 2:668 ff. (what was left out was the section on *usus*).

79. See French 1994, 326 ff.

80. On this reaction to Harvey's *Generation of Animals* see Goltz 1986.

81. Schrader's *praefatio* to **Harvey** 1674. For a similar treatment of Aristotle's *Parts of Animals* by Caesar Odonus see MACLEAN in this volume, 164.

82. Agrimi and Crisciani 1988, 216–217. On *practica* in medieval medicine see also Demaitre 1975, 1976; Wear 1985, Crisciani 1990. On Savonarola see CRISCIANI in this volume.

83. Laín Entralgo 1950, 68–104.

84. Agrimi and Crisciani 1994.

85. Crisciani 1996, 18–19. See also Murray Jones 1991. It is probably an overstatement, however, to argue as Owsei Temkin did, that the *consilia* literature shows that medieval physicians were incapable of "thinking the individual" (Temkin 1929, 42–66).

86. Crisciani 2001.

87. **Champier** 1532. On Champier (1472?–ca 1535) see Vasoli 1960, Roger 1973, and especially Copenhaver 1978.

88. **Champier** 1516c, 1516b (a satire of Scholastic medicine), 1503 (on which see Tracconaglia 1922), 1516a, 1510, 1884. For a detailed (but not always reliable) list of his work see Allut 1859.

89. **Champier** 1532, *praefatio*.

90. *De ingenio sanitatis* was the title of *Methodus medendi* as translated by Gerard of Cremona: see Kibre and Durling 1991, 117–122.

91. For instance lib. I hist. 38: the extract from Galen is a *casus*, while Champier's comment is on the disease in general. In lib. IV hist. 5 and 6, Champier discusses a case from Hippocrates' *Epidemics III*, as reported by Galen. In Lib. IV, hist. 25, he reports a case of his own.

92. Laín Entralgo 1950, 67.

93. First published posthumously in 1568, the *Theatrum Galeni* was reprinted in 1587 as *Loci communes medicinae universae* (see **Mondella** 1587, s.v. "historia"). A protagonist of the new humanist genre of the *epistolae medicinales*, a correspondent of Fracastoro, Fuchs, and Gessner, Mondella (?–1553) practiced in his native Brescia, taught botany in Rome, and was involved in the founding of the Botanical Garden in Padua (evidence on this is unclear: see Minelli 1995). On his correspondence with Fracastoro and Fuchs see **Mondella** 1543; on his friendship with Gessner see **Mondella** 1551, 93r. He describes his antiquarian activities (the study of ancient coins) in **Mondella** 1551, Dialogus VII.

94. *Epistola Jo. Cratonis, qua recte Galenum legendi ratio breviter ostenditur* in **Crato** 1560. A very influential medical and religious reformer, personally connected with Luther and Melanchthon, and the center of a wide network of correspondents, Crato (1519–1585) studied with Giovanni Battista da Monte in Padua, practiced in his native Breslau and then at the courts of Ferdinand I and Maximilian II in Vienna, and later of Rudolph II in Prague (see Gillet 1860–61 and Louthan 1994, both of which, however, focus on Crato's religious rather than medical activities). He edited the second volume of da Monte's *Consultationes* and Falloppia's *Observationes anatomicae*. His massive output of medical correspondence, in the form of *Consilia* and *epistolae medicinales*, was published posthumously by Laurentius Scholzius in seven volumes (see **Crato** 1591–1611).

95. On the structure of the clinical cases in *Epidemics* see Potter 1989.

96. Laín Entralgo 1950, 108.

97. According to Nutton there are two types of "case histories" in Galen: (a) in *Methodus medendi* and *De locis affectis*, one case is singled out as an illustration of a more general argument. The case serves to legitimize from experience the advice already offered in general terms. (b) In *On Prognosis*, we find especially the self-advertising *curatio* (Nutton 1991b, 9–11).

98. Siraisi 1997, 207.

99. The last *Centuria* was published posthumously two years after his death (**Amatus Lusitanus** 1570). The seven *Centuriae* were first published together in Barcelona (**Amatus Lusitanus** 1628). On Amatus (1511–1568), one of the most prominent Jewish physicians of the late Renaissance, see Friedenwald 1944, 1:332–380.

100. On the recipe as the typical format of the *experimenta* and *secreta* literature not only in medicine but also in medieval and Renaissance manuals on arts and crafts, see Keil and Assion, 1974; and especially Eamon 1994, 83–90, 112–120.

101. See the texts reported in appendix to McVaugh 1971, 107–118.

102. On this shift from recipe to case narrative see Pomata 1996, 184–192.

103. On the two sources of legitimization of medical practice see Pomata 1998, 50–52.

104. **Saccardini** 1621; **Cardano** 1643, 188 ff. On Saccardini, a Paracelsian healer who was executed for heresy in Bologna in 1622, see Ginzburg and Ferrari 1978.

105. Nutton 1991a gives examples from Ambroise Paré and William Clowes.

106. Ruland (1532–1602) was court physician to Rudolph II in Vienna. He also wrote several works on the Greek language (**Ruland** 1556, 1567) and on the lexicon of alchemy (**Ruland** 1612). His *Curationes* were still read and used in the seventeenth century (they were excerpted in **Bonet** 1679).

107. **Ruland** 1580, *praefatio*, α 7 r–v.

108. See Lonie 1985, 155–174, 318–326; Nutton 1989a, 420–439, and in general Smith 1979, 13–31.

109. **Da Monte** (Montanus) 1558, 543. On da Monte see Bylebyl 1979, 335–370; Bylebyl 1991, 157–189; Bylebyl 1993, 40–68, and *Dizionario biografico degli italiani*, 32:365–368.

110. **Da Monte** 1554. These are notes from da Monte's lectures, collected and published two years after his death by one of his students, the Polish Valentinus Lublinus.

111. **Da Monte** 1554, 40v: "You will have in a single case both the theoretical and the practical [parts of medicine]."

112. **Da Monte** 1558, 542–543.

113. **Da Monte** 1558.

114. **Valleriola** 1605, *praefatio* (first edition Lyon, 1573). Valleriola (1504–1580) also wrote an often reprinted compilation of *Loci medicinae communes* (Lyon, 1562) from Galenic and Hippocratic texts, plus several commentaries on Galen. See *Biographie médicale* 1855, 1:391.

115. **Amatus Lusitanus** 1570, 9v. In his lectures on *Epidemics*, da Monte refers to the cases as *casus*, but in the *Consultationes* he uses "historia casus," "historia morbi," or simply *historia* to refer to a narrative account of a case (**Da Monte** 1583, cols. 135, 503, 118).

116. **Vallés** 1589, "Ad lectorem." Cf. **Selvatico** 1605, *praefatio*: "Sanitas autem non in communi, sed Socrati aut Calliae comparatur." Francisco Vallés (1524–1592) was court physician to Philip II of Spain. His *Controversiae medicae* were edited by Crato: on this work see López Piñero and Calero 1988. On his medical skepticism see Pittion 1987, 121–123.

117. **Foreest** 1584. The 32 books of his *Observationes et curationes* came out 1584–1619. On van Foreest (1521–1597) see Houtzager 1989; Nutton 1989b; Müller 1991, 49–52. The *Medicinales observationes* of Jodocus Lommius (see **Lommius** 1560), the first medical text, as far as I know, to be published under the name of *Observationes*, is also the work of a town physician and is dedicated to the municipality he served (Brussels in Lommius's case). But in contrast to van Foreest's and other later *observationes*, those by Lommius are general descriptions of symptoms and rules of prognosis in various diseases, and not records of individual cases.

118. **Foreest** 1584, 12–13.

119. **Cornarius** 1599. On Cornarius (ca. 1535–1600) see the biographical entry in Hirsch 1884–1889, s.v. A similar list of his rare cases in chronological order from 1561 to 1610 was published by a practitioner in Antwerp, Henricus Smetius, in his *Miscellanea medica* (see **Smetius** 1611, 523–593). These are not

only rare cases he treated but also rare anatomical conditions he observed in dissections. Many thanks to Ian Maclean for referring me to this source.

120. Laín Entralgo (1950) had already drawn attention to this feature of the *observationes* (chap. 3, esp. 110–121).

121. As already noted by Laín Entralgo 1950, 112, 121, 130–131.

122. On the epistemological significance of the observation of rare and strange cases in early modern natural philosophy see Daston 1991b; Daston and Park 1998, chap. 6.

123. **Donati** 1586. Donati (1538?–1602) was court physician for the Gonzaga in Mantua (for biographical information see **Castellani** 1788 and *Dizionario biografico degli italiani*, 41:49–51). Donati also wrote a volume of commentary on several ancient historians, including Livy and Tacitus (see **Donati** 1604). Cornarius's *Historiae admirandae rarae* were published together with his *Observationes medicinales* and *Consilia* in **Cornarius** 1599.

124. **Dodoens** 1581, *praefatio*. On Dodoens (1516–1585) see *Dictionary of Scientific Biography*, s.v. Like Cornarius he was also physician to Maximilian II in Vienna. A prominent botanist, he was called to teach practical medicine at the newly founded university of Leiden in 1582, a year after the publication of his *Observationes*.

125. On **Velasco de Tarenta** 1490, see Demaitre 1976, 87. Benivieni's text had been published posthumously from his manuscript notes. See Giorgio Weber's introduction to **Benivieni** 1994, 7–33.

126. **Schenck** 1609, "Praefatio ad lectorem." This work came out in seven volumes in Basel and Freiburg 1584–97 and was reprinted in Frankfurt 1600, Freiburg 1604, Frankfurt 1609, Lyon 1644, and Frankfurt 1665, with additions by Laurentius Straussius (see **Lipenius** 1679, 309). *Paraterēseon* (from the Greek *paraterēsis*, observation) was added to the title in the 1609 and following editions. On J. Schenck von Grafenberg (1530–1598), a town physician in Freiburg, see Hirsch 1884–1889, s.v.

127. Blair 1997, 65–77. On the significance of the practice of excerpting in the rise of the early modern notion of "fact" as a detached morsel of experience see Daston 2001b, 758–759. On the long tradition of medical *florilegia* see Wallis 1995.

128. **Schenck** 1609, "Praefatio." According to Schenck, however, the first to excerpt rare cases from Galen was Maimonides (the reference is to **Maimonides** 1579, 473–491, "De quibusdam natura miraculosis rebus apud medicos lectis").

129. It should be noted in this respect that already in the first half of the sixteenth century the genre of *epistolae medicinales* is one of the *loci* for the detailed description of *casus* (as pointed out by Nutton 1991a). On the exchange of specimens through medical correspondence in this period see Olmi 1991.

130. Petroni's *observationes* were never published (he died in 1585, a year after Schenck began publishing his work). On Petroni see SIRAISI in this volume, 328, 334–335, 344.

131. See the entries "observationes" and "historia medica" in **Lipenius** 1679, **Linden** 1651, and **Mercklin** 1686.

132. The expression "res publica medica" is used by Theodor Zwinger in his prefatory letter (dated 19 August 1584) to Schenck's *Paratereseon*.

133. See for instance **Ferdinando** 1621, but examples could be multiplied (Pomata 1996, 173 ff.). The medical dictionaries register the two terms' equivalence with some delay. The 1642 edition of Castelli's *Lexicon medicum* includes *observatio* but not *historia*. The 1700 edition has both, cross-listed, and under *observatio* it is noted that "hodieque utplurimum aequipollet Historiae Medicae" ("today it is mostly equivalent to *historia medica*").

134. See Daston 2003 (and the bibliography cited therein, 16 n. 3) on the *Ephemerides* of the Academia naturae curiosorum. The *Acta eruditorum* were published in Leipzig from 1682 to 1731. The *Acta medica et philosophica Hafniensia*, edited by Thomas Bartholin, were issued in Copenhagen from 1671 to 1679. A strong interest in rare cases is displayed in Bartholin's *Historiae anatomicae rariores*, which presents firsthand accounts of rare anatomical conditions observed in practice and dissection (see **Bartholin** 1654). A remarkable example of the penchant for the bizarre in mid-seventeenth-century medical literature is the much-cited anthology of near-miraculous *historiae* excerpted and collected by the Portuguese physician Gaspar de los Reyes Franco (see **Reyes Franco** 1661, reprinted in Frankfurt 1670).

135. **Selvatico** 1605, "Praefatio."

136. **Zacutus Lusitanus** 1636–42.

137. **Perlini** 1613a, 10 (see also **Perlini** 1610). He taught in Rome; see **Perlini** 1613b. Some biographical information on him in **Marchesi** 1726, 202. He is praised in **Morgagni** 1719, preface to Adversaria VI.

138. It is worth noting that the authors of *curationes* and *observationes* were eminently practitioners, rather than academic doctors: itinerant practitioners like Amatus Lusitanus, court physicians (like Martin Ruland the Elder, Dodoens, Cornarius, and Donati), or town physicians (like Valleriola, van Foreest, and Schenck).

139. **Dodoens** 1585, "Praefatio."

140. They often quote *De optima secta*, chap. 8 (for instance **Selvatico** 1605, "Praefatio"); **Castelli** 1700, s.v. *observatio*).

141. See **Castelli** 1642, s.v. *observatio*: "Empyricorum vox est, qui quicquid observassent fieri in morbis, id *athroisma*, et quasi axioma nuncupabant." Cf. **Castelli** 1700, "vocabulum Empiricae sectae proprium, significans cognitionem rei alicujus sensuum fide comparatam."

142. **Ruland** 1580, β r–v; **Foreest** 1584, 14–15.

4

White Crows, Graying Hair, and Eyelashes: Problems for Natural Historians in the Reception of Aristotelian Logic and Biology from Pomponazzi to Bacon

Ian Maclean

The natural historians who produced works on botany and zoology in the course of the sixteenth century did not do so in a void. There was a rich encyclopedic literature about plants and animals from Pliny to the Middle Ages, and the Aristotelian tradition contained, as well as a complex discussion of the logical issues that arise when individual observations are incorporated into conceptual schemes of one kind or another, an extensive set of biological and zoological texts. These natural historians had (almost without exception) been trained in the mental habits of Aristotelianism; while it would be misleading to suggest that this proved to be an impediment to their investigations (for they showed themselves to be capable of recognizing the limitations under which such mental habits placed them, and of finding ingenious solutions to such limitations, as we shall see), it would be equally misleading to say that they found it easy to liberate themselves from all the conceptual baggage which came with a training in Aristotelian thought. When the English mathematician Thomas Harriot (1560–1621) was working on an aspect of Scholastic philosophy, he set down all the metaphysical presuppositions to which he saw himself committed; having read through this list (consisting of twenty-nine principles and almost as many differentiae), we can understand why he wrote to Kepler in 1608 that thinkers in England were "stuck in the mud" and desired only to "philosophize freely."[1] Not all early modern natural historians engaged explicitly with these conceptual issues; but some, notably Andrea Cesalpino, were deeply interested, and tried to reconcile their innovative work with them. There were also Aristotelians of various hues who looked at the theory either with the revision of peripatetic philosophy in mind, or with other ideological (principally theological) concerns. The prominence of medically trained scholars in both these groups is no coincidence, as Galen was an important channel for methodology and Aristotelian biological ideas, as well as a source for anatomical and physiological research in his own right.

In this paper I shall attempt to show what effect these habits of mind and metaphysical commitments had on the field of enquiry into nature by examining the theoretical discussion of classification from Pomponazzi to Bacon. To do this, I shall have to look at some of the elements of the arts course, as well as the commentary tradition on the zoological works of Aristotle and the botanical works of Theophrastus, before passing to the texts of a selection of natural historians. (The Aristotelianism I shall describe is that accepted by most scholars in the late Renaissance and incorporated by them into their own thought; I am not concerned here to adjudicate whether their interpretations concord with the most recent scholarship, although I shall on occasions refer to this.) I shall end with a famous passage in Bacon's *Advancement of Learning* which seems to me to mark a decisive shift in the terms in which this issue is discussed.[2]

Orders of Knowledge and the Logic of Definition: White Crows and Graying Hair

What is known as "scientia" in the sixteenth century is causal knowledge, which is convertible into syllogisms of the first figure. Below this is knowledge based on sensory perception ("experientia") and the best available authority ("opinio").[3] In various contexts, Aristotle sets down the areas in which the highest degree of accuracy and certainty can be obtained (notably *Physics* 1.1 and *Nicomachean Ethics* 6.3–7); and in *Metaphysics* 2.2 (995a15–20), he relates this explicitly to his biology: "mathematical accuracy is not to be demanded in everything, but only in things that do not contain matter. Hence this method is not that of natural science, because presumably all nature is concerned with matter." This is why he opens the *De partibus animalium* (639a1–12) with a distinction between *episteme* (scientific knowledge) and *paideia* ("eruditio," "peritia"), the latter being the general exercise of judgment, albeit guided by a set of rules.[4] As the text makes clear, he has not abandoned the pursuit of causes, but he has placed it in the context of the difficulties of the subject matter: its diversity, variability, and propensity to change over time. He acknowledges that the senses are the source of all knowledge about nature, but insists in *Posterior Analytics* (2.19, 99b20 ff.) that for the intellect to grasp its subjects, general concepts have to be produced inductively from our experience of particular perceptions.

In earlier chapters of the same book (2.13–14), he demonstrates how to arrive at a definition of one's subject, whether this is real or exists only in intelligible form;[5] we are here only interested in what he has to say about the definition of the real, which is illustrated with the same examples as are found in the *De partibus animalium*. One must begin with the highest genus (e.g., animal) and proceed by division through all the relevant differentiae (e.g., rational/irrational, mortal/immortal) until one arrives at a convertible definition of the species in question (man). This will produce also intermediate or subalternate genera.[6]

In the *Topics* (1.5–8), Aristotle examines the question of definition further, and discusses it in the context of the predicables genus, species, proprium (peculiarity or property), and accident.[7] The proprium is convertible with its species: man (alone) has the property of being able to laugh, and being able to laugh defines man and no other creature. An accident, on the other hand, which does not exist apart from the subject in which it inheres, can be shared with many other subjects, whether separable (e.g., "sitting" predicated of a person) or inseparable ("whiteness").[8] It is thus not convertible with a species; Aristotle concedes (1.5, 102b25–26) that "nothing prevents the accident from becoming both a relative and a temporary property," but he adds that "it will never be a property absolutely." This point is made in rather different terms in *Prior Analytics* 1.13 (32b4–10), in a discussion about possibility; it generally happens that human beings as they age go gray, but this is not necessary (it depends on the individual living long enough). So "going gray" is to be treated as an accident, or one of the pathemata of parts of animals (*De generatione animalium* 5.3–5, 778a24 ff., 782a14 ff.). This example is picked up by later commentators, as we shall see.

One of the most important channels for the logical component of this doctrine is Porphyry's *Isagoge*, which addresses the related questions: What is a thing? and What kind of thing is it? and discusses the predicables (genus, differentia, species, proprium, and accident: the "quinque voces"). After a classification of genera by the ten predicaments of the

Categories, the *Isagoge* engages in an exhaustive separate comparison of each of the five predicables with the others. The best known comparison is that between genus and differentia, in which the famous "tree" is set out which supplies the scientific definitions of the genera subalterna or species "stone" (corporeal, inanimate substance), "plant" (corporeal, animate, insensitive substance), "animal" (corporeal, animate, sensitive, irrational substance), and "man" (corporeal, animate, sensitive, rational, mortal substance). Perhaps misleadingly, Porphyry chooses here an example from the natural world. In these definitions, the genus inheres in the species, and the species in the genus; but the differentiae of the species do not all inhere in the genus or the genera subalterna (one cannot predicate "rational" of "animal," for example).[9]

Porphyry's discussion of differentia, proprium, and accident is also very pertinent here. An essential differentia cannot admit of the more or the less (a man is unquantifiably rational and mortal), but an accidental differentia (a man is rich) can. There are separable accidents (the state of sleeping predicated of a given man) and inseparable accidents (the blackness of a crow). A problem arises with the distinction between an inseparable accident (the blackness of a crow) and proprium (the capacity of laughter in man). Other things (coal, Ethiopians) share blackness with the crow; but the only creature capable of laughter is man. According to Porphyry, a crow can lose its blackness without its essential nature being altered; it is possible to conceive of a white crow. Propria resemble accidents in four different ways: they may be accidental to a species, but not to every member of the species (e.g., the property of studying geometry in a man); they may be accidental to the whole of the species, but not to that species alone (e.g., two-footedness in man and in birds); they may be accidental both to one species alone and to all members of the species, but not at all times (e.g., the grayness of the hair of aging human beings); or they may be always convertible with the whole species (e.g., the capacity for laughter in man). According to this doctrine, species can only be distinguished by the last of these versions of propria.

Boethius, whose various pedagogical works on logic were widely read and studied through the Middle Ages and the Renaissance, expatiated on a different part of Aristotelian logic. To him is ascribed a book on division in which he defends Porphyrian dichotomy or bipartite division and suggests that all division is reducible to it; but he concedes also that there are lesser (topical) forms of division, which he describes in the same work.[10] He wrote moreover an exposition of Porphyry, which makes yet more explicit the doctrine of the *Isagoge*, and commentaries on various topical texts, in which he makes an important distinction between definition, (which belongs to demonstrative logic), and description, by which he denotes a less formal level of definition able to deal with open classes of subjects nonconvertibly. He also makes explicit the relationship of description to visual representation.[11]

Zoology: Horns and Eyelashes

In various parts of the Aristotelian corpus many, but not all, of the presuppositions that relate to biology are made explicit. Nature makes nothing in vain, and seeks always to achieve the best possible result; all living things consist of matter and form; all require nutrition, and all reproduce; there is a hierarchy in nature, at the top of which stands the human being; and in all genera, the hottest, most active, and most complete beings (in the case of human beings, these are the males) are set above the rest.[12] It is important, in picking out observed characteristics of animals, to consider to what subject they belong, and what properties they entail: this should lead to a causal explanation. Of these, that which adduces the final (as opposed to the material, formal, or efficient) cause is the most to be desired. In the example Aristotle gives—horned animals that possess a single row of teeth and a third stomach—we can determine that because they have horns, they lack sufficient matter to produce two sets of teeth, and they therefore need a third stomach to help to break down the food they ingest (*De partibus animalium* 3.3, 631b1 ff.; 3.14, 674a22 ff.). The priority given to the final cause and to purpose in nature (*De partibus animalium* 1.1, 641b12–13) also helps to answer the questions of why animals have eyelashes (it is to protect their eyes) and whether human beings are intelligent because they have versatile hands, or whether human beings have versatile hands because they are intelligent (*De partibus animalium* 4.10, 687a7 ff.).[13]

Definition arises in biology in various ways; one of these is in the distinction of species. In *Topics* 6.6, 144b12 ff., Aristotle has recourse, in a more general discussion, to this distinction when he asks whether the same differentia (e.g., two-footed) can attach to two genera; the example given concerns man (two-footed pedestrians) and birds (two-footed winged creatures). In various passages in the *Metaphysics* (books 4 and 6) the doctrine of definition is amplified and reiterated in relation not only to ontological but also epistemological concerns. Aristotle says that it is impossible to enumerate all the accidents of an individual (so in that sense, an individual is unknowable: 4.4, 1007a15), and that as a consequence there is no science of the accidental (6.2, 1027a20–21). Species, therefore, is the limit of the intelligible. A very important passage of *Metaphysics* 7.12 (as we shall see in the work of Cesalpino), discusses definitions consisting only in genus and ultimate differentia, which are reached by the process of division; here Aristotle is said by most Renaissance commentators to expose the redundancy of definitions such as "a footed two-footed animal" and to reject the use of accidental qualities in definition.[14]

The *Historia animalium*, which is traditionally considered before the *De partibus*, is Aristotle's extensive record of the animal kingdom based on sense observation (1.1, 491a23), invoking fifteen differentiae and seven classes of animate beings;[15] the most important areas of investigation, which give rise to comparisons between species, are manner of life, activities, dispositions, and parts (407a10f.). The *De partibus* also deals with this last, but it begins with a discussion of methodology and presupposition. Aristotle argues that one must deal not with ultimate species but with genera (e.g., birds, fishes),[16] and one must base these genera on commonly held classifications (1.4, 644b1–2), a practice that is made explicit throughout the *Historia animalium* in the form of reference to information obtained from such sources as fishermen and shepherds. He rejects an approach (implicitly associated with Plato) through dichotomy—definition by genus and ultimate differentia—because it splits natural kinds; he asserts the doctrine that a privative differentia (e.g., "not winged" or "not footed") cannot be associated with a genus to obtain a species; and he recommends that a number of differentiae be taken together, bearing in mind that the same subalternate genus (e.g., ants) may be defined by opposite differentiae (winged and not-winged), and that some subalternate genera may have the characteristics of other subalternate genera (e.g., two-footedness of man and birds). He recognizes the primacy of the final cause, through which it is possible to distinguish two sorts of necessity: absolute (material) necessity and conditional necessity, the latter determining the nature of a subject through its purpose in being.[17] As a general approach to zoology, the *De partibus* contains identifications both between genera by analogy (e.g., fish spine and human bone) and by "the more and the less," through which members of the same genus differ in attributes (e.g., in color or size of beak: 644b10 ff.). In both the *Historia animalium* (7 (8).1, 588b4 ff.) and the *De partibus* (4.5, 681a25 ff.), the point is made that it is impossible to determine the exact line of demarcation between animate and inanimate things and between various animate things; nature takes the form of a continuous ascent from plants to the highest animals.[18] Alluded to in the *De partibus*, but not developed, is the doctrine of the three souls: vegetative, sensitive, and rational (*De anima*, 2.2, 413b2 ff.); the first of these is found in plants as well as animals and together with the second is materially present in the body; the last is different in nature, and found only in man (413b28–30).[19] This doctrine is usually referred to as essentialist;[20] genera and species are defined by their substantial form, in whose matter inheres their principle of life.

Theophrastus's *De plantis* and *De causis plantarum* are also important sources for the Renaissance. The first book of the former work discusses how plants are to be classified and makes a number of points about the difficulty of precise definition of species and the possibility that classes may overlap. The division into tree, shrub, undershrub, and herb is said to comprise "all or nearly all" plants. Among the bases for further classification that are used are the distinctions wild/cultivated, fruit-bearing/fruitless, flowering/flowerless, and evergreen/deciduous; differences in appearance and habitat; and differences in parts, some of which are essential, others not. To this list, other descriptive features (roots, knots in trees, habit, leaves, seeds, taste, flowers, fruit) are added. The *De causis plantarum* deals with the life cycle of plants, their diseases, their flavors and odors, seasonal occurrence, and the distinction between spontaneous phenomena and those brought about by human intervention. The emphasis in these works (as also in Dioscorides) is much less on what a thing is (the essentialist question) than what kind of thing it is (the question of description).

154   Ian Maclean

Commentary on the Zoological Texts

Before we consider the range of responses to the Aristotelian doctrine of classification in the Renaissance, it is pertinent to mention the most important channels through which it reaches its sixteenth-century readers. Pliny is not one of these, as he firmly sets aside any ambition to give a causal account of the diverse natural phenomena he records;[21] Galen, on the other hand, is. His *De usu partium corporis, De placitis Hippocratis et Platonis, De methodo medendi*, and *Quod animi mores corporis temperamenta sequuntur* contain a number of tributes to the *De partibus animalium*, as well as some critiques.[22] Galen is himself very interested in logic, as passages in a number of his works reveal, and various attempts were made during the Renaissance to reconstruct his lost work on inference, the *De demonstratione*.[23] He shows himself to be familiar with the Platonic logical procedure of division, and he knows the objections which can be made to it.[24] He develops the technique of analogy (borrowed from Stoic rather than Aristotelian logic) used in the *De partibus* for his own purposes.[25] He records the hierarchy of being from vegetable to man, and adopts the Aristotelian view that the soul is to some degree inseparable from the body.[26] Most importantly, he, like Aristotle, is fully committed to a teleological view of nature: for him, all animals have been provided with the best possible organization, and it is through a study of function relative to the whole organism that one can determine the purposes, actions, and uses of parts of the body (the subject matter of the *De usu partium corporis*). Like Aristotle, he believes that because man is intelligent, he has been endowed with versatile hands; and like Aristotle, he argues that eyelashes have been provided to protect the eyes.[27]

Averroes, who was looked upon as an authoritative expositor of Aristotle even in the Renaissance, also writes on the biological works. In expounding systematically the doctrine of the *De partibus animalium*, book 1, he confesses that he finds it very difficult, wonders whether his text is not deficient, and promises to return to it if he has time; his comments are still cited with respect by his Renaissance successors. He is also cited by later writers for his division of real definition into two classes, "formaliter" and "effective," the latter being scientifically less demanding than the former, as it can accommodate the claim that "some things change their essence in generation, mutation, and alteration; some things retain their substance."[28] This means in effect that definition can be a combination of genus and accident.

Through the patronage of pope Nicholas V, the zoological texts of Aristotle and the botanical texts of Theophrastus were translated more than once into Latin in the fifteenth century; the version of Theodore Gaza soon established itself as the standard text.[29] The first thing to note about their reception is the small number of reactions that are found to them in comparison to other parts of the Aristotelian corpus. This is in part because they were rarely taught in the arts course, and even considered by some as "not difficult enough" to require a teacher.[30] All translators and commentators agree with Averroes that they are obscure and difficult, and most especially that the acceptance of multiple differentiae in definitions found in the *De partibus animalium* is hard to reconcile with the doctrine about definition found in the logical works and the *Metaphysics*.[31] It is notable that a number of the figures involved—Theodore Gaza, Niccolò Leonico Tomeo, Julius Caesar Scaliger, and Daniel Furlanus—are known for their expertise in Greek: it seems that these translators and exponents (together with Agostino Nifo) thought that philological study was one way to improve the texts' comprehensibility. Their knowledge of Greek also gave them access to the expository paraphrase of Michael of Ephesus, which was not translated into Latin until 1559. Many of the contributors to the debate were medically trained and may have formed a special interest in these treatises through Galen's engagement with them; these include Nifo, Scaliger, Jakob Schegk, Pietro Pomponazzi, Andrea Cesalpino, Nicolaus Taurellus, and Cristoforo Guarinoni.[32] It is possible also that the publication in 1521 of Averroes's commentary may have inspired some, notably Pomponazzi, to engage in exegesis of the Aristotelian texts; Stefano Perfetti argues that these commentators aimed at the production of nothing less than a "scientia de animalibus."[33]

The discussion of these zoological texts reveals a number of interrelated difficulties with which these commentators grapple. These include the instruments and orders of knowledge brought to bear on zoology; the determination of genera; the predicables proprium and accident; the relationship between genera and differentiae; gradation, redundancy, and chance in nature.[34] Through the collocation of loci, these last two problems are discussed also in the context of *Metaphysics* 7.12, and hence also in that of theology, as we shall see.

Instruments and Orders of Knowledge

The first issue to be considered arises from the opening distinction in *De partibus animalium* between "peritia" or "eruditio" on the one hand (described as an inductive method by which evidence is selected and rejected)[35] and "scientia" on the other. These modes of knowing can be seen in a hierarchy, with "peritia" occupying either a higher or a lower place.[36] For Leonico Tomeo, the reliance on "peritia" places a limit on the pretensions of zoology to "scientia"; it characterizes the natural history of the *Historia animalium*, with its critical attitude to hearsay evidence and its record of effects, not causes: a point made also by Gaza in the introduction to his translation.[37] For Furlanus, "peritia" is the first step to an "explicatio animalium" that looks not at "remote" but at "primary and proximate causes," including formal causes; this is implicitly not as elevated a causal enquiry as is found in the *Physics*.[38] A different tradition sees both "peritia" and "scientia" as necessary to any philosophical enterprise; this is the position eventually adopted by Pomponazzi.[39] Nifo may also espouse this view in his reference to a sequence of scientific tasks, beginning with the description of the nature of the object being investigated, passing to the rehearsal of the causes of that nature, and ending with an explanation of them: Stefano Perfetti terms this approach "apodictic zoology."[40] Jacopo Zabarella suggests that Nifo's sequence can be perceived in Aristotle's overall zoological program, beginning as it does with the noncausal and "crude" ("pingui Minerva") practice of accumulating observations in the *Historia animalium*, passing to the more sophisticated approach of the *De partibus*, and ending in the formal analyses of the *De anima*.[41] It is clear from these remarks that not only the instrument of knowledge is in question but also its transmission to others. This leads a number of commentators to relate the opening of *De partibus animalium* to the four pedagogical methods or "viae" ("definitiva," "divisiva," "resolutiva," "demonstrativa").[42] One (Guarinoni) sets out this sequence (or rather, as he characterizes it, these interdependent approaches) as "historia," "[via] divisiva," "resolutio," and "compositio," by which in the end the *Historia* itself can be made to yield up demonstrative syllogisms.[43] His is the most affirmative treatise on the scientific status of Aristotle's zoology and the most optimistic presentation of the passage from observation to scientific knowledge.

The Determination of Genera

A second issue these commentators all mention is the unscientific source of the general classes of animate beings, and the use of hearsay evidence from country-dwellers, fishermen, hunters, shepherds, and the like to establish them.[44] It is moreover not clear whether these genera are clearly separated from each other, or whether they comprehend all of nature; as we have noted, this would compromise the science of animals, for in Aristotelian terms there can be no scientific knowledge of an incomplete class. Leonico Tomeo points to the inconsistency and variability of corruptible things, which reduces the status of classifications built upon them.[45] Commentators also note that the system described by Aristotle contains other compromising features, including limitrophic classes (such as zoophytes) and species that belong to more than one class.[46] The description of individual members of genera is said to come from the senses and what is "better known to us";[47] it makes reference to qualities such as color, shape, texture, taste, smell, even though these are excluded *qua* accidents from the definition of the essence of a thing. According to Aristotelians, as we have seen, all sensible objects are unintelligible in the sense that the sensible and the intelligible belong to different orders of knowledge;[48] for genera of animals and plants to be more than mental constructs, however, they must relate to the individual sense perceptions through which they were formed.[49] These problems, which arise elsewhere in Aristotelian philosophy, are particularly acute in his zoology, as this relies so heavily on lower orders of knowledge, incomplete classes, sensory information, and accidents.

Proprium and Accident

In these commentaries, a distinction is made between the enquiry into the essence (the "principia" and "substantia") of animals and the account given of their attributes and operations.[50] This bears on the pretensions of the discipline of zoology (as it asks both the question What is an animal? and What kind of thing is an animal?) and on the distinction between proprium (which relates to the essence) and accident (which is only qualitative).⁵¹ The former question may require a species to be defined in defiance of what is commonly accepted (the white crow as a crow, for example) and may lead to a qualitative accidental feature that none the less seems to be a property being ignored (e.g., the fact that man's hair goes gray). Three of the four categories of evidence cited in *Historia animalium*—"vita," "actiones," "mores"—are acknowledged by most commentators to yield only accidental information.⁵² Kristian Jensen points out that Cesalpino thought that he had found a way of saving the Aristotelianism of his approach by claiming that the physical parts of plants (the fourth category of evidence) are instruments of their vegetative soul, and hence may be claimed to be essential differentiae (or, to put it differently, he made some accidents of plants into propria),⁵³ but most other commentators admit that reliance on physical appearance transgresses the Aristotelian interdict on definition by accidents.⁵⁴ Because it is not always given to man to know the essence of things, he may have to be content with accidental differences, as Furlanus and Taurellus point out.⁵⁵

This lower expectation of knowledge leads, I believe, to an acceptance of the evidence of accidents and a merging of proprium and accident in the course of the century, notably in the work of Zabarella. In the account of Aristotle's zoological works in his *De methodis*, Zabarella distinguished between knowledge of the essence of a thing and knowledge of its "propria accidentia," which relate either to the genus or to the species. Although he argues that it is generally necessary first to know universals in order to know particulars, he claims that it is not necessary to know generic proper accidents before those relating to species, and he develops this point at some length.⁵⁶ In this development, "accident" begins to lose its status as a separate predicable and becomes a descriptive term for any form of contingency.

Genus and Differentiae

One of the most important issues is that of definition by genus and ultimate differentia. As we have seen, there is a potential contradiction in Aristotelian loci between *De partibus* 1 and *Metaphysics* 7.12. Guarinoni does not refer to the problematic claims about definition by multiple differentiae in the *De partibus*; he even produces a dichotomous table from a passage in the *Historia animalium* that seems to bear out his claim that scientific definition can be achieved from the evidence set out in that work.[57] Others (the author of the Ambrosiana manuscript and Leonico Tomeo) simply record the objections to dichotomy (which they attribute to Plato and Porphyry)[58] as given in the *De partibus*, and make no reference to *Metaphysics* 7.12.[59] Several commentators (Nifo, Furlanus, Schegk, Cesalpino, Taurellus) explicitly acknowledge the difficulty raised by the collocation of these texts. Some commentators simply allow the inconsistency to stand. In one commentary Nifo makes the suggestion that there are two orders of division: Platonic, which is dichotomous; and the "vulgar division" of the *De partibus*, which is implicitly not scientific and can be derived from multiple and even accidental differentiae.[60] But in another context, Nifo makes the claim that all division is reducible to dichotomy, and that accidents cannot be part of any definition.[61] Cesalpino's adversary Nicolaus Taurellus also makes Nifo's first distinction between Platonic and vulgar division, and claims in a similar way that definition in its proper sense can only be derived from genus and ultimate differentia; "vulgar division" leads to no more than description.[62] Others try more actively to save the coherence of Aristotelian doctrine. Schegk, for example, uses *Topics* 6.6 to defend "composite differentiae," again implicitly accepting that what will emerge from the process will not have the status of demonstrative knowledge.[63] By far the most ingenious defense of multiple differentiae comes from Cesalpino: he explicitly raises the problem of *Metaphysics* 7.12, and claims that in this text Aristotle did not intend this passage as a "true conclusion from true premises" but rather "as an absurd consequence of a flawed mode of definition"; for Cesalpino, the problematic chapter in the *Metaphysics* is concerned with unity in definition, not dichotomy, and the passage about genus and ultimate differentia is an argument ex absurdo in favor not of scientific definition but of multiple divisions of the genus, including accidental divisions.[64] Cesalpino's attempt to justify multiple differentiae and accidental qualities in definitions leads him to a novel interpretation of Aristotle, demonstrating the lengths to which scholars of his generation were prepared to go to save the coherence of peripatetic doctrine by the conciliation of texts.[65]

Gradation, Plethora, Redundancy, Chance, and Variation

These issues are all connected to formal and final causes in Aristotelian philosophy and their relation to purpose and to variation in and between species. Of the four causes, only the efficient and the material are subject to change and time; the final and the formal are not. Scaliger says that animals and plants can be classified in different ways according to which causes are in question; they can even be grouped by a principle—their usefulness to man—wholly extrinsic to their natures (a principle explicitly rejected by Cesalpino).[66] It is accepted by these commentators that of all causes, the final is the most significant for the study of zoology; the formal is also very important, as the vital principle of an animate body is said to be its form.[67] The insistence on these two causes is connected with the near-universal rejection of the Democritic version of atomism; the consequences of the theory that the universe was purposeless and the soul material were well understood, but execrated, at least in public.[68] Even Cesalpino, who tries to incorporate the study of material and accidental features of plants into his taxonomy, gives precedence to the higher causes in the form of the vegetative soul,[69] and he makes no mention of the doctrine of the latitude of forms, to which I shall return below.

The issues of variation and redundancy arise also. In the *De partibus*, it is acknowledged that one of the means by which a species is identified is through the excess or defect of a given feature (a beak, for example).[70] Plethora and excess are also part of natural processes and natural design; individual natures can have plethoras or superfluous humors that contribute to the variability inside species. Redundancy is also an issue. Even though Aristotle claims, as we have noted, that nature does nothing superfluous, his pupil Theophrastus provides a locus for the contradictory claim, pointing out that nipples in males, hairs in the nose and horns on the heads of stags all constitute examples of natural redundancy. The whole issue of exceptions in nature, which operates only "for the most part" in respect to its own rules, is linked also to the excess and defect of the moment of generation of natural beings. These questions were much debated in the Renaissance.[71]

Theology, Medicine, and Dialectics

These debates relate closely to theological discussions of the period about substantial form and the plurality of souls in man (Thomas Aquinas is of note here through his discussion of the issue in *Summa theologiae* 1a76.3), about the theory of a purposeless universe, and about the nature of the Eucharist. Even before the Reformation, the Parisian philosopher Joannes Dullaert's discussion of the white crow and accidentality led to the issue of the accidents of the host (an issue much debated in the later Middle Ages), which he approaches from both a realist and a nominalist position: he mentions a nominalist "neotericus" (not identified) who argues that blackness can be a property of a crow, whose argument he refutes by claiming that "blackness" is not convertible with "crow."[72] The conflict between various protestant views of the Eucharist and that propounded by the Council of Trent generated a vast literature involving the discussion inter alia of proprium and accident, in which the discussion of books 7 and 8 of the *Metaphysics* plays an important part. There is a revival of the study of metaphysics across the whole spectrum of religious opinion, in which the relationship of propria to accidents, the connection of both of these to matter and to substantial forms, the plurality of souls (which natural philosophers such as Crippa also discuss), the distinction between imperfect and perfect definitions, and the question of ultimae differentiae, genus, subalternate genera, and species all arise.[73] One text (by the Scotsman Thomas Rhaedus, of the gnesiolutheran University of Rostock) is even entitled "on the accidental proprium," although no such laxity with terms is found here as was found in the work of Zabarella. Rhaedus is engaged in a polemic with theological writers of various confessions (Bellarmino, Timpler, Goclenius, Keckermann) over the issue of "the communication of properties," which concerns the double (divine and human) nature of Christ, and the accidents of the host.[74] Theologians were aware of the biological parallels drawn in the Aristotelian texts, and they even refer to medical discussion of vital principles.[75] It seems very likely that both the commentators on Aristotelian zoology and natural historians of the mid and late sixteenth century were aware of these ideological debates and their relationship to their own preoccupations.

It seems equally likely that the same commentators and natural historians were aware of the medical implications of their enquiries, not least because many of them were trained as doctors. Medical teaching at the time stressed that the best methodology for medicine was the "via divisiva" (the creation of species from genera, or in medical terms, the identification of disease through its accidents or symptoms);[76] its definition of disease was not, however, wholly ontological, as it could arise inter alia from the dyscrasis of the individual. Nancy Siraisi has shown how difficult it was to distinguish between cause, symptom, and disease and to produce a reliable classification.[77] It is interesting to note that Galen himself has recourse to book 1 of the *De partibus animalium* in his introduction to the *De methodo medendi* in setting out the problems of defining and identifying disease, thereby providing a direct link with the issue of zoological species; several doctors pick this link up, and relate it to the "via divisiva."[78] A further feature of the scholarly discourse of medicine at this time is its development of a lax form of logic in which all of its demonstrative tools—division, genus, differentia, inference through signs—are given a less precise status. Such a relaxed attitude to reasoning would be very suitable to sixteenth-century natural history also, as we shall see.[79]

At the end of the fifteenth century, professors of medicine were also responsible for reviving the medieval investigation into the intension and remission of forms, no doubt as a tool with which to deal with the medical problem of continuous change. In Galenic medical theory, the logical form of contrary known as "mediate" as opposed to immediate (e.g., white/[gray]/black as opposed to odd/even) is employed to account for this continuous change, sometimes in contradiction of the doctrine of substantial forms, which would seem to determine the nature of any physical object in a fixed way.[80] Both Aristotelian and Galenic loci speak of change as occurring through the medium:[81] this change presupposes a continuity which comes to be known as the "latitudo formarum."[82] Latitude of forms is a doctrine particularly pertinent to medicine, "for medicine is the knowledge of everything in degrees, from the smallest to the largest," as one prominent professor of medicine put it. The logical issues are here closely entwined with such ontological problems as whether the passage from "salubritas" to "insalubritas" is continuous, or gradual through degrees (a "medium transitus"); whether the middle term between health and sickness participates in both extremes, or excludes both extremes (a "medium formae" or "medium in genere"); whether an overlap is possible; whether the latitude of health refers only negatively to resistance to disease; whether it refers both negatively and positively (to degrees of health); and whether the neutrum which is perceptible to sense (such as stomach pains, headaches, a bitter taste in the mouth) does not exist in the order of the intelligible (and hence of demonstrative truth).[83] These discussions all bear on the living body as a process, and the perception of that process through accidents; these are issues which arise also in natural history at this time.

Also of importance is the revival of dialectics, and especially of topical reasoning as found in the *Topics* of Aristotle, the rhetorical works of Cicero and Quintilian, and the relevant commentaries by Boethius. Argumentation from loci in the Renaissance is often associated with the name of Rudolph Agricola and his influential *De inventione dialectica*; chapter 5 of its first book contains, as well as a comprehensive list of such loci, what seems to be an original contribution to the practice of definition which Agricola terms "descriptio" after Boethius: this "expresses the thing more verbosely . . . and sets it as it were before the eyes for inspection." The example he gives is the following: an ass is a animal with non-cloven feet, long ears, and capable of reproduction; a mule and a horse are not cloven-footed, a mule and a hare have long ears, a mule is not capable of reproduction. The horse, the hare, and the mule all possess only two of the three attributes of the ass; therefore the definition is adequate. Later chapters deal with genus and species and proprium in a logically informal way and list the places of argument that are taken up by certain Renaissance natural historians, as we shall see.[84]

Species in Natural Histories

To attempt systematically to demonstrate the influence of the discussions surveyed above in Renaissance works of natural history would be a dauntingly broad-ranging enterprise, especially as some of the prominent writers in this field (Cesalpino, Bauhin, Aldrovandi) are themselves also commentators.[85] All I can do here is to sketch out a few connections that seem to me to be significant. It seems to me uncontroversial to claim that Aristotle's project to produce a general zoology was an important, if not the most important, model for natural-historical enquiry at this time. Andrew Cunningham has persuasively argued that this was taken up by Hieronymus Fabricius of Acquapendente, who followed the sequence implicit in the Stagyrite's zoological works (observation and inspection, followed by a systematic account from which causes could be derived and a general etiological theory developed).[86] This combined inductive and analogical with deductive processes of thought in a very similar way to the Aristotelian project described above. I believe that this approach, together with those of earlier and contemporary anatomists, botanists, and zoologists, evinces the same conceptual problems as those with which Aristotle's commentators at this time were grappling; it reflects the difficult of proceeding beyond crude generic classes, to which Gessner refers in a letter to Joachim Camerarius of 27 January 1565.[87]

The first of these problems is the (vulgar) source of the initial division into genera and the attitude to hearsay evidence.[88] There is good Aristotelian precedent for relying on the testimony of others (in the *Historia animalium*, he not infrequently cites evidence from reliable sources such as fishermen) as well as that of one's own eyes.[89] Conrad Gessner comments directly on this: he declares that he has included all testimonies even if he didn't himself necessarily believe all of them ("fidem meam in pluribus non astringo"), because he decided that it would be premature to exercise judgment before having assembled all the evidence available; his protégé Otto Werdmüller is happy to accept in a similar spirit the traditional classification of animals.[90] Later in the century, this liberal spirit seems to have hardened in some quarters: Ulisse Aldrovandi warns his reader about the dangers of hearsay with an example, and demands that the natural historian be the "oculatus testis" of all that he records (although, as Laurent Pinon points out, he also proudly lists the textual evidence he cites, which is not to be found in Gessner).[91] Another indication of the growing importance of sensory evidence is to be found in the work of Caesar Odonus, who cuts out the methodological sections of the *De partibus animalium* and reduces it to an alphabetical series of *historiae* about individual animals.[92]

The issue of proprium and accident emerges most forcefully in the debate about the wisdom of illustrating herbals and anatomical textbooks. This has been lucidly set out by Sachiko Kusukawa, in respect of the exchange of views between Leonhart Fuchs and Sébastien de Monteux; at issue is whether a plant can be known by its accidents, and whether any image of a plant should restrict itself to features which occur simultaneously, or whether it is permissible to produce a composite picture showing buds and leaves as well as fruit and seeds. If accidents are shown, then the record of the plant will be (in Renaissance terms) a description rather than a definition; this issue is another version of that addressed in the debate described above about instruments and orders of knowledge and propria and accidents.[93] Guillaume Rondelet makes an explicit connection with the loose logic of medicine in his book on fishes, in speaking of understanding differentiae in a "broader way" and permitting accidents as well as propria in the determination of species.[94] It would seem that, as in the case of the preference of eye-witnessing over hearsay, there is a tendency as the century progresses to rely more on evidence of the senses which had traditionally been seen as accidental and uninformative about propria, and a blurring of the line between propria and accidents similar to that found in the works of Zabarella.[95]

The question of genus and differentiae also arises. Classification, as Karen Reeds points out, can proceed by alphabet, by habitat, by similarity of form or action, by organs of generation, by uses, and by yet other criteria. Behind many of the classifications lies the fourfold comparative description of genera (by "vita," "actiones," "mores," and "partes") which is employed in the *Historia animalium*.[96] Benedictus Textor's *Stirpium differentiae* of 1534 specifically associates his work with the commonplace rather than logical tradition; he applies dialectical loci—quantity, quality, place, time, action, and use—to the general classification of plants, and thereafter analyzes them by parts (root, stalk, branch, leaf, flower, fruit, seed). He is followed in this by the Swiss zoologist Joannes Fabricius, who in 1555 echoes his title and also employs, together with medical terminology, terms in a topical rather than logical register ("subiectum," "adiuncta," "comparata") to generate his field of differences.[97] Gessner, as is well known, presents his descriptions of animals in a series of sections with letters rather than numbers (to allow for omissions in the set; he thought that readers would be tolerant of an incomplete alphabetical sequence, but not of an incomplete numerical one); in this, location and habitat are given an important place, perhaps reflecting the growing interest in regional difference celebrated in such newly discussed works as Hippocrates' *Airs, Waters, Places*.[98] Cesalpino's genera are based only on organs of generation (fruit and seeds), as these could be argued to be indications of the vegetative soul and hence privilege the formal cause.[99] Contemporaries did not find the result to be very satisfactory, and Bauhin's *Pinax* of 1623 was organized on rather different lines, in which single outstanding features of plants in relation to their structure, habit, and habitat were used to generate a taxonomy based on resemblances of form.[100] It is also worthy of mention that classifications tend increasingly to be open-ended and even to permit of unclassifiable residues; both Cesalpino's and Bauhin's plant taxonomies were designed to accommodate as yet undiscovered plants.[101]

The last issue to be mentioned is the variability of nature. Even though theories such as that of idiosyncrasy and the use of extended lists of variable factors in diagnosis and therapy are in themselves admissions of the diversity of forms in the physical world, the abandonment of strict taxonomy and a postcanonical approach to the constituents of the human body greatly accentuate the loosening of the conceptual bonds of what is taken to be natural, as Nancy Siraisi has shown in an important article on Vesalian anatomy.[102] A later example is afforded by Bartholomaeus Eustachius, who in his investigation into kidneys refers on more than one occasion to the "variable and surprising productive craft [*ars*] of nature," and suggests that "she [the personification is explicit] is not subject to any law and does what she pleases."[103] This poses a considerable challenge to those who wish to classify by number or size or shape, and again predisposes natural historians cautiously to engage in frequent and precise observations.

Bacon: Back to Eyelashes

I come finally to Francis Bacon. He himself shows no great interest in the establishment of species except where he sees the attempt as misleading (possibly because he thought that in the state of knowledge current in his time, it would be premature to determine biological classification of any sort), but he does engage in reflections on some of the issues set out above. The *Novum organum* includes a consideration of genera which one can afford to accept provisionally and deviant and borderline instances, which include living creatures who challenge classifications (such as flying fishes).[104] In the *Advancement of Learning*, Bacon sets out a grand map of all the disciplines, identifying those areas that are in his view deficient, which is relevant here. He divides natural history into three parts, in which he sees the study of "nature erring" or "irregulars as nature" (which would presumably include the study of the white crow) as neglected, because it will not be possible to produce a general etiology of nature without taking them into account. He divides philosophy into divine, natural, and humane, putting "primitive philosophy" (later called metaphysick)—that is, the accumulation of axioms, which must precede any enquiry into nature—into the second category, which is further subdivided into the "inquisition of causes" and the "production of effects." The former of these (natural science or phisick) is dedicated to the enquiry into "variable or respective" (i.e., material and efficient) causes, while metaphysick enquires into "fixed and constant" (i.e., final and formal) causes. Bacon records the view that these may lie outside the reach of man, but thinks them worthy of pursuit provided that this produces theories and promotes man's capacity to manipulate nature. The particular deficiency he finds in metaphysick lies in its pursuit of final causes, which he identifies as a feature not only of Platonic but also of Aristotelian and Galenic philosophy:

For to say *that the haires of the Eyeliddes are for a quic-sette and fence about the sight: Or, That the firmness of the Skinnes and Hides of liuing creatures is to defend them from the extremities of heate and cold: Or, that the bones are for the columnes or beames, whereupon the Frames of the bodies of liuing creatures are built; Or, That the leaues of trees are for protecting the Fruite; Or, that the cloudes are for watering the Earth; Or, That the soldinesse of the earth is for the station and mansion of liuing creatures*: and the like, is well inquired and collected in METAPHYSICKE, but in PHYSICKE they are impertinent.[105]

The note to the most recent edition says that the example of eyelashes is taken from Xenophon *Memorabilia Socratis* 1.4.4–6, which was indeed a much published text in the Renaissance; but the *De partibus* also contains this and several other of these examples of reasoning by final cause cited by Bacon (653a30; 654a30 ff.; 658b1 ff.).[106] I am inclined to believe that he has the latter text in mind—not least because he chooses to rehabilitate a view of nature evoked in the *De partibus* (1.1, 640b10) and associated with the names of Empedocles and Democritus, which Aristotle specifically refutes there and in the *De coelo* (3.4, 303a2 ff.):

The Natural Philosophie of *Democritus*, and some others who did not suppose a *Minde* or *Reason* in the frame of things, but attributed *the form thereof able to maintain it self to infinite essaies or proofes of Nature*, which they tearme *fortune*: seemeth to mee (as farre as I can iudge by the recitall and fragments which remain vnto vs) in particularities of Phisicall causes more real and better enquired then that of Aristotle and Plato.[107]

168 Ian Maclean

Bacon is aware of the association of atomism with atheism, and answers the charge with a (to me) surprising parallel between God and the devious Renaissance prince of Machiavelli and Guicciardini who keeps his ministers and ambassadors in the dark:

> Neither does this call in question or derogate from diuine Prouidence, but highly confirme and exalt it. For as in ciuill actions he is the greater and deeper pollitique, that can make other men the Instruments of his will and endes, and yet neuer acquaint them with his purpose: So as they shall doe it, and yet not knowe what they doe, then hee imparteth his meaning to those he employeth: So is the wisdome of God more admirable, when Nature intendeth one thing, and Prouidence draweth forth another; then if hee had communicated to particular Creatures and Motions the Characters and Impressions of his Prouidence.[108]

Bacon recommends the study of only material and efficient causes, and he offers an exemplary translation of the etiology of the eyelash into a form which does not rely on the determination of purpose ("*Pilsotie is incident to Orifices of Moisture*"). The radical nature of this suggestion seems to me to lie in the decomposition of an integrated field of etiology: whereas before him, the final cause, albeit distinct in sense, was intimately linked to the other members of the Aristotelian quartet, Bacon here suggests a divorce that will do much to promote the scholarly status of accidents and particulars and will allow white crows, gray hairs, horns, and eyelashes to be subject to new modes of assessment by subsequent natural historians.

Notes

I have been much helped in the preparation of this paper by the published work of Cunningham, Jensen, Kusukawa, Perfetti, Pinon, Moraux, and Siraisi. I should like also to thank Nancy Siraisi, Gianna Pomata, Brian Ogilvie, Laurent Pinon, and Sachiko Kusukawa for their additions, comments, and suggestions.

Sources cited in boldface type appear in the Primary Sources of the bibliography.

1. See British Library, Add MS 6789, fol. 511r–v; **Kepler** 1954, 172 (letter of Harriot to Kepler, 13 July 1608): "ita se res habent apud nos ut non liceat mihi adhuc libere philosophari. Haeremus adhuc in luto."

2. I have not included material from the commentaries on Theophrastus; it should be said, however, that many botanical writers (notably Cesalpino, Scaliger, and Bauhin) owe much to his texts. It should be stressed in particular that for the scholars and natural historians whose work is discussed in this paper,

the meaning of "genus" and "species" is not that found in modern zoology or botany; see Atran 1990.

3. See Dear 1995 and Maclean 2002, 114ff.

4. It is interesting to note the similarity of this concept and Galen's "endeixis" (indication): see Maclean 2002, 306–315.

5. Scholastic philosophy here distinguishes between "ens reale" or "res ipsa" and "ens rationis."

6. Convertibility betokens the interchangeability of the subject and predicate: a rational, mortal animal is man (and nothing else); man (and nothing else) is a rational, mortal animal. The ultimate differentia in this case is "rational"; the subalternate genus is "mortal animal."

7. In *Topics* 1.8 (103b3ff.), these predicables are distinguished as follows: an accident, while belonging to a species, is neither its definition, nor its genus, nor its proprium; a proprium is convertible with its species, but does not indicate its essence; a definition is convertible, and does indicate its essence; a genus is not convertible with its species (man, being rational, is not convertible with the genus animal). In *Posterior Analytics* 1.4, 75a20ff., Aristotle makes the point that there is no "scientia" of the accident.

8. The most important examples of accidents that Aristotle cites are for our purposes sensory qualities. See also *De sophisticis elenchus* 5 (167a8–14) on the much-cited fallacy of an Ethiopian being both black and not-black, because he has white teeth (whose whiteness is accidental to the Ethiopian's substance).

9. **Martini** 1605, 5: "species genere est perfectior, quia addit semper aliquid ad genus." But it is sometimes claimed that all individuals inhere in universals, and universals in all individuals; see Maclean 2002, 122n.

10. *De divisione*, PL 64.875–891 (883). In this work division is broken down into four classes: genus into species (e.g., animals into rational and irrational animals); whole into parts (also known as enumeration, e.g., house into roof, walls, and foundations); polysemic words or propositions into single senses (which in the Scholastic system becomes the "sic distinguitur" by which contradictions are resolved by appeal to *ratio*, *ordo*, *modus*, etc.); and finally the species of "divisio secundum accidens," which is itself divided into three classes: subjects divided by accidents (e.g., some men are black, some white); accidents divided by subject (things desired are either desired by the body or the soul); and accidents divided by accidents (some white things are solid [pearls], others liquid [milk]).

11. "Ex proprietatibus informatio quaedam rei et tanquam coloribus quibusdam depictio." Boethius's doctrine of description is taken up by jurists; see Maclean 1992, 108.

12. See *De generatione animalium* 2.4, 739b19–20 (the locus classicus); Maclean 2002, 234ff. On arguments against purpose in nature, see Maclean 2002, 247–248, and *De partibus animalium*, passim, on plethora and on relative uselessness (e.g., horns swept back are not of use for self-defense, but do not impede grazing: 3.2, 663a8; cf. 648a16). On hierarchy in nature (including the passage

from larvipara to vivipara) see *De generatione animalium*, esp. books 2 and 3; also **Wotton** 1552, 1: "natura continue ab inanimatis ad animalia transit." On nature striving to achieve the best result, see *De coelo* 2.5, 288a3–4, and below, note 13.

13. "Anaxagoras indeed asserts that it is his possession of hands that makes man the most intelligent of the animals; but surely the reasonable point of view is that it is because he is the most intelligent animal that he has got hands. Hands are an instrument; and nature, like a sensible human being, always assigns an organ to the animal that can use it (it is more in keeping to give flutes to a man who is already a flute-player than to provide a man who possesses flutes with the skill to play them); thus nature has provided that which is less as an addition to that which is greater and superior; not vice versa. We may conclude, then, that, if this is the better way, and if nature always does the best she can in the circumstances, it is not true to say that man is the most intelligent animal because he possesses hands, but he has hands because he is the most intelligent animal." On the comparison of nature to a human creator or artisan, see *De partibus animalium* 1.1, 639b15ff.

14. Cf. *De partibus animalium* 1.3, 643a28; 644b7.

15. Van den Broek 1991, 60–61. Laurent Pinon has kindly told me that these differentiae and genera, although used for terminological purposes, were not themselves influential in Renaissance natural history.

16. Cf. Albertus Magnus and Zabarella on the doctrine of beginning with the "more universal" and descending to the particular: Reeds 1991, 174; **Zabarella** 1586–87, 1:64.

17. *De partibus animalium* 1.1, 642a10: "a hatchet, in order to split wood, must, of necessity, be hard; if so, then it must, of necessity, be made of bronze or iron." It is not clear what sort of necessity is being referred to in *Progression of Animals*, book 8 (708a21): "every animal which has feet must necessarily have a even number of feet." Aristotle also refers to "coercive" necessity (*Metaphysics* 5.5, 1015a20 ff.). See **Leonico Tomeo** 1540, 179; also Balme 1987b (a fuller account).

18. Gaza's translation of the passage in the *Historia animalium* runs as follows: "sed adeo de inanimatis paulatim sensimque ad animata natura transit, ut continuatione ipsa lateat eorum confinium et medium utrisnam sit extremi."

19. The doctrine of *De anima* is associated by Aristotle in the opening chapter of the first book with a higher order of enquiry, both because of the nobility of the subject matter and because of the degree of certainty to be obtained from its investigation. On this see **Furlanus** 1574, 43.

20. Pinon 1995; Balme 1987a dissents from the view that Aristotle's biology is essentialist; see also Atran 1990, 89–122.

21. *Historia naturalis* 11.2.8: "nobis propositum est naturae rerum manifestas indicare, non causas indagare dubias"; see also POMATA in this volume, 109–110, 138 n. 22, citing Ermolao Barbaro on the distinction between Pliny's "particular and sensory reasons" and Aristotle's "universal and speculative reasons."

22. Moraux 1985. Moraux does not discuss, however, the use of the *De partibus animalium* in Galen's *De methodo medendi*, book 1.

23. On the importance of this in the Renaissance, see Maclean 2002, 115.

24. *De methodo medendi* 1.3, in Galen 1821–33, ed. Kühn (hereafter K), 10.18 ff.; Moraux 1985, 330–331.

25. On "analogismus" see Maclean 2002, 159 ff. **Guarinoni** 1601, 56 makes some interesting points about analogy (including the distinction between Euclidean "similitudo rationum" and Aristotelian "aequalitas rationum metaphorice") in respect of observation and the recording of parts of animals, which I shall not have space to develop here.

26. *Quod animi mores corporis temperamenta sequuntur*, K4.767–822; also *De semine* 2.6, K4.611.

27. Moraux 1985, 330–335, citing *De usu partium* 3.1.3, K3.5 ff.; *De instrumento odoratus* book 2, K5.879; *De partibus animalium* 2.13, 657a32 ff.

28. See **Aristotle** 1550, 6.66–67; Maclean 2002, 145.

29. See Siraisi 1993, 173–174.

30. For a list of the sources here used, see Biblioteca Ambrosiana, B Ambr. N 26 Sup. (hereafter MS Ambrosiana); **Cesalpino** 1593, **Crippa** 1566, **Furlanus** 1574, **Guarinoni** 1601, **Leonico Tomeo** 1540, **Fabricius** 1555, **Nifo** 1546, 1547, **Odonus** 1563, **Scaliger** 1592, **Taurellus** 1596, 1597, **Werdmüller** 1555, **Wotton** 1552, **Zabarella** 1586–87. For biographical details of Gaza, Cesalpino, Furlanus, and Leonico Tomeo, see Perfetti 1999a, 1999b, 2000; Cranz and Schmitt 1984. Bernardino Crippa taught philosophy at Bologna from 1561 to 1566 and later moved to Rome (see Mazzetti 1988, 102); the Oxonian Edward Wotton (1492/3–1555) was trained at Padua; Nicolaus Taurellus (1547–1606) taught at Altdorf. There are two further works of great rarity which I have not been able to consult: Arcangelo Mercenario, *Iudicium super rationibus Aristotelis primo de partibus animalium capite primo* (Padua, apud Simonem Galignanum de Karera, 1570, 4to), written, according to Perfetti (2000, 186), to "open a path to Aristotle on psychology"; and Paolo Belmissero, *Libri duo primi Aristotelis de animalibus elegantissime et ad amussim in elegias triginti sex translati* (Rome, 1534, 4to). See also Blair (1997, 35), citing Gilbert Jacchaeus on the facility of the zoological texts.

31. Pomponazzi, quoted by Perfetti 1999a, 113, 117; **Schegk** 1556, 112 (referring to the "obscuris et perplexis rationibus et argumentis" of the *De partibus*, book 1); **Zabarella** 1586–87, 5.50 (referring to its "prolixum prooemium").

32. The author of the Ambrosiana manuscript is unknown; there is no indication that he knew Greek, and some signs that he had read the Averroes commentary. Crippa's commentary is on the *De motu animalium*, but the same points arise in it as in commentaries on the *De partibus* and the *Historia animalium*.

33. The suggestion about Averroes is found in Perfetti 1999, 301. Perfetti 2002, 432–434 connects the revived interest in these texts with that evinced in the same

period for Aristotle's *Parva naturalia*, and makes the claim about the ambition to produce a "scientia de animalibus."

34. I have set aside the problem of the relationship between genera and subalternate genera discussed in *De partibus animalium*, book 1, which engages several commentators. See, for example, MS Ambrosiana 3r–v.: "Aristoteles dicit ibi q[uod] qu[ando] una species sub duobus generibus est alterum sub altero est"; "diversa esse genera in aliquo pr[a]edicamento quae non sibi invicem subalternantur."

35. MS Ambrosiana 1r: "methodum quandam et modum docendi qua sciamus quid tractandum quidve non tractandum sit"; **Crippa** (1566, 2) suggests that it is an inductive method. Perfetti (2000, 213) quotes Cesare Cremonini's manuscript treatise on the passage, and his characterization of "paideia" as "an instrumental habitus" deduced from what the "peritus" does.

36. **Leonico Tomeo** 1540, A7r: "naturalem scientiam non proprie scientiam appellari debere... sed peritiam potius, et eruditionem quam [Aristoteles] *paideian* appellat, cum non ex propriis immediatisque ostendendae rei procedat causis, sed ut in pluribus communibus utatur principiis, quae multis aeque quadrare possunt"; also **Leonico Tomeo** 1540, A4r–v, where he claims that as in ethics, there are no "certissimae demonstrationes" in the studies on animals. Perfetti 2000, 213 refers to the "irreconcilable gap" between paideia and episteme in the eyes of Leonico Tomeo. See also below, note 54.

37. Perfetti 1999a, 458, quoting Pomponazzi on Aristotle accepting hearsay evidence in the way Christians accept that Christ really lived.

38. **Furlanus** 1574, 37, 43 (on the passage from a "narratio quaedam rudis" to the discovery of causes); also **Crippa** 1566, 1–2 (who attempts to link the zoological treatises to the pursuit of truth as defined in *Metaphysics* 2.3, 2.11, and 4.2); and Du Val in **Aristotle** 1619, 1.13, where "peritia" is described as a "vim et facultatem bene ac perspicienter iudicandi, quae quid recte aut non recte detur tradet."

39. Perfetti 1999a, 449–453, quoting Pomponazzi: "videtur quod Aristoteles vult quos isti habitus, scilicet scientia et peritia, sunt inseparabiles"; see also **Crippa** 1566, 5.

40. See Perfetti 1999b, 311–312. Zoology's apodicticity rests on the claim that Aristotle's demonstrations are universal and can be wrong in detail without their conceptual schema being impugned.

41. **Zabarella** 1586–87, 1.64–66; the same claim is made in **Crippa** 1566, 1–2. See also POMATA in this volume, 110–111.

42. Perfetti 2000, 132 ff. cites Simon Porzio's definition of "paideia" as "modus intelligendi, docendi et sciendi"; see also Perfetti 2002, 439.

43. POMATA (in this volume, 110) quotes Gaza's definition of "historia" as "an exposition of the thing as it is, which the schools of philosophers of our time used to call the *quia*." See also **Leonico Tomeo** 1540, E4-5v, C4v: "physicas demonstrationes quia esse et non propter quid"; **Guarinoni** 1601, 20,

324–325, where it is claimed that observation leads to taxonomy, that the compositive, resolutive, and divisive methods "inter se cohaerent et una alteri inhaerere vel potius inserta esse videntur," and that the history of animals leads to demonstrative syllogisms. The methods here referred to seem to be those set out in Galen's *Ars parva*, book 1, K1.305, on which see Maclean 2002, 201–203; also Gilbert 1960; Jardine 1988 (on regressus). Perfetti 2000, 99 points out that **Nifo** reduces these texts to syllogisms; although he concedes (1546, 4r) that "notitia animalium omnium perfecte non potest habere [quia] semper aliquid novi fert Africa [et propter] animalia quae non cadunt sub sensum," he claims nonetheless that Aristotle developed a sufficient theoretical structure for zoology and for the classification of animals into "species ultimae."

44. **Nifo** 1546, 19 concedes that the status of the "scientia de animalibus" is reduced because it is derived not from "metaphysicis et logicis" but from "vulgaribus, . . . piscatoribus, venatoribus, pastoribus, iis qui mundum et sylvas et loca ferina peragrarunt"; **Furlanus** 1574, 242 relates this to Galen *De methodo medendi* 1.5, K10.40 f.; **Zabarella** 1586–87, 5.54 refers to the unhappy marriage in *Historia animalium* of those things which "ad principia et ad substantiam animalis pertinent" and those which relate to "accidentia et operationes, sed ruditer nec ita distincte, ut in aliis sequentibus libris, quos scientiales esse diximus, siquidem rudis et historica narratio eam exquisitam partium distinctionem, quam demonstrativa contemplatio habet, habere minime potuit."

45. **Leonico Tomeo** 1540, A4r: "in iis libellis [*De partibus animalium*; *Historia animalium*] non exactae illae et primae immediataeque rerum explicantur causae, de quibus in Analyticis plenius [Aristoteles] nos docuit; neque demonstrationum ea sane exercetur ratio, quae in primo certitudinis, ut aiunt, gradu esse censetur. Naturalia enim eum non admittunt doctrinae modum, et quod minus etiam id fieri possit, varias plerumque et inconstans natura et corruptibilium rerum nonnihil fragilis obstat conditio. Sed is certae docendi traditur modus, qui subiectae conveniens est materiae, et qui si non exactam et incommutabilem rerum scientiam facit, peritiam tamen in nobis et elegantem solertiae cuiusdam et eruditionis efficere potest habitum." This habitus is compared to the prudence described in *Nicomachean Ethics* 6.1. **Crippa** 1566, 6 acknowleges a similar point by including as the fifth "membrum" of the doctrine of animal movement a catchall category ("reliqua quae ad has causas et ad totam ipsam motus doctrinam spectare videntur"), which shows that the doctrine cannot be reduced to scientific demonstration.

46. *Historia animalium* 1.6, 490b8 ff.; **Guarinoni** 1601, 312; **Leonico Tomeo** 1540, B1r: (extrapolating on Aristotle): "quaecunque eadem animalibus insunt specie diversis, ipsa vero nullam habent differentiam: [Aristoteles] asserens quaedam ex accidentibus illis communibus, et affectionibus nullam inter se habere differentiam, cum de diversis tamen dicantur speciebus, ut univocam de illis praedicationem nobis ostendat." See also **Fabricius** 1555, 13.

47. Perfetti 1999b, 313 (citing Leonico Tomeo).

48. **Taurellus** 1596, 13: "de genere specie et differentia propre omnia sibi vendicat logicus" (not the natural historian).

49. **Scaliger** 1592, 139 (140.1) (on Cardano's failure to distinguish plants by divisions relating either to "stirpes" or to "perennatio"; his divisions are said to be neither "subtiliter inventae" nor "philosophicae legibus enarratae"). Scaliger's own classification ("divisio") of plants is taken from Theophrastus, see **Scaliger** 1592, 463 (139). **Cardano** seems implicitly to accept that classification will be the result of an inductive process (1663, 9.487: "ex experimentis pluribus fit experientia").

50. See **Crippa** 1566, 1; **Zabarella** 1586–87, 1.64.

51. On Melanchthon's more modest position on these questions see Kusukawa 1997, 422–423.

52. See e.g. MS Ambrosiana 8v: "nos vero meliori modo agemus, facilioreque ac magis copiosa via procedemus: quandoquidem per accidentia neque semper per opposita animal dividemus. Cernimus enim differentias per vitas, per actiones, per mores ac per partes, quorum quaedam sunt accidentia"; also **Guarinoni** 1601, 32.

53. Jensen 2000, 195; see also Atran 1990, 18.

54. E.g. **Nifo** 1547, 407: "in substantia genus et differentia sunt eiusdem naturae in specie; in accidentibus genus et differentiae sunt diversarum naturarum aliquando, quia genus est in praedicamento substantiae, differentiae in praedicamento accidentis." **Leonico Tomeo** (1540, A3v) links this directly to description: "in primo enim res ut sunt et sensibus substernuntur, exponi manifeste videntur, nullis earum expressis causis, nullisve rationum persuasionibus concinnatae, sed nudis tantum verbis simpliciter descriptae, et veluti lineamentis quibusdam ruditer effigiatae. Quae certe pars non ab re de animalium historia inscribitur, cum narrantis quasi historici personam in illis commentariis philosophus plane agat." In physiognomy, it is even disputed whether visible accidents are informative of character or not; see Maclean 2002, 317n (quoting **Fontanus** 1611). On the link between physiognomy and zoology, see Perfetti 2000, 103–104 and F. **Bacon** 2000a, 94–95.

55. **Furlanus** 1574, 258–259; "perspicuum [est] Aristotelem non damnare divisionem, quam et septimo Metaphysicorum libro, et secundo Posteriorum Analyticorum in praedicatis, quae rei naturam explicant, inveniendis indicat necessitatem"; **Taurellus** 1597, 79: "nos rerum essentias non videmus: idcirco saepe cogimur differentiis accidentalibus esse contenti"; see also MS Ambrosiana, 8r–v.

56. **Zabarella** 1586–87, 1.64–66, where the phrase "propria accidentia" is used to describe both what traditionally had been classed as propria (e.g., the ability to laugh in man) and accidents (movement). Galen refers in his works to an *idion symptoma* which has much the same implications as "proprium accidens." See also Aquinas, *ST* 1a 45.1 (on the distinction of "forma accidentalis" and "forma substantialis"), and Thomas Harriot cited by Clucas 2001, 200 on whiteness as "accidental form."

57. **Guarinoni** 1601, 119. The dichotomy is constructed as follows (I have used a different but equivalent textual layout; see Maclean 1992, 40n):

1 animantia ob escam

1.1 partim aquatica quae aquam

1.1.1 recipiunt, ut pisces branchiis muniti

1.1.2 non recipiunt, quaeque aera

1.1.2.1 captant

1.1.2.2 non captant

1.2 partim terrestria quae aera

1.2.1 admittunt

1.2.1.1 alia ab aqua

1.2.1.2 alia a terra

1.2.2 non admittunt {ves[p]ae, apes} insecta omnia

This dichotomy can be contrasted with the topical divisions of **Textor** 1534 and **Fabricius** 1555 (see above, 165).

58. Of all these commentators, Pomponazzi is the only one to question the attribution of dichotomy to Plato; see Perfetti 1999a, 449.

59. **Leonico Tomeo** 1540, E4v-6r; but at F3v-4r, he produces an argument against accidental differences: "animadvertendum autem hoc in loco esse censeo, et alibi, cum de differentiis genus dividentibus verba facimus de essentialibus semper intelligere differentiis. In his enim absurdum est neque fieri potest, unam et eandem speciem in contrariis differentiis reperiri, et ab illis constitui. Nam in accidentalibus differentiarum modis hoc esse constat, et nullum sequitur inconveniens. Si enim animal per album et nigrum dividitur, sub utraque sane differentia et hominem et equum, multasque alias animalium species sine controversia reponi clarissimum est." "Sine controversia" here means, I believe, "per experientiam."

60. **Nifo** 1546, 19.

61. **Nifo** 1547, 406–408.

62. **Taurellus** 1597, 78–91. By claiming (90) that "ut enim species aliqua recte definiatur: necesse est ultimam haberi differentiam. Si namque centum pluresve differentiae sint: et desit ultima: definitio nulla erit. Haec nempe sola proprium rei definitae essentiam ostendit," he implicitly accepts that other forms of descriptions are possible, but cannot be scientific. Cf. **Crippa** 1566, 2: "animalium genus ex variis, diversis ac mutiplicis partibus, quae non omnibus notae plane sunt, tanquam ex materia, et ex anima, tanquam ex forma constitu[it]ur."

63. **Schegk** 1556, 105–120; he wonders (120) whether Aristotle may simply be contradicting himself on what is meant by "ultimate" difference. See also **Schegk** 1584, 874–878.

64. **Cesalpino** 1593, 16–20 (not "tanquam verum ex veris" but rather "tanquam absurdum ex pravo definiendi modo"), lucidly set out by Jensen 2000. Jensen does not note that Cesalpino also refers to the often-invoked three logical conditions set down in *Posterior analytics* 1.4, 73a28 ff., which relate to the types

of proposition that can appear in a demonstration; they help determine the difference between the scientific and topical orders of division and provide here the link back to definition (of universals) by genus and differentiae illustrated in Porphyry's tree.

65. On these questions of interpetation, see Maclean 2002, 206–233.

66. **Scaliger** 1592, 597–599 (182); Jensen 2000, 194 (on Cesalpino), 189 (on Monteux).

67. **Leonico Tomeo** 1540, B3v-4v: he refers to the four (material, formal, efficient, final) causes as relations ("in quo"; "quod"; "a quo"; "propter quod"), declares the formal cause in animals to be the soul, and "finis causarum omnium causa." The material cause is identified with the passage in the *De partibus* on (absolute) necessity (B6v): "appellat [Aristoteles] causam hanc [materialem] necessitatem, quoniam necessario substernitur omnibus; et cum aliae vel fiant, vel faciant, haec a principio perstat, inestque, et tum ante finem et formam, tum post illa semper consistit et remanet." See also **Furlanus** 1574, 83–84.

68. **Du Laurens** 1600, 4–5: "Epicurus...hominum corpora casu et fortuito facta affirmabat"; **Leonico Tomeo** 1540, B6v, C6r-7v (mentioning Democritus on *nous*, atoms and form), C3v: "in naturalibus igitur nihil temere fit, sed omnia praecedentes habent causas."

69. Jensen 2000, 195.

70. See **Leonico Tomeo** 1540, G7v; **Guarinoni** 1601, 66.

71. Theophrastus *Metaphysics* 6, 10b5 ff. (the editio princeps is in the fourth volume of the Aldine Aristotle of 1497), cited by **Valleriola** 1577, 45. On this passage see Lennox 1985. There is an entry in **Brasavola** 1556, 312v which reads "natura aliquid fecit frustra" (referring to *De anatomicis adminstrationibus*, book 9), but ten other loci are cited that refute this. Galen opposed the view that there is redundancy in nature, which he associates with Erasistratus: see **Galen** 1991, 95 (Hankinson, citing *De naturalibus facultatibus* 2). On Aristotle's support for the view that there is nothing superfluous in nature, see *De generatione animalium* 2.4, 739b19 f. and 4.4, 770b28 ff.; on Aristotle's acceptance of attenuated purposelessness in nature (in respect of horns), see above, note 12. See also *De generatione animalium* 4.4, 77035 ff. See also Maclean 2002, 247–251.

72. **Dullaert** 1521, 28r: "dico quod accidens non dicitur inseparabile: quia non possit separari sed quia difficulter separari potest: quare nigredo corvi est accidens inseparabile ad illum sensum: et sic cepit Porphyrius.... coloratum dico ipsum accidens et non proprium: licet quod neotericus in sua libri Porphyrii expositione dicat nigrum proprium esse respectu corvi et coloratum respectu animalis quemadmodum et ly risibile est proprium respectu hominis. Sed contra hoc argumentum proprium et sua propria passio debent converti (loquimur secundum viam nominalium de qua ipse est) sed supernaturaliter stat esse corvum non nigrum: ergo illi termini non convertuntur: scilicet nigrum et corvus. Consequentia tenet. Etiam non omne nigrum est corvus, nec omne coloratum est animal: ergo illi termini non sunt proprium quarto modo quod tamen prefata opinio dicit. Similiter risibilitas est res risibilis; ergo color est res colorata. Quare

aliter dico ly coloratum est accidens et dicitur inseparabile ad sensum prius declaratum." On the accidents of the host, see **Dullaert** 1521, 27v. Dullaert died in 1513 in his forties.

73. On metaphysics in Catholic contexts, see Lohr 1988; in Protestant contexts, see Sparn 1976; on debates about the Eucharist, see Leijenhorst and Lüthy 2002. See also **Crippa** 1566, 4v (on the plurality of souls); **Fonseca** 1604, 195, 224, 228–229, 263, 484 (on the relation of accidents and propria to substantial forms and the Eucharist), 355 ff. (on definition, division, genus, and species, comparing *Metaphysics* 7.12, with *Topics* 1.5 and *Posterior Analytics* 2.12–14, but not *De partibus*); **Rhaedus** 1609; **Cramerus** 1601, 65–66, 94, 115 (on species and differentia, multiple differentiae, and latitude of forms); **Rubeus** 1618, 64–68, 266–267.

74. **Rhaedus** 1609.

75. See **Fonseca** (1604, 365) on the Spanish medical professor Christophorus a Vega, and whether hair and nails have souls.

76. See Maclean 2002, 140–145.

77. Siraisi 2002a; see also Julius Caesar Scaliger in **Aristotle** 1584, 19: "omnis morbus, ut nunc loquuntur Philosophi recentiores, aut est per essentiam, aut per communicationem. Galenus in primo de locis affectis male affectis *idiopatheian* et *sympatheian* appellari iubet. Nos ita dicere possumus, morbum esse secundum causas membrum afficientes in quo is morbus est: alium autem per consensionem alterius membri, qui cum membrum aegrotum naturae alicuius vi coaptatum sit." For a more general account of the difficulties of classifying disease, see Maclean 2002, 259–269.

78. *De methodo medendi* 1.3, K10.20 ff.; **Argenterio** 1610, 1463: "quantum enim difficile sit bene dividere, docet Arist. *initio Operis de partibus animalium*, tantoque profecto difficilior nobis est dividendi ratio: quid nemo hactenus ex antiquis scriptoribus enim satis explicare sit conatus"; **Cardano** 1663, 9.487. Perfetti (2000, 146 ff.) points out that Francesco Vimercate also makes the link between these Aristotelian texts and Galen's *De usu partium* and his *De placitis Hippocratis et Platonis*.

79. See Maclean 2002, 143 ff.

80. Maclean 2002, 137–138.

81. Aristotle *Physics* 3.1, 200b29–201a9; Galen *De sectis*, K1.64 105.

82. This involves the distinctions "simpliciter" and "ut nunc," potency and act, and the four qualities described in *Categories* 8, 8b25 f. (state and condition, natural capacity, affective qualities, and shape or form); see Maclean 2002, 139–140; 256–258.

83. Joutsivuo 1999; cf also *Categories*, 12a10–12; *De generatione animalium* 2.1, 733a34 ff.; the analogy with shades of color in Galen *De sanitate tuenda* 1, K6.14; Blair 1999, 139 ff. (on Bodin).

84. **Agricola** 1967, 26 ff.: in the chapter on division, Agricola reproduces the standard passages from Aristotle and Porphyry, and in the chapter on genus

and species there is a brief mention of animals: "ut si dividere brutum velimus, dicamus aliud in terra, aliud in aere, aliud in aqua vitam ducere, vel item dicamus, bruta omnia aut natare, aut volare, aut ingredi, aut repere." In his commentary on this passage, Alardus (**Agricola** 1967, 30) suggests that the source of this might be *Topics* 5.4, 133b6 ff., where the same elimination is implied (man is biped and pedestrian; the genus bird is biped and winged; there are animals which are quadruped and pedestrian); but the point is not made there explicitly. Jensen (2000, 188–189) points out that Agricola is cited by Fuchs as a source on "adiacentia" and accidents (**Agricola** 1967, 106 has a definition used by Fuchs ["quod abesse rei, et adesse, ut non corrumpatur res, potest"]); but a phrase almost identical to this is also found in Porphyry, who is most likely the common source. Kusukawa 1997, 422–423 suggests with greater force that Melanchthon is a more likely source for Fuchs, citing persuasive textual echoes. See also Moss 1996.

85. Such an enterprise would also have to take into account the change over time that enquiries into nature undergo. Pinon 1995, 4 ff. gives a useful periodization ("la période des correspondances"; "l'enregistrement de la nature par l'image"; "la période d'exploration; la période académique").

86. Cunningham (1985, 198) argues that Fabricius was engaged on "an open-ended research programme on animals devoted to the acquisition of true causal knowledge ('scientia') on certain types of topic ... such as parts, organs and processes, and employing a thought-through and consistent methodology and epistemology, a suitable technical vocabulary and the like."

87. "Res certe infinita est: et quae augeri perpetuo possit, praesertim si quis generibus non contentus species quoque omnes persequi velit": cited by Rath 1950–51, 159–160.

88. For the relationship of the following points to definitions of "historia" see POMATA in this volume, 111–112.

89. E.g., *Historia animalium* 6.12, 566b23 (on the longevity of dolphins); *On Breathing*, 478b1–2, quoted by Cunningham 1985, 219.

90. **Gessner** 1551, β2v; **Werdmüller** 1555.

91. **Aldrovandi** 1599, 5r. The distinction between "demonstratio ostensiva" and "scientia demonstrativa" is clearly relevant here. See also **Dioscorides** 1518, AA3r (on physicians being "oculati testes" of medicinal plants).

92. **Odonus** 1563, and Perfetti 2002, 436–437. See also POMATA (in this volume, 121–122) for the similar treatment of Harvey's *De generatione animalium*.

93. Kusukawa 1997; also Jensen 2000, 189.

94. **Rondelet** 1554–55, 3: "cum perutilis necessariaque sit piscium cognitio, et illa ex eorum differentiis pendeat, quae qualesque sint nunc recensere oportet, si prius admonuerimus differentiae nomen hic latius patere, ut omne id quod aliud ab alio quoquomodo differre facit, significet, sive id accidens sit, quod possit, vel non possit separari, sive proprium. Cum enim verae differentiae, maximeque propriae paucissimae sint et ad inventum difficilimae, in tanta penuria, antiquorum philosophorum exemplo ad alias nobis confugiendum fuit, quibus ita perspicue

et dilucide res indicantur, ut earum distinctio pernosci possit, etiamsi differentiis maxime propriis careamus: nam ut scribit Arist[oteles] etiam accidentia ad cognoscendum quis res sit, valde conferunt."

95. See van den Broek 1991, 67 (citing L'Obel).

96. Reeds 1991; see also **Wotton** 1552; Cunningham 1985, 198 (on Fabricius).

97. **Textor** 1534; **Fabricius** 1555. His medical terms come from the list of non-naturals (food, drink, sleep and wakefulness). Textor (d. 1556) was a pupil of the Parisian anatomist Jacques Dubois (Sylvius).

98. **Gessner** 1551, γ1–3 (the sequence, after an illustration of the animal, runs: "nomina; regio; actiones; mores; usus; alimenta; remedia; philologica et grammatica"); Siraisi 1994b, 83 (on Hippocrates). **Gerard** 1633 is very similarly disposed; even **Aldrovandi** 1599 has strong echoes of this disposition. It should however be noted that the title of Gessner's *Icones animalium* (1553) refers to them being "per certos ordines digestae."

99. Jensen 2000; Reeds 1991, 19.

100. Reeds 1991, 120–127. The only allusion Bauhin makes in the preface to his method refers to a previous version of the *Pinax* of which he says: "plantae, methodice secundum genera et species, habita potissimum ratione formae externae, distributae habentur" (**Bauhin** 1623, *4v).

101. Jensen 2000, 192–194; Maclean 2002, 249–250. **Crippa** (1566, 6) has a residual category of causes of movement.

102. Siraisi 1994b; see also **Du Laurens** 1600, 6.

103. **Eustachius** 1564, 51, 123, 129, 131 "natura in hominibus autem frequenter quasi omni lege soluta libera voluntate utens"; "variam et admirabilem eiusdem naturae artem." The phrase "lege soluta" is adapted from *Digest* 1.3.31, on which see Maclean 1992, 91.

104. F. Bacon 1858–74, 1.159, 171–172, 173–179, 225, 282, 283 (i.16, 60, 61, 66; ii.2, 29, 30). Also in **F. Bacon** 1878, 56–57; F. Bacon 2000b, xxxviii–xxxix.

105. F. Bacon 2000a, 86–87.

106. On Bacon's knowledge of Aristotle and Boethius at Cambridge, see Durel 1998; on his knowledge of the zoological works, see Wolff 1977, 178–180. The reference to bones as beams is found in Galen *De anatomicis administrationibus* 1.2, K2.220, 226; it may be that Bacon is quoting from an intermediary such as Andreas Vesalius, the first sentence of whose *Humani corporis fabrica* of 1543 repeats the analogy.

107. See also **F. Bacon** 1858–74, 168–170 (*Novum Organum* 1.51, 57).

108. F. Bacon 2000a, 87; Biow 2002, 144.

5

Antiquarianism and Idolatry: The *Historia* of Religions in the Seventeenth Century

Martin Mulsow

"Most Experts usually agree that all history has its beginnings in fables and that they enjoy a close relationship. Fables, however, as everybody knows, received their name from beans, or *fabae*."[1] This is why it seems fitting to begin a discussion about history with beans and, more precisely, with the very sage who had advised against their consumption: Pythagoras. Whether Pythagoras did in fact spurn beans remains doubtful; arguments can be put forward that contradict this. The ancient Egyptians, after all, viewed them as divine.[2]

Baldassarre Bonifacio, who began his *Historia ludicra* wittily with this story about beans, was not the only one who had something to say about the religious history of beans. Gerhard Johannes Vossius, the famous polyhistor from Amsterdam, also mentioned them in his monumental *De theologia gentili*. He talks about them in book V, where he treats the cultic significance of plants. Under the scornful heading "abusus" he tells us about "faba refriva," the bean that the Romans brought home at the beginning of the year in order to obtain a positive omen.[3]

Vossius and Bonifacio composed their books at approximately the same time, during the 1640s. Both were the product of an overdeveloped culture of excerpting, which was then widely practiced by scholars, especially the historically-philologically learned polyhistors. But the historical material on the religious history of beans plays different roles in the two books. Bonifacio's work is an example of erudite pursuit of curious learning. His work emulates the *Lectiones antiquae* of his fellow countryman Coelius Rhodiginus, an entertaining compilation of the type that developed in the times of Aulus Gellius.[4] Bonifacio's beans then are in a true sense "fruits of reading." The "histories" that draw on these fruits are odd fragments of the erudite tradition; they are rich in punch lines, and occasionally they were laid out in a frivolous way deliberately intended to amuse and at the same time to educate the reader.

Vossius's treatment of beans seems different because at first sight it appears to fit the larger framework of *historia naturalis*. With respect to its overall structure, Vossius's monumental work is based on a passage in Paul's letter to the Romans, 1.22–23: "Claiming to be wise, they became fools, and exchanged the glory of the immortal God for images resembling mortal man or birds or animals or reptiles." Paul's text inspired Vossius to organize his encyclopedia of false religious worship according to the different parts of nature. Starting with the worship of the celestial bodies, it continues with the worship of animals and plants until it arrives at the treatment of the worship of elements.[5] The basic thesis is that idolatry constitutes the erroneous worship of a second cause in place of the worship of the first cause, God. Idolatry means worshipping something created rather than its creator. An accurate understanding of nature can correct this error and by doing so restore the original possibility of a knowledge of God. In that way, history of religion coincides with *historia naturalis*.

In what follows, I intend to examine the contexts within which scholars pursued the history of religions in the sixteenth and early seventeenth centuries—a period in which the scholarly obsession with curious facts, individual observations, and descriptions supposedly unburdened by theory reached a high point. I will examine the ways in which this study was connected to both natural history and antiquarianism. The essay begins by examining the problematic status of scholarly interest in paganism in an age of confessional strife. It then traces the ways in which antiquarianism replaced mythography and examines the special problems of reference that affected historians of the pagan religions.

Historia, Fabula, Genealogia

Taking a closer look at how Vossius wedded history of religion with *historia naturalis* results in disappointment. Though Vossius lived in a time when natural history and its subdisciplines, botany and zoology, were developing rapidly, he quotes mainly from Pliny.[6] Sometimes he draws on the authority of Aristotle, thereby primarily relying on the interpretations of Julius Caesar Scaliger, as he does in the case of beans.[7] Hence the work is like Bonifacio's: a product of bookish culture, not of the new empirical sciences. But a major difference remains: Vossius treats the history of religion within the framework of a specific discourse on idolatry that, largely influenced by his own work, developed during the seventeenth century.[8] By contrast, the connection of religious myths with *fabulae*, which was favored by Bonifacio, was still motivated by traditions that reach back much further: in particular, the kind of history of religion that used the corpus of the ancient traditions about the gods as a treasury of stories. Depending on one's own position, one could either discard these as fairytales or interpret them in a moral or allegorical way.

How did the status of *historia* change in the course of time, as the new method replaced the old? Our initial glimpse has shown that it is too simple to say that the older variant treated *historia* in a narrative sense, whereas the more recent form viewed *historia* as textually independent knowledge of individual facts. The relationship appears to be much more complicated, and the reason for this difficulty lies not least in the precarious status of the study of pagan religions.

Christian mythographers had always been well advised to apologize to their readers in advance when they intended to write about pagan gods. Lilio Gregorio Giraldi, a humanist scholar in Ferrara, begins his comprehensive *De deis gentium varia et multiplex historia*, written in 1548, with a declaration that his report of the "superstitiones" of the ancients should not be accused of impiety, and he emphasizes: "unus inquam, unus Deus est."[9] Mythographic *historia* in Giraldi's sense belonged to a long tradition that stretched back to texts from late antiquity and the Middle Ages by authors such as Fulgentius, Albricus, Phornutus, and Palaephatus[10] and that gained a new form above all in the work by Boccaccio. Boccaccio called his mythography *Genealogiae deorum*. With this title he indicated that the descent and relationships of the ancient gods served him as a guiding principle, according to which he intended to instill order into confusingly diverse textual traditions.[11] Within the discourse about paganism, "genealogy" is a comparatively neutral semantics. Other, less neutral types of semantics at the mythographer's disposal included the *superstitio* of the ancients, their *idolatria*, and finally their *fabulae*. Bonifacio drew on this last. Emphasizing the "fabulous" element was sometimes quite appropriate, especially when an author wanted to stress his orthodoxy in this minefield of research, as the Jesuit François Antoine Pomey of Lyon made clear with the title of his work of 1659: *Historia deorum gentilium fabulosa*.[12] All of this seems to suggest that mythography deals above all with the inventions of the human imagination.

On the other hand, this kind of apologetic gesture could give the impression that the study of mythography was insignificant in comparison to more serious disciplines such as law or the natural sciences. Traces of rivalry among scholars gathering around their patrons are apparent—for example, in the case of Girolamo Aleandro the younger, an antiquarian and philologist from Rome. He had to defend himself in the dedication of one of his books to Cardinal Odoardo Farnese against the accusation that his study of pagan "fables" was nothing but a form of idle play, a "res ludicra." Bonifacio adopted this same terminology in a more assertive way. Aleandro also tried to regain his initiative and to go on the offensive, not by practicing mythography as a playful game, but rather by emphasizing its underlying sincerity. It would be equally possible, he claimed, to turn the tables on his rivals and to hold that any study that does not directly deal with the highest creator of all things would be just a game.[13] This is exactly what he focused on in his book *Antiquae tabulae marmoreae Solis effigie . . . explicatio* of 1616.

Here was the other possible strategy: not to stress the differences between pagan religions and Christianity and play the former down as fables, but to emphasize their similarities with Christian monotheism. It was especially helpful in this case if one studied the sun, as Aleandro did in his book. For sun worship—especially in Rome during these years and even more when the Barberini, whom Aleandro would serve as a secretary, came into power—tended to be considered as an indication of an esoteric monotheism practiced by the pagan peoples of antiquity. Aleandro proceeded even more aggressively at the end of his dedication, where he alluded complacently to Galileo's work *Delle macchie solari*, which had been published two years earlier by the Accademia dei Lincei ("Academy of the Lynxes"). "Human beings in our time are so lynxlike, that they even discover spots on the surface of the sun. Why then should I not be allowed to treat the sun in my writings as well?"[14] More precisely: If Federico Cesi's Academy sponsors the study of the sun, why then not the Farnese?[15]

The spectrum from playfulness to profound religious sincerity was very broad—or at least it appears broad today, though the concettistic spirit of the Roman academies and the openness of baroque Catholicism (especially as practiced by the Barberini) concealed it from contemporaries. Anyone who applied the categories *historia sacra* and *historia profana* had to draw a sharp line between access to cognition by means of natural knowledge and through revelation. This distinction became blurred, however, since the *vestigia* of divine knowledge appeared among the pagans. This is why during the Renaissance, the Sibyls, Hermes Trismegistus, or Orpheus enjoyed high esteem as representatives of a *prisca theologia*. Aleandro was still able to refer to that tradition, although it had become subject to increasing criticism from the late sixteenth century on.[16]

In northern Europe, philosophers reflected on the correct division of the sciences. Given that one accepted God, man, and nature as the basic areas of existence—as Jean Bodin did—a division of the sciences into *historia divina*, *historia humana*, and *historia naturalis* suggested itself.[17] The ontologically defined *historia divina* could then be identified with the gnoseological *historia sacra*, but it could also be understood in a more neutral way and not only refer to the realm of Scripture and theology.[18] Hence the historian and philologist Johann Jakob Beurer of Freiburg attempted to explicate *historia divina* by means of Zwinger's definition of *historia* as *cognitio singularium*. He eventually arrived at the conclusion that *historia divina* takes the statements ("sententias") of ancient writers, sacred ones as well as profane, about God and divine affairs as the subject of its study.[19] The one particular thing ("singulare") that becomes the subject of *historia* is thus a statement, and not a fact as it is in the *historia naturalis*. For classificatory reasons, Beurer admits in addition to the Bible pagan texts as well, and accepts statements about God as well as about "divine affairs." The scope of his definition seems at first glance to fit into the gray area of the history of religion as it developed around 1600.

To be sure, *historia divina*, even as Beurer understood it, does not add up to a *historia religionum*. What category would such a *historia* belong in? Would it form part of *historia humana*, since pagan religions were only fables of human origin? No one investigated this set of problems explicitly. Rather, they crystallized in these decades out of widespread scholarly practices, not from efforts to classify the sciences. The research program, which Beurer's methodological essay of 1594 could not really define, was in fact created by practices such as antiquarianism, travel accounts and philological investigations. Even if scholars still carefully distinguished between *religio vera* and *religio falsa*, they took the totality of religious phenomena more and more into account. A look at their practices reveals that they were already going beyond Beurer's distinction between the individual facts of the *historia naturalis* and the individual statements of the *historia divina*. The antiquarians now brought into play not only the *sententiae veterum* on the gods, but also statues, reliefs, and medals, all of which had to be interpreted. When travelers wrote about the religions of the New World, they offered direct observation of cults based on autopsy, not only *sententiae*, and their information no longer came entirely from the *veteres*.[20] The relationship of this developing scholarly study of religion, which still took very disparate appearances, to *historia naturalis*, which was also developing and changing,[21] remained unclear—too much so for contemporary terminological distinctions to express it. In both, elements of transmitted textual knowledge continued to exist, and in both there was increasing discussion of individual facts,[22] as well as of material objects and pictures.

Historia played a complex role in the transformation of the study of religion from the mythographers' "Genealogiae deorum" to the scholars' monographs "De origine et progressu idolatriae," and here I can only sketch it briefly. The change took a direction which was intended only by few, but which had been feared by many. Increasing scholarly study of foreign religions in their cultic contexts and attempts to present them as a corruption of the true original religion—a failed cultural transfer from Judaism—created a sense of distance toward Christianity and made it seem relative, since it was increasingly viewed as just one among other religions.[23] If a scholar could unmask ancient oracles as priestly deceptions, then he could just as well ask if his own religion were not based on frauds perpetrated by the priesthood, too. Once there was no longer a clear line between the texts of the *historia sacra* and the *historia profana*, scholars could come to believe that the sacred writings also originated in a profane context as well. But these are long-term consequences, which became apparent predominantly only during the second half of the seventeenth century.

Antiquarianism and Allegory

In what follows, I prefer not to dwell any further on beans; instead, I would like to move from natural objects of worship at a very low level to the highest level, the worship of the sun and of heavenly bodies (fig. 5.1). In the course of the Renaissance, the reception of Neoplatonism and Hermeticism as well as the new theory of Copernicus had provided the sun with an exalted position.[24] When Italian scholars such as Pirro Ligorio, Girolamo Aleandro, Lorenzo Pignoria, or Giacomo Filippo Tommasini began studying the cults of Isis and Osiris, the Attis myth or Mithraism, for the most part in the company of collectors such as Cassiano Dal Pozzo or as *familiares* at court, they were aware of the connection between these cults and the practice of worshipping the sun.[25] How did these scholars interpret their material? Mythographers before them had not hesitated to repeat over and over—though with an increasing apparatus of knowledge about antiquity—late antique allegorical interpretations of the myths about the gods. Around 1600, philological and antiquarian knowledge tended more and more to destroy these old allegories.

One antiquarian who specialized in the history of religion was Lorenzo Pignoria, a scholar from Padua. In 1605, when he interpreted the "Tabula Bembina," a metal cult object dedicated to the goddess Isis, he abstained from using the common allegorical speculations. "I approach this matter under your best auspices," he wrote to his friend Markus Welser, "and I do not want to try to explain the panel allegorically [*allegorikos*], but to do so far as I can on the basis of the narratives of the ancients [*ad veterum narrationem*]. More than anyone I hate the exaggerated interpretations of these things, which are mostly entirely inappropriate to them and which the Platonists have introduced as confirmation of their uncertain fables [*labantium Fabularum*] (and with too little regard to their master): I prefer admitting my ignorance to being a nuisance to my learned reader."[26] When Pignoria musters *narrationes* against *fabulae*, he means in effect texts and stories from antiquity, which he sets in opposition to allegorical interpretations. *Historia* as *vera narratio* stands in opposition to *fabula* here not because of the difference between sacred and profane history, but as a source and simpler form of writing as opposed to overinterpretation. Pignoria belonged to the long

Figure 5.1
Gerhard Johannes Vossius, *De theologia gentili* (Frankfurt, 1668): worship of the heavenly bodies.

line of scholars from Padua who were trained as Aristotelians and distanced themselves from the methods and views of the Platonists. His friends Gianvincenzo Pinelli, Paolo Aicardi, Prospero Alpino, and Melchior Guilandini were equally interested in Egyptian culture. Alpino had traveled to Egypt, and in his book *De medicina Aegyptiorum* he described different cures that he had encountered there, instead of relying on the fables transmitted in the books.[27] A particular "culture of fact" developed in that milieu. Like those in Ferrara (see CRISCIANI and GRAFTON in this volume) and in Rome (see SIRAISI in this volume), this milieu still needs a close description.[28]

Autopsy was Pignoria's basic principle as well. He often uses the first person in his writings. "I have seen a large piece of marble on the Piazza Campidoglio in Rome in 1606," he says in one instance where he describes a Mithraic inscription;[29] or he mentions the names of such informants as Peiresc, Aleandro, Herwart von Hohenburg, and Markus Welser. This personalization, localization and temporalization of knowledge and testimony evidently corresponds to the new style of *historia*, in which the credibility and verifiability of individual observations are essential. Pignoria's interpretation of the Isis table abstains from attaching any deeper meaning to it. Instead he proceeds, through digression after digression, in the manner of a detective. By using comparisons[30] he identifies the gods portrayed, and he follows clues such as the position of their feet, which he then illuminates with passages from the Aristotelian *Problemata* on the curved feet of the Egyptians, quoting from the learned exchange with Antonio Querenghi on the same problem.[31] In the study of his residence Pignoria kept a model of an ancient foot.[32] Indeed, Pignoria even draws on Prospero Alpino's eyewitness account of the commands that the Egyptians used for their dogs and which he identifies as "Tao, Tao" rather than the "To, To" that Pignoria knew.[33] This difference not only suggested a change of the diphthong "au" to "o" and a change of the name of the god from "Taaut" to "Thoth," but it also helped to verify Plutarch's account that dogs were called by the name of the god Hermes ("Thot"). This again illuminated a curious fact reported by Aelian, that the dogs in Memphis shared their prey and—as Horapollo said—gazed with open eyes on the images of the gods.[34] The site of Isis's death was in Memphis, and Pignoria suspected that the flooding of the Nile, dogs, and Isis's search for Osiris were connected. To prove this he produced the evidence of a coin from the collection of the Venetian patrician Giovanni Mocenigo (fig. 5.2). This supported Heliodorus's statement that the river was revered as a god. According to Aelian, moreover, a dog helped Isis search for Osiris, and it was well known that the rise of Sirius, the dog star, caused the Nile to rise. Finally, a Phoenician report transmitted by Pausanias claimed that the tears of Isis explained the flooding of the Nile. These materials yielded both the conclusion that Isis and the dog were connected with the Nile and an explanation for the fact that dogs were at the head of the procession of Isis.[35]

In this sort of *historia* of religion, which combined textual evidence with numismatics and eyewitness descriptions, scholars had to be aware of the hypothetical nature of their own theses. Leonard Barkan has demonstrated how the use of ancient fragments that had been unearthed—parts of statues, remains of reliefs—created a space of uncertainty and sparked insoluble discussions. This is the sphere in which we have to place the *cognitio singularium* of Pignoria or Peiresc (see MILLER in this volume).[36]

Is it possible to identify the appearance of this new form of *historia* in the development of the history of religion with the end of allegorism? Does it consist of a hypothetical discussion of individual facts with no further "grand narrative"? These statements pose too sharp a break, and the difficulties—as well as the charm—of our subject lie particularly in this circumstance. Allegorism did not come entirely to an end in the era of antiquarianism,[37] nor did antiquarians such as Pignoria abstain from putting individual things into larger perspectives. Rather, the antiquarian transformation of the old mythographers unleashed a new set of theories and hypotheses. The origins of pagan unbelief, polytheism, and idolatry now became urgent problems.

The Origins of Idolatry

One day in the summer of 1571, the natural philosopher Bernardino Telesio and his student Antonio Persio were engaged in a conversation in Telesio's home in Naples on the nature of light and sun when Telesio's colleague Quinzio Buongiovanni entered. Buongiovanni was a quarrelsome Aristotelian and very quickly a heated debate developed about

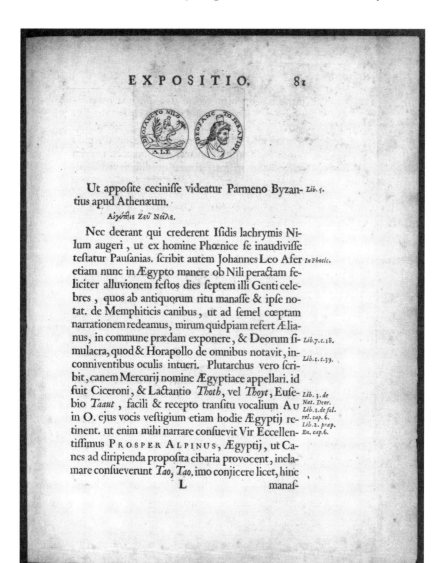

Figure 5.2
Lorenzo Pignoria, *Mensa Isiaca* (1670).

whether the sun is warm or not. According to the Aristotelian Buongiovanni, it could not be warm, because it belonged to the supralunary sphere beyond the sensible qualities; while Telesio claimed that the sun was a hot body, whose rays of light transmit heat directly to the earth. When the discussion eventually reached a dead end, Telesio took Buongiovanni's arm and led him outside in front of the house into the heat of midday. "Can't you feel the heat of the sun?" he yelled at Buongiovanni. But the latter refused to accept the evidence.

What matters to us at this point is less how argument, evidence, and the theory-ladenness of observation interacted here than how this dispute was solved. The student Persio thought that it was about time to diffuse the tension with a gallant quotation, which he threw into the debate. He quoted the verses from Ovid that describe the human being as a creature that looks up to the stars—and he earned relaxed laughter from both sides by doing so.[38]

During these very years, Vincenzo Cartari had developed an "anthropological" awareness of the basic human disposition to religion. His *Imagini delli dei de gl'antici* appeared in a second edition in the year of the dispute at Telesio's house. Right at the beginning of the book Cartari says, "Man lifts his eyes up to the sky as if he sensed naturally that everything good comes from there."[39] What does he see in the sky? He sees the sun, the moon and all the other celestial bodies. Though he knows that God is infinite and invisible, it is all too easy for him to mistake the heavenly bodies for the gods themselves. This explains, as Cartari concludes, agreeing with Plato,[40] why the celestial bodies were objects of worship among the early Greeks and among other peoples as well.[41] Man's inclination toward religion, which originated in his upright posture, is at the same time the origin of his inclination to idolatry.

Lorenzo Pignoria commented dryly on Cartari's anthropological speculation: animals, he said, have occasionally been ascribed a sense of religion as well.[42] He may have been thinking of Camerarius's emblem book, where elephants are accorded a sense of God and the stars (fig. 5.3).[43] But he certainly had also in mind the Egyptian religion, which considered several animals as sacred. Does this then mean that Pignoria was just a destructive critic, and that he rejected any assumptions about the origin of religious cults? In fact, carefully and in remarks scattered throughout the work, he develops a very comprehensive and innovative

The Historia of Religions in the Seventeenth Century 193

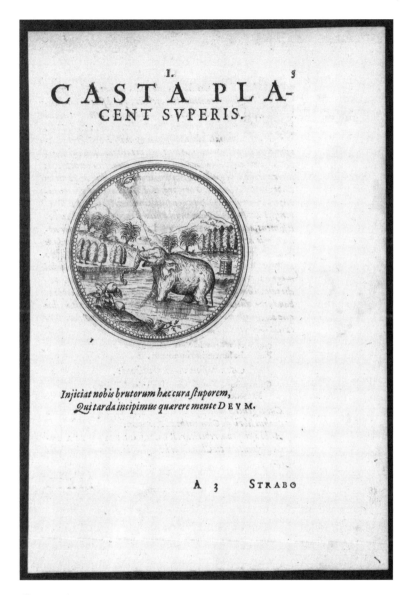

Figure 5.3
Joachim Camerarius, *Symbolorum & emblematum ex animalibus quadrupedibus desumptorum centuria* (1595): emblem with a "pious" elephant.

theory of the history of religion.[44] Pignoria believed that Egypt was the place of origin of superstition. This learned man, who was educated by the Jesuits and had portraits of cardinals Baronio, Bellarmino, and Cobelluccio in his study,[45] applies the terminology of *superstitio* and uses a conservative semantics along orthodox lines, especially when he makes drastic judgments and even talks about the "deceptions of the devil."[46] Without explicitly mentioning Lactantius (as Athanasius Kircher did after him), who saw the origin of idolatry in the curse made by Ham, the son of Noah, which had separated him from the roots of the true religion,[47] Pignoria nonetheless agreed with him on the negative evaluation of Egypt. The Egyptians, who were deprived of "the shining splendor of divine things," were for him a people who had fallen into darkness.[48] He found the beginnings of all religion in crude mystery religions such as the cult of Isis, the Bacchanalia or the cult of Eleusis, and he speculated that many authors had passed over this early period in silence because these cults were "immoral and cruel."[49] For Pignoria, the level of civilization as a normative entity became a criterion of evaluation. His study, in which statues of the gods and inscriptions, gems and crystals, Lombardic manuscripts, Indian plants, and marvelous seashells were heaped up on every surface, also contained bronze tablets showing the fights that took place during the Bacchanalia.[50]

From Egypt, so the theory goes, "this Egyptian superstition spread throughout almost the entire world, even though the sages frowned upon and attacked it."[51] Taking up Benito Arias Montano's speculations about the journeys of the ancient Hebrews to Peru, allegedly the legendary land of gold of Ophir, Pignoria even hypothesizes a possible spread of the Egyptian religion to Central America and East Asia.[52] Similarities between depictions of exotic gods and Egyptian gods seemed to confirm his view.[53] But this was only one part of Pignoria's allusive history of idolatry. The other part referred to the continued life of Egyptian superstition in the post-Christian world. Pignoria knew that cardinal Baronio had used a gem for illustrative purposes in the second volume of his *Annales ecclesiastici*, where he treats the year 120 and discusses gnostic sectarians, mainly Basilides and Carpocrates.[54] Pignoria reproduced a number of other gnostic gems as well to support his thesis that Egyptian gnosis was nothing more than a muddled continuation of the original Egyptian superstition.[55] A few years earlier, these gems had led

Peiresc as well to conclude that gnosis was part of a long and continuous history of heresy.[56]

The case of Pignoria illuminates early modern "histories" of religion. His diffusionism results from his efforts to keep the history of religion within the framework of *historia sacra*: the distinction between true and false religion preestablishes the conditions according to which he develops his views. This then leads almost inevitably to a diffusionist model, both for Hebrew culture with regards to true religion and for Egyptian culture with regards to false religion. Models such as the one later put forward by Huet, which views "false" religions as corruptions of the one true Hebrew religion, fill out this picture.[57] On the other hand, as we have seen, Pignoria retained a sharp distinction between *narratio* and *fabula* in the sense of the relationship of sources to interpretations. This distinction poses problems for the larger framework. In this case, the distinction between biblical and pagan sources does not initially play a role. This is the elementary practice of the antiquarian, who examines clues without engaging in any speculations.

Pignoria's frame of orientation, which draws a clear line between true and false religion,[58] could be called "exclusive." In addition to this framework, there was another, more "inclusive" one: it softened this strict distinction by pursuing the traces of "true" religion—that is, the underlying monotheism—within the realm of "false" religion. Pignoria's friend Girolamo Aleandro pursued this perspective further, and in 1616 he demonstrated that an antiquarian history of religion could also be written on the basis of belief in the *philosophia perennis*. This meant the return of Platonism and the return to a controlled form of allegorism, which would rely above all on ancient models of allegorization. Hence, this was not a mere speculative interpretation of antiquities, but an attempt to put them in relation to contemporary texts.

This approach was exemplified by interpretations of the excavations of ancient Mithraea in Rome,[59] which became more comprehensible only when the correct texts, such as Porphyry or Macrobius, were used as keys to illuminate them. The use of these works on cults of the sun and on Homeric allegory made earlier allegorical misinterpretations obsolete.[60] The interaction of the excavations with the culture of the Barberini court also played an important role. Since the Barberini had chosen the sun as the representative symbol of their power,[61] it was natural that scholars such as Holstenius and Aleandro highlighted the solar hymns of Julian or Proclus.[62] This gave Aleandro's form of antiquarian *historia* a greater degree of relevance; as suggested above, he could now claim with reason that he was dealing with the most important thing in the world: the worship of the one and only God, in the guise of the sun.

A good friend of Aleandro and Pignoria was the young John Selden, a lawyer, politician, and expert in oriental languages, who published his book *De diis Syris* in 1617. Here he examined on a broad philological basis (including Arabic and Rabbinic sources) the Syriac-Canaanite gods mentioned in the Bible.[63] By listing individual "syntagmas," in each of which a certain complex of names of gods was discussed, Selden modeled the structure of his book on Giraldi's approach. Like Giraldi, Selden used a basically philological method. Unlike Pignoria, he relied more on texts than on artifacts. But he avoided the title *Historia*, which Giraldi had still used. Selden avoided laying too much stress on the exclusive or the inclusive semantics. Making efforts to maintain his neutrality, he claimed that he wanted to examine the "origins of polytheism." He contrasted the "exclusive" observance of the polytheistic concept among the Syrians and Egyptians with an "inclusive" reconstruction of the esoteric monotheistic religion among the sages. But this no longer included the statement that pagans had received divine revelation in the sense of a *prisca theologia*. Instead, Selden held that the esoteric religion had developed through the "study of the secrets of nature."[64] Astronomical observations and reflections on the circularity of the celestial spheres had, according to Selden, led these sages to develop a philosophical conception of God as the One, who remains unchanged in all change. But they had not disseminated these complex secrets, which were too difficult for the common man to understand.

This is how Selden created a theory of religious development from the esoteric ideas of the neoplatonically inspired *philosophia perennis*, which he somehow secularized. In addition, in Selden the influence of Maimonides becomes apparent for the first time. This complements the view that pagan religion originated in the cult of the celestial bodies, which was widely held. Eusebius had already speculated in his church history that after the Fall and an initial period of primeval atheism, true religion developed among the Hebrews, whereas an astral cult developed on the other side as well.[65] Maimonides by contrast provided more precise historical arguments by naming the religion of the Sabians as the original form of astral religion.[66] Apart from that, he set the origins of idolatry as early as the antediluvian times of Enosh, thereby claiming even more basic relevance for this problem. The influence of Maimonides became more powerful in 1641, when Dionysius Vossius translated the section *Avodah Zarah* of the *Mishne Torah* as *De idololatria* and his father utilized these theories in his *De theologia gentili*.[67] From that point on, the semantics of "idolatry" takes over in the works on the history of religion.

Problems of Reference

Scholars like Aleandro, Pignoria, or Selden pose a twofold problem with their study of idolatry. First of all, we have to ask, to whom did the idea of a pagan solar monotheism refer, from a historical point of view? Secondly, to whom or what did pagan worship refer if an idol such as the sun were venerated?

Though the scholars of the early seventeenth century were aware of and intensively discussed the second problem, they were only vaguely aware that the historical point of reference was also problematic. Even in the beginning of his interpretation of the marble tablets, which featured the sun god with numerous attributes, Aleandro uses Homer as his source of reference. The "progenitor of all philosophers and poets" had taught that the highest numen, which we call God, not only dwelled in heaven but was distributed throughout the entire "universitas rerum," and that he would fill everything with his majesty and presence.[68] This reference to Homer as an authority of course breathes the spirit of Homeric allegorism, as favored by Stoics and Middle- and Neoplatonists. Aleandro also quotes Maximus of Tyre and, to mention a Christian writer as well, Justin Martyr as witnesses that Homer had indeed believed in one single God as supreme "moderator." After that, however, appears the central reference for the identification of this God with the sun: the first book of Macrobius's *Saturnalia*, and especially the speech that Macrobius puts into the consul Praetextatus's mouth.[69]

This speech seemed to show that the Greco-Roman gods, especially Apollo, Bacchus, Hercules, or Mercury, referred to the sun. The attributes of these deities and semideities were depicted on the marble relief.

According to Aleandro, they referred to the elements over which the single numen presides: Bacchus to water, Apollo to fire, Hercules to earth, Mercury to air. They referred also to the four seasons, which again were determined by the sun.

Yet Aleandro did not become uncritical and unhistorical. In the second part of his book, he distances himself from allegorism and, by citing Tertullian, makes clear that it was precisely those pagans who felt uneasy about their fables ("fabularum puderet") who had developed the natural-philosophical interpretation of the gods. Like Pignoria, he tried to be on the safe side, calling this type of sublimation of paganism into a philosophical religion the devil's work—a judgment that reflects the rejection of monotheistic tendencies among pagans by many early Christians.[70]

Aleandro thus did not simply continue the allegorical tradition, but used it for an understanding of cultural documents.[71] The allegorism and sun worship that flourished in the Barberini circle had elements of reflection and criticism. The antiquarians of this circle pointed out that the creator of the relief was guided by philosophical-allegorical convictions; but evidently they remained fascinated themselves by the philosophical perspectives on the discoveries of the relief and became inspired by the Neoplatonic heritage.

The more monuments like the solar relief were discovered, the more objects of the cults of Isis, Cybele, or Mithras surfaced, the more the complex world of late antique intersection of religions became the focus of attention and posed new mysteries to its interpreters.[72] The Rome of late antiquity had been "a place for all deities": imperial claims and conquests had created a form of universalism, which again influenced Greco-Roman religion.[73] When Aurelian, for example, conquered Palmyra in 272–273, the statues of Sol and Bel were brought to Rome to adorn the new temple of Sol. Whether Porphyry already propagated a cult of the sun scholars continue to debate,[74] but it is certain that both Julian and Sallustius cultivated a comprehensive solar theology.[75] This same theology appears some decades later in Praetextatus's speech in Macrobius.[76]

The Roman scholars of the third and fourth centuries themselves found it hard to grasp the multitude of deities and their complex relationships with each other. The antiquarianism of the third century was all about cultural monuments, which could no longer be understood prima facie and whose meaning had to be reconstructed. Philological and philosophical authors of late antiquity dealt with texts such as the Orphic hymns, which were many centuries old and from which they tried to extract philosophical meanings, or with what were actually newer texts such as the *Chaldaean Oracles*, to which they, however, assigned the dignity of deep antiquity.[77]

If scholars from the sixteenth and seventeenth centuries predominantly used testimonies of late antiquity such as those of Macrobius, then they exposed themselves to the danger of falling into a "reception of reception":[78] they referred to documents that themselves interpreted a much earlier past. This created such a complex muddle of indirect relationships that it rarely became clear where the primary evidence lay, in early or in late antiquity, especially since the common practice of citing authorities as sources without evaluating them critically tended to conceal rather than to reveal the different layers of reference. John Selden developed special theories to explain the multiplication of names of deities so that he could at least partially disentangle this muddle. Diffusionists assumed that the deities of some cultures derived from those of other cultures, which could mean that the worship of Moses was in fact worship of Bacchus, which again, however, was actually the worship of Osiris, etc.[79] These hypotheses of cultural translation could send any given reference on a complex journey.

But despite all of these dynamic elements, a static element remained in these theories, one that resisted any attempt at a critical disentanglement of the historical layers. Those who accepted, as Aleandro and Selden did, that an allegorically concealed, esoteric monotheism had flourished among educated pagans[80] saw this monotheism as a static phenomenon that stretched over the entire ancient period, rather than as the result of a historical development or one that had occurred at a late stage of antiquity, forming a competing alternative to Christianity. Selden's views bound monotheism to early developments in astronomy and philosophy, but not to the last representatives of paganism. In Aleandro's work on pagan religion, it is not so much natural science that served as a norm. Despite the antiquarian method with which he approaches the matter—or maybe just because of it—the interests of late antiquity are mingled with those of baroque Catholicism, and the religious intersections of late antiquity are mingled with the distant mirror of these intersections in Renaissance philosophy.[81]

These complex circumstances make it hard to determine the status of the *historia* of religion around 1600. There is another difficulty, a problem of reference of no less complexity that needs to be understood before we can bring some light into the jungle of seventeenth-century treatises about the history of deities and idols. What constitutes the point of reference for the worship of a given idol, such as an image of the Egyptian god Anubis, which is partly canine and partly human? The Old Testament did not use a uniform term for "idols." Several of the expressions used signify "nothing" or "something insignificant." The Septuagint translates many of these expressions with "eidolon," which has the flavor of the "void" of a treacherous being.[82]

Carlo Ginzburg and others have emphasized that the late antique distinction between "idols" (*idola*) and "images" (*similitudines*), which made the medieval cult of images (as opposed to a cult of idols) possible, was influenced by contemporary Aristotelian and Stoic attempts to organize the universe in accordance with the categories "Being" and "Non-being."[83] A certain reference to a particular idol can be only an apparent reference, as is the case with imaginary compounds (those, for example, which are partly human and partly canine). Their individual parts have a point of reference, but the wholes do not.

A related problem was that occasionally what seemed to be idols were in fact only bearers or servants of the invisible deity. Since one could not see an invisible god, one could easily misunderstand these representatives as substitutes for the real God. This means that the golden calves that Jeroboam built in Dan and Bethel were possibly just pedestals for the invisible God, who sat enthroned upon them.[84] Hence one seventeenth-century scholar defended Aaron's worship of the golden calf as orthodox. In his view, by some act of mental reservation Aaron was able to worship the true God in the calf, which the mass had forced him to venerate.[85]

If these were the problems of reference, which opened up an abyss behind the entire discussion on idolatry, then what in fact served as a point of reference for sun worship, which so fascinated antiquarians during the early seventeenth century? Nothing? The natural, created body of the sun? Or God, for whom the sun constitutes a meaningful representation? This ambiguity was not confined to the interpretation of the past. It also manifested itself in the neopagan practices of some Renaissance intellectuals, such as Pomponio Leto or Campanella, as well as in several New World religions.

Aleandro pondered all of these questions as well. The sun, he claimed, constituted only an image for the intellectuals and sculptors he studied; in reality, it referred to God. The Greeks had had to capture the one numen, which Plato called the father of all gods, not only with their intellect, but with their visual sense as well. "Therefore, since nothing more beautiful or more useful appeared to the sight, they mistook the sun itself for God."[86] Even the apologist Minucius Felix, who had still ridiculed pagan idolatry in his *Octavius*, pointed to the special position of the sun, which is present everywhere from heaven down to the earth. Even Pliny, who made fun of the fables about the gods, appeared to doubt if there were a true God besides the sun.

Having covered himself against possible objections, Aleandro followed the Catholic theology of imagery, which permits the veneration of "simulacra" if linked to God in the correct way.[87] On the other hand, he also followed the syncretism of the *philosophia perennis*, which assumed a universal knowledge of God in all cultures, even among pagans. Worship could then be a complex symbolic act, and its true meaning could only be uncovered within its particular cultural context.[88]

Surprisingly, these very "baroque" and "Catholic" interpretations exercised a significant influence on deism, which was then taking shape. While Gregorio Panzani, nuncio of Pope Urban VIII, stayed in London in 1635, he met Herbert of Cherbury and talked with him about his philosophy and the Catholic faith. Herbert had heard that Urban liked his work *De veritate*.[89] When Herbert then read Vossius's *De theologia gentili* in 1641, he realized that Vossius's presentation of the ancient celestial cults could very well be linked with the solar cult of the Barberini. One only needed, as Aleandro had indicated before him and as was understood in the Barberini circle, to view sun worship as an attempt to refer to the one God by means of some visible form or manifestation: an attempt that might have its limitations, but that could be appreciated in the face of human imperfection.[90] For that indirect relationship, Vossius had introduced the term *cultus symbolicus*. This served to distinguish it from the *cultus proprius*, which referred to the practice where only the object per se was venerated and not the invisible matter it represented.[91]

But Vossius did not regarded sun worship as a merely symbolic cult; the examples that he cites make this clear: "I define as *cultus proprius* what happens when what is worshipped is viewed by itself and in itself as God. The cult of the sun constituted such a cult, whether it be the cult of Hercules or the Theban cult or one springing from a different lineage. I define *cultus symbolicus* as when something is worshipped not because it is perceived as God, but because it denotes God: for example, the cult of the sun in the Vestal fire, or Hercules as depicted in a statue."[92]

Herbert, then, needed to reverse Vossius's arguments if he wanted to interpret the sun cult as symbolical. His reasons for doing so need not concern us here; he struggled not to abandon the pagans as eternally condemned by an unjust God to hellfire, but to present them as capable practitioners of natural religion. What concerns us here is the way in which he reconfigures the references of the individual facts of *historia* of religion. The same historical-philological facts about sun worship that Vossius had accumulated are now rearranged and reinterpreted. Everywhere that Vossius saw an idolatrous cult of Baal or Horus, Herbert uncovered symbolic worship of the true God.

Herbert, who wrote his own hymn to the sun, in a certain way combines Macrobius, Cartari, and Maimonides: he combines the solar syncretism of late antiquity with the investigation of the origin of religion. And he varies the anthropological approach, which we examined in Cartari, and according to which looking up toward the sky constituted the origin of all religion, the true as well as the false ones. Yet Herbert remained more sympathetic to the pagans than Cartari: since they could not believe that God was hiding from them, they worshipped the sun as his noblest representative.[93] Herbert, in contrast to Aleandro, remains close to astronomy and astrology; his demonstration that all deities could be reduced to the God of the sun needs to be understood above all as a reduction of the planetary gods to the sun. Since the planets were identified at a relatively late stage, Herbert argued, their names were derived from the name of the sun—as were their cults from the cult of the sun.[94]

In addition, Herbert tries to make the symbolic character of the worship more plausible by endowing the celestial bodies with souls, thus raising their status. If the sun and stars had souls, so this theory went, then the old argument against idolatry that idolaters venerate low and lifeless matter, which was unworthy of its worship, no longer holds. For Herbert, as D. P. Walker has shown, the stars certainly possessed souls. He does not seem to mind having to manipulate some quotations from Vossius rather massively in order to prove his orthodoxy in this point. He also quotes—without mentioning any names—the work of Adam Tanner, a Jesuit from Ingolstadt, who had discussed the connections between the worship of the stars and the stars' possession of a soul in his *Dissertatio peripatetico-theologica de coelis*. Tanner, however, had rejected the idea that the stars possessed souls, whereas Herbert used this Jesuit's authority in order to suggest that Catholics were on his side in this point.[95]

How can these manipulative strategies be judged? Obviously they spring from a twofold desire to built consensus: consensus with the contemporary Catholic and Protestant literature and, above all, consensus between Christians and pagans in antiquity. Herbert's argumentative strategy accumulates authorities; it perceives in the *consensus gentium* an essential criterion for truth. For philosophical and theological reasons, Herbert reverts to a method older than the comparative approach of *historia* of religion. He looks for consensus rather than emphasizing the differences between Christians and pagans. As a consequence, he abandons the skeptical-hypothetical method that scrutinized each fact individually, and instead subjected his sources to a method that leveled them by filing away apparent differences. This turns out to be crucial, for only the clear distinction between true and false religion had made the comparative approach possible and opened up the field for comparisons in the *historia* of religion.

My goal in this article has been to show that the efforts to create a *historia* of religion, one in which individual observations and individual facts played a major role, fell within a time of dramatic intellectual change. There was no single established discourse that united travel literature, antiquarian decipherment, and philological investigations, and there were different possible terminologies (*fabula, superstitio, idololatria*). For the same reason the term *historia* was rarely used, although the practices of inquiry into religion clearly belong to the early modern culture of fact and description designated now by *historia*. Above all, however, the conceptual frame and the reference of the descriptions of individual facts remained unclear: after discarding allegorical, moralizing mythography, a free space emerged in which it became possible to debate whether idols referred to a single God, whether the idolatrous popular religion of pagans was subject to a monotheistic religion of learned men, and whether the allegorism of antiquity in fact referred to that kind of monotheism. But at the same time, even as the *cognitio singularium* became more precise, the reference of these *singularia* became blurred, and its framework became more ambiguous and multilayered. This lack of clarity had massive consequences for deism and libertinism, but that is another story.

Notes

I am grateful to Ulrich Groetsch and Tony Grafton for the translation of this paper from German, as well as to Peter Miller, Jonathan Sheehan, Glen Bowersock, and the Historia group for discussions. Ralph Häfner was so kind as to give me the opportunity to read his *Habilitationsschrift* (see note 62) before its publication, through which I became acquainted with Aleandro and Pignoria.

Sources cited in boldface type appear in the Primary Sources of the bibliography.

1. See **Bonifacio** 1656, 1. Bonifacio (1585–1659) was a learned jurist and papal ambassador from Rovigo. This passage follows Caput primum: "De Pythagora." On the relation between *historia* and *fabula* see Bietenholz 1994, though he does not mention Bonifacio. Bonifacio plays with the meaning of "historia" as "quicquid vere graviterque explicatur" (see POMATA in this volume, 108–109) when he adds "ludicra" to his title, thus producing an oxymoron.

2. **Bonifacio** 1656, 36. On the religious history of the bean see Klauser 1954.

3. **Vossius** 1668, vol. II, lib. V, 96: "quasi faba esse pignus frugum, ac quia fabam retulissent, etiam frugus relaturos domum." On Vossius (1577–1649) see Rademaker 1981.

4. See **Bonifacio** 1656, 1; **Coelius Rhodiginus** 1542. On the reception of Aulus Gellius, see Häfner (forthcoming).

5. See Graevenitz 1987.

6. On the reception of Pliny see Borst 1994; Nauert 1979; and PINON and OGILVIE in this volume.

7. Scaliger 1557.
8. As an introduction is useful Schmid 1990; Stroumsa 2001.
9. Giraldi 1580, 2 (separate pagination).
10. See Seznec 1953.
11. See the recent edition, **Boccaccio** 1998. See in general Kany 1987.
12. **Pomey** 1659.
13. **Aleandro** 1617. The original edition appeared in 1616. On Aleandro (1574–1629) see Pélissier 1888; Lhote and Joyal 1995.
14. **Aleandro** 1617, fol. Aiii v: "Sunt homines hac tempestate adeo lyncei, ut in Sole maculas deprehendant. quid ni meis in scriptis quamvis de Sole loquentibus?"
15. On the Farnese see Zapperi 1994; on the antiquarianism of the Farnese circle see Herklotz 1999, 214 ff. On the Accademia dei Lincei see Redondi 1983; Olmi 1992.
16. See Wind 1958. On the criticism of this tradition see Mulsow 2002.
17. See **Bodin** 1566.
18. See Seifert 1976, 111 f.
19. **Beurer** 1594, 16: "Historia divina intelligatur, quatenus ea litteris prodita sententias veterum tum divinorum hominum, tum prophanorum de Deo rebusque divinis in sese continet, non quatenus ea in formam scientiae et facultatis ex eadem tandem per certa quaedam theoremata redacta est." On Zwinger see the contribution of Blair in this volume.
20. See Rubiés 2000.
21. See in general Findlen 1994; Grafton and Siraisi 1999; Olmi 1992.
22. See Daston 2001b.
23. See Ginzburg 1998b. The process in its full-fledged extension in the eighteenth century is described by Manuel 1959.
24. See *Soleil à la Renaissance* 1965.
25. For the details of these connections see Herklotz 1999; on Ligorio see Schreurs 2000. On antiquarianism in general see the classical account in Momigliano 1950; see further Da Costa Kaufmann 2001.
26. **Pignoria** 1670, 1 f. The work appeared for the first time in Venice in 1605. On Pignoria (1571–1631) see Volpi 1992; the contemporary biography is **Tommasini** 1669.
27. **Pignoria** 1670, 53 mentions Alpino's journey to Egypt. On Alpino see Siraisi 2002b.
28. See Shapiro 2000. Shapiro limits her thesis of the juridical origins of the attention to "matters of fact" to England, with its Common Law system. But it might be fruitful to explore the Aristotelian university culture in Italy in connection with its juridical milieu at specific places (Pignoria, e.g., was trained in law by Guido Pancirolo, Angelo Matteacio, Ottonello Discalcio, and

Marcantonio Ottelio, and in philosophy by Francesco Piccolomini and Jacopo Zabarella) looking for other kinds of cultures of fact. See in this respect De Renzi 2001.

29. **Pignoria** 1647, 293: "Io vidi in Roma l'anno 1606 un gran pezzo di marmo, nella piazza di Campidoglio."

30. On the new culture of comparison see Miller 2001a.

31. **Pignoria** 1670, 52: "In proxime sequenti Osiridis simulacro status pedum notandus, ceteris quoque huius Tabulae Imaginibus communis. sunt enim omnes varis pedibus, & in se contractis. cuius rei ratio ab Aristotele petenda, qui in problematis quaerit, cur Aethiopes, & Aegypti blaesi sunt." It follows a philological discussion about the meaning of the Greek term *to blaisoi*, an eyewitness report by Prospero Alpino on feet in Egyptian imagery, and on 54–60 the comments of Querenghi on this subject. See Aristotle *Problemata* 14.4. On the importance of the *Problemata* see Blair 1999. See as well the account that Carlo Ginzburg has given of the history of the "clue paradigm" (Ginzburg 1989).

32. See **Tommasini** 1669, 87.

33. **Pignoria** 1670, 81 f.

34. Aelian *De natura animalium* 7.19.

35. Aelian *De natura animalium* 10.45; Heliodorus *Aethiopiaca* 9; Pausanias *Phocica*. See in general Krauss 1980; Fischer 1980.

36. Barkan 1999, 131: "historia" in the old sense is replaced by autopsy. On Peiresc see Miller 2000.

37. Don Cameron Allen has described those interpretations that replaced the older allegorical method, while still ascribing meanings to symbols, in the "Symbolic Interpretations of Renaissance Antiquarians"; see Allen 1970, 249–278.

38. Biblioteca Corsiniana, Rome, MS. Linceo VII, fol. 351r ff.: Antonio Persio, *De natura ignis et caloris*, XI.29. See Mulsow (forthcoming): "male igitur Poeta ille? ego in excandescentem Quintium protuli ridens, Pronaque cum spectent animalia caetera Terram Os homini sublime dedit coelumque tueri Iussit, et erecto et ad sydera tollere vultus." See Ovid *Metamorphoses* 1.84–86. On the scientific contexts see Mulsow 1998.

39. **Cartari** 1647, 1: "l'homo alza gli occhi al Cielo, & spesso anco le mani insieme giunte, quasi che naturalmente senta, che di là su viene ogni bene." On Cartari see Volpi 1996.

40. Plato *Kratylos* 397c-d.

41. **Cartari** 1647, 2.

42. **Pignoria** 1647, 291.

43. **Camerarius** 1595, 9. On this book, see the contribution of OGILVIE in this volume. I am grateful to Brian Ogilvie for the reference.

44. See Seznec 1931.

45. See **Tommasini** 1669, 83.

46. **Pignoria** 1670, 10, where the successes in healing in the cult of Isis are ascribed to a cunning of the devil: the devil did this in order to get the cult dis-

seminated. The "conservative" semantics are typical for the seventeenth century. While in the Middle Ages from Isidore of Seville on there was a considerable tolerance toward the pagan gods—as heroes that brought culture or as benefiters of humankind (see Seznec 1953, 14 ff.)—in early modern times it was replaced by a much more critical attitude. Seventeenth-century apologetics revived the apologetical arguments of the second to fifth centuries.

47. See Lactantius *Institutiones divinae* 2.14. According to Lactantius, Cham fled to Canaan; one could, however, as Kircher did, see in Cham or his sons the founding fathers of Egypt. On Kircher see Pastine 1978; Findlen 2004.

48. **Pignoria** 1670, 7: "Fateor id quidem, & ingenti Dei beneficio accidisse credo, ut prae oculis homines haberent, in quas tenebras inciderent illi, divinarum rerum illustri splendore destituti."

49. **Pignoria** 1670, 8: "Hujus silentii ea caussa erat, quod haec vel turpia vel crudelia essent, qualia Eleusinia, Pessinuntia, Adonia, Isaica, Bacchanalia, Ityphallica, Omophagia, & Mithratica fuerunt, quae exagitare consuerverunt Ecclesiae Christianae indefessi Praedicatores." As a recent view on the origins of religion in cruel cults see Burkert 1983. The ancient source is Cicero.

50. **Tommasini** 1669, 88: "Tabellae aeneae, quae typos exhibent pugnarum Bacchanalium."

51. **Pignoria** 1670, 10: "notandum est, Aegyptiacum superstitionem, exagitatam licet, & sapientiam contumeliis vexatam, per Orbem fere universum diffusam fuisse." As source, Diodorus Siculus *Historia* 1.2 is cited: "ubique fere terrarum Isidem cultum."

52. Pignoria, "Seconda parte delle imagini de gli dei indiani," in **Cartari** 1647, 361–400, esp. 361 f. On the debates on Solomon's journey to Ophir see Gliozzi 1977.

53. See Miller 2001b, esp. 203.

54. **Baronio** 1597.

55. **Pignoria** 1670, 73 f. and the Auctarium 84 f.; the illustrations are on 86 ff. The collection of gems was run by Natalino Benedetti, and Peiresc had helped Pignoria to establish contact with him.

56. See Miller 2001a, 74 f. on Benedetti and Pignoria, and 89 on Peiresc.

57. See **Huet** 1679; on Huet see Dupront 1930.

58. Jan Assmann has labeled this distinction the "Mosaic" one; see Assmann 1998.

59. See Lanciani 1989, 238.

60. On the misinterpretations see Panofsky 1960, 96 ff.; Saxl 1957, 13–44.

61. See Scott 1991.

62. See Häfner 2003.

63. **Selden** 1668. The first edition appeared London 1617. On Selden (1584–1654) see Mulsow 2001; Miller 2001b, esp. 193–200.

64. **Selden** 1668, 72.

65. Eusebius *Historia ecclesiastica* 7.2.1; *Demonstratio evangelica* 4.8.2.

66. On this hypothesis see Elukin 2002.

67. **Maimonides** 1641. See on the context of this translation Katchen 1984.

68. Aleandro 1617, 3.

69. Aleandro 1617, 4: "At prae ceteris Macrobius primo Saturnaliorum pluribus argumentis evincere contendit, deos omnes ad Solem referri, idque unum esse numen, quod sub varia nominum appellatione gentes venerentur." See Macrobius *Saturnalia* 1.2.

70. Aleandro 1617, 53: "Scitum est, quod animadvertit Tertullianus, Ethnicos, ubi fabularum puderet, quas de Diis commenti fuerant, ad interpretationem rerum naturalium confugisse, ad dedecus suum ingenio obumbrasse. Diaboli ea fuit versutia, uti specie lucis excaecaret, quemadmodum aiebat Lactantius."

71. So Pignoria in his letters to Aleandro, e.g., about the bees, which according to him had served in antiquity as a symbol for the Muses. See **Pignoria** 1628. At the same time, the Accademia dei Lincei was concerned with the natural history of the bee; see Freedberg 2002.

72. See Barkan 1999, 168 ff.; on Peiresc see Aufrère 1990; on Mithras as Sol Invictus see Bibliothèque Nationale, Paris, MS. Dupuy 746, Girolamo Aleandro, *Sol Invictus*, fol. 218 ff. On Cybele see **Pignoria** 1669. See as well Miller 2001a.

73. See Fowden 1993, 37–60.

74. See Buffière 1956, 535 ff.; Lamberton 1986, 108 ff.

75. See Athanassiadi 1992; the polytheist aspects of the solar theology, however, are emphasized by Smith 1995.

76. See Liebeschütz 1999.

77. Even from today's perspective it is difficult to discern whether the late antique projections of the past hit something right. After all, early documents such as the Jupiter hymn already display a certain amount of "monotheism." See West 1999.

78. I take the concept "Rezeptionsrezeption" from Enno Rudolph.

79. See, e.g., **Tenison** 1678, 126: "First, That Moses was the ancient Egyptian or Arabian Bacchus. Secondly, That Bacchus was the Egyptian Osiris. Thirdly, That the ancient Egyptian Bacchus and Osiris was no other than Apis." I am grateful to Jonathan Sheehan for the reference.

80. See **Selden** 1668, 67 ff.

81. On Renaissance "syncretism" see, e.g., Farmer 1998; Schmidt-Biggemann 1998. On the problem of the use of antichristian neoplatonic thinkers in this "syncretism", see Häfner 2003, 114: "Man brauchte kein Libertin zu sein, um derart scharfe Kritiker des frühen Christentums wie Porphyrios oder Salustios im Horizont dieses Humanismus zu promulgieren, solange die integrierende Kraft des römischen Katholizismus noch so mächtig fortwirkte, wie sie sich unter der besonderen intellektuellen Konfiguration der Herrschaft der Barberini erwies."

82. See Fredouille 1981, 846 ff.

83. See Ginzburg 1998b. See also Ladner 1983; Freedberg 1989.

84. See Albright 1949, 298 f.

85. **Monceaux** 1605. I am grateful to Jonathan Sheehan for this reference. On the problem of mental reservation see Zagorin 1990.

86. **Aleandro** 1617, 2 f.

87. See Hecht 1997. Hecht refers to works such as **Gregorio de Valencia** 1580.

88. Even today it is open to discussion how in each case sun worship was meant. In antiquity it certainly could be both worship of the sun itself and worship of the highest god by means of the sun; see Fauth 1995, 32; Liebeschütz 1999, 186.

89. Rossi 1947, 2:490–499.

90. **Herbert of Cherbury** 1663, 19: "Adeo ut quamvis superius Sole numen sub hisce praesertim vocabulis coluerunt Hebraei, Solem, neque aliud numen, intellexerunt Gentiles, nisi fortasse in Sole, tanquam praeclaro Dei summi specimine, & sensibili eius, ut Plato vocat, simulachro, Deum summum ab illis cultum fuisse censeas: quod non facile abnuerem, praesertim quum Symbolica fuerit omnis fere religio eternum; uti qui non solummodo hoc in illo, sed aliud ex alio colerent."

91. Vossius's terminology seems to have been adapted by **Casali** 1646.

92. **Vossius** 1668; I cite here from the edition Frankfurt 1668, 1:30.

93. **Herbert of Cherbury** 1663, 20.

94. **Herbert of Cherbury** 1663, cap. V: De nominibus Planetarum ad Solem ductis, 29: "Quum non valde antiqua circa quosdam Planetas fuerit Gentilium scientia, vel saltem non valde universalis, aliquot Planetarum nomina ad Solem reducta olim fuisse ostendemus."

95. Walker 1972, 178 f. and 186 f.; **Herbert of Cherbury** 1663, 40; **Vossius** 1668, 170; **Tanner** 1621, 65 f.

6

Between History and System

Donald R. Kelley

> Man, in a word, has no nature; what he has is... history. Expressed differently, what nature is to things, history, *res gestae*, is to man.
> —José Ortega y Gasset (1941)

"History" as the study of past experience has usually been set in the perspective of the humanities, especially of Renaissance humanism and the "art of history" (*ars historica*) associated with the liberal arts (*studia humanitatis*), but this is not the only story to be told about the fortunes of the term and concept *historia* and its semantic neighbors, and so of the associated genre and discipline of history in a modern sense.[1] In the early modern period, "history" had a variety of meanings drawn from different sources. In grammar, "history" referred to the interpretation of authors (*enarratio auctorum*), as contrasted with the order and arrangement of words and the art of speaking correctly (*ratio loquendi*).[2] This was a distant linguistic echo of the old opposition between content and form, which has haunted the idea and the semantics of history across the ages and most especially in the early modern period. In some contexts history was virtually identified with rhetoric, and it also had close associations with poetry, from which indeed, according to Cicero, it was descended. "History" also referred to the most elementary, literal mode of medieval exegesis, as distinguished from figurative interpretations.[3]

From the present point of view, *historia* occupies a vast semantic field between the Herodotean coinage of the term (or the appearance of the root in Homer) and modern derivations, vernacular spillovers, and innovations, such as "historicism," whose champions and critics accept (though very selectively) much the same perspective. Although it has generated its own literary and critical genres, *historia* has invaded many areas of knowledge, memory, and imagination, serving to indicate facts, acts, examples, cases, anecdotes, and conglomerates of information on particular—or general, but anyway definable—subjects. It has often accommodated, though it has not required attention to, the dimension of time. In any case it should be kept in mind that the term is largely a phenomenon of print culture and so of the nearly modern period, for only in books has "history" left significant traces for posterity.

What did "history" mean in the sixteenth and seventeenth centuries? Though it had accumulated a large mass of semantic baggage, it was the most neutral and adaptable of terms. According to the standard dictionary published by Rodolphus Goclenius in 1613, history had four main significations: it was, citing Gaza on Aristotle's history of animals, a simple description without demonstration (*descriptio sine demonstratione*); it was, citing Theodor Zwinger's *Theater of Human Life*, knowledge of the *singulorum notitia, particularis cognitio*; it was, following Cicero, a historical unit (*syntagma seu corpus historicum*) that required more than one narrative, such as the commemoration of antiquity (*Antiquitatis ipsa commemoratio*), the history of the Turkish people, or the history of Cicero's consulate; and finally, history was understood, for example by philosophers like Galen, as observation, knowledge derived from personal experience or faith in the senses.[4] Another more neglected semantic area is the association of *historia* with religious ritual and ceremonial, and in this connection Du Cange, who ignores the classical precedents, notes the equivalency of the term with "figure" and "image."[5]

This represents a very rough charting of a semantic field stretching back over two millennia in Greek and Latin, but the later meanings were even more various, as history was both opposed to "science," which dealt with causes, and yet also regarded as a particularly rich form of knowledge rising to philosophical heights, as reflected in the endlessly repeated rhetorical aphorism of Dionysius of Halicarnassus that history is "philosophy teaching by example."[6] It is in this connection that history was regarded as the "mistress of life" (*magistra vitae*), whether in moral or in political terms.

From a modern scholarly perspective, however, the Western story really begins with Aristotle's famous and controversial contrast between poetry and history, which maintains that the latter stands in relation to philosophy as the particular to the general, so that, as an early modern commentator on Aristotle (and author as well of an essay in the *ars historica*) put it, "poetry speaks more of universals, history of singulars."[7] By implication then, as Goclenius wrote, "history is the study of particulars, theory of universals" (*historia particularis notitia est, theoria universalis*). This assumption continues with the early modern usage comprehensively surveyed and analyzed a quarter of a century ago by Arno Seifert in his ground-breaking book on the onomastic and terminological applications of *historia*, which dealt above all with scientific and proto-scientific usage.[8]

The term enjoyed a remarkable semantic expansion from narrative to perception of singular things to experience to other forms of probable knowledge, and yet in a long perspective and in a broad sense the practice and theory of historiography has remained the central and abiding concern, even for Bacon. Classically, *historia* meant inquiry into specific but unspecified things and actions (*res, res gestae*), and soon the distinction between such things and the memories and reports thereof (*narratio rerum gestarum*) became confused, especially in English and the Romance languages. The same occurred in scientific usage, with the ambivalence between things and facts, between observations and descriptions, and between cognition and narration. Yet all these associates of *historia* logically occupied a place under the Herodotean rubric of inquiry.

Historia has its vernacular histories, too, as Karl Keuck showed in his history of the term and such derivatives as "histoire" and "story."[9] In German areas the extraordinary semantic expansion of the term has been traced by Joachim Knape in his history of "Historie" and its affiliates in the medieval and early modern periods in artistic and liturgical as well as literary terms.[10] In general, the ambivalences of "history" are apparent in two often-confused traditions, which may be called the philosophical or theological and the philological. For Augustine history was ultimately the substance of the providential plan as set down in Holy Scripture (itself a "history," forming the basis, for example, of the thirteenth-century *Historia scholastica* and Vincent of Beauvais's *Speculum historiale*).[11] Other less theologically oriented scholars limited the term to narratives of human experience, with Isidore of Seville adding a plausible but spurious etymological reinforcement (*a videre vel cognoscere*), associating "history" with the faculty of sight.[12]

In the first instance history was already embedded in a system of interpretation, explanation, and even prediction; in the latter incarnation "history," contrasted by Quintilian to "fable" and "argument," was a literary art that could be extended to all sorts of phenomena, natural as well as human.[13] Indeed, one remarkable thing about *historia* in the wake of the classical revival was the shift of interest from things (*res*) or even the narrative of things done (*narratio rerum gestarum*) to things written (*res literaria*) and more broadly cultural, as in Polydore Vergil's book on the inventors of things *De rerum inventoribus* (1499), which was an encyclopedic and chronologically arranged survey of all the arts and sciences, starting with the foundational vehicle, "letters" or "literature" (*literatura*), and including history itself.

So the flexible and generic phrase "history of [whatever]" was joined to other disciplines, including philosophy, art, science (and the particular sciences), and literature, with the latter, *historia literaria*, encompassing efforts throughout the encyclopedia and beyond and serving as a bibliographical locus of everything preserved in writing, of course including history itself. Perhaps the most influential of these was D. G. Morhof's *Polyhistor*, which associated the history of literature with Milieu's "history of the universe of things" as well as with Bacon's scheme of knowledge and Gabriel Naudé's plan for a library, and which listed some forty varieties of learned "history," scientific, sacred, and secular.[14]

By the seventeenth century, then, *historia*, with its vernacular derivatives, had become a common principle of organization according to the "natural" order of chronology both for particular disciplines and for the whole encyclopedia of Western learning. Of course the practice of disciplinary histories long preceded the idea of historical progression, as in the lives of artists and writers, the "origins" or "antiquities" of wisdom and philosophy, and other topical or analytical commentaries and monographs; but these scholarly efforts *ante literam* were joined to the nominal traditions under the rubric *historia* and in bibliographical and historiographical retrospect were carried down to the electropresent.

In the sixteenth century the term *historia* also expanded in a scientific context in connection with "natural history," on the model of Pliny's work, as well as the history of medicine, anatomy, and other naturalist disciplines, where history as the "knowledge of what exists" (*cognitio quod est*) and "knowledge of singulars" (*cognitio singularium*) reinforced the notion of neutral "things" and "facts" and "particulars," independent of theoretical or even formal considerations. Within this tradition, too, "history" expanded from empirical fact to connected narrative in the form of what were later called "case histories," which were the medical equivalent of *exempla*.[15] This represents an aspect of the radical empiricist tradition that is largely set apart from the dimension of time and human values. It underlies the famous Rankean definition of history as "what really happened" beyond the domain of the immediate observer and the *histoire événementielle* practiced by positivist—and "scientific"—historians at the turn of the last century and deplored by their later critics.

"Natural history" was popularized by Bacon in his program to reform knowledge and to root out the powerful remnants of Aristotelian philosophy, which still gave form and technical vocabulary to the field. The phrase was common in many areas of natural science, including zoology, botany, medicine, and anatomy, but with exceptions such as Oviedo's *Natural History of the Indies* (1523), it did not achieve wide currency until the 1540s, beginning especially with the work of Gessner (1541) and Fuchs (1542).[16] There was also significant semantic interchange in the work of physicians who wrote history both in the field of medicine and those of human affairs and antiquities (fig. 6.1).[17] Of course, not every author employing the vocabulary of "history" was aware of the whole semantic range that could be found in the sixteenth century.

Terminologically, *historia* came of age in the fifteenth and sixteenth centuries with the modern tradition of "arts of history" (*ars historica*, a genre of literary criticism parallel to the *artes poetica*, *rhetorica*, and *logica* and later *critica* and *hermeneutica*), which, drawing on ancient precedents, was explored first by Italian scholars, who were concerned above all with how to write history in classical style. The dominant themes were pleasure (*voluptas*), utility (*utilitas*), and especially truth

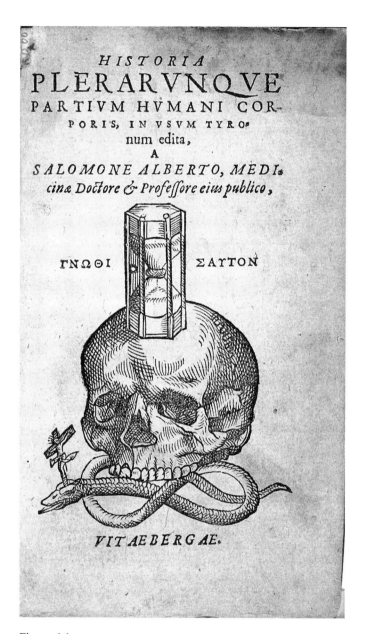

Figure 6.1

Title page, Salomon Alberti, *Historia plerarunque partium humani corporis* (Wittenberg, 1583): moralized anatomy links anatomical *historiae* with symbols of the passage of time and human life.

(*veritas*); but of these truth, or accuracy, was primary: this was Cicero's "first law of history" (*prima lex historiae*), which distinguished it from both rhetoric and poetry. To these may be added memorability, although even in the "arts of memory" this did not really come in for critical discussion and was in effect relegated to authorial choice. This is a genre which has still not been explored fully, perhaps because, however truthful, it has neither utility nor pleasure.[18]

The conceptual limitations of history arose from its concerns with the particulars of human time and space, but humanists made a virtue of this philosophical failing, so that Lorenzo Valla could argue that "the discourse of historians exhibits more substance, more practical knowledge, more political wisdom, more customs and more learning of every sort than the precepts of any of the philosophers."[19] This is a literary— a grammatical and rhetorical—variation on the philosophical theme that history was restricted to particulars, but with a favourable twist.

In the mid-sixteenth century, scholars, especially French and German, shifted emphasis from the writing to the reading of history, from the "art" to the "method" of history (in a pedagogical sense), and from its literary appeal to its interdisciplinary utility and even, paradoxically, philosophical value. Within fifteen years three pioneering books were published on this subject: Christophe Milieu's *Writing the History of the Universe of Things* (1551), François Baudouin's *Institution of Universal History and Its Conjunction with Jurisprudence* (1561), and Jean Bodin's *Method for the Easy Comprehension of History* (1566).[20] These were all included, along with other ancient and modern contributions, in Johannes Wolf's anthology.

The little-known Swiss scholar Milieu, who brought a Platonic sort of holism to the scattered materials of history, was probably the first author to employ the term "history of literature" (*historia literaturae*) and to have a critical conception of this genre. He repeated the commonplaces of the *artes historicae* but transcended the tradition by combining them with a historicized and encyclopedic vision of Vergil's proto-cultural history. "Moreover, the use of history is spread through every branch of learning," he wrote; and he concluded that the best way to organize knowledge was not through the reason or ordering of philosophy but through the reason of history (*historica ratio*), which in effect meant chronology. Thus history was extended not only to human society but also to the whole hierarchical "universe" of nature and culture.

In his book Milieu distinguished five grades of human experience, subjecting each to a historical "narrative" (*narratio*) with the following historicized formulas: *historia naturae, historia prudentiae, historia principatus, historia sapientiae,* and *historia literaturae*. For Milieu these rubrics indicted the structure of the story of what he called "the progression to wisdom" (*progressio ad sapientiam*) from the lower stage of unreflective and instrumentalist "prudence," originating in wonder at the heavens, to the subsequent "invention" of the first disciplines, beginning with astronomy, which he called "the principal cause of the investigation of Philosophy," and followed by the invention of writing, poetry, history, including natural history, geography, geometry, optics, architecture, music, and the practical arts.[21] This ordering followed "the historical progress of the disciplines" (*historica disciplinarum progressio*) from the age of fear and wonder down to the age of reason, erudition, and civilization. The last "transition" in Milieu's scheme was the historical development of the disciplines from the history of wisdom embodied in the late medieval university to the level of literature in the age of humanism. "To wisdom is joined, as in a partnership," he wrote, "the history of literature, by which we pass from the times of brilliant and learned men to the state of civil society."

Baudouin also displayed an encyclopedic impulse, but his specific agenda was to defend a particular alliance between history and the science of law, with the further "pragmatic" purpose of contributing to an ecumenical and irenic program of reconciling the contending religious confessions on the threshold of the wars of religion.[22] For Baudouin the value of jurisprudence arose not only from its critical view of evidence but also from its ideals of a universal humanity based on pious legal traditions, which gained legitimacy in the early Church and which suggested ways of bringing confessional peace to sixteenth-century Europe. From this perspective the course of *historia*, and especially *historia perfecta*, gained the appearance of continuity, unity, and universality, encompassing as it did not merely the empire of Rome, which constituted Polybius's "universal history" (*historia katholike*), but the whole world of nations, pagan and infidel as well as Christian. In this context the "things" (*res*) that were the subjects of history were nor so much human deeds as human institutions: *res publica* and *res privata*, that is, the state and private property. Yet impressive and prophetic as was Baudouin's book, it was hardly more influential than Milieu's, both achieving renown mainly through Wolf's anthology.[23]

With the much more influential work of Bodin history encounters formal philosophy as well as law and politics, and with it the tendencies toward categorization and organized "science." Bodin begins with the normal Scholastic questions, *historia quid et quotuplex*. Following the humanist tradition of the *ars historica* he answers first that history is the "true narration of things" but, employing the trichotomizing conventions of philosophy, posits that it is threefold, that is, human, natural (including mathematical), and divine, corresponding to human will (and the individual acts flowing therefrom), natural causes, and God's plan. Moreover history is to be studied in that order, though with some overlap. Turning not (as in Italianate *artes historicae*) to the writing but rather to the reading of history and to the knowledge rather than the narration of materials, Bodin arranges in a hierarchical way the various classes of historical narration, from world history to national and local histories down to biography. In effect, then, he makes historical study a branch of literary history, where it has remained ever since in the handbooks of historical method, of which Bodin's book is *primum in genere*.

Following the reform of logical and rhetorical method, the projects defined by Bodin were continued by other theorists of the *ars historica* (if not practitioners of *historiographia*), which often proceeded with scholastic and critical listings of *opinions* (itself, it may be suggested, a rudimentary form of literary history). This intersection of history and philosophy was an early sign of the search for a systematic framework for the formless empiricism on which historical inquiry was apparently founded. Thus rational categories and topics were imposed on the literary tradition of historical writing, not only the trichotomy of divine, human, and natural species of history (adapted from the law), but also the distribution of the substance of history from the singular (biography) to the general ("universal history"). For the arrangement of historical materials Bodin drew on Ramist topical rhetoric, which was dichotomizing not trichotomizing.[24] In retrospect it is not surprising that Bodin's book has been placed in the tradition not only of handbooks of historical method but also of treatises on the "philosophy of history."

This philosophical impulse is already clear in Bodin's *Methodus*, which was often regarded as a method not of history but of political science and which represented a step on the way to his *Republic*. Despite his political orientation, Bodin was indeed an encyclopedic scholar driven to seek system, which finally indeed included the whole corpus of nature. Yet Bodin's system was not quite that of the Roman jurists whom he, often carelessly, exploited, though it did owe much to the comparative jurisprudence of French jurists like François Connan and Eguinaire Baron. Bodin was also a determined and virtuoso interdisciplinary scholar, as apparent in his usages, or coinages, of such portmanteau words as "geographistoricus" and "philosophistoricus" (one who combined the narration of facts with the precepts of wisdom). The old question of whether history was an art or a science Bodin avoided altogether by concluding that it was "above all sciences."[25] Bodin had a number of followers in the tradition of the reading of history, and of these Pierre Droit de Gaillard took an even more elevated view of the value of history. For him history, like law, was a form of wisdom (*sophia, sapientia*, the "knowledge of things divine and human," according to Augustine as well as Cicero), and not merely the desire for wisdom (*philo-sophia*). Wisdom, or self-knowledge, according to Gaillard, came from the reading of history, "both universal and particular."[26] This was a theme of historical thought down to the time of Croce and R. G. Collingwood, who likewise paid deference to Bodin.

Francis Bacon followed the divisions of Bodin, though he expressed them as history "naturall, civile [including antiquities], and ecclesiasticall," adding a number of post-Scholastic subdivisions which need not be discussed here.[27] Bacon was of course mainly concerned with natural ("and experimental") history, and here he deviated from the customary view, noting that "natural history is not about single objects."[28] It should be recalled, too, that Bacon himself not only contributed to "civil history"—history in terms of human time—but also assigned at least a marginal or preliminary role to old-fashioned literary history as a part of the advancement of learning, since it afforded a way of supplementing research through the vicarious experience of creditable authors and so reinforcing reason with memory. It was this receptivity that gave Bacon a posthumous place in the tradition of modern Eclectic philosophy.[29]

Following (though not citing) Bodin, G. J. Vossius brings us back to the genre of *ars historica*. Vossius's first aim was to distinguish between history as a literary art (*historices*) and history as a form of knowledge, especially of contingencies (*historia versatur circa res contingentes*). He accepted the formula of Aristotle, which made reason universal and sense experience particular, though he was interested only in particulars (*res singulares*) that were "worthy of memory" and not merely those set down in "annals," the lowest form of history.[30] Vossius noted ancient usage in natural science, such as Theophrastus's *historia plantarum*, Pliny's *historia mundi* (sic), and Aristotle's *historia animalium*; and he duly noted the opinions that associated history with sight and "autopsy." He was, however, much more interested in the literary tradition of the term, invoking the earlier *artes historicae*, including (after the ancients) Pontano, Barbaro, Patrizi, Viperano, Robortello, Foglietta, Fox, and Coccejus. As for the purpose, or final cause of history, this was, as many authors had said, utility, including not only private virtue but also "civil philosophy" (*philosophia civilis*).

But since history was based on human particulars, on sense rather than reason, it could never be stable and immune from error; and moreover it was tied to contingencies, including person, cause, place, time, and action, which suggested an early sort of historicism and scholarly relativism—or so it has been argued.[31] Vossius also celebrated the commemorative value of history as the knowledge of antiquity (*cognitio rerum antiquarum*), adapting here the periodization of Varro (and Augustine), which recognized an "obscure time" of myth and fable before historical writing. Here he cited the famous verse of Horace (later applied by Lord Acton to the historians before the nineteenth-century historical school): "Many heroes lived before Agamemnon, but all are overwhelmed in unending night, unwept, unknown, because they lacked a bard."[32] For the most part, however, he was concerned with literary style and "structure," pursued with a wealth of illustrations from classical authors and objections to modern critics.

In the work of Vossius, as in that of Bodin, came the importation of ideas, such as causation and regularity, from natural science into historical narrative and the introduction of philosophical standards and goals.

Indeed, the main focus of this paper is precisely on this striving of the study of history, in the work of some reflective authors, to rise above human particulars to collective and perhaps natural patterns, and to a synthetic and systematic vision—moving from "idiographic" to "nomothetic" goals. Going beyond the tired old exemplarist topos of Dionysius of Halicarnassus, this meant both world history and "philosophy of history," which by the eighteenth century were often exchangeable terms. In the wake of this modern encounter between history and neo-Aristotelian philosophy, history came also into contact with the notion of system, illustrated especially by the Ramist or Ramoid diagram offered as a synopsis of Vossius's book, following the lead of Bodin. One problem created by this confrontation between history, defined as the product of human will, and system was another version of the dilemma of free will and determinism, which has hounded historians across the centuries.

The early practice of history was interdisciplinary, too, as the alliance with chronology and geography—the "two lights of history," as Vossius called them—was sealed, and as other "auxiliary sciences" joined this alliance, making up the "encyclopedia" of history, parallel to that of law, medicine, philology, etc. What history also shared with these disciplines (if not with philosophy in general) was the faculty of judgment (*de judicio historico*), meaning not only ordinary interpretations but also value judgments, since history was supposed to offer moral and political lessons. To these were added other "auxiliary sciences" pertaining to the identification and interpretation of sources, all arranged into a methodological and pedagogical encyclopedia. Of course, these interdisciplinary connections were all established within the old tradition of learning, which itself constituted the largest "encyclopedia," along with its baggage of old opinions and prejudices; and in this context history, though classically bound to ideals of "truth," could make claims not on philosophical truth but only on a probable sort of knowledge.

The encyclopedic expansion of history was also apparent in the emergence of the field of the history of learning and the disciplines developed especially by seventeenth- and eighteenth-century polyhistors like Alsted and Morhof. For example, J. A. Fabricius's *Historie der Gelehrsamkeit* (1752–1754) built on a rich tradition of scholarly and bibliographical works, including those of Bodin and Milieu and La Popelinière's *Histoire des histoires* (1599); and it included the history not only of the liberal arts, including history, but also of the "three so-called higher faculties of the university": *Gottesgelehrsamkeit oder Kirchenhistorie* and (over two centuries before Ernst Troeltsch) *theologia historica*; *Arzneykunst*, *Arzneygelehrsamkeit oder Medicin*; and *Rechtsgelehrsamkeit*.³³ In this way one of the "liberal arts" became, from a certain point of view, queen—or at least, in effect, secretary—of the sciences.

Another interdisciplinary enterprise associated with the university as well as classical tradition was the new field called the history of philosophy (*historia philosophica*), based ultimately on Diogenes Laertius's careless and anecdotal survey of ancient philosophers. Despite its almost oxymoronic name, this self-proclaimed "science" affected to join the scattered and contingent tradition of philosophers and their "lives and opinions" (in the formula of Diogenes Laertius) with the higher truth of their quest. Many philosophers stayed with the modernized Aristotelian tradition, making history a department of an encyclopedic philosophy defined, in the words of the polyhistor Johann Alsted, as "the knowledge of everything that is intelligible" (*philosophia est omnis scibilis cognitio*).³⁴ However, the new discipline of the history of philosophy, joined to the "Eclectic Method," sought wisdom in a different form than the scholastic systems devised by individual thinkers. According to J. G. Heineccius (editor of Baudouin's works), "one should not seek truth by oneself, nor accept or reject everything written by ancients and moderns; therefore, no other method of philosophizing is more reasonable than the Eclectic Method."³⁵ The summa of Eclectic philosophy and founder of the modern philosophical canon was J. J. Brucker's "critical history of philosophy" (1742–1744).³⁶ Still another example of the currency of eclecticism and historical learning may be seen in the largely unexplored mass of academic theses that treated a wide range of questions and keywords—including *historia*, *historia literaria*, and *historia philosophica*—relevant to the present discussion.³⁷

As one anonymous enthusiast wrote in 1702, "This Method, as I take it, is preferable to that of culling one General Systeme of philosophy out of all their writings, and to quoting them by scraps scattered here and there."³⁸ A crucial problem was that, like historians of religion who had to include theological error in their research, Eclectic philosophers, in their search for truth, found it necessary to include philosophical error and folly (*stultitia*) as well as wisdom (*sapientia*), as in Christian Thomasius's *Historie der Weisheit und Torheit* (1693). This set them apart both from champions of the new science and from advocates of "critical philosophy" in a Kantian sense. Moreover, this antiquarian and pedantic inclination eventually led them to diverge from philosophy into historical scholarship, and specifically from the history of philosophy into cultural history.[39]

The problem of "history" from the very beginning was its detachment from form and structure, and this contrasted sharply with the early modern search for proper pedagogical "methods" of disciplinary knowledge, including history.[40] Research, description, and collecting did not need a rigid plan, as the unfocused (or many-focused) and omnivorous approach of scholars like Peiresc made plain; but this was not satisfactory to the ambitious authors of the Republic of Letters, who sought a modern container for the cornucopia of Renaissance learning, employing metaphors, such as that of "theater," not only of "life" (*pace* Zwinger) but also the "theater of history."[41] But here the key word was "system" (closely followed by "syntagma"), which, likewise going back to Greek sources, had a somewhat simpler semantic history than *historia*.

"System" represented one radical way of containing, and perhaps organizing, the infinite number of particulars offered by historical study.[42] In the early seventeenth century there was a veritable avalanche of "systems," sometimes joined to "method," distributed over the human and natural sciences, beginning with systems of logic (1600), theology (1602), politics and economics (1603), metaphysics (1604), rhetoric (1606), ethics (1607), law (1608), astronomy (1611), geography (1612), chronology (1613), grammar (1615), cosmology (1615), medicine (1618), and the world itself, *Systema mundi* (1638), in the context of the discussions of Copernicus and Galileo (1738). Inevitably there was also a "system of systems," *Systema systematum* (1613), by the master systematizer, Bartholomaeus Keckermann (1613). By 1700 over a hundred fifty works were published by polyhistorical scholars under the title of "Systema," associated with all the disciplines of the old encyclopedia.

In any case, in the seventeenth century the "spirit of system," reflected in the works of ancient and medieval philosophers, had returned with a vengeance, along with the subversive idea of a "new philosophy." Here history became an impediment rather than a benefit for members of the "party of nature." Galileo contrasted philosophers like himself with those benighted "historians" or "memory experts" who countered his rational arguments with mere textual authority. Descartes's *Discourse on Method*—which was originally to have been entitled "history of my mind," which is to say interior reflection rather than a history of philosophy from other authors' minds—was even more systematic in its exclusion of history as a useless repository of outdated knowledge. Descartes was disillusioned with philosophy because although "it has been cultivated for many centuries by the most excellent minds, and yet there is still no point in it which is not disputed and hence doubtful"; and so it was with the other human sciences.[43]

Descartes's solution was to separate reason from memory and in effect to ban history from his method, his third "rule for the direction of the mind" being to attend "not to what others have thought, nor to what we ourselves conjecture, but to what we can clearly and perspicuously behold and with certainty deduce."[44] The model of Cartesian wisdom is that of the solitary mathematician working with simple logical and analytical concepts, unencumbered by past learning and enchanted by the vision of an enlightened future purged of prejudice and error.

This attitude may be contrasted with the view of Descartes's nemesis, Pierre-Daniel Huet, who (like Descartes) prefixed an autobiographical declaration (1692) to his profession of philosophical faith: "From my school reading so ardent a love and esteem of ancient philosophy had pervaded my inmost being, that from that period classical literature seemed to me the handmaid of this science. By this passion I was led to obtain a knowledge of the sects of those ancient high-priests of philosophy which are treated of in Diogenes Laertius..., [which], compared with the observations of others gave me an intimate acquaintance with the history of philosophy."[45] For Huet philosophical understanding was inseparable from this "boundless" science founded by Diogenes Laertius, whose work (he confessed) he took as his "companion" in all his travels. In this opposition there appears the contrast between unreflective doxography and an essentialized or spiritualized "philosophy," contrasted with accumulated erudition, which was made more explicit in Leibniz's praise of his mentor Jakob Thomasius. "Most of the others are skilled rather in antiquity than in science and have given us lives rather than doctrines," Leibniz wrote in 1669. "You will give us the history of philosophy, not of philosophers" (*non philosophorum sed philosophiae historia*).⁴⁶

Here again is the old contrast between philosophy and history—of universals and singulars—with the latter connoting atomized information, content without form as it were, except what could be provided from ideas outside this semantic tradition. For historiography itself, including the history of disciplines, where one scientist built on the achievements of others, this meant especially the importation of arguments from rhetoric and especially emphasis on causation: that is, the efficient if not final causes. For the theory of history it meant the ordering of historical materials according to logical method (in a Bodinian rather than a Baconian or Cartesian sense). Yet the problem of chronology remained, and this was a permanent obstacle to the idea of history as a true and systematic "science."

Here is the way the problem was expressed by one historian of science, J. S. Bailly, writing about astronomy: "A science is a sum of truths. To link [*enchaîner*] them, to present them in their order from the simplest to the most complex is the object of elementary exposition; but the chain of these truths is not the order of their discovery."⁴⁷ This insight was also present in the French *Encyclopédie*, which likewise added a historical dimension to the "chain" of sciences envisioned by D'Alembert in his "preliminary discourse"; and it received definitive expression in Comte's distinction between a "historical" and a "dogmatic" or systematic approach.⁴⁸ Here is another way of contrasting the diachronic arrangement of history, conceived as one damn thing after another, and the synchronic one of philosophy, which is pursued in a "presentist" and ahistorical mode. Rivaling both the chronological and the logical or systematic modes of organization came, within the same encyclopedic tradition, the more practical but also more arbitrary way of arranging, which was that of alphabetization, which begs all theoretical questions.⁴⁹

Meanwhile the discipline of history, gaining a foothold in the universities from the sixteenth century, was forming its own encyclopedic and systematic form. According to Bartholomeus Keckermann, "methodus" was equivalent to "systema," by which, however, he meant not a logical structure (dialectic being only one minor part of the system), but rather an encyclopedia or universal library, accountable to all the other arts and sciences as well as to logic and to reason.⁵⁰ In Germany in particular, history ascended in scientific status by drawing on this medieval tradition and creating its own "encyclopedia"—as indeed did medicine, law, and philology.

Once again the context of this was the old genre of the "art of history," which received new life in the eighteenth century, especially in the German universities, with dozens of treatments—lectures, prefaces to books, essays, articles, reviews of books, chapters of textbooks, and systematic and "encyclopedic" treatises—devoted to the praise and analysis of history.⁵¹ Such publications, in Latin as well as German and ranging from high erudition to elementary pedagogy, carried on the old Ciceronian commonplaces defining history ("the witness of time, the light of truth, the life of memory, the mistress of life, and the messenger of antiquity") and laid down its (moral not natural) laws, "first that an author must not dare to tell anything but the truth and second that he must make bold to tell the whole truth."⁵² So truth and utility continued to be the essential aims of historical study, but in an increasingly elaborate form in keeping with the standards of modern scholarship and criticism and adapted to the conditions and requirements of modern society and the state. As in the Renaissance, but more systematically and professionally, the art of history (*historische Kunst*) was promoted to the level of science (*Geschichtswissenschaft*) and of a "theory."

Following the conventions of the rhetorical view of historical narrative, these new and vernacularized *artes historicae* had an acute sense of the difference between the "material," or sources, of history and the literary form (*Nachrichten*; *Erzählungen*; *Darstellung*) by which it was made intelligible, communicable, and perhaps useful. To emphasize this distinction between content and form, German scholars turned to the original Greek form "Historie," which became the standard term of the new arts of history. As S. J. Baumgarten wrote in his preface to the German translation of the great collective English world history ("Über die eigentliche Beschaffenheit und Nutzbarkeit der Historie," 1744), "History [*Historie*] is a well-grounded account of past events" (*geschehene Begebenheiten*—the "happenings," *res gestae*, of unreconstructed narrative *Geschichte*). In recounting such events, truth (*Wahrheit*) or certainty (*Gewissheit*) continued to be the most important value of historical narrative.⁵³ Tacitean impartiality, reinforced by the nonpartisan stance of Gottfried Arnold, and avoidance of prejudice (*Vorurteil*) were important; but as many scholars acknowledged, the certainty attainable by history was of a quite different order than that of mathematics. Thus history had to be factual and empirical rather than logical and demonstrative, since its brand of truth depended on the scholarly and critical aids produced by three centuries of Renaissance learning and the medium of print.

Along with the professionalization of historical studies in the universities, especially in eighteenth-century Germany, came the development and systematic training in the so-called auxiliary sciences (*Hilfswissenschaften*), whose tradition is still alive. History itself had performed such an ancillary role, especially with respect to law, theology, and philosophy, and indeed in eighteenth-century professorships "history" was often combined with other disciplines, including metaphysics, logic, rhetoric, poetry, law, literature, and languages. In the eighteenth century, however, history reigned sovereign over the subdisciplines of geography, chronology, genealogy, heraldry, diplomatics, paleography, numismatics, sphragistics, statistics, and of course bibliography, all of which broadened and deepened the concept of "history" as well as rendering it more technical, professional, and "scientific" (*Geschichtswissenschaft*). Reinforcing this structure was a mounting mass of works of reference, editions of standard works and documentary sources, pedagogical manifestos, scholarly exchanges, textbooks, maps, scholarly journals, bibliographical listings of the "literature" of particular fields of study, and specialized books in the history of philosophy, art, the sciences, and history itself.

In the age of the "new science," history had an ambivalent position, seeking as it was to establish its own scientific status while setting itself apart from natural science—the opposition between *res humanae* and *res naturales*, the realms of free will and determinism, corresponding to that between *Geschichtswissenschaft* and *Naturwissenschaft*. As F. W. Bierling put it, "The truths of history cannot easily be compared with those of the natural scientists."[54] History was still tied to an empirical base, still concerned with particulars, and it could hope not for certainty but only for probable and practical knowledge. History was becoming "critical," but in a sense closer to philology and textual criticism than to Kantian philosophy and general categories.

Yet history did reach out to larger horizons in two parallel ways, which may be called the encyclopedic and the global. Pedagogically, history constituted itself as an "encyclopedia," surrounded by a set of auxiliary sciences (*Hilfswissenschaften*), which were received into the German universities.[55] History also extended itself by incorporating natural history and (at least implicitly) eliminating *historia divina* or *historia sacra*, as the biblical story and attendant chronology were undermined and replaced by the findings and speculations of modern astronomy, geology, geography, archaeology, and natural history as a background to conventional human history drawn from written records. This is the chief difference between traditional narratives of universal history, such as that of Bossuet, and new models like Herder's and the textbooks of world history published by German scholars.

All this could be invoked in support of history's own quest for "system." As Carl Hausen wrote in 1766, "History [*Geschichte*] in and of itself has no general principles. Only the historian knows how to give it systematic expression. This he must seek to realize through oral and written discourse, and he can accomplish this when he understands the theory of history [*Theorie der Geschichte*], [which] is expressed in the judgment, choice, criticism, and application of materials."[56]

The quest of historical study for some kind of "system" beyond its narrow empirical base went beyond disciplinary alliances and the "encyclopedia" formed by the auxiliary sciences, for it also involved temporal and geographic expansion.[57] The career of world or universal history (*historia universalis*; *Weltgeschichte*; *Universalhistorie*) goes back to the Christian world chronicle tradition of Eusebius and his medieval and Reformation epigones, including Schedel, Sleidan, and Conrad Cellarius, who established the ancient-medieval-modern periodization of Western history, and was continued by Lutheran scholars, especially University of Göttingen scholars like Gatterer, Meiners, Eichhorn, Schlözer, Schlosser, and other authors both major and minor (listed in the bibliography complied by H. J. Pandel for Blanke and Fleischer's anthology of the theory of history).[58]

Such world histories ranged from small textbooks, for example Johann Colmar's little *World in a Nutshell* (1730), a catechistic question-and-answer summary, to multivolume narratives, such as the translation by Baumgarten of the cooperative English universal history published in

38 volumes (1736–1765), on which Gatterer, Heeren, and Schlözer worked. "In this edition," announces the advertisement of 1779, "the plan is methodized; inaccuracies corrected; and the style improved; whereby, it is presumed, the work will be rendered a system of History, hitherto unequalled in extent of useful information, and agreeable entertainment."[59] Even Niebuhr, reader of both Kant and the classics, tried to make his historical narratives "systematic," following modern critical historians: "The wish of these historians was to gain the whole of the mythical age for history: their assumption, that the poetical stories always contained a core of dry historical truth: and their system, to bring this core to light by stripping it of everything marvellous."[60]

The later history of this historiographical drive toward "system" in the eighteenth and nineteenth centuries includes the further elaboration of world or universal history, extending beyond particular nations, individuals, and the conventions of strict Biblical chronology, the Four World Monarchies, and the Translation of Empire; while the history of religion came also to accommodate the pagan gods and the history of—or history as—myth.[61] This genre, universal history, had been placed at the head of the reading list in Bodin's *Methodus*, at the opposite end from particular histories, meaning especially freestanding biography. Here indeed was the Renaissance root of the Enlightenment idea of the history of humanity and the "philosophy of history." Not only Voltaire's *La philosophie de l'histoire* but also Herder's *Auch eine Philosophie der Geschichte* as well as the even better-known works of Hegel and Schlegel derive from this tradition. This romantic genre attracted hundreds of scholarly and speculative efforts down to the present, culminating in such systems as that of Croce (building on Vico's "new science"), which in effect wholly subverted the Aristotelian convention by equating history and philosophy.

For Enlightenment scholars history was not merely philosophy "teaching by example," in the aphorism of Dionysius of Halicarnassus still being cited by authors like Bolingbroke; it was a "system" that reflected not only the past and the potential of human reason but also its future promise for social perfection. "Conjectural history" (a phrase applied by Dugald Stewart to Adam Smith) inferred patterns from and imposed schemes of periodization on the historical process, but aspired to end the conventional series of "epochs" in a transcendent age of moral, social, and political perfection.[62] The so-called "four-stage theory" of cultural development was first publicized by Turgot, Goguet, and Smith, who distinguished the ages of gathering, fishing and hunting, agriculture, and commerce underlying the shift from "barbarism" to "civilization."[63]

In the early nineteenth century a larger system came into play in order to formulate the idea of human progress—"archaeology's first paradigm"—which was accomplished through the efforts of Scandinavian scholars, who distinguished the three stages of savagery, barbarism, and agriculture before the emergence of the fourth phase, the domain of history, which is to say, that of civilization.[64] Here is the domain of another neologism and progeny of *historia*, which is the field of "prehistory."

Another of the novelties of the eighteenth century was the emergence of a field called "history of ideas," a phrase introduced by the Eclectic scholar J. J. Brucker and, in this connection, by Vico. Vico often used the word *storia* in his *Scienza nuova*, where he adapted his *storia delle idee* from the Platonic tradition (which was the subject of Brucker's survey) to a larger and interdisciplinary field of investigation. In the "heyday of ideas," as Ian Hacking has called it, "history" was an alien force, but it finally gained at least marginal acceptance. For Descartes and other early modern philosophers ideas were rational and universally valid concepts independent of time, and the *Cogito* neither had nor had any use for memory or imagination. For Vico on the other hand, in self-conscious opposition to Descartes, ideas lived in time; memory—individual and collective—was absolutely essential to true science; and imagination was not the threat that Bacon thought it posed but rather a creative and synthetic form of memory. Vico's conclusion, therefore, was that modern students should draw on the whole legacy of Western arts and sciences that embodied this memory. Descartes would purge classical literature and history from his program of studies, while Vico made them not only foundational but also socially useful in ways that for Descartes were irrelevant.[65]

Not geometry but Latin and Greek were the keys to true understanding, Vico argued, and yet the privileging of philology rather than philosophy did not mean surrendering the ideal of systematic "science." "The ancients should be read first," Vico advised, and added, "I would suggest that our professors should so coordinate all disciplines into a single system so as to harmonize them with our religion and with the spirit of the political form under which we live. In this way, a coherent body of learning having been established, it will be possible to teach it according to the genius of our public polity." In his *Scienza nuova* Vico became a pioneer in the project of combining singulars and universals, philology and philosophy, history and "sistema" (a word he reserved, however, not for metaphysics but for the hybrid constructions of predecessors like Bodin, Grotius, and Pufendorf, or for the law of nations).[66]

One of the essential aims of Vico's *New Science* was to trace both an "ideal universal history" and what he called the "history of human ideas" (*storia dell'umane idee*), and to encompass—like Polydore Vergil, Vives, Milieu, Louis Le Roy, and others, but more systematically—the whole historical continuum and encyclopedia of learning that Descartes wanted to discard. "This history of ideas," Vico wrote, "will present the rough origins both of the practical sciences in use among the nations and of the speculative sciences which are now cultivated among the learned."[67] Vico's project was continued in the next century, especially by Victor Cousin and by his older colleague the baron de Gérando, who published a vast survey of the history of Western philosophical "systems."

Of course what I have called the drive toward system did not affect every scholar, for many historians continued to take pride in their myopic commitment to particularity; and indeed empirical-minded philosophers like Condillac (who also wrote history) warned against the abuse of abstraction and "system."[68] Baron d'Holbach, too, though he attacked the "stupid respect for antiquity," warned against the neglect of experience in favor of "systems and conjectures," which led not only to error and prejudice but also to "religious terror."[69] It may be noted that the late eighteenth century, despite its affection for philosophy, was the age of enthusiasm for "fragments" as well as the genre of history. As Friedrich Schlegel wrote, "It is equally disastrous for the mind to have a system and to have none. Surely, then, it will have to decide to combine the two"—which is perhaps the logic behind the coinage of the term "historicism."⁷⁰ Moreover, scholars were well aware that philosophical systems were individual constructs, as indeed were histories (of whatever); and this insight lay at the roots of another "art" that emerged in the context of Renaissance learning, which was hermeneutics (*ars hermeneutica*, as Leibniz called it). The aim of hermeneutics was first comprehensively expounded by the author of another "art of history," J. M. Chladenius, who introduced the concept of "point of view" (*Sehepunkt*) into critical scholarship, the argument being that there was no observation without an observer, no narrative without a narrator, and so, at least implicitly, no history without an individual historian. In this sense, too, history was a matter of particularities.

In the late eighteenth century there arose one of the last and most troublesome of the offspring of *historia*, which showed its full reception into the world of system. This was "historicism" (*Historismus*), which was introduced by Schlegel and Novalis during the 1790s and found concrete expression in the "historical school," which opposed itself self-consciously to the "philosophical school" associated with Hegel. Novalis associated "historism," no doubt pejoratively, with mysticism and "the system of confusion," while offering a miscellaneous listing of systematic "methods" (Fichte's, Kant's, chemical, mathematical, artistic, etc.). Later the term was occasionally used in philosophical polemics, especially regarding the question of the value of history for philosophy—a question often answered in the negative by academic philosophers. Yet despite the contentious elaborations of "historicisms," old and new, the nucleus of focus on a single thing or fact had not been lost, as evident from the motto chosen by Friedrich Meinecke for his great book on the subject, that is, Goethe's aphorism, *Individuum est ineffabile*.⁷¹ No philosophy or system can cancel out this human truth, from which, as Goethe wrote, a whole world may be derived.

History and system met again in the Enlightenment practice and theory of the history of culture and civilization by French and British as well as German scholars, for example François Guizot, who devoted his academic teaching to the notion of "civilization." He was followed in this by the younger generation of "new historians," most notably Jules Michelet, who, under the influence of Vico, conceived the idea of a "history of the world as a system."⁷² As he told students in 1825, "Science is one: languages, literature and history, mathematics and philosophy, and knowledges apparently most remote are actually joined, or rather form a system, of which we in our weakness [can only] consider as separate and successive parts." This was only one of the precedents of Henri Berr's idea of history as "synthesis," which underlay the work of what has been called the "Annales paradigm," with its continuing scorn for the individual facts and events—the mindless narrative of *histoire événementielle* confined to *res gestae*, incidental happenings, or *geschehene Begebenheiten*, denounced originally by the Durkheimian sociologist François Simiand.

Another more philosophical effort in this direction was made by José Ortega y Gasset, who wrote extensively on "historical reason"—in particular contrast to natural reason—and explicitly on "history as a system."[73] For him, history is "the *systematic* explanation of events or conditions past, though tied to time and interests of successive generations." It is in this essay of 1941 that Ortega y Gasset—echoing Michelet echoing Vico echoing Augustine—made his famous declaration that "*Man, in a word, has no nature; what he has is history*. Expressed differently, what nature is to things, history, *res gestae*, is to man. Man, likewise (like God), finds that he has no other nature than what he has himself done."

There are many stories to be told about *historia*, including the story of "story" itself. The one I have chosen to introduce focuses on the onomastic trajectory between the extremes of history defined as a single thing or things, fact or facts, event or events, figuring in empirical science as well as concrete human narrative, and history as the past in its broadest, Toynbeean range and even identical, as conceived by Croce, with its old rival philosophy. At either end of this gigantic semantic spectrum, running from the individual and the local to the collective and the global, history has been taken as either "objective," the "thing in itself," or as a human inference or representation thereof; and this conceptual or linguistic duality, recognized ever since antiquity, continues to trouble some scholars. But this, too, of course is not the fault of Herodotus or even Ranke but just another by-product of history's millennia-long love-hate relationship with philosophy.

Notes

Sources cited in boldface type appear in the Primary Sources of the bibliography.

1. In this paper I draw on a number of earlier publications listed in the bibliography.
2. **Quintilian** 1989, 1:9.1.
3. **Origen** 1984, 100–104.
4. **Goclenius** 1613, 626.
5. **Du Cange** 1938.
6. **Dionysius of Halicarnassus** 1895, 11:2.
7. **Riccobono** 1995, 744.
8. Seifert 1976, reviewed by me in the *Journal of Modern History* 54 (1982): 320–326. See also my review of Kessler, Landfester, and Cotroneo on the *ars historica* in the *Journal of Modern History* 47 (1975): 679–690.
9. Keuck 1934.
10. Knape 1984, reviewed by me in *Speculum* 60 (1985): 197.
11. Smalley 1974, 12, 175.
12. **Isidore of Seville** 1962, 1:1.41–44.
13. **Quintilian** 1989, 2:4.2.
14. **Morhof** 1732.
15. See MACLEAN in this volume.
16. See OGILVIE, POMATA, and PINON in this volume.
17. See SIRAISI and CRISCIANI in this volume.
18. But see GRAFTON in this volume.
19. **L. Valla** 1962, 2:6.
20. Kelley 1970, 1998.
21. Kelley 1999, 342–365.
22. Kelley 1964.
23. **[Wolf]** 1576.
24. McRae 1955.
25. Kelley 1973.
26. **Gaillard** 1576.
27. **F. Bacon** 2000a; and see Fattori 1980.
28. **F. Bacon** 1996.
29. Kelley 1973.
30. **Vossius** 1623; and see Wickenden 1993.
31. Kelley 1970.

32. Horace *Odes* 4.9.
33. **Fabricius** 1752–54.
34. **Alsted** 1615, 10.
35. **Heineccius** 1756: "Folglich kein anderer Methodos philosophandi raisonable sey, als: METHODOS ECLECTICA."
36. **Brucker** 1742–44.
37. Marti 1982.
38. **Diogenes Laertius** 1702, preface.
39. Kelley 2003a.
40. Franklin 1963 and Gilbert 1956.
41. **Helvicus** 1618. On Zwinger see BLAIR in this volume.
42. Ritschl 1906.
43. **Descartes** 1985, 1:114–115.
44. **Descartes** 1977. The translation is that of Elizabeth S. Haldane and G. R. T. Ross, *The Philosophical Works of Descartes* (Cambridge: Cambridge University Press, 1931, reprinted 1978), 1:5.
45. **Huet** 1810, 2:203.
46. **Leibniz** 1670, fol. 2v.
47. **Bailly** 1969, 143.
48. **Comte** 1949, 1:145.
49. *Encyclopédie* 1751–80, Preliminary Discourse.
50. Zedelmaier 1992; Schmidt-Biggemann 1983; Loemker 1972; Freedman 1988; Yeo 2001.
51. Blanke 1991; Pandel 1990, 156–180.
52. **Cicero** 1988, 2.9, 2.15.
53. Blanke 1991, 178. On this world history, see further 229–230 here.
54. Blanke 1991, 159.
55. Marino 1975.
56. Blanke 1991, 130.
57. Kelley 2003a.
58. Marino 1975 and Boockmann 1987; **Blanke and Fleischer** 1990.
59. *Universal History* 1779, 1.
60. Walther 1993.
61. See MULSOW in this volume.
62. Hulliung 1994, 52–78.
63. **Turgot** 1973; and see Meek 1976, 1977.
64. Kelley 2003b.
65. Kelley 1984.

66. Vico seldom used the term *historia* in his Latin works, but *storia* occurs over 200 times in the *Scienza nuova* (Concordance).

67. Vico 1968, 391.

68. Condillac 1798, 1; and see Diemer 1968.

69. **Holbach** 1994a, 1:1, and cf. **Holbach** 1994b.

70. *Athenäum-Fragment*, 53, **Athenäum** 1798–1800.

71. **Novalis** 1907, 118. Meinecke 1959: "Habe ich dir das Wort Individuum est ineffabile, woraus eine Welt ableite, schon geschrieben? Goethe an Lavater 1780."

72. Michelet 1959, 217.

73. **Ortega y Gasset** 1941.

II
The Working Practices of Learned Empiricism

7

Conrad Gessner and the Historical Depth of Renaissance Natural History

Laurent Pinon

What does Renaissance *Historia naturalis* have in common with human or civil history? Is the term *historia*, as it appears in many sixteenth-century naturalist works, much more than a recollection of the title of Aristotle's *Historia animalium*? In this paper I shall attempt to analyze the effective use of the concept of *historia* in the zoological work of Conrad Gessner and, conversely, to show that natural history as practiced during the Renaissance has intrinsically a strong historical dimension. Not surprisingly then, natural history shares some of the methodologies and practices of human history, including procedures of proof.

Conrad Gessner's work appears to be most relevant for this inquiry because this polygraph author, apart from his fundamental zoological work, was also involved in the debate concerning the classification of sciences. Trained in the Protestant reformed educational system of Zurich, Conrad Gessner (1516–1565) was one of the most learned humanists of his time.[1] After a few medical writings and many philological and erudite editions, he published in 1545 his monumental *Bibliotheca universalis*, which gives an account of every book published in Latin, Greek, or Hebrew. The second part of this bibliography, organized according the main branches of knowledge, offers concrete testimony of Gessner's opinion about the term *historia*. The *Historia animalium*, published from 1551 onward, is his masterwork of the following decade. From those two monuments of erudition, it is possible to compare Gessner's theoretical definitions of history and natural history with his actual practice of natural history. Drawing also on a few examples taken from Pierre Belon and Ippolito Salviani, I will argue that this strong relationship between history and natural history is not specific to Conrad Gessner but represents a general characteristic of Renaissance natural science.

History and Natural History in Gessner's Classification of Sciences

Gessner's classification of sciences appears in the *Pandectarum libri*, the second part of the *Bibliotheca universalis*, organized according to the divisions of philosophy. This classification clearly suggests that he did not consider the history of animals a part of history. In this bibliographical encyclopedia, Gessner considers history as one of the propaedeutical disciplines (*praeparantes*) that serves as an introduction to the "substantial" disciplines, which are the parts of philosophy (physics, metaphysics, ethics) and the three higher university faculties (law, medicine, theology).[2] These propaedeutical disciplines are the well-known seven *artes liberales* and five "ornamental arts" that may embellish the literary culture: history, poetics, geography, magic, and mechanical arts. In discussing the boundaries of history with neighboring disciplines in his bibliography, Gessner cannot find any clear boundary between history and geography and often hesitates over the best category to which to assign a book. He even wonders whether history and geography should not be unified in a single field. If he does not go so far, he nevertheless chooses the same order in both disciplines, following the different geographical areas.

The other disciplines that might share boundaries with history according to Gessner are ethics, for the moral usefulness of historical examples, and theology, because of ecclesiastical history. He proposes a special treatment for fabulous histories that might belong either to poetics (if they are written in verse) or to history, although they must be separated from true history. He also makes a special case for doubtful histories, which surely can be included neither in true nor in fabulous history. History, as Gessner conceives it, seems to be related to truthfulness, but not exclusively.

Quite surprisingly, we do not find in Gessner's discourse about history any links to natural history, as one would have expected from an author so deeply involved in that field.[3] Some other authors do not hesitate to consider natural history as a section of history. Poliziano, for instance, in his *Panepistemon* divides "true history" into four parts, including natural history.[4] As Jean-Marc Mandosio notes, the two disciplines actually have a lot in common. Both of them are constructed as narratives with an important element of enumeration; annals or chronicles may be compared to botanical or zoological inventories. Both of them are also constituted of single events bound together by weighty methods of compilation. The lack of a chronological dimension could be, according to Mandosio, the critical point that might separate natural history from history. Some Renaissance authors like Jean Bodin also differentiate human and natural history by degrees of certainty. While human history, being dependent on human will, is very unpredictable, natural history follows regular rules and every single fact may be verified by observation of the corresponding natural being.

At this first step of analysis, it appears that in his classification of science, Gessner makes a strong distinction between history and natural history, although those two disciplines are linked together by other Renaissance scholars according to the important characteristics they have in common. I should like in the following sections of this paper to discuss the proximity between natural and human history by examining Gessner's project for the *Historia animalium*, the methods he employed for compiling this huge work, and his actual practice of narration, which is very close to a historical one. In addition, some of the common characterizations of natural history may have to be revised in the case of Gessner's practice. The species he describes, for instance, always have a chronological dimension; some of them, like monsters, do not follow the regular laws of nature and require a specific analytical treatment.

While Gessner clearly separates history and natural history, one may wonder what meaning the word *historia* actually has in Gessner's *Historia animalium*. The next section discusses his effective use of the word and, more generally, the overall intellectual project of the *Historia animalium*.

An Ambitious Project

In the *Historia animalium* one can find at least two different meanings of the word *historia*, which is employed either in the singular or in the plural. The title of the work suggests an influence from Aristotle's own

Historia animalium, by far the main zoological source from antiquity.[5] But it is worth noticing that in the case of Aristotle, *Historia animalium* was only a part of a more comprehensive reflection about animals that also included the *De partibus animalium* and the *De generatione animalium*. As Aristotle himself explains, the *Historia* describes the different parts of animals while the *De partibus* intends to explain the causes of the phenomena described.[6] Very few passages indeed in the *Historia* contain any such explanation.[7]

Theophrastus's botanical opera, the *Historia plantarum*, also very familiar to Gessner, bears a similar title.[8] In this case, some modern translators and commentators suggest translating *historia* as "research," thus keeping the ambiguity between a research in progress and the result of research, which is knowledge.[9] In this case again, the *Historia plantarum* is coupled with another treatise, the *De causis plantarum*, where causal analysis is much more developed.

Unlike his prestigious predecessors, Gessner uses the term *historia* for the core of his discourse: there is almost nothing that is not *historia* in his zoology. While the *Icones animalium* published later do not bear the word *historia* in their title, Gessner explicitly presents them as extracts of the main work, containing mainly images and nomenclatures.[10] Does that mean that Gessner's *Historia animalium* is purely descriptive?

The complete title of the first volume, devoted to live-bearing quadrupeds, clearly indicates Gessner's intended audience: philosophers, physicians, grammarians, philologists, poets and, as he says, anyone interested in matters related to language.[11] This expectation of a wide readership distinguishes the book from other contemporary Renaissance zoological works, which do not have such an insistence on language. Gessner's *Historia animalium* is not a simple description of the natural world, as I will argue more precisely below. The word *historia* also appears in the broad sense of a complete discourse about animals in its preliminary pages, when Gessner thanks all the people who helped him compile zoological knowledge. "Here is established, as an acknowledgment, the list of the scholars of our century who provided us with objects for the history of animals (pictures, names in various languages, descriptions)."[12] Here again, *historia* seems to include images and nomenclatures and not only the descriptions or properties of animals. *Historia* is also used in this fuller sense but on a smaller scale when it is employed for a single animal. In this context, it refers to the chapter devoted to an animal or a single group of animals. This appears, for instance, when Gessner presents the internal organization of each chapter, in a section of the foreword entitled "Explanation of the order we followed for almost every animal history."[13]

Elsewhere however, *historia* is used in a more restrictive way, to contrast with other material related to animals. When Gessner explains how he managed to get zoological information without the support of a patron, he significantly separates images, names and histories: "I did what I could, and managed to acquire some friends in various parts of Europe, who furnished me kindly, frankly and generously with many illustrations of all kinds of animals, depicted from life, and some of them provided with the animals' names in various languages and with their histories."[14] In this more restricted acceptation, *historia* appears to be what there is to tell about an animal previously identified by its image and various names.[15]

The advice to the reader on the title page (fig. 7.1) comments on the very wide contents of the book: "Not only the simple history of animals is included, but also everything we were able to see, for instance very rich commentaries and many emendations on ancient or recent authors on this subject."[16] *Historia* is here distinguished from the commentaries Gessner frequently added and from the philological corrections he might have made to the sources he cites. This leads to a striking point: what Gessner calls *historia* seems to be quite external to his judgment. The history of animals, in that sense, appears as a pure collection of zoological knowledge that has nothing to do with the author's opinion or subjectivity. If Gessner keeps himself in the background, he nevertheless has a fundamental role: he is the critical authority able to decipher to which animal every particular knowledge is related. In the preliminary pages of the *Historia animalium*, he explicitly mentions the restoration of the links between the words (*nomina*) and the things (*res*).[17]

Even if the term *historia* is not used univocally in the *Historia animalium*, the fact that Gessner's encyclopedic work claims to be a collection of all the available data about every animal suggests that the actual meaning of the word is quite close to its sense in Aristotle's or Theophrastus's titles. The Gessnerian *Historia animalium* is indeed an extraordinarily wide and complete collection of particular zoological facts. The

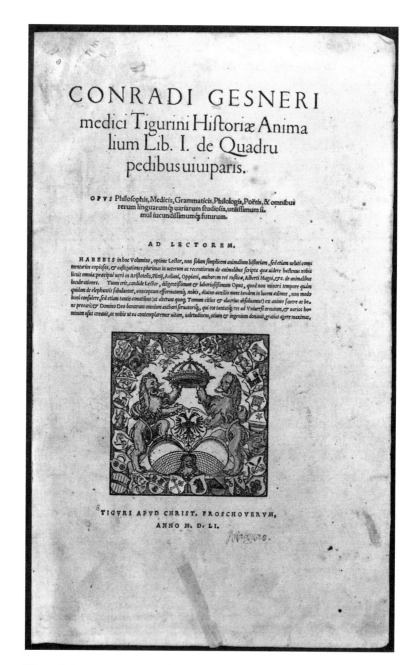

Figure 7.1
Title page of Conrad Gessner's volume on quadrupeds, *Historiae animalium liber I, de quadrupedibus viviparis* (Zurich: Ch. Froschauer, 1551).

size and alphabetical organization of the four successive volumes printed from 1551 to 1558 are perfectly convenient for that purpose. Responding to potential criticism, Gessner explains in his bibliography of his own works that his books are not supposed to be read extensively, but rather be used as dictionaries:

> Everything is arranged in a handy and permanent order, so that, whatever one wanted to know about whichever animal, one would find it immediately and without any trouble. So this is not conceived for an uninterrupted reading, but for being used as any dictionary, chiefly to investigate: those who accused us of being too prolix in this work did not understand that.[18]

The comparison with the *Icones animalium* is enlightening here, and reinforces this interpretation. Although they are presented as mere illustrated summaries, the *Icones animalium* contain an important innovation: an organization of the described animals following the natural order of nature. The *Historia animalium*, by contrast, is essentially an accumulation of single *historiae* that has almost nothing to do with causal explanations or taxonomic considerations. The philosophical order of the *Icones* contrasts with the alphabetical order of the *Historia*.

When Gessner explains the utility of his book,[19] he never evokes explanations or causal interpretation of the collected facts. Instead, he mentions medicine, cookery, other unliterary *artes*, morals, and domestic economy, and strongly emphasizes the fact that natural knowledge necessarily leads man to admire the work of God.[20] In this conception of science, compiling particular information about every animal is a goal in itself with immediate utility, not a preliminary step for a philosophical approach to the natural world. This double ambition of comprehensiveness—in the number of species described and in the information attributed to each of them—is elevated in Gessner's case to the rank of an epistemology. It would be misleading to take it for a kind of theoretical goal that no one seriously expects to achieve. Although the number of known plants grows greater and greater, Gessner is optimistic about the ability of his contemporaries to succeed in botany: in a letter to Fuchs sent in October 1556, he estimates that if the information already collected is correctly shared among scholars, they would be able by the end of the century to complete the unique and definitive botanical book (*opus unum absolutum*).[21] The French author Pierre Belon shares this optimism in his *Histoire de la nature des Oyseaux* when he estimates that the number of bird species cannot go beyond five hundred: though God could easily have created many more, He would not have chosen a number of animal species exceeding the human ability of description: "Car si quelcun maintenoit deux mil especes d'oyseaux, feroit comme celuy qui diroit qu'il est plusieurs mondes, et qu'il y a un Soleil et une Lune en chascun monde, qui est chose du tout incroyable."[22] This epistemology is perfectly consistent with the main goal attributed to natural history: the glorification of God by man's admiration of His creation, even down to the smallest details.

This method of knowledge based on compilation, and the very concrete matter which it treats, has an obvious influence on the style of the *Historia animalium*. As a good humanist, and as Zwinger does in his *Theatrum* (BLAIR in this volume), Gessner feels obliged to apologize for the absence of an elevated style in his books. He reproduces for that purpose the similar apology made by Pliny, who was himself a fine stylist.[23] Should he be more dependent on the demands of rhetoric than Pliny was? Gessner will not shrink from putting the right word in the right place, or digressing, simply for stylistic reasons: "In those writings where a knowledge of facts is looked for, not the charm of a beautiful style, but the real truth should be expressed."[24] He also justifies some repetition, noting that it arises from the logical order of the book and not from his negligence. Even inside a chapter, he reproduces very similar assertions from different sources, because the accumulation of testimonies is a crucial element of truthfulness: "If something is said in the same words by many witnesses, it is much more reliable."[25] But Gessner's logorrhea also faced practical and financial constraints, and after the first massive volume of the *Historia animalium* he tried to reduce the size of the following ones.

The Content of the *Historia animalium*

The actual content of the work corresponds quite well to Gessner's preliminary announcements. Each animal history normally begins with an illustration engraved on wood, followed by a text divided into eight main sections, lettered from A to H. The knowledge gleaned by Gessner is methodically distributed among the following topics:

A. Names of the animal in ancient and modern languages
B. Geographic distribution and regional differences, outer and inner morphology
C. The animal's way of living and behavior
D. Character, ingenuity, vices and virtues, sympathies and antipathies
E. Utility for man, except in cookery and medicine
F. The animal and its parts as food
G. Medical uses obtained from the animal
H. Philological questions

The last section, very often the most voluminous one, is divided again into eight subsections, identified by lower case letters, that respond to the previous ones but are related to philological topics. For instance, subsection "a" is devoted to the meanings of the names, proper names, epithets and metaphors associated with an animal. The last subsection, containing all the remaining knowledge, is again divided into five parts: history, veracious or fabulous; predictions, wonders, and monsters; religion; proverbs; and emblems, following Alciati or other authors. Very few animal descriptions, of course, include material to fill all these sections and subsections, but this general organization of the discourse is methodically applied to the whole book on live-bearing quadrupeds.

Two points in particular may surprise the modern scholar on reading Gessner's *Historia animalium* for the first time: the heterogeneous assemblage of facts gathered from a very wide range of books and the presence of wonders and imaginary animals. At first sight, the book looks like a juxtaposition of data without any criticism or logic. But the coherence comes when one remembers that Gessner wants to report every little thing said about any animal. This includes, of course, unknown or fictitious animals. For those cases, the compilation of information is not so easy. Gessner is puzzled, for instance, about divergent opinions and descriptions concerning the unicorn, an animal to which he dedicates seven pages. Although there is much to say about the unicorn, it is difficult to find a convenient image: "The picture is such as this animal is commonly painted nowadays, about which I do not have any certainty."[26] He gives the traditional image of the animal, a white horselike animal with a single, spiral horn that corresponds to the narwhal's teeth held in several royal collections in Europe (fig. 7.2). Concerning the appearance of the unicorn, Gessner explains that he can only gather the various descriptions he found and let other scholars more erudite than

Figure 7.2
How should the unicorn be depicted? From Conrad Gessner, *Historiae animalium liber I, de quadrupedibus viviparis* (Zurich: Ch. Froschauer, 1551), 781.

himself form their own opinion. He was not able to verify the wonderful medical properties accorded to the horn as an antidote because of its very excessive price: "I leave this way of experimenting to the rich, for the price of what is called the real stuff is no less today than that of gold."[27] One can understand, then, that even if Gessner has doubts (though he does not explicitly express them) about the existence of the unicorn, this would not be a reason at all to suppress the corresponding chapter: whatever one may think about the animal, Gessner gives real textual references and describes real medical practices, both of which do exist.[28]

This conception of zoological knowledge, much wider that what one would expect for natural history, accounts for some serious misunderstandings by historians about Gessner's scientific practice. Many historians of science have blamed him for compiling all kinds of information without criticism. This is to ignore Gessner's explicit goals and the fact that his natural history is much more than a simple study of nature: it is actually an inventory of every element of natural knowledge since antiquity. Consequently, the real basis of Gessner's zoological discourse is not the animals themselves but the animal's names, to which the zoological knowledge is related. While Gessner's zoology could exist in a world without animals, it is not conceivable in a world without books.

This revised reading of Gessner's zoology is increasingly widespread among scholars today, but some of them may have gone too far in opposing the classical conception of natural history to the practices of Renaissance scholars. William Ashworth Jr., for instance, in a remarkable presentation of Gessner's work, calls it "emblematic natural history."[29] He proposes a detailed analysis of Gessner's history of the fox that may indeed suggest at first sight that "emblematic history dominates his attitude toward nature." But one may answer that the fox is not the most representative animal from this point of view: many other species have a much weaker emblematic meaning or none at all. While the emblematic perspective is obviously an important part of Gessner's discourse on animals, it is excessive to take it for the main goal of the work.

The Status of Description

The way Gessner organizes every chapter of his books and the kind of zoological description he practices provide the best evidence for the kind of natural history he strives for. We saw how the unobservable unicorn is the subject of a chapter in the first book of the *Historia animalium*. But this is not an exception: many other animals that figure in Gessner's books are very difficult to observe or to identify with certainty. Among those, one may think of the three animals described by Caesar in his *De bello Gallico*.[30] In the Renaissance, the *De bello Gallico* was primarily read from a historical point of view, but a few passages required a technical understanding of what Caesar saw and what exactly he described. Among those technical parts of the work is the description of three strange beasts from the Hercinian forest, in book 6.

The main point of this discussion is the fact that, according to Caesar, these animals had never been seen elsewhere. The first one is described as a *bos cervi figura*, that is an ox with the appearance of a stag.[31] Caesar asserts that this animal has a single horn between the ears, divided into two webbed antlers. The second one, *alces*, has the form and coat of a goat, but a little taller, and has legs without knees. It is hunted, says Caesar, by sawing in advance most of the tree on which it has to lean in order to sleep. When the tree is knocked over, the animal falls over with it and cannot get up. The third animal, called *urus*, appears to be a very savage ox that cannot be domesticated and can be caught only by digging a large pit into which it will fall. His horns, once set in silver, are used in banquets, adds Caesar. The zoological importance of Caesar's descriptions comes from the fact that no other ancient writer mentions those animals—which might have been expected, since they were not supposed to have been seen anywhere else.

This is the reason why Renaissance natural historians looked so closely at Caesar's descriptions and tried to identify the three animals. We find traces of this quest in Gessner's work. He refuses the identification of the first animal as a bison, implicitly proposed by the French author Pierre Gilles, who referred to a unicorn bison in his adaptation of Aelian's *De natura animalium*.[32] In the same way, Calepino's *Dictionary* mentions the existence of unicorn bisons. Against those two authors, Gessner argues that webbed antlers were not convenient for a bison, and recognizes a reindeer in Caesar's *bos cervi figura*. He also discusses the two other animals and proposes to see an elk in Caesar's *alces*, even if the elk does not have, as far as he knows, the described nonbending legs. The identification of the *urus* is complicated by the fact that it can be related to two different wild oxen that lived then in central Europe, either the bison or the aurochs, the latter of which definitely became extinct during the seventeenth century.

Gessner is involved in a kind of historical inquiry in order to determine what animals Caesar had seen or heard about, and to compare them to the known local fauna.[33] As he notes, the best way to decide between possible identifications is to investigate among the people who live in the places visited by Caesar, and in particular to know what names they give to the suspected animals. He was able to get information from his correspondents in central Europe: one of them, a Polish nobleman, even sent him horns of a bison and of an elk (fig. 7.3).

Gessner also made use of the books published by cosmographers and other people who visited those countries. He duplicated two engravings from Sebastian Münster's *Cosmographia*, first printed in 1544, that represent animals very close to those mentioned by Caesar: an elk, *ellend* in German, and a long-bearded wild ox.[34] He also took information from the description of Muscovy by Sigmund von Herberstein, itself heavily indebted to Caesar's text.[35] In his discussion of Lithuania, Herberstein

Figure 7.3
Elk horn obtained by Gessner as part of his research into animals described by Caesar. From Conrad Gessner, *Icones animalium quadrupedum viviparorum et oviparorum, quae in historia animalium describuntur* (Zurich: Ch. Froschauer, 1553), 27.

describes three savage beasts, unknown in Germany, that are precisely the aurochs, the bison, and the elk. He explains for instance that the *urus*, called *thur* by Lithuanians, lives only in Masovia and is a black wild ox with a white line on its back. Herberstein even saw the skin of a recently hunted one.

It is striking that, in this particular case, a large number of observed facts, witnessed either by Gessner or by his contemporaries, are linked to a literary investigation on an ancient classical text. Gessner still had an open mind when he published his first volume and he would later write several addenda, in an appendix published in 1553 with the second book of the *Historia animalium* and in the reissues of the *Icones animalium*.[36] One may assume in this case that natural observation is constrained by a historical approach: Renaissance scholars are observing the fauna of this area not exactly with Caesar's eyes, but at least with his words in mind. I would be inclined to describe this as the "historical filter" through which Renaissance natural historians examine the natural world.

We find quite similar methods of inquiry into wonders or monsters, whose occurrence is, by definition, exceptional. In most cases, their description and interpretation rely on a single event. But the sea monster described in the fourth volume of the *Historia animalium*—with the head and general shape of a monk—appears to be a more complicated case.[37] Gessner's description is included in a chapter devoted to mermen (fig. 7.4). He quotes extensively from the long chapter of Rondelet's book describing this monk-fish and includes many details about how it was first observed and how the information about it was transmitted to Rondelet.[38] It was found on the Norwegian seaside, in a place named Diezus, near a town called Den Elepoch. Rondelet describes in detail all the intermediaries between the monster and himself: a Spanish nobleman saw the animal and gave an image of it to Queen Marguerite de Navarre (sister of the French king François I), who passed it on to Rondelet. The uninterrupted link between the monster and Rondelet; the nobility of the Spanish man; the exceptional virtue, knowledge and nobility of the queen; the way she guarantees the description ("this great Queen maintained..."): all these elements endorse the observational narrative. In his book, Gessner carefully duplicates the entire chain of proof. He also borrows from Rondelet the references to mentions of mermen in Pliny.[39] However, if Gessner supplies arguments for believing in this monster, he also quotes without changing a word the next chapter of Rondelet's book about the bishop-fish, together with Rondelet's doubts about its credibility. Gessner does not act here as the credulous writer he is often accused of being, but simply as a faithful narrator of whatever had been written before him. He finishes with a quotation from Pierre Belon, who sets out to account for such monsters through the playfulness of nature (*ludus naturae*) and her ability to transgress the usual rules of generation.[40]

After these descriptions taken from the two modern reference books about fish, we need to look at Gessner's *corollarium*, the part of the chapter for which he claims responsibility.[41] He adds new information about the monk-fish, in a manner that clearly reveals his methods of critical analysis. First, he explains how he received an identical image from Georgius Fabricius, this monster being found later, in 1546, in Denmark near Copenhagen. It was four cubits long with a black face like an Ethiopian's. Another correspondent, the young Hector Mythobius, gave

Gessner and the Historical Depth of Renaissance Natural History   255

Figure 7.4
Gessner's version of the monk-fish. From Conrad Gessner, *Historiae animalium liber IIII, qui est de piscium et aquatilium animantium natura* (Zurich: Ch. Froschauer, 1558), 519.

additional details about the second monster, such as the fact that it was caught while trawling for herring, which Gessner also quotes in his *corollarium*. There are moreover a few additional descriptions from Belon together with the report that a Frenchman told Gessner that such a monster had been captured near Bordeaux. The *corollarium* ends with an excerpt from Albertus Magnus about the *monachus maris* from the Atlantic Ocean.

We may note that Gessner's voice or opinion never appears in this chapter: there is no critical position or opinion about this kind of monster; Gessner simply informs the reader about all the known occurrences. In such preternatural cases, every detail about description, date or place is crucial, for it may indicate how many occurrences have been recorded and perhaps explain the meaning of a wonder.[42] Under the particular conditions of a natural observation that cannot be repeated, the methods of natural historians are precisely those of historians. The goal is to understand what kinds of animals someone saw many years or centuries ago. For such rare events, natural history has to follow the methods of historical enquiry.

A different type of enquiry is employed by Gessner when he tries to get information about the animals of the New World. For those animals that have no equivalent in the Old World, what Gessner is able to say is quite limited. The difficulty of obtaining information about such animals clearly appears from their slow appearance in zoological books. Some of them, like the armadillo or the guinea pig, appear in the appendices of Gessner's books: this suggests that Gessner was informed about them while he was already printing his *Historia animalium*. The short chapters devoted to these two animals contain images and precise descriptions, but no ancient references, no discussion about their names, no philological section. For the guinea pig, compared to the rabbit or a small pig and named *cuniculus vel porcellus Indicus*, Gessner supplies an image made from the live animal he received from a Parisian friend.[43] Later, as he explains, he obtained a pair of animals from another correspondent and could thus give precise information about their color variations, feeding habits, and spectacular fertility. There is no bookish information in this additional chapter, except for a final reference to Peter Martyr, who in his *Historiae navigationum ad novas insulas* mentions three kinds of animals close to the European rabbit, among which this species might figure. But Gessner's empirical approach toward the guinea pig has its limits: he confesses he did not taste its flesh as a dish.

In this case, access to an American animal is much facilitated by its easy breeding and high fertility. By becoming commonplace in this way, the guinea pig is no longer spectacular when Gessner describes it, and he need not justify its existence or attest the reliability of his information. The description for the armadillo is quite similar: Gessner received an image and a dried specimen from a German apothecary and subjects Belon's engraving and description to a critique.[44] But he gives little more information than a vernacular name, *tatu*, the ability of the animal to roll up like a hedgehog, and the precise number of its digits. In a sense, one may say that these new animals are without any history: they enter quite naked in Gessner's books and there is little more to say about them beyond describing them and explaining the origin of information about them. Gessner's attitude toward these animals is very close to that which one might imagine a civil historian to have when describing a newly discovered kingdom whose language he did not know. Fortunately in a way, these "naked" animals are still few enough in number in the middle of the sixteenth century not to disturb the general order of knowledge adopted by Gessner; but their increasing number would soon be fatal to humanist zoology.

Truth and Proof in Gessner's Descriptions

Many difficulties of Gessner's task result from his ambition of absolute exhaustiveness about the knowledge related to animals. As we explained, he did not intend to describe only the animals he might have under his eyes; he also wanted to get information concerning the fauna far away from his home. This humanist town physician, settled in Zurich by his professional obligations, often wished he could have traveled across the world. He had to rely on the testimony of others in order to complete his inventory of the natural world. Like his contemporaries, as we previously saw, he tried to glean the little information that was available about American animals and other exotic species. Analyzing the distribution of species around the world, Gessner's work has a real geographical dimension. But Gessner also had to deal with chronology.

If the species described have, not surprisingly, a spatial dimension (they are implicitly supposed to be homogeneous in all the different places where they are recorded), they also have an unexpected chronological depth: the species observed in Gessner's days are systematically compared to those previously described by the ancients. When his correspondent John Caius describes a newly discovered species he calls *leuncia*, Gessner answers that it is nothing else than the lynx, named Θώς by the Greeks, and thus does not need a new name.[45] Any modern species not recorded in ancient reference works requires special investigation and explanation. One has to decide whether it is a new animal or one whose ancient description has not yet been discovered. Gessner's task thus presupposes a historical analysis: in order to understand what the ancients said about animals, to translate correctly the name they gave to beasts, birds, or fish, he has to reconstitute, in a way, the world in which they lived.

But the relationship between Gessner's natural history and human history is further enhanced by the strong interest the naturalist shows toward human history. History of knowledge seems important enough from Gessner's point of view to justify the presence, in the preliminary pages of the *Historia animalium*, of a list of the lost zoological treatises of antiquity, with about twenty titles.[46] Moreover, the striking way in which Gessner keeps in the new issues of his books old engravings that should have been replaced by the new and better ones he had obtained later might make more sense in light of this historical sensibility. When he keeps, for instance, the old image of a mongoose he found in a Venetian manuscript of Oppian, this is of course a way to use once more an expensively engraved woodblock, but this is also a way to keep alive a previous stage of zoological knowledge.[47]

We may also consider that Gessner's ambition of exhaustiveness leads to a method of knowledge and investigation quite close to the one commonly used in human history. Human history also fundamentally has the same ambition of telling everything worth noticing on a specific topic or area. From this point of view, Gessner is exactly in the same position as the civil historian who has not witnessed all the facts he mentions. Neither the historian nor the natural historian can follow the ancient definition of history mentioned by Isidore of Seville: an account supported by direct examination.[48] The impossibility of general autopsy leads Gessner to develop principles of validation for the images made from natural models he did not see himself. In the first volume of the *Historia animalium* he explains that all the images are made *ad vivum*, either under his own supervision, or sent by perfectly reliable friends.[49] This may be possible for live-bearing quadrupeds, not so numerous and quite well known. But for the following volumes, devoted to birds and fish, Gessner cannot maintain this requirement for accuracy. Almost every image, he says at the beginning of the fourth volume devoted to fish, is true.[50] He means that he systematically indicates the origin of his images and that their validity rests on the sources from which they are taken.[51]

Gessner has a very strong opinion about the veracity of the information contained in his books. But the veracity claimed by Gessner is actually restricted to his own part of the work. The point on which he might be wrong is his effort to associate every available fragment of information with the right animal. The remaining truth claim is just a matter of trust toward the various sources of his information. This specific status of truth in Gessner's work appears to be a point of difference from a historical narrative. Gessner's task is to record faithfully in his books every element of knowledge related to animals, whether true or not. Each particular element of knowledge, even if false, is significant as a small part of the written world. Thus, even if Gessner had had the ability to differentiate true from false knowledge, he would nevertheless have kept the latter in his book, as a testimony to the diversity of human opinions. No ultimate truthfulness is supposed to emerge from Gessner's completely achieved work. This may be the main difference with human history.

Contemporaneous Practices of Zoology

Except for its incomparable scale, what Gessner did was not unique in the middle of the sixteenth century. A few examples taken from contemporary zoological works should demonstrate the extent to which Gessner's practices were widespread.

First, the word "history," either in Latin or in vernacular, appears quite often in book titles: fourteen modern authors used it for purely zoological works printed during the sixteenth century.[52] History is thus a widely shared appellation for Renaissance zoology. The more or less equivalent terms employed in other books include "differences," "nature," "diversity," and "description." The associations between those appellations—for instance, "the nature and properties of . . . ,"—and the contents of the corresponding books suggest, as a first step of analysis, that the term *natura* might be employed as a kind of synonym to *historia*. This is suggested as well by the titles of early editions of Aristotle's *Historia animalium*.[53]

Camerarius, for instance, uses the term in contrast with the properties of animals.[54] But the effective value of *historia* in book titles is certainly not unified, and one may easily find counterexamples to this interpretation. More significantly perhaps, the French title of Jean Bauhin's book about butterflies, published in 1593, opposes *traité* and *histoire*.[55] While the *traité* is an inquiry into the available literature, the *histoire* is the fruit of Bauhin's own investigations. This opposition, if we take it as deliberate, might indicate a shift at the end of the century toward an empirical meaning of the term *historia* that it did not have, a few decades before, in Gessner's works.[56] It is worth noticing that the development of empiricism is related to a renunciation of exhaustivity. One cannot be both exhaustive and empiricist: Gessner made his choice, Bauhin made another one for his *Traicté des animauls aians aisles*. Those two authors represent two possible, but not really compatible, epistemologies of natural history.

Two other examples of contemporaneous natural history allow a discussion of the representativeness and specificities of Gessner's scientific practices and their relation to history. It is worth comparing Pierre Belon with Gessner because he was a great traveler and acute observer, but a lesser expert in philology and ancient languages than Gessner was. In his remarkable book about dolphins and cetaceans, Belon makes an interesting use of ancient knowledge and Roman testimonies. Having characterized cetaceans as aquatic animals with quadruped-like internal organization, he logically concludes that the hippopotamus is one of them and tries to find the right representation of this mostly unknown animal.[57] To solve the confusion due to divergent descriptions, he makes use of two ancient representations, a golden medal from the famous collection of Jean Grolier and an antique marble he saw in the papal garden of the Belvedere (fig. 7.5).[58] Both of them are allegorical representations

Le pourtraict du Cheual marin, prins du reuers de la medale de l'Empereur Adrian, tel que nous a communiqué monseigneur le Treforier. Grolier.

Figure 7.5
The hippopotamus from an antique marble of the Belvedere's garden. From Pierre Belon, *L'histoire naturelle des estranges poissons marins* (Paris: R. Chaudière, 1551), 50.

of the Nile river that portray a hippopotamus and a crocodile. After discussing the reliability of Roman imagery, Belon concludes that they give the right image of the animal. This use of archaeological sources should of course be interpreted in the light of the specific Roman context, as described in this volume.[59] Using antique iconographical sources to identify an animal he never saw himself and knew only through ancient literature, Belon's method of knowledge is very like Gessner's in the way it borrows objects and methods from history.

Exactly contemporary to Gessner's and Belon's works, Ippolito Salviani's *Aquatilium animalium historia* offers unusual evidence of the value of an extensive record of ancient zoological references in an extraordinary table containing all the available bookish references concerning fish (fig. 7.6)[60]. Divided in ten columns, this table occupies no fewer than 120 pages in Salviani's book. The first three columns give the names of fish in Latin, Greek, and Italian. The next one indicates the

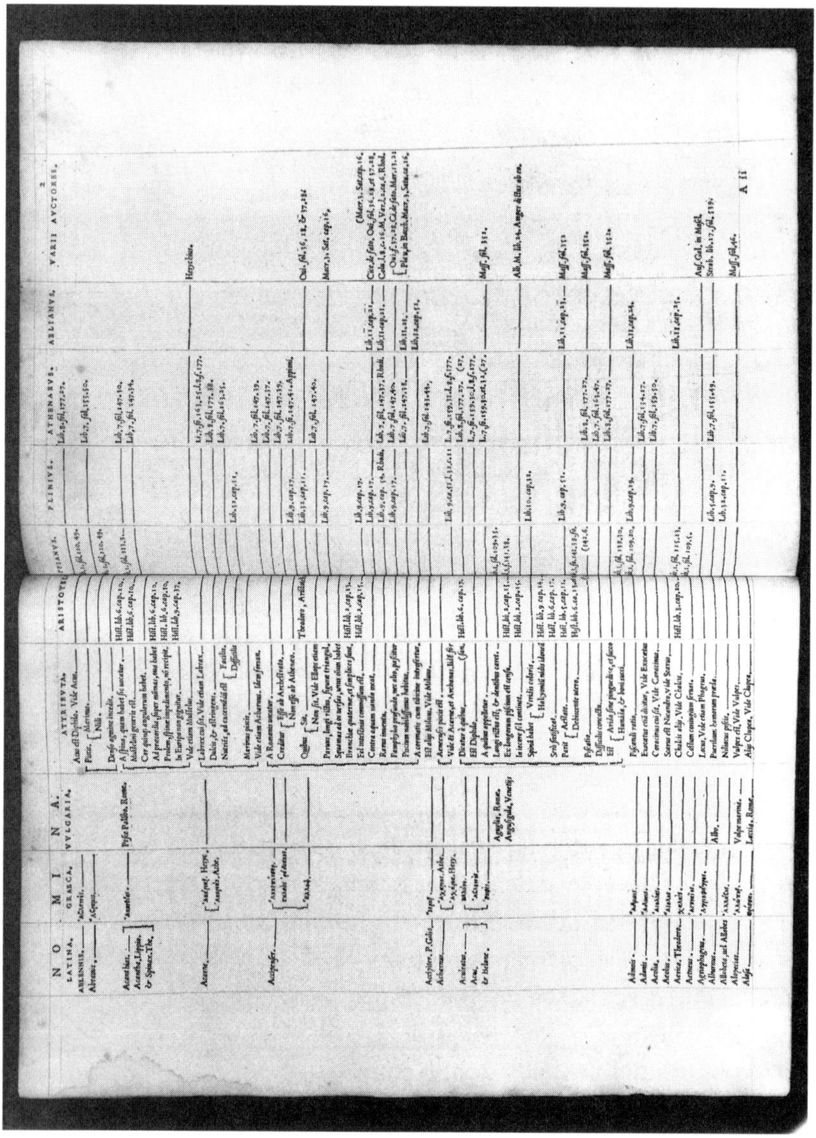

Figure 7.6
Table of references concerning fish. From Ippolito Salviani, *Aquatilium animalium historiae* (Rome: I. Salviani, 1557), fols. 10v–11r.

part or character of a fish being considered, while the remaining columns are devoted to the main writers: Aristotle, Oppian, Pliny, Athenaeus, and Aelian (the last column records textual references from other authors, among which very few are modern). Every fact about a fish is related to one or several textual references, without quotation. The large predominance of ancient writers (about 90 percent of the 5,640 references) gives an idea of the practical importance of a historical survey of zoological knowledge. Salviani shared Gessner's difficulties in associating Greek and Latin names and descriptions with the fish he observed: several addenda and even emendations during printing show that he frequently changed his conclusions. Salviani's table appears to be a preliminary aid to natural history, indicating all the authorities to take into account for the history of each animal. It is a reference frame for the whole book, whose chapters describe fish he observed himself. In this case again, bookish references—that is, the "historical" part of the enquiry—directly support empirical observations.

Thus, the use of ancient knowledge to classify and to describe the natural world is a widely shared practice, even for a less well-educated author like Pierre Belon. Every Renaissance naturalist has to master the ancient knowledge and somehow act as a historian. But the specific purposes of Gessner's *Historia animalium* strongly enhance this general feature. With his onomastic and exhaustive approach to the natural world, with his interest in zoological knowledge more than in the animals themselves, Gessner is much closer than one might have expected to the historian. Not surprisingly then, the *Historia animalium* involves methods and forms of truthfulness and veracity very close or perhaps even identical to those of human history. This practice of zoology gives sense to Poliziano's conception of a "true history" divided into four parts, including natural history.[61] Gessner certainly is the most "historian" of all natural historians. No one after him tried to do more, except Ulisse Aldrovandi, the Italian Catholic author who competed with his Protestant predecessor.[62]

If we try to think about Gessner's motivations for such a long-term work, the historical dimension is still important. Gessner ends his preface to the reader with the wish that his work will last for eternity, not for the sake of his own knowledge, which, he says, is insignificant, but for the sake of so many works accumulated through the centuries and carefully stored in a single volume. Through this ambition of keeping knowledge for posterity, Gessner clearly works for history. With his printed books, he does for zoological knowledge what Noah did before him for animals. His *Historia animalium* claims to be a milestone in human history.

This historical depth of natural history survives to the end of the Renaissance, even if it becomes weaker and weaker as time goes by. In a brilliant analysis of Linnaeus's new natural history, Staffan Müller-Wille concludes that stones, plants and animals "acquired depth and irreversibility in the dimension of time" and considers this acquisition to be a novelty.[63] But Gessner's *Historia animalium* as we understand it suggests that we should consider Linnaeus's intervention more as a reconstruction of the historical depth of natural history than as its first invention. One may also notice that this historical depth still exists in today's natural sciences through the modalities of species references that rely on their first historical description. Thus, although the chronological dimension has sometimes been used to differentiate natural history from human history, it actually seems to be a constitutive and irreducible part of natural history.

Notes

Many thanks to Gianna Pomata and Nancy Siraisi for their kind invitation in Berlin. I would also like to specially thank Ian Maclean, Ann Blair, and Candice Delisle for their very useful emendations to this paper.

Sources cited in boldface type appear in the Primary Sources of the bibliography.

1. Among the works related to Conrad Gessner, see Serrai 1990; Wellisch 1984; Braun 1990 for its iconography; Leu 1990 on Gessner as a theologian. Gmelig-Nijboer 1977 is a specific essay about Gessner's *Historia animalium*, and Friedrich 1995, more general than what its title suggests, gives clever interpretations about history.

2. I closely follow here the enlightening presentation of history in the Renaissance by Mandosio 1995, 47–48.

3. Even though Gessner's zoological works were printed only in the following decade, he had already done a lot of work, as is evident in the section of the *Pandectae* devoted to natural history (**Gessner 1548**).

4. **Poliziano** 1971, 1:471. I follow Mandosio 1995, 46.

5. See Pomata in this volume for a general survey of the meanings of *historia* and Maclean in this volume regarding Aristotle.

6. *De partibus animalium* 2.1, 646a8–12 and *Historia animalium* 1.6, 491a6. Louis (1955) mentions the main occurrences of the word ἱστορία in Aristotle's work and concludes that, except history, it only has the sense of knowledge or science.

7. The explanation that the seal's dentition looks like a saw because the seal has essential characteristics in common with the genus of fish and many fish have dentition of this kind is a remarkable exception (*Historia animalium* 2.1, 501a22).

8. I use the common Latin titles, but it is worth noticing that they have many variants in Greek.

9. See for instance the discussion in the introduction to Théophraste, *Recherches sur les plantes*, ed. Suzanne Amigues (Paris: Belles Lettres, vol. 1, 1988).

10. The complete titles are of this type: *Icones . . . cum nomenclaturis* (**Gessner** 1553 and 1555b). Those two volumes were republished and enlarged in 1560 so as to come out with the third volume about fish (**Gessner** 1560a and 1560b).

11. **Gessner 1551**. The full title is *Opus philosophis, medicis, grammaticis, philologis, poetis et omnibus rerum linguarumque variarum studiosis, utilissimum simul iucundissimumque futurum*.

12. **Gessner** 1551, sig. α6v, "Catalogus doctorum virorum nostri saeculi, qui aliquid ad animalium historiam (imagines, nomina in diversis linguis, descriptiones) nobiscum communicarunt, gratitudinis ergo institutus."

13. **Gessner** 1551, sig. γ1v, "Ordinis ratio, quem per singulas fere animalium historias secuti sumus."

14. **Gessner** 1551, sig. α2v, dedication to the Senate of Zurich: "feci quod potui, aliquot in diversis Europae regionibus mihi comparavi amicos, qui benigne, candide, liberalitater, multas animantium omne genus effigies ad vivum repraesentatas, quarumdam etiam nomina in diversis linguis & historias, mecum communicarunt."

15. This distinction frequently appears in Gessner's forewords. See for instance **Gessner** 1551, sig. β3r.

16. **Gessner** 1551, title page, "non modo simplex animalium historia continetur, sed etiam veluti commentarii locupletissimi, et castigationes plurimae in veterum ac recentiorum eiusdem argumenti, quae videre nobis licuit omnia."

17. See for instance **Gessner** 1551, sig. β3r, "Quid vero laudabilius ex omni humanitatis studio quàm suscipere tam liberale tamque laudabile munus renovandae vetustatis, vel ab interitu potius vendicandae, & nomina rebus, & res nominibus redendi."

18. **Gessner** 1562, n. 37, "Ita digestis commodo perpetuoque ordine omnibus, ut quicquid de uniquoque animali cognoscere libuerit, nullo negocio statim inveniatur. Neque enim ad continuam lectionem à nobis haec condita sunt, sed ut eorum qui Dictionariorum esse solet, ad inquirendum maximè, usus sit: quod non intellexerunt, qui prolixitatis in hoc opere nos accusarunt."

19. **Gessner** 1551, sig. α2v, "The fruits and value of this work."

20. **Gessner** 1551, sig. α3r, "The dignity and eminence of the history of animals toward a contemplation and admiration of the works of God." See on that point Siraisi 2001, 276–277, and Friedrich 1995, 44–47.

21. **Gessner** 1577, 137v, "Sunt enim infinitae plantarum species, quarum magnam partem singulos ignorare necesse est, propter regionum diversitatem. Quòd si suas quisque observationes in commune protulerit, spes est aliquando fore, ut ex omnibus opus unum absolutum ab aliquo Colophonem addituro perficiatur: quod ut nostro seculo fieri optarim, ita vix sperare ausim." See MACLEAN in this volume, 164, for a less optimistic position of Gessner on this topic, in a later letter.

22. **Belon** 1555, 66.

23. **Gessner** 1551, sig. β2r.

24. **Gessner** 1551, sig. β3r, translation from Gmelig-Nijboer 1977.

25. **Gessner** 1551, sig. β3r.

26. **Gessner** 1551, 781, "On the unicorn."

27. **Gessner** 1551, 785 (section 'G'). See SIRAISI in this volume, 330, about Andrea Bacci's treatise on the unicorn.

28. Gmelig-Nijboer (1977, 101) nicely summarizes Gessner's position toward doubtful animals: "whether Gessner himself left their existence undecided, doubted it, or denied it altogether, did not alter the fact that they belonged in an inventory of animal-knowledge as he understood it."

29. Ashworth 1996.

30. This argument was presented at the 2001 annual meeting of the Renaissance Society of America, in a session entitled "Natural History and Human History in Renaissance France," organized by Ann M. Blair with Nancy G. Siraisi as respondent.

31. Caesar *De bello Gallico* 6.26–28.

32. **Gilles** 1533, 136, "Bison cervi speciem similitudinemque gerit: cuius à media fronte inter aures unum cornu in excelsitatem magis dirigitur, quàm ea quae nobis nota sunt, cornua."

33. See for instance the chapter devoted to the elk (*alces*) in **Gessner** 1551.

34. **Münster** 1544.

35. **Herberstein** 1556 (first printed in 1549).

36. **Gessner** 1553, 20, 27, 60, 63.

37. **Gessner** 1558, 519–521.

38. **Rondelet** 1554–55, 492–493.

39. Pliny *Natural History* 32.11 and 9.5.

40. **Belon** 1553.

41. **Gessner** 1558, 520–521.

42. On wonders, see the seminal studies of Lorraine Daston: Daston 1991b and Daston and Park 1998, among others.

43. **Gessner** 1554, appendix, 19.
44. **Gessner** 1554, appendix, 19–20, and **Belon** 1553b, 210r.
45. **Gessner** 1577, 135v.
46. "Catalogus veterum non extantium, qui de animalibus scripserunt," **Gessner** 1551, sig. γ4v.
47. **Gessner** 1560a, 102.
48. Isidore of Seville, *Etymologiarum libri XX*, 1.41.
49. "Nostrae vero icones, quas omnes ad vivum fieri aut ipse curavi, aut ab amicis fide dignis ita factas accepi, (nisi aliter admonuerim, quod rarum est) quovis tempore et perpetuò se spectandas volentibus, absque labore, absque periculo, offerent." **Gessner** 1551, sig. γ1v, "De picturis animalium in hoc Opere."
50. **Gessner** 1558, sig. β1v.
51. "Qualescunque sint autem, verae sunt, hoc est, vel ad naturam factae : vel ad archetypum alterius authoris qui semper nominatur." **Gessner** 1558, sig. β1v.
52. **Aldrovandi** 1599, **Bauhin** 1593, **Belon** 1551 and 1555, **Gilles** 1533, **Boussuet** 1558, **Emiliani** 1584, **Caius** [Kay] 1570, **Marschalk** 1517–20, **Menabeni** 1581, **Rondelet** 1558, **Salviani** 1557, **Schaller** 1579, **Turner** 1544, **Verville** 1600.
53. See Ogilvie in this volume.
54. **Camerarius** 1595 is subtitled: *Exponuntur in hoc libro rariores tum animalium proprietates tum historiae ac sententiae memorabiles.*
55. **Bauhin** 1593. The full title is: *Traicté des animauls aians aisles, qui nuisent par leurs piqueures ou morsures, avec les remèdes; oultre plus une histoire de quelques mousches ou papillons non vulgaires apparues en l'an 1590, qu'on a estimé fort venimeuses.*
56. Concerning *historia* and empiricism, see Pomata 1996.
57. **Belon** 1551, 47v.
58. **Belon** 1551, 50r.
59. See Siraisi in this volume.
60. **Salviani** 1557.
61. Angelo Poliziano, *Panepistemon*, in **Poliziano** 1971, 1:471.
62. See Pinon 2003 for an example of the way Aldrovandi indicates how he surpasses Gessner by the number of cited bookish references. About the crow, he adds more than a hundred to Gessner's 132 references.
63. Müller-Wille 2001, 36.

8
Historia in Zwinger's *Theatrum humanae vitae*

Ann Blair

One of the ways in which *historia* reached its widest audience in the Renaissance was in collections of short passages excerpted from various historical sources and sorted under topical headings, often called *exempla*. The number of works printed in this genre, broadly conceived as collections of short segments of anecdotal historical material, is considerable.[1] They include Latin works, ranging from ancient to Renaissance compositions and typically sorted either alphabetically or systematically, but also vernacular ones, most famously the *Silva de varia lección* of Pedro Mexía, which is miscellaneously arranged and was the object of numerous editions, translations, and imitations in six languages, from 1540 to roughly 1682.[2] In Latin the genre also ranges in size from portable octavo editions, e.g., of Valerius Maximus's *Factorum ac dictorum memorabilium libri IX*, to the largest folios, such as Theodor Zwinger's *Theatrum humanae vitae* (from 1565 to 1604) and its sequel in eight volumes, the *Magnum theatrum vitae humanae* (from 1631 to 1707). The number and extent of these collections of historical excerpts is evidence of an ample early modern readership, both learned and less learned, for the stuff of exemplar history presented ready-made for easy consumption.

The notion that history should serve as a storehouse of vicarious experience from which to derive examples of behaviors both to imitate and to avoid (the "exemplar theory" of history) was dominant from antiquity until the eighteenth century.[3] In the fifteenth and sixteenth centuries this pragmatic interest in learning from history spread well beyond the religious themes traditional in medieval *exempla* and was applied to a new range of ancient sources and to a broader set of worldly concerns. The wide variety of works of this kind that were available by 1555 can be sampled in the compilation of eleven collections of *exempla* published by Johannes Herold as the *Exempla virtutum et vitiorum* (1555). The religious theme suggested by the title of Herold's compilation is borne out by two works in the collection, including the only medieval text in the book,[4] while other works in Herold's compilation specifically address domestic economy, love, government, and military strategy.[5] The largest portion of the compilation is devoted to examples derived from human history, in two ancient collections and three near-contemporary ones, totaling 916 pages.[6] The latter highlight the historical figures involved in each *exemplum* (as is typical in Renaissance collections) and create separate subsections for pagan examples or "more recent" ones. Herold provides access to this massive collection of *exempla* through a large alphabetical index that spans the whole volume.

The interest of early modern printers and readers in the compilation of historical particulars, even on a large and expensive scale, is clear. What is less clear is the role that these collections were expected to play—how they were meant to be used and how they were actually used. Many collections, including Herold's, contain little or no prefatory material.[7] But Theodor Zwinger, whose *Theatrum humanae vitae* was a most impressive contribution to the genre both for its size and its longevity, provides a detailed discussion of the subject, form, and goals of his work. In this brief study focused on Zwinger's preface (in the edition of 1586) I hope to shed light on the theory and practice of *historia* as it was consumed by many in the Renaissance—as distributed into a myriad discrete examples of human behavior. In closing I will address analogies between Zwinger's collection and the gathering of "facts" of a Baconian kind. I will suggest that the accumulation and potential for recombination of particulars drawn from historical accounts, on a scale beyond any individual's interests, sets a precedent for the manipulation of discrete observations associated with the Baconian use of the "fact."

The *Fortuna* and Context of the *Theatrum humanae vitae*

The *Theatrum humanae vitae* (hereafter *THV*), first published by Theodor Zwinger in 1565 in one volume of 1,400 pages, grew through two more editions in Zwinger's lifetime to 4,500 pages in 1586 and was printed again posthumously in 1604. In 1631 the *Theatrum* was reworked and much augmented by the Flemish prelate Laurentius Beyerlinck, as the *Magnum theatrum vitae humanae*; this work appeared in eight folio volumes totaling 7,468 pages plus an index of 600 pages and was printed without major changes in five editions down to 1707. The *THV* offers a rich example through which to examine methods of composition and text management common to many early modern reference works. The text passed through a number of hands in the course of its 140-year career; its composition was in this sense collective over time. Although the work was always attributed to a single author, the composition of each new edition also involved the help of amanuenses (barely acknowledged) and practical shortcuts, such as cutting and pasting from printed books.[8] The problem of order is especially important in the *THV*, given its unusual bulk but also the great care that Zwinger devoted to an elaborate systematic presentation of the material. Zwinger's deeply layered hierarchical subdivisions and multipage dichotomous diagrams were all abandoned in Beyerlinck's *Magnum theatrum* in favor of an alphabetical arrangement of headings, since, as the preface explains, "many approved less of the studied [i.e., systematic] order for history."[9] The increasing numbers of indexes appended to the *THV*, which were unified into a single alphabetical index in the *Magnum theatrum*, also testify to concerns about improving access to the text.

Though typical in these ways of other encyclopedic compilations of the time, Zwinger's *THV* was different in its intellectual conception. The *THV* was not designed as just another compilation of material from which to pad one's prose or oratory, such as the long-lived *Polyanthea* of Domenico Nani Mirabelli (26 editions from 1503 to 1686). Of course, the *THV* may well have been used in the same way as the florilegium, primarily as an engine for Latin *copia*. Indeed, early modern critics of these genres tended not to distinguish between Zwinger and other compilations in complaining about the way contemporaries used them to gather excerpts that they misunderstood and misused.[10] In the shift from Zwinger to Beyerlinck, in fact, many of the features of the *THV* that distinguished it from other compilations and that Zwinger probably most valued were abandoned: not only the systematic arrangement and hierarchical tables of headings, but also Zwinger's long theoretical preface about the work and its philosophical ambitions. Thus the title page of the *Magnum theatrum* boasts that the work "is arranged according to the norm of a universal polyanthea," associating the work with a genre that Zwinger himself had shunned.[11] Zwinger's own conception of the *THV* was thus not as long-lived as the work itself, but it offers a rich set of claims about the nature and role of *historia* that was admired at least in its original context of late sixteenth-century Basel and diffused through the multiple, though constantly revised, editions of the *THV*.

Theodor Zwinger (1533–1588) was well connected to the Basel elite, especially through his mother (Christine Herbster, sister of the printer and professor Johannes Herbster, known as Oporinus) and his wife, Valeria Rüdin, daughter of a wealthy guild master. His father was a tanner from Bischofszell who settled in Basel in 1500 and acquired citizenship in 1526. After his father's death when Zwinger was thirteen, his mother married Conrad Wolffhart, or Lycosthenes, of Strasbourg. Although it has been suggested that this remarriage triggered Zwinger's departure for years of travel abroad, Zwinger used Lycosthenes' collection of *exempla* (left unpublished at his death) as the point of departure for the *THV*, as he acknowledges in the title pages of the editions of 1565 and 1571 and in the preface (but not the title page) of 1586.[12] After working for a bookseller in Lyon, studying classical languages and with Petrus Ramus in Paris, and acquiring a medical degree in Padua (in 1559), Zwinger returned to Basel as professor at the university, successively of rhetoric, ethics, and theoretical medicine.[13] He was friendly with other Basel humanists like Felix Platter and Basil Amerbach (who was also his brother-in-law), and he accumulated a large correspondence with scholars throughout Europe.[14] His publications include commentaries and tabular presentations of Galen (*In artem medicinalem Galeni* and *In librum de constitutione artis medicae*, 1561), Hippocrates (1579), and Aristotle's *Nicomachean Ethics* (1566 and 1582) and *Politics* (1582), as well as treatises on travel (*Methodus apodemica*, 1577), *rustica* (*Methodus rustica*, 1576), ethics (*Morum philosophia poetica*, 1575), and history (published in the *Artis historicae penus*, 1579). Zwinger is also noted for his edition (the only one ever published) of the work of the Platonist Francesco Cattani da Diaccetto,[15] and for his interest in Paracelsianism.[16] The *Theatrum vitae humanae* in its three editions frames a lifetime of exceptionally open-minded and wide-ranging teaching, reading, and writing, especially informed by medicine, ethics, and history.

Historia and *Theoria*

Zwinger's long introduction to the work (and the shorter preface to his treatment of *historia* in the *THV*, volume VI, book 2) elaborates on a notion of exemplar history favored by many humanists. History offered, at no risk, lessons for action, particularly for those in public office. Zwinger refers specifically to the needs of those who "because of serious occupations cannot rely on an assiduous reading of histories."[17] But the scale of Zwinger's *THV*, with its tens of thousands of *exempla* and its universal scope, took exemplar history well beyond the purpose of advising those in government. Zwinger had gathered the material from which anyone could be instructed on any question. "Thus, having brought together in one place all the actions and passions of all men who have lived from the beginning of the world to its end, we may be allowed to wish rather than to hope that this *Theatrum*, finally perfected, would resemble in some way that censorious display before the most just Judge."[18] Zwinger did not shy away from immodest ambition, and registers only a slight discomfort at comparing his task of gathering all human behaviors with that of God at the Last Judgment.[19] Zwinger carries out his ambitions with great diligence and thoroughness, starting with a long preface (33 pages in the edition of 1586) explaining the origins and purpose of the work.

Knowledge stands on two legs, Zwinger begins, relying now more on one, now more on the other, according to the nature of the subject: these are "history, that is the knowledge of singulars, with the help of the senses among which sight holds the principal place; and theory or the nature of universals, with the help of reason. The latter are called precepts, the former *exempla*."[20] The distinction is traditional, but Zwinger notes with unusual emphasis the equal status of the historical and empirical on the one hand and the theoretical grounds for knowledge on the other.[21] The status of the knowledge of particulars guarantees that of his whole project, which rests on observations of human behavior. "There are two kinds of philosophical studies, the one theoretical, the other historical. But in this work, which, as the monument of our attentiveness, can also serve to help others, we pursue not precepts but *exempla*."[22] In the preface to the book on *historia* (volume VI, book 2) Zwinger even articulates the superiority of examples over principles: "Certainly ethical philosophy contains general precepts of the virtues and vices; but history brings forth particular examples of them, which are more efficacious and suitable for moving [people to action] insofar as sense experience seems to be superior to reason in clarity and certainty."[23] Therefore examples of human behavior can accomplish more effectively the same goals as an ethics taught by precepts, echoing the Senecan tag that "the way through precepts is long but that through examples is short and effective."[24]

The goal of the *THV*, Zwinger concludes in a final section "on the purpose or use of this work," is a double one, aimed both at contemplation and at action, and examples are well suited to accomplish both ends.

Therefore the purpose of history is the same as the purpose of *theoria*: leading in part to the apprehension of truth, in part to the possession of the good. The examples of human history gathered in the 29 volumes of this theater serve *firstly* and in themselves philosophical contemplation, either physical and medical, metaphysical or theological, mathematical, mechanical, and for the most part ethical, political, economic: whence those experienced [in these fields] get their name; just as the wise are called so from the knowledge of precepts. But *secondly* to action: insofar as, instructed by these examples, they are led and inflamed through some kind of examination [*epilogismum*] to similar actions, physical, mathematical, theological, ethical, mechanical, and from these themselves they are pronounced doers of good and evil. Thus human history is useful to experience in knowledge and to art in action. These are the twin goals, defined briefly, but very great and broad in their use.[25]

Earlier in the preface Zwinger had explained how each kind of *historia* corresponds to a philosophy or theory of precepts: natural history corresponds to physiology, medical history to medicine, and so on.[26] With its accumulation of the full range of human behaviors, the *THV* is "a human natural history."[27] The wide-ranging examples of the *THV* can be applied to many disciplines, but ethics is the principal field for which Zwinger destined the *THV*. The examples serve not only ethical theory, but also and especially inspire ethical behavior, as the reader reasons (by *epilogismus*) from the particulars accumulated on a theme.[28] Zwinger is not explicit about the process by which the study of historical particulars leads to action, but it would seem to involve some kind of tacit induction.

History and Rhapsody

Zwinger distinguishes his practice of history from that of more ordinary "historici," both in his use of sources and in his order of presentation. Zwinger describes three methods of historical observation: *autopsia* relying on sight, which Zwinger calls the principal sense for history; *akroasis*, which relies on oral reports; and *anagnosis* grounded in written sources." We leave *autopsia* to the historians and we use *akroasis* rarely and sparingly; we use mostly *anagnosis*."[29] Acknowledging that his sources are mostly the texts of others, Zwinger devotes ten pages to a discussion of his choice of authors and *exempla*. Among his sources, for example, Zwinger contrasts the historians who "follow a continuous narrative of *res gestae*" from the "historical rhapsodists" who select from histories some *exempla* as emblems or little flowers.[30] Zwinger not only discusses at considerable length the individual rhapsodists "whose labors we have used now frequently, now sparingly in the *Theatrum*"; he also concludes by identifying the *THV* as a rhapsody of examples.[31]

The *THV* thus offers a particular kind of history, based neither on eyewitness nor oral reports nor on chronological order, but which rather distills through selected examples arranged thematically the utility of history for action: "The historian follows the truth of *res gestae* circumscribed by the bounds of persons, places, and times. The Rhapsodus studies [how] to accommodate the utility of *res gestae* to himself both in public and in private."[32] The rhapsodist is not bound by chronological order, nor by the injunction to historians not to interject their own judgment (or to do so only modestly). Instead the rhapsodist enjoys greater liberty in seeking the knowledge of vices and virtues.[33]

In Zwinger's practice of history or historical rhapsody, the truth claim concerns not the events described but only the reporting of them (see PINON in this volume). "We say nothing that was not said before: and if you condemn the authority of those whom we follow, this is done capriciously [*pro arbitrio*] since you admit that they were reported faithfully from others, truthfully, that is, produced by us in good faith."

276   *Ann Blair*

Zwinger contrasts history with medicine in emphasizing that in history one must rely on the authority of others.

> We are here in a historical argument, which is different from the medical activity [*medica professio*]: therefore we must embrace things that others have said, collected and observed. If these things were to prove false, our credulity would merit reprehension, but our faithfulness is worthy of pardon. We cannot all do everything. The task of the collector is to report in good faith the words and writings of others and to watch and follow the truth of the report, as I say, rather than of the event [itself].[34]

In history, unlike in medicine, there seems to be no external criterion of judgment from which to evaluate the veracity of reports.

Whereas in medicine (I am extrapolating here) there may be grounds for dismissing a report based on personal experience or a sense of what is possible in nature, human history knows no such criteria. This attitude, confirmed by Zwinger's sense of the enormous range of human behaviors as recorded in the *THV*, also helps to explain how a contemporary usually known for his skepticism, Montaigne, could object to excessive *incredulity* applied in evaluating the reports of historians. Montaigne criticizes Jean Bodin, for example, for dismissing as incredible Plutarch's report of a Spartan boy who hid a young fox he had stolen under his shirt and kept it there unto death, even while the fox devoured his entrails. Montaigne objects:

> I find this example badly chosen, since it is very hard to assign bounds to the achievements of the faculties of the soul, whereas we have more chance to assign limits to physical powers and to know them.... We must not judge what is possible and what is not according to what is credible and incredible to our sense.[35]

In a similar spirit Zwinger has accumulated *exempla* in the *THV* without seeking to justify or to query their verisimilitude. Zwinger's justification seems in part one of overload ("we cannot all do everything"), and in part inherent in history, since no one person can report directly from all times and places.

Given his explanation of his procedure (of reporting in good faith what others have said but without critiquing the veracity of their statements), Zwinger acknowledges the importance of his selection of sources and devotes many pages to defending his choices. Zwinger first justifies his decision to include poets, even though they describe fabulous things: "Indeed there is no fable which does not contain a trace or appearance of truth. . . . There is hidden [in them] a mystical meaning, whether physical, ethical or theological, which becomes apparent only to the initiate, those imbued with solid philosophy." Poets are studied in school, after all, and also add pleasant variety to the work.[36] Next Zwinger defends his decision to draw on obscure and new authors. Just as the lees of the wine help to preserve it, and ancient medicine was preserved in the Arabic language, though it is barbaric and unknown to us, so too obscurity is no strike against the quality of an author. "Humble and popular historians often propose more faithfully the truth of *res gestae* than those who were induced to write with great rewards."[37] Zwinger articulates here a principle made explicit in later scientific reporting (and dear to historians of it), but he is most proud of his practice of openly citing these lesser-known sources. Earlier compilers used them too, Zwinger points out, but made little or no mention of them, to avoid being accused of doing so and to acquire a greater authority of polyhistory; but "in this omission itself almost everyone condemns them, and rightly so. Now with equal acerbity some are irritated by our candor, because we cite authors faithfully as we should, but not the authors that they would like."[38] In practice Zwinger does cite other compilers, although he also leaves a small portion (about five percent) of his *exempla* without attribution.

Obscure authors are often the more recent ones, and Zwinger justifies especially his use of new authors. "If the authority of Campofulgosus or Sabellicus, as new men, seems base to you then why not also that of Valerius and Gellius and Macrobius and Laertius? who, it is clear enough, were new with respect to those from whom they described their material."[39] Zwinger echoes many other humanists in arguing that moderns should have the same prerogatives as the ancients did in their day, including that of being "new." "The authority of antiquity comes not so much from the prerogative of time as from the gravity of the authors. Things which are now old were once new: and what is now scorned as new, posterity will venerate with the name of antiquity."[40] Indeed Zwinger's list of authors includes all those in Herold's *Exempla virtutum et vitiorum* (published in Basel in 1555) which likely served as a major source of material for Zwinger and of inspiration for his printer, Zwinger's uncle Oporinus, whom Zwinger credits with both initiating and carrying through the first edition.[41]

As further justification for the vast diversity of authors and examples gathered in the *THV*, Zwinger closes with a kind of natural theological argument. Just as in nature there is nothing so bad that good cannot come from it (for example, theriac can be made from the viper's poison as an antidote to it), so too in the *THV* there is no example, however bad, from which good cannot be derived. Examples of all kinds each serve their purpose. "Old [*exempla*], because of the prerogative of age, bring with them no small majesty. Recent ones strike our senses more and are more capable of moving us, particularly if they are local. Those which are rare are recommended by their very novelty and strangeness. Those which are common or even proverbial acquire authority from habit or use."[42] In the *vivaria* of princes there are not only noble animals, but also rabbits and hares; in the botanical garden of Padua there are the most noble plants, but also, for the sake of the students, ignoble and common ones, "for there is no herb so vile that it does not contain something useful."[43] Similarly in the *THV* the vile and the obscene examples serve a purpose as deterrents, and the variety of *exempla* overall makes the *copia* less fastidious, just as the magnificent buildings augment the beauty of a city and the lowly buildings contribute by making the fine buildings seem more magnificent.

Even more than pleasure, the diversity of examples provides utility, of an explicitly public kind. "Examples which you may reject as exotic, will please another most. Similarly do not pick from a public field plants which will benefit another, even if they are not pleasing to you: unless perhaps you think that all the others were grown just for you."[44] Like the world itself, Zwinger's *Theatrum* offers something for everyone, in which personal displeasure is subsumed under the common good of all. Zwinger acknowledges that, like a mapmaker, he has illustrated only the major headings and invites the studious reader to supplement them by pointing out smaller places or empty ones.[45] The *THV* is thus open to public contributions as well as public use.

History and Rhetoric

Zwinger devotes a section of his preface to the form of the *Theatrum* and specifically to the diction or prose style appropriate for *historia*.

It is the task of history to observe diligently the circumstances of things and the series of time, not to care too much about ornamentation but to rely on moderation, so that the style is neither uncultured nor horrid, but not altogether embellished and excessively adorned with rhetorical trappings which would cast doubt on the reliability of the history, as if the writer cared more about eloquence than about truth.[46]

Zwinger is first concerned to follow a middle course between the excesses of brevity and prolixity. Some rhapsodists, like Valerius and Egnatius, are too prolix and ornate, he opines, while others are too brief: in providing an example without the relevant circumstances they do not allow the inexperienced reader to make good use of it. "Brevity is useful to those who know the histories, to refresh their memories, and it contributes to a synopsis: but it is useless and odious for those not learned or skilled in history, who are the greater part, and even, I would add, for the learned who because of serious studies cannot attend to histories."[47] Thus Zwinger promises to supply the necessary background to understand, for example, an *exemplum* like the "softness of Sardanapalus": Sardanapalus was the last king of Assyria and because of his laziness he was destroyed by Arbax, king of the Medes.

Zwinger follows here the advice of Polybius and other theorists of exemplar history who thought that "contributory conditions of events" should be included in *exempla* to facilitate a knowledge of motives and causes.[48] Zwinger does indeed generally include essential elements of historical context with each *exemplum*, though he warns that as a result in some cases it may appear that he has strayed from the heading under discussion: "this is done so that the whole history can be told under one heading [*titulus*] and then only alluded to succinctly under other headings, and often the reader is sent to the earlier heading (and that nonetheless rather rarely lest frequent cross references become tiresome)."[49] Zwinger does not attend, however, to the intentions or context of the author or source from which the *exemplum* has been taken. In 1685 Adrien Baillet would complain about the use of reference books, mentioning the "theater of human life" in particular: "Those who read disconnected bits in these large reference books cannot know the intention of their original author, and it is hard for them not to apply them badly and against the use for which they were made."[50] Baillet's complaint stems from a different conception of historical correctness than that articulated in the *THV*. Instead of recovering the past context of each passage, Zwinger's intention is to make it possible for readers to use *exempla* in ways appropriate to their needs, without concern for the original authorial intention behind the passage. The *THV* is a storehouse where *exempla* are removed from their original sources, associated with commonplace headings and held ready for reuse by the reader. But one hundred years later Baillet was concerned, much as we are today, to acknowledge and respect the intellectual origins of a piece of text that one cites. I am not sure how widespread Baillet's concern was and whether it was not simply articulated as part of a general attack on reference books, which had become a common refrain.

Most significant perhaps is the way Zwinger distances himself from the rhetorical ambitions of historians. Renaissance historiography was often indebted to rhetoric and might involve the composition of suitably polished orations to put in the mouths of historical figures.[51] But Zwinger explicitly rejects the rhetorical emphasis of narrative history in favor of what he calls "naked things."

> Those who have written of certain kinds of men and states are in the habit of pursuing diligently each smallest detail, even often to intermix some foreign matters for the sake of ornament. But we who consider naked things [*nudas res*] and study in summary insofar as possible, we reject all ornaments: we must embrace only that which is necessary to the knowledge of things.[52]

Of course, as Zwinger has already explained, the "things" of the *THV* are reports taken from other authors. Zwinger indeed goes on to acknowledge that "the style of this Theatrum is varied, since the writers whose authority we have followed used [each] a different style. Indeed when we retained the *exempla* of Valerius ... [and others], changing certain few things, we neither could nor wanted to change the syle, thinking that we should attend to things rather than words."[53] Discussions among Renaissance scholars of the proper relations between *res* and *verba* were widespread and complex.[54] Zwinger's antirhetorical statements, combined with his obsession with dichotomous diagrams and the methodical disposition of topics (e.g., in his *Methodus apodemica* and throughout the *THV*), can probably be attributed to his admiration for Ramist principles of disposition.[55]

The Arrangement of the *Theatrum*

After the theoretical preface the reader is plunged into the first of the twenty-nine volumes of the *Theatrum*, with its own dedication (to Julius Count of Salma and Neoburg, counselor to emperors Maximilian and Rudolph II) and a diagram outlining its principal headings laid out over ten pages.[56] The text itself is composed of short paragraphs arranged in two columns under headings of various levels of generality (identified by their font: large capital letters, small capitals, italics) and various dingbats (variants of the *hedera* or ivy leaf). Between 1565 and 1586 editions of the *THV* developed a more sophisticated and legible page layout, but the visual impact of the *THV* consists throughout in the manipulation of bits of text mostly displayed in paragraphs or in roughly dichotomous diagrams; there are no images.[57] Despite Zwinger's use of the theater metaphor as a place for "watching," introduced in the preface of 1586, the role of the spectator of the *THV* is to read and manipulate bits of text. These take the form of a paragraph, usually of two to ten lines but occasionally running to forty or more lines, neatly demarcated with a line break and an "undent" (the first line beginning earlier than the later lines of a paragraph).

Each paragraph is one *exemplum* and is primarily identified by the name of a person (Claudius Imperator, Georgius Trapezuntius) or people (Iudaei, Druydarum discipuli), occasionally anonymous (Quidam, Vir Atheniensis). The designation of the person is capitalized in the text and is the element that is alphabetized in the *index exemplorum*, which first appeared in 1571. The *exemplum* thus *is* the human (or humans) whose behavior is described along with (as Zwinger notes) essential elements of historical context and often culminating in a short quotation (not typographically distinguished from the rest of the paragraph). The short paragraph usually ends with an indication of the source: almost always an author, often a title, occasionally a book or chapter number, and more rarely still an indication of an earlier source.[58] These short paragraphs are the stuff of the *THV*, lined up and sorted under headings. Thus "Oblivio" appears as a subheading of MEMORIA and is subdivided into two sections: considering its subject—forgetfulness of letters, names, a mandate; and considering its causes—old age, a blow, a fall, wine, poison, disease, study.[59] These subheadings have only one

282   Ann Blair

or two examples apiece; others, such as "great doctors" run for up to ten pages. Occasionally (especially in 1565) Zwinger even devised some headings that were left empty, evidently designed to show the logical range of the concept and hoping to elicit contributions of readers (as mentioned in the preface) or future additions of his own to flesh them out.[60]

Each *exemplum* is independent of those around it and can be shifted anywhere in the *Theatrum*, as one can trace through the major revisions made in the editions of 1571 and 1586.[61] In 1586, for example, the contents of two sections on *oblivio* that had been quite separate in the first edition (located on pages 18 and 1154) were brought together, merged under somewhat different headings. One example disappeared from the section altogether[62] and one was added. About a quarter of the examples were modified in the revision: not in their content, which remains identical, but in their source citation—citations were dropped, added where there had been none or alongside a preexisting citation, or in some cases simply changed.[63] I have yet to determine whether these changes represent improvements—i.e., corrections or omissions of errors. It is clear from this brief comparison of the treatment of "oblivio" in 1565 and 1586 that Zwinger (no doubt with the help of others, including a cousin whom he thanks by name) devoted great care to optimizing the position of each *exemplum*.[64]

The process of revision was not mechanical but complex and carefully studied, based on the mobility of each unit of text within a shifting grid of headings. Zwinger's structural revisions tended toward streamlining the divisions of different topics to highlight their similarities. The number of "books" in the *THV* seems to increase modestly from the initial 19 to 20 in 1571 to 29 in 1586, but this number masks the fact that the 29 divisions of 1586 were not books as in 1565, but volumes, most of which were further divided into books (from 2 to 11 in each volume), so that the 1586 edition actually comprises a total of 112 books. Books that in 1565 were labeled with a general virtue were given more detailed titles and subdivisions in 1586. Books 6 and 7 entitled "de liberalitate" and "de magnanimitate" in 1565 were called "on moderation and immoderation with respect to money and wealth" and "on moderation and immoderation concerning glory and honor," each with multiple subdivisions, in 1586. The structure and headings are more unwieldy but bring out the similarities between different topics, e.g., the Aristotelian theme that virtues come from moderation and vices from the absence or the excess of a quality.[65]

The 1586 edition also adds whole new topics, often with a practical bent, such as volumes on the solitary, academic, religious or political life, and an enormously expanded treatment of the mechanical arts, which grows from 25 pages in 1565 to 464 pages in 1586.[66] One of the additions is a book on the discipline of history in a volume entitled "on the practical aspects of philosophy leading chiefly to both the knowledge and possession of the good."[67] After a prefatory explanation of the principles of exemplar history, Zwinger provides both a narrative outline of the discipline (heavily indented), with an emphasis on its causes, material (time and place) and formal, and a tabular division of the discipline over two pages that highlights the headings that appear in the text. The text itself is quite brief, extending only from pages 1583 to 1594. The first half comprises short descriptions of different authors of history and what they wrote: inventors of history, eccesiastical historians, universal historians, then those who wrote the histories of particular peoples, those who wrote lives, and finally rhapsodists. The second half considers the practice of history more generally: the uses of history (for kings, military leaders, women; to provide virtue, courage, tranquillity, or health); the exercise of history, as true or as mendacious narration (the latter treated at greater length than the former), as explanation, recitation, debate, and judgment; finally the love of history and its negative, the contempt of history. As in a number of other sections (e.g., on medicine, with its very long list of doctors classified chronologically, or in tables of universal chronology), Zwinger brings together material which is not exemplary in any clear way, but rather derives from an encyclopedic ambition to represent all of human experience in an enumeration of historians and the topics they treated.

The *THV* and the Emergence of the "Fact"

Many years ago Walter Ong commented on Zwinger's *Theatrum*: "'Induction' is being encouraged here, subtly but really, by typographically supported developments within the commonplace tradition. The units are not individual observations of experiments, but bits of text."[68]

Ong drew an explicit parallel with Baconian induction; other scholars have also commented on Zwinger's influence on Bacon and Comenius.[69] More recently, both historians of science (including Steven Shapin and Simon Schaffer, Lorraine Daston, and Peter Dear) and scholars of literature (Mary Poovey and Barbara Shapiro) have initiated a discussion of the origins of the "fact" in early modern Europe.[70] This is a question (indebted to Foucault's project of an archaeology of knowledge) asked not about the content of ideas or theories or their influence or diffusion, but about the categories of thought presupposed in the way people in many different fields constructed arguments—by referring to "matters of fact," which were presumed neutral and uncontentious, worthy of agreement. Despite differences in their accounts, these scholars all agree to locate the crucial development of the "fact" as a basis for argument in the early modern period. I am not aware of any discussions of the emergence of the fact in periods prior to the early modern.

My purpose is certainly not to venture a definition of a "fact." Although we all take many of them for granted, the term continues to elicit controversy, as in the arguments used by creationists against evolution: they argue that evolution is a hypothesis, a theory, not a fact. In this case, a "fact" might be said to be a theory on which there is agreement–evolution is a fact for evolutionists, but not for creationists. How much evidence and what kind of evidence it takes to make a theory into a "fact" is the underlying issue, and it is not clear that these two particular camps could ever reach agreement on those basic premises. Similarly, as Shapin and Schaffer have shown, there were plenty of disagreements about what constituted "matters of fact" within the Royal Society. Although the "fact" was especially favored as a kind of argument because it was expected to command assent better than theoretical statements, it is no simple matter to dissociate "fact" from "theory."

Yet it is precisely this claim, to offer matters of fact without the biases of theory, that was made increasingly frequently in early modern arguments, not only in natural philosophy (as Daston emphasizes) but also (as Barbara Shapiro argues) in law, history, and religious controversy— hence the claim that the modern fact was born in the seventeenth century. Of course, the term "fact" (from Latin *factum*) and other vernacular equivalents are all attested long before the seventeenth century; but these terms meant "deed," as in the *faits et dits* of some great leader, which would be recorded in a history (or excerpted and collected in works like Valerius Maximus's or indeed Zwinger's). Empirical observations and arguments from them also certainly existed prior to the seventeenth century. What is new about the "fact" in the argument of Lorraine Daston (or the "modern fact" as Mary Poovey specifies) is a double peculiarity: first, the fact is a "nugget of experience detached from theory"; second, the fact is used not as empirical observations traditionally were, to illustrate or exemplify some general principle already agreed on, but rather as a corrective to and best foundation of natural philosophy—in other words, as evidence from which to build a new generalization.[71]

In this development Lorraine Daston has emphasized the role of the rare and exceptional specimens collected in cabinets of curiosities: because they did not fit received theories, they could become free-floating facts, detached from theory. Although they were not actually theorized very successfully or at all, the monstrous and the unusual formed a crucial part of Francis Bacon's program for the gathering of prerogative instances from which one could hope some day to reach sound generalizations, Daston argues.[72] Most recently Daston has considered why early modern facts were short (e.g., taking the form of numbered observations in Bacon's *Sylva sylvarum*): short facts seemed easier to apprehend (demanding less attention of the reader), they were better suited to command the assent of the widest number of people, and they offered possibilities for interesting recombination without involving the snares of theoretical commitments.[73]

To what extent can Zwinger's *THV* be related to the developments emphasized by this recent historiography on new modes of "factual" reasoning? Zwinger accumulated a record number of *exempla*. These are, like textual "facts," short segments or nuggets of text that have been purposefully detached from the contexts in which they first appeared. However, the *exempla* are then assigned a topical heading and a position within a complex systematic arrangement. There is no material that is extraneous to the existing categories or resists classification, as the rare and the marvelous do in Baconian natural history. *Exempla* are short perhaps for much the same reason that facts were, to command attention more effectively. Both ancient and medieval theorists of the *exemplum* discussed its role in rhetoric and, in the Middle Ages, in the sermon in particular: *exempla* made a point easier to grasp and to retain in memory and were thought to be especially effective in motivating the will, hence their utility for sermons and other hortatory genres.[74] The *exemplum* generally resembles the fact in its formal qualities: it is a brief description, which can be gathered in compilations where it is available for reuse in a variety of ways.

Zwinger's *exempla*, which are predominantly drawn from histories on the model of Valerius Maximus, also resemble "facts" in providing details about the actions and sayings of a particular person or people in a particular time and place. Zwinger regularly includes *exempla* from his own experience and social context, from the poverty of Conrad Gessner as a child, to the generosity of Erasmus toward Zwinger's compatriots Froben, Episcopius, and Basil Amerbach.[75] While some medieval *exempla* also took the form of excerpts from ancient or recent history, many were drawn from bestiaries, folk legends, or religious sources, and offered universal statements of moral lessons.[76] Whatever their origin, medieval *exempla* were more explicitly moralized than Zwinger's. The primary indication of the message that Zwinger expects to be drawn from an example is the heading under which it appears; he only rarely includes a moral conclusion of his own, and not all the headings in the *THV* are moral in character.

In an unusual case in which Zwinger does offer a moralization of his own, it comes as a surprising conclusion not related to the heading under which the example appears. Under "great hunters" Zwinger tells the story of Acteon, who sought to rest in a grove but was turned into a deer for staring at the naked Diana and was killed by his own dogs. Zwinger concludes the account, "This happens even today. When they want to please the lords, they spend all their patrimony in hunts and other trifles and so much that at the end they are shredded and cooked by their own dogs."[77] The moral Zwinger proposes here is unexpected: "do not be a spendthrift." Nor is it brought to the reader's attention, since it appears in a section which is more encyclopedic than moral in character, devoted to an enumeration of famous hunters. In this rare instance Zwinger has moralized his *exemplum*, but in such an idiosyncratic way that readers need not have accepted the moralization along with the story of Acteon.

Zwinger acknowledges openly that one *exemplum* can be used in different ways and that he therefore repeats passages under different headings in the work.[78] I have traced one passage to at least five entries in the *THV* of 1586, with eight points of access from the indexes. The terrible line attributed to Vitellius at the battlefield where his forces had defeated those of Otho—clearing the path to Vitellius's rise to emperor—that "the smell of a dead enemy is sweet, but that of a dead compatriot is sweeter" appears under "cruelty," "cruelty—leaders toward soldiers," "cruelty—delight in the calamities of others," "smell," and "pleasure from smell," each indexed in the *index titulorum*. "Vitellius Caesar" also appears in the *index exemplorum* and is the most complete index entry, with references to four of the five passages.[79] In the *index rerum et verborum* (also new in 1586) the remaining variables are singled out: "enemy" and "compatriot." Zwinger maximizes the ways in which each *exemplum* can be accessed and dictates no single use of a passage. As an example of the power of smell this *exemplum* is not moralized at all.

Zwinger's motives are not primarily rhetorical. These *exempla* may be used to ornament an exercise in oratory or to illustrate a precept, Zwinger acknowledges, but he strives for a higher, philosophical purpose: to lead the reader to ethical action.

> If you want to contemplate the actions and passions of human life in the histories of others, you will find described here the headings or some markers to follow: and if you would rather use our collections, you will be supplied with examples to illustrate all sorts of philosophical precepts. But if from the example of others (which everyone should very much strive for) you will be driven to act and work, theoretical, practical, and mechanical [examples] will be offered to you, which you may express by imitation if they are good, or detest and flee if they are bad.[80]

Personal judgment is central to this active use of *exempla*, and Zwinger has done his utmost to serve the needs of as many different readers as possible, to cater to every possible set of interests, mental categories, and motivations, just as a "public field" provides plants for different tastes that cannot please everyone.[81] Given the role of personal judgment, these historical "facts" for moral use are probably not as widely shared or shareable as the natural facts on which the Royal Society hoped to build consensus within its ranks and beyond. Instead, in the moral realm, with its nearly infinite supply of indefinitely recombinable *exempla*, readers apply their own sets of interests and reach their own conclusions.

Whether the nuggets of experience presented in the *THV* were ever used as the grounds of new generalizations—the second hallmark of the modern "fact"—is hard to determine. The marginal annotations left in the *THV* offer evidence of thoughtful reading, but not of further stages of reflection. Marginal annotations I have found in the *THV* are typical of learned reading: they offer cross references within the work, make additions and corrections to the index and source citations, and add further examples from personal and reading experience. In one such case a reader copies out in the margin the moral assigned to a story of Augustus, who chose justice over friendship in condemning a friend.[82] Another strategy for assessing how the *THV* was used is to identify in contemporary works passages that are likely to have come from Zwinger, although the *THV* is almost never cited explicitly. Fausta Garavini has argued convincingly through a close analysis of Montaigne's "Histoire de Spurina" (*Essais*, II:33) that Zwinger was Montaigne's source for this *exemplum* concerning a handsome Etruscan young man who disfigured himself to avoid being the object of desire: Zwinger had classified it under resistance to desire (rather than, as other collections did, as an example of chastity) and placed it alongside the story of Xenocrates, which Montaigne also included in this essay.[83] In this case Montaigne borrowed the anecdotes along with the heading Zwinger had assigned them, and that combination was the key to identifying their likely source. Cases of *exempla* reused independently of their original heading (and more similar to "facts" for becoming thus "free-floating") would be even harder to trace convincingly to their source.

Whether or not Montaigne always accepted Zwinger's heading assignments for the *exempla* he borrowed, his purpose was precisely not to derive rules for action from historical experience, as Zwinger hoped his readers would. Neither was Montaigne's purpose to illustrate preexisting moral truths, in a traditional rhetorical use of *exempla*. Instead, in his unique way, Montaigne used the abundant material he gathered through commonplacing to scorn the notion that lessons could be drawn from the variety of human experience and to question the instructive value of the *exemplum* altogether: "Tout exemple cloche et la relation qui se tire de l'expérience est toujours défaillante et imparfaite."[84] From his consumption of *historia* as nuggets of human experience Montaigne derived neither traditional *exempla* that illustrate received opinion nor "modern facts" designed to serve as evidence for a new generalization. Instead, Montaigne challenged the notion that something like "facts" drawn from experience could lead to sound knowledge.

Zwinger's massive collection of *exempla* is, on the one hand, the high-water mark of a practice of exemplar history that originated in Roman antiquity and maintained a presence in the Middle Ages before experiencing a great resurgence in the Renaissance.[85] On the other hand, Zwinger's *Theatrum* is also comparable in some ways to later works full of "facts," such as the *Sylva sylvarum* (1627) of Francis Bacon. The *THV* is a collection of particulars, mostly drawn from reading rather than direct experience, that were painstakingly collected and sorted into headings in the hope that they would motivate both philosophical contemplation and ethical action. The scale and the variety of the compilation, which was meant to include something for every reader and on every topic and went far beyond the needs of any individual, was designed as a contribution to a public field from which readers would pick and choose as they pleased. Zwinger's work was no doubt used to impress by its massive presence on a bookshelf and its *exempla* employed to embellish prose in predictable ways, by illustrating preestablished generalizations. But the *THV* also offered such an array of material that it could prompt unexpected and original reflections—not only those of Montaigne on the impossibility of deriving knowledge from experience, but also generalizations of a more constructive kind, such as the ones Francis Bacon envisioned deriving some day from the unsorted mass of particulars that he left in manuscript at his death. The collecting of and generalizing from the nuggets of experience called "facts" in seventeenth-century England may well owe something to traditional practices of excerpting, compiling, and sorting, especially when these were applied to the short excerpts of historical particulars that held such appeal in the Renaissance and on such a massive and impersonal scale as in Zwinger's *Theatrum humanae vitae*.

Appendix

Theodor Zwinger, *Theatrum vitae humanae . . . in XIX libros digesta . . . cum gemino indice*. Basel: J. Oporinus and Froben brothers, 1565. 19 books and 1428 pages; series titulorum (in order of appearance) and apology for lack of alphabetical index. Typically bound in one large folio volume.

———. *Theatrum vitae humanae* . . . *in XIX libros digesta* . . . *a multis haeresibus et erroribus* . . . *vindicatum et repurgatum*. Paris: Nicolas Chesneau, 1571. 19 books and 2316 pages; expurgated for Catholic readership; series titulorum and index rerum et verborum.

———. *Theatrum vitae humanae* . . . *accessit elenchus triplex*. Basel: Froben, 1571; 20 books and 3455 pages; series titulorum; elenchus titulorum (alphabetical); index exemplorum; catalogus autorum. Designed for binding in two volumes.

———. *Theatrum humanae vitae* [note title change] . . . *cum tergemino elencho*. Basel: Eusebius Episcopius, 1586; 29 books and 4373 pages; series titulorum; elenchus titulorum; index exemplorum; auctorum magis illustrium nomenclator. Bound in 5 vols.

———. *Theatrum humanae vitae* . . . *cum quadrigemino elencho*, ed. Jakob Zwinger. Basel: Sebastian Henricpetri, 1604; 29 books and 4374 pages; identical to 1586 edition; with addition of a new index: titulorum catalogus; elenchus titulorum; index exemplorum; rerum et verborum memorabilium index; auctorum magis illustrium nomenclator. Variously bound in 8, 9, and 10 vols.

Laurentius Beyerlinck, *Magnum theatrum vitae humanae*. Cologne: Hieratus brothers, 1631; Lyon: Huguetan and Ravaud, 1656 and 1666; Lyon: Huguetan, 1678; Venice: Paulus Balleonius, 1707. 26 sections, each paginated separately; designed to be bound in 8 vols. I: A-B 997pp; II: C-D 1080pp; III: E-G 907pp; IV: H-L 1094pp; V: M-O 1056pp; VI: P-R 1247pp; VII: S-Z 1087pp; VIII: index, 604pp, for a total of 7468pp + index of 604pp.

Notes

I am grateful for insightful readings to all the participants in the workshop and especially to the editors of this volume and the two commentators on my paper, Ian Maclean and Brian Ogilvie.

Sources cited in boldface type appear in the Primary Sources of the bibliography.

1. The number becomes staggering if one also considers (as I will not attempt to do here) the closely related genres of collections of *sententiae*, apophthegms, and proverbs, which include Erasmus's *Adages* and numerous florilegia and polyantheae. Pierre Villey emphasizes the common features of collections of *exempla* and *sententiae* as both sources and models for Montaigne's *Essays* (Villey 1908, 2:27–31); the complex interactions between these genres would reward further study.

2. For a brief overview see **Mexía** 1989, 1:54–59.

3. See Nadel 1964. The exemplar theory was dominant, but not universal: Guicciardini for one questioned whether one could draw examples from history; see Gilbert 1965, 299–300.

4. Nicolas Hanappus, *Virtutum et vitiorum exempla* (whose title Herold borrowed). Hanappus was Patriarch of Jerusalem and died in the siege of Acre in

1291; see Welter 1973, 230–233. The other work on a religious theme is near-contemporary: *De vita religiose per exempla instituenda* by Marcus Marulus Spalatensis (from Split, Croatia, 1450–1524).

5. These are: *Aristotelis oeconomicarum dispenationem exempla*, a brief text (658–666), which is unattributed and may be the work of Johannes Herold himself; *Erotica* by Parthenius (1st century BCE, previously published in Basel, 1531); *De politiis* by Heraclides Lembus (2nd century BCE, previously published in Rome, 1545, with Aelian [see note 6]; 10pp); Sextus Julius Frontinus, *Exemplorum sive stratagematorum libri IV* (1st century CE, probably first published here).

6. Valerius Maximus, *Factorum ac dictorum memorabilium libri IX* (1st century CE; 172pp); Aelian, *Variae historiae* (2nd century CE, previously published in Rome, 1545; 152pp), which includes some natural historical topics; Marcus Antonius Sabellicus, *Libri exemplorum* (first published Venice, 1507; 177pp); Baptista Campofulgosus, *De dictis factisque memorabilibus* (first published Milan, 1509); and *De rebus humanis variorum exemplorum librum* (preface dated Paris 1516, probably published here for the first time; 75pp) by one Guy de Fontenay of Bourges (Guido Defonte Nayo Bituricensis).

7. The copy of Herold that I have used, at Houghton Library, lacks most of the first two leaves, but any prefatory material that it contains is less than one page long, given that the first leaf was devoted to the title page and the verso of the second leaf to a list of authorities used (still partially visible). On Herold, see Burckhardt 1967.

8. This practice is explicitly noted in the preface to **Beyerlinck** 1666, sig. [e3]v. Samuel Hartlib also attributes it to Zwinger himself, on what evidence I do not know. **Hartlib Papers Project** 1995, Ephemerides, 1641, "ars excerpendi." For more discussion of this phenomenon, see Blair 2003.

9. **Beyerlinck** 1666, sig. [e3]v. Nonetheless, the articles in Beyerlinck can be very long ("Bellum" runs to 106 pages) and often maintain Zwinger's subdivisions within them.

10. Baillet 1685, I:468. See also **Liberius** 1681, sig. A2r.

11. *Magnum theatrum vitae humanae: hoc est rerum divinarum, humanarumque syntagma catholicum, philosophicum, historicum et dogmaticum, ad normam Polyantheae universalis dispositum* (1631).

12. Dufournier 1936; Karcher 1956. Zwinger and Amerbach also drew up the inventory of Oporinus's manuscripts after his death in 1568, including a number of works by Lycosthenes, as studied in Gilly 2001.

13. For good evidence about his high standing in this community, see **Reusner** 1589, which features a portrait and some eleven odes of praise, one of which is ten pages long.

14. Zwinger and Basil Amerbach made bequests to one another of handsome items in their collections; see *Das Amerbach Kabinett* 1991, 5:95. Felix Platter wrote a life of Theodor Zwinger for the 1604 edition of *THV*. Some 2,300 letters to and 400 letters from Zwinger are extant in the Universitätsbibliothek in Basel;

more letters from Zwinger can presumably be found scattered around European libraries. Carlos Gilly of Basel is planning an edition of Zwinger's correspondence. Some of Zwinger's correspondence is discussed in Rotondò 1974.

15. See Gilly 2002. I am grateful to Carlos Gilly for a copy of this article and other articles in press.

16. Oporinus had been amanuensis to Paracelsus. Zwinger was at first hostile to Paracelsus, complaining of the confusion he caused students; see Berchtold 1990, 2:669. But Zwinger introduced Paracelsus into his lectures and composed a preface to one of his works, which was published anonymously; see Gilly 1985, 23, 126; Gilly 1977–79.

17. Zwinger 1586, sig.[***7]v.

18. Zwinger 1586, sig. **3v.

19. See also the portrait Zwinger commissioned by Hans Bock in which Zwinger is holding a skull as a *memento mori*, but the skull is adorned with a crown of laurels, sign of immortal glory. In the background the fall of Bellerophon is depicted, an *exemplum* of human ambition struck down by the gods, and yet the *exemplum* also calls attention to that ambition itself. Following the common practice of the Basel elite (a few murals have survived to this day, thanks to restorations over the centuries) Zwinger had plans made for murals for the outside of his house. The drafts by Hans Bock emphasize the same theme, adding depictions of Icarus and Phaethon to that of Bellorophon. The paintings were to be inscribed with appropriate sentences, such as: "Hast du vil/rum undt er/erjagt das/als weltt von/dier eracht/huett/dich steig/nit gar/zu hoch/das du/nit aber/fallst her/noch." See *Spätrenaissance am Oberrhein. Tobias Stimmer (1539–1584). Ausstellung im Kunstmuseum Basel* (1984), 79–81. For records of the inscriptions inside Zwinger's house, see **Tonjola** 1661, 400–403; **Gross** 1625, 475–480.

20. Zwinger 1586, sig. *3r.

21. For an insightful analysis of the place of Zwinger's conception of history as "sensata cognitio" in the context of contemporary works by Bodin and Patrizi, see Seifert 1976, 79–88.

22. Zwinger 1586, sig. [*6v].

23. Zwinger 1586, 1580.

24. Seneca *Epistles* 1.6.

25. Zwinger 1586, sig. [***6]r. Emphasis mine.

26. Zwinger 1586, sig. [*6]v.

27. Zwinger 1586, sig. [*7r].

28. "Epilogismus," coined from the Greek ἐπιλογισμός, reckoning, calculation, examination, is found in medical texts to mean specifically arguing from effects, or "a comparison of like with like, which is proper to empirics"; see Maclean 2002, 160.

29. Zwinger 1586, sig. [*7]r. I have not succeeded in finding a source for this tripartite distinction. Ἀνάγνωσις and ἀκρόασις appear in the list of spiritual exercises enumerated by Philo of Alexandria, as quoted in Hadot 1981, 35 and

are discussed in chapters 13 and 14 of the *Exercises* of Aelius Theon (ca. 1st century CE); see **Kennedy** 2003, 66–70.

30. **Zwinger** 1586, sig [*7]r. On rhapsody, see Ong 1976 and Chatelain 1997.

31. **Zwinger** 1586, sig. **2r, [***7]v.

32. **Zwinger** 1586, sig. [***7]r.

33. **Zwinger** 1586, sig. ***6v.

34. **Zwinger** 1586, sig. [**5]v. "Non omnia possumus omnes" is an accurate quotation from Virgil's *Eclogues* 8.63; I am grateful to Ian Maclean for pointing this out. Zwinger's own practice here is a fine example of the creative reuse for his own purposes of a quotation that had acquired proverbial status.

35. **Montaigne** 1988, II, ch. 32, pp. 722–725; **Montaigne** 1965, 546–548. Cf. **Bodin** 1566, 66–67.

36. **Zwinger** 1586, sig. **2v.

37. **Zwinger** 1586, sig **3r.

38. **Zwinger** 1586, sig. **3r.

39. **Zwinger** 1586, sig. **3r. Marco Antonio Sabellico (1436?–1506) wrote on the history of Venice and published commentaries and editions of ancient texts (including Suetonius and Justin Martyr) as well as the work Zwinger probably has in mind here: *Exemplorum libri X* (Strasbourg, 1511, reedited at least 1555). Campofulgosus (also Fulgosus) is Battista Fregoso (1453–1504), author of *De dictis factisque memorabilibus* (Milan, 1509, with at least seven reeditions between 1541 and 1604).

40. **Zwinger** 1586, sig. **3v.

41. **Zwinger** 1565, 19 on Oporinus. Zwinger names his sources in **Zwinger** 1586, sig **v.

42. **Zwinger** 1586, sig. **4r.

43. **Zwinger** 1586, sig. **4v. Cf. the saying attributed to Pliny that there is no book so bad that some good could not be got out of it; Pliny the Younger *Letters* 3.5.8–12.

44. **Zwinger** 1586, sig. **3v.

45. **Zwinger** 1586, sig. **3v–**4r.

46. **Zwinger** 1586, sig. ***4r.

47. **Zwinger** 1586, sig. ***4r.

48. Nadel 1964, 302, paraphrasing Polybius and Tacitus.

49. **Zwinger** 1586, sig. ***4v. Zwinger reserves a special symbol, the pointing finger, to highlight cross references from one section to another.

50. **Baillet** 1685, 1:467.

51. See Burke 1969.

52. **Zwinger** 1586, sig. ***4v.

53. **Zwinger** 1586, sig. ***4v.

54. See Kessler and Maclean 2002.

55. **Zwinger** welcomed Ramus on his visit to Basel in 1568 (Berchtold 1990, 2:679) and chose Ramus as godfather for his son Jakob (see Dufournier 1936, 325). Ramus praised Zwinger in his *Basilea* 1571; see Ong 1977, 177. The most recent expositions of Ramus's life and influence include Skalnik 2002 and Feingold, Freedman, and Rother 2001. The classic exposition of Ramism by Ong (1958) has been supplemented by Meerhoff 1986 and Bruyère 1984.

56. Although the work as a whole is dedicated to the "One and Triune Lord," the twenty-nine separate volumes of the 1586 *THV* give Zwinger an opportunity to line up an impressive number of human dedicatees, ranging from humanists in Basel and across Europe, from Spain to Poland, to Baron Hartmann of Liechtenstein or Count George of Thuringia. The editions of 1565 and 1571 are dedicated to Christophorus, Ioannes, and Esaias Weidmoser, lords of Winckel and Falckenstain, the sons of Christophorus Weidmoser, whose exemplary fortitude in the face of death is included in the *THV*; see **Zwinger** 1565, 408.

57. See also on this point West 2002, 242 n37 and 52.

58. The citations for the examples of "memoria debilis" are: no source (in 1565 the same excerpt is attributed to Ravisius Textor); Plut. (in 1565 no source was provided); Plutarchus in Theseo; Macrobius lib. 2. cap. 4. Saturnalium; Ludo. Vives de Causis corrupt. artium lib. 2; Suetonius; Ex Seneca, Caelius [Rhodiginus] libro decimotertio, capite trigesimoprimo Antiq. Lect. (a rare equivalent to our "as quoted in"); Aristot. in 3. Rhetoricorum; Cic. in Bruto. **Zwinger** 1586, 36.

59. This final, paradoxical category includes two exempla: Socrates' complaint that study is the ruin of memory because men confident in what they know exercise their memory less (no source given) and "Antisthenes of Athens to a servant complaining of the loss of his commentaries, said: You should have written them on your mind, not on paper. Laertius." **Zwinger** 1586, 36.

60. E.g., in Zwinger's attempt to classify the natural resources around the world. In 1565 headings for "Regions fertile in wine," and "in milk" are left with no entries at all, while in 1586 this section is much expanded and includes headings on the presence of minerals, metals, trees, and animals of many kinds, none of which are left blank. Compare **Zwinger** 1565, vol. XIX, pp. 1313 ff, and **Zwinger** 1586, vol. XXIII, book 2, pp. 3969 ff.

61. Since I have had limited access to the edition of 1571, I am only able to offer comparisons of the editions of 1565 and 1586; some of the features that I describe from the edition of 1586 may also be present in 1571.

62. An example featuring Bartolus, in **Zwinger** 1565, 18; compare with **Zwinger** 1586, 35–36.

63. For example, "Dion. Nicaeus et Xiphilinus in Antonino" becomes "Dion. et Suidas"; see **Zwinger** 1565, 18, and **Zwinger** 1586, 35. A citation attributed to "Brusonius, lib 4 cap 4. Eras. Rot. lib. 4 Apoph." is attributed to "Macrobius lib. 2. cap 4. Saturnalium"; see **Zwinger** 1565, 1154, and **Zwinger** 1586, 36. See also note 58 above.

64. Zwinger acknowledges help in preparing the third edition from the "faithful and elegant hand of Basilius Lucius alone, a dearest cousin [on his mother's side] . . . in copying and in bringing together things that belonged together for three years and more." **Zwinger** 1586, sig. ***5r.

65. **Zwinger**'s teaching of ethics warrants further study. His commentaries on Aristotle's *Nicomachean Ethics* of 1566 and 1582 offer increasingly more detailed tabular presentations of the subject. Although he focused on Aristotle, Zwinger also admired Plato and recommended choosing from pagan ethics what was compatible with Christianity; see Kraye 1988, 320, 329.

66. A point made in Gilly 2002.

67. **Zwinger** 1586, vol. VI, book 2, p. 1554; this volume title is already present in book 4 of the edition of 1571.

68. Ong 1976, 109.

69. Berchtold 1990, II:656, mentions Comenius's citation of Zwinger in his project for a theater, which was never realized. Gilly 2002 surmises that Bacon drew from Zwinger for his division of history in *De augmentis scientiarum*, 2.2–12.

70. Shapin and Schaffer 1985; Dear 1995; Poovey 1998; Shapiro 2000; Cerutti and Pomata 2001a. Lorraine Daston's many articles on the topic include Daston 1991a, 1991b, 1996, 1997, and 2001b.

71. Daston 1991a, 343; for a critique by Poovey, who argues that Bacon overemphasized for rhetorical effect the difference between his new regime of fact gathering and those which had preceded it, see Poovey 1998, 9. Poovey focuses instead on the way in which numbers came to hold special appeal as facts, because they rejected the "ornamental excesses associated with rhetoric" while remaining amenable to generalization, to serve as evidence for something. Though she traces the origins of this interest in number back to double-entry bookkeeping, her story focuses on late seventeenth- and eighteenth-century England.

72. See Daston and Park 1998.

73. Daston 2001b. I am grateful to the author for an English version of this article.

74. For references to ancient rhetorical theory, see David 1998, 10; on medieval rhetorical theory, see Bremond, Le Goff, and Schmitt 1982, 31.

75. **Zwinger** 1586, 3853 on Gessner and 912 on Erasmus's gifts of vases of silver and gold that he had received from various princes.

76. Bremond, Le Goff, and Schmitt 1982, ch. 2.

77. **Zwinger** 1565, 200.

78. **Zwinger** 1586, sig. ***4v.

79. The various topical headings in **Zwinger** 1586 refer to pp. 70, 120, 2755, 2738, 2787. The index entry for "Vitellius Caesar" lists them all except the reference to 2787.

80. **Zwinger** 1586, sig. ***6v.

81. Hampton 1990 has emphasized the role of the exemplar in developing a notion of selfhood in the Renaissance.

82. **Zwinger** 1565, 260 (Houghton copy, *2000-512F).

83. Garavini 1992.

84. **Montaigne** 1988, 3:1070. Montaigne builds his argument here around a proverbial Latin expression "omnis comparatio claudicat." On Montaigne's attitudes toward commonplacing, see Goyet 1986–87.

85. Valerius Maximus was one of a number of authors to establish the genre of the compilation of *exempla* in the first century CE; David 1998, 10. On the medieval fortuna of Valerius Maximus, which included excerpts and a poetic adaptation in the eleventh century, full manuscript copies in the thirteenth and commentaries in the fourteenth century, see Reynolds 1983, 428–430.

9

Histories, Stories, *Exempla*, and Anecdotes: Michele Savonarola from Latin to Vernacular

Chiara Crisciani

Michele Savonarola and His Works

In general, the Scholastic thought of the thirteenth to fifteenth centuries shared—and indeed reinforced—the Aristotelian concept of knowledge (*scientia*) as deduction of necessary conclusions from principles and as knowledge of causes. This is true even of those currents of Scholastic thought—for example, the various schools of nominalism—that from a logical, gnoseological, and above all ontological perspective most strongly emphasized sense evidence and the perception of unique reality, that is, of the individual.[1] Those who adopted such an orientation immediately encountered the problem of how it was possible to acquire *scientia* that maintained the usual Aristotelian characteristics. For this reason, one can really speak, as regards the later Middle Ages, of "a natural philosophy without nature," of a displacement of truth exclusively to the level of language, with either skeptical or Platonic/Augustinian consequences.[2]

In this context empirical knowledge (not always, however, *experientia*, which in Scholastic Aristotelian epistemology has a special meaning),[3] that is, direct sense experience, is often understood as a very confused and unstable kind of knowledge, in which items of information come at random from a low level. This type of information is, to be sure, regarded as an indispensable part of the process of acquiring knowledge and is even recognized as unavoidable in certain disciplines (for example, medicine). But it is perceived as needing to be integrated as rapidly as possible into more readily controlled and ordered forms and levels of knowledge.[4] Such an approach was found even in medicine, which was of all the disciplines taught in universities the most involved with observation and operative techniques.[5] Nevertheless, in certain areas of medieval thought one encounters a valorization of empirical knowledge as a form of *experientia* that not only sets forth empirical information but is also productive of special forms of knowledge that do not necessarily need further elaboration and that do not, on that account, fall into irrationality (perhaps they fall in nonrational fields) or complete failure to communicate. The areas I have in mind are some forms of mysticism, some trends in alchemy (a branch of knowledge very strongly rooted in validation of empirically and experientially acquired information), and even some phases of crisis and transition in medieval medicine.

This evaluation of empiricism is seldom discussed in medieval texts;[6] very often it is simply found in actual practice. But it is also possible that in most instances this "epistemology of the empirical" (as I term it) can only be implicit, given that it is found in contexts in which the need to "know, teach, and learn by doing" is affirmed. Hence this epistemology may be hidden in technical directions and accounts of mystical experiences, or it may underlie a recipe or an account of one person's life. Thus, it is obvious that even the legitimacy and possibility of research into medieval epistemology of empiricism raise many theoretical and historical problems. Nevertheless this paper will, I hope, be a contribution to such research.[7] In any case, research into these issues, supposing it is possible, certainly implies attention to facts and the uses made of facts (although I agree with Ann Blair[8] in considering the term "fact" anachronistic in pre-Baconian contexts).

I do not want to repropose here old-fashioned and historically misleading ideas about precursors. The philosophical and cultural context of the late Middle Ages and the fifteenth century only occasionally presents what can be defined as a prehistory of *historia* as discussed in this volume; even if, for the most part, the same terms are used, it is with very different meanings and intentions. Moreover, it should be noted that *historia* and empiricism are not always linked in a direct relationship, as some of the essays in this volume clearly show.[9] In some instances, indeed, *historia* (conceived of as a collection of facts, cases, and data) is present, but the facts collected are merely textual facts, although considered true. In other texts, such as those of Savonarola, there is a strong emphasis on the value of empirical data, even those emerging from everyday experience. Such data are carefully enumerated and described, but evidently without the author having any conscious program to collect them under the heading of *historia*. The latter situation might, however, be considered a starting point toward more deliberate and programmatic ways of collecting data, facts, and *exempla*, as Gianna Pomata's essay shows.

All these considerations provide, I think, the more reason to pay attention both to continuities and to differences, to take note of slow changes in terminology and concepts. To be sure, an increasing quantity of occurrences may eventually amount to a change in quality—such, indeed, may be the case with the use of *exemplum*. By being attentive to these processes, we can recognize the "explosive novelty" of revolutionary epistemological ideas, the acceptance of which was perhaps rendered more urgent also by the impasses that the natural sciences—including medicine—had reached in the late Middle Ages.

In this context, Michele Savonarola is one of the most interesting representatives of medicine in the fifteenth century, a period that was certainly a critical and eventful phase in the development of medical learning and indeed an epoch of transition. Michele was also one of the earliest of that substantial group of physicians who over the next two centuries found it easy and a natural connection to produce works on both medicine and history.[10] In Savonarola's eyes, as we shall see, the two fields were linked both by the conceptual structure that they required[11] and, above all, by values and obligations that were not only intellectual but also ethical. More importantly, however, for Savonarola history and medicine were both fundamental parts of the court commitment of a physician-scholar who conceived of his position in regard to the prince as very similar to the respectful relation, paternal on one side and filial on the other, that bound together Aristotle the Philosopher and Alexander the Great throughout life.[12]

Michele Savonarola has been defined as "an atypical Renaissance physician."[13] In reality, he perhaps had a fuller awareness than other contemporary physicians of some new features of a new profession diffusing itself in northern Italy: that of the court physician. That a physician could act as a counselor and a pedagogue was already possible, though of rare occurrence, during the Middle Ages.[14] This counseling function is rooted in the very definition of medical knowledge, which was carefully analyzed by medieval physicians, particularly in their *orationes* in praise of medicine.[15] According to them, medicine is a specialized form of knowledge with a unified and defined *subiectum*: the human body. But the human body is an organism that exists within and is conditioned by a dense net of influences and relations (above all, the link between body and soul), which the physician also has to take into account. Therefore, his specialized knowledge becomes the core of many different but necessary competences and presents itself most fully as a sort of encyclopedic knowledge that brings together all the information necessary to promote the well-being of the *subiectum*-organism. Hence, it encompasses the liberal arts, of course (defined as *subservientes*), but also oneiromancy, physiognomy, alchemy, meteorology, astrology, cookery, and, above all, psychology. This articulated and broad competence enables the physician to give advice on all kinds of topics. They range from sexual and affective behavior[16] to eugenic practices designed to produce vigorous progeny; from cautions in organizing a military campaign[17] to suggestions regarding use of leisure and appropriate reading for a convalescent; from problems of travel in exotic lands[18] to the choice between studies or military life for an adolescent. Therefore, it is no wonder that a physician with this range of competences should become a prominent adviser with many abilities and duties (always based upon the doctrinal value of his learned knowledge). This was especially the case in the courts of northern Italy in the Quattrocento, where court relationships overlapped friend and family links. Here the physician often became a functionary as well as a trusted "friend" and indispensable adviser. The various activities of Guido Parato, Giovanni Matteo Ferrari da Grado, Benedetto Reguardati at the Sforza court;[19] of Pantaleone da Confienza in Piedmont;[20] and of Ugolino da Montecatini[21] and Pierleone da Spoleto[22] in Florence testify very clearly to this situation.[23]

In 1440, when Michele Savonarola moved from Padua to Ferrara and became a physician at the d'Este court, he assumed such differentiated responsibilities.[24] Before then, he had been a famous professor at the University of Padua (where he had previously studied) and had been lecturing on practical medicine.[25] During his preuniversity education, he was perhaps among the pupils of the humanist Giovanni Conversini da Ravenna, another of whose students was the famous Vittorino da Feltre (if so, these early connections with noted humanist educators might explain Michele's lasting interest in pedagogical problems). Among his teachers, colleagues, and friends we can find illustrious names of Paduan medicine and philosophy: Paolo Veneto, Giacomo della Torre da Forlì, Galeazzo Santasofia, Biagio Pelacani, Antonio Cermisone, Bartolomeo Montagnana. The motivation for appointing Savonarola to Ferrara was described thus: "He gave great lustre to medicine with his singular ingenuity, his care and skill in curing human bodies, and his volumes and books, of which he composed many."[26] Indeed, at Padua Michele was not only a famous professor but also a much-appreciated practitioner. There he wrote his main medical works, all of which are in the field of practical medicine (the physician is in fact, according to Savonarola, first and foremost an *artifex sensualis*):[27] *Directorium ad actum practicum*, *De vermibus*, *Summa de pulsibus, urinis et egestionibus*, *Practica canonica de febribus*, and *Practica major*.[28]

At Ferrara, Michele continued to teach for another ten years and finished his *Practica major*; subsequently, however, he was completely involved in his court duties.[29] He does not seem to have been engaged in diplomatic tasks, as other court physicians were; instead, the range and the relevance of his political, ethical, and religious interventions are impressive. This enlargement of interests was doubtless stimulated by the lively and exciting cultural context promoted by Nicolò and Leonello d'Este in Ferrara.[30] In my view, Michele's multifaceted production in Ferrara has to be seen as a whole, as a unified ethical/pedagogical program, in which dietetic advice, suggestions about the best form of education for the young princes, and devotional recommendations are all integrated. Certainly, the whole of this program is influenced by the characteristic trends of "Padanian humanism,"[31] but its primary model is an ancient one, now renewed, namely the pseudo-Aristotelian *Secretum secretorum*.[32] Indeed, the whole of Savonarola's writings in Ferrara can be seen as a Christian overlay of the contents of the *Secretum secretorum*, which itself is a successful combination of an encyclopedia with a *speculum principis*.[33] Astrological physiognomy, dietetics, hygiene, eugenic advice and education of children, marvelous remedies, court anthropology, wise administration, practice of virtue—these are the themes treated in the *Secretum* and they are also the subjects of the texts (medical and nonmedical) written by Michele in Ferrara. However, as a

Christian he went beyond these topics and also produced pastoral and devotional works.

Therefore, it is true that Michele's interests and writings are in a certain sense atypical. They are atypical especially in their number and variety, and especially because of the range of genres they cover. They range from an academic treatise on practical medicine and scientific monographs (as we have seen), to criticism of the vices of the court[34] and even a devotional treatise (in Latin) for a suffering friend that exhorts him to practice the virtue of patience, salutary for both soul and body.[35] Finally, Savonarola's bilingualism, in both his medical and his ethical and political writings, is also very significant.[36] Indeed, the most important ethical, political, and religious writings produced at Ferrara are all either in the vernacular or exist in both vernacular and Latin versions. The same bilingualism is also true of some of his medical writings. The teaching contained in the vernacular treatises written at Ferrara—*De regimine pregnantium* (in the vernacular despite its Latin title), *De gotta*, *Libretto di tutte le cose che se manzano*, and *Sulla peste*—is all to be found in the Latin *Practica major* or *Practica canonica de febribus*, both written at Padua in the first part of Savonarola's career. A comparison of these four vernacular texts to the content of Savonarola's Paduan textbooks helps in understanding how Michele differently structured and expounded the same doctrinal subjects in each of the two languages. The analysis that follows is focused precisely on this comparison.

Michele's passage from Padua to Ferrara, from the university to the court, involved not only the enlargement of his interests and a new range of topics on which he developed treatises: it also involved a change in the format of his texts. Although some of the works he wrote at Ferrara are best described as treatises (*Del felice progresso*, *Speculum phisionomie*,[37] *De vera et digna seculari militia*, *De balneis*[38]), the four medical works listed in the previous paragraph all fall into the genre of *regimina sanitatis*. The move from the university to the court was doubtless also responsible for the decision to use the vernacular, which was certainly more appropriate—as Michele knew—for a courtly audience. The following questions then arise, especially as regards his medical writings: Are the four vernacular medical texts translations, or were they composed in the vernacular? And do they imply a conscious program of broader divulgation of medical knowledge or not?[39] These are questions of particular interest in the context of current enthusiasm for the study of the vernacularization of scientific texts in early modern times.[40]

History and Other Stories

Taking into account Savonarola's entire production, I intend now to sketch out the various meanings of history he presents. There are at least two general literary forms of history to be found in texts written by Savonarola; I shall first describe these before returning to consider the medical works more analytically.

At the end of his Paduan period, Michele wrote a *Libellus* in Latin in order to describe and praise his town. Works of this kind formed a widespread genre during the Middle Ages and continued to be produced, although on different lines, by humanists as well (for example, Decembrio, Bruni, and Aeneas Sylvius Piccolomini).[41] Savonarola's history of Padua is rather traditional.[42] He first praised the excellent location of Padua (according to the scheme of the four elements), then narrated its mythological foundation. Having thus located the town in space and time, he next turned to its ornaments, that is, its beautiful buildings, its relics and books, its paintings and statues, and its prominent citizens. These are listed and described following a criterion that placed features of religious interest above secular ones ("de divinis et spiritualibus a nobis primo loco agendum est" [9], and "terrena spiritualibus semper cedere debent" [21]). The description is traditionally exhaustive and rhetorical. But there are no anecdotes about contemporaries and only two about earlier personalities; one of them concerns a Neapolitan student who was fond of art (55) and the other is a brief narrative about Gentile da Foligno. The latter story recounts how this learned physician, on entering the study of Pietro d'Abano, was moved to tears and with appropriate reverence laid down relics: the handwritten notes of the dead physician-philosopher (27). Savonarola sometimes states how he gained his documentary data and indicates his sources. His list of famous physicians is a useful source of information about his preference for practical medicine and about the teachers who most influenced him in learning.

It is only in this work that Michele introduces and follows the concept (so widely discussed among humanists)[43] of the superiority of poetry to history.[44] However, his praise for Pietro d'Abano (listed among the philosophers) relies on the fact that Pietro devoted himself with enthusiasm and success to "moral and natural history" ("morali atque naturali historie" [27]), which in this context has to be understood as philosophical research in general. Savonarola introduces here two concepts of *historia* (history and philosophical research), without emphasizing and carefully discussing their differences. He provides no further analysis of the relationship between history and poetry, nor any considerations about the peculiarity of this *naturalis et moralis historia* of Pietro d'Abano. It seems as if Savonarola uses the difference between history and poetry only to establish the correct order of his list and description of illustrious Paduan men. Unfortunately, we do not have the text of Savonarola's *De la decorante Ferrara*, which would enable us to make a comparison between his traditional encomium to and description of Padua and his representation of the city that was his new home. It seems likely that he might have praised and described Ferrara not only in a different language but perhaps also using different tones and categories.

Compared with the traditional and ordered exposition of the *Libellus*, the *Del felice progresso*, a political-historical encomium[45] written in Ferrara for Borso d'Este,[46] is much more articulated and multifaceted. Michele does not define this text as *historia*; as a matter of fact, this work is not purely historical. It has been considered by its editor (Mastronardi) as a "singular panegyric built up with a systematic interlacing of different genres."[47] Michele begins by underlining the pleasure (a typical category that, certainly for Michele, characterizes court literature) that Borso will obtain from reading this book. In fact, in this work (which exists in both Latin and vernacular versions) Michele deals both with Borso's seizure of power[48] and the best form of education for a prince. Traditional themes and topoi[49] of humanistic historiography come together here in an apparently eclectic treatise,[50] which however in reality displays an innovative type of structural unity.

To grasp this unity it is necessary in the first place to stress that "history" here has various meanings. In the first place it has the humanistic-Ciceronian sense of *historia magistra vitae*.[51] Thus, Savonarola several times tells Borso that the prince should know "as example" the histories and deeds of the past in order to be able to foresee and prepare for future events.[52] For this purpose it would be useful if the deeds of prudent men of old were translated for the ruler; if a learned man were continually to read aloud to him; and that he should memorize these *historie* which are also "advice and records" ("consegli e documenti").[53] Finally, it would be useful if the entire court were involved in this immersion in the exemplary past; it would be very advantageous if, during banquets, there were some reading aloud at table, following the monastic practice, but using historical works.[54] In addition to providing advice for future action, history is panegyric and above all a monument; indeed, only history records the prince's deeds and keeps the memory of events alive. Princes survive in later times only if the historian and literature (*litterae*) keep their memory alive beyond the chances of fortune and the brevity of life.[55]

Magistra and monument: these characteristics belong both to "grand history," for example that of Julius Caesar,[56] and to the closer to hand (and certainly more modest) history of the house of Este. The rule and death of Nicolò, the deeds of Leonello and Borso as young men, the sudden and unforeseen death of the former: Michele repeatedly narrates these events precisely as a eulogy and memorial of the dead and in order to legitimate—but also to exhort and guide—Borso, the survivor and new signore. Yet this history, with such a high ethical and political and above all educational purpose, unfolds in the manner of a chronicle. It is attentive to particulars, to details; it includes precise descriptions provided by someone—Savonarola himself—who "saw what happened." Ambassadors, festivals, ceremonies of mourning and of investiture, the palaces in which the events took place, the hangings on the walls, the clothes of the participants, the varied hues of colored silks: all are described in detail and objectively.

In this context, which Savonarola defined as "description of the truth,"[57] "made-up histories" are also included, as he himself noted. They are not fables or myths or allegories. Rather, they have to do with history, which always teaches and exhorts, understood primarily as eloquence: that is, as wisdom in narrating events and ideas with persuasive efficacy. This function and this eloquence are expressed in a series of *orationi*[58] in the middle of the treatise on the best form of government and on the just choice of Borso as an ideal prince in every way. Savonarola admits that, in reality, these orations were never delivered; but they could have been, because the people who speak in these literary fictions are real, eminent citizens of Ferrara who are alive and well known. In every case, the orations have a legitimating function reinforced by the verisimilitude of deliberate and insistent objectivity of the details described (the orator's posture, clothes, official position are all precisely indicated). Indeed, Savonarola announced at the beginning that alongside the narration of truth he would permit himself some fabrication "for adornment." But these orations, rather than being purely ornamental, are, to a large extent, historical commentaries on the events of the history of the house of Este inserted in a (relatively modest) context of political theory.

The multiplicity of themes and the multiform meanings of "history" in *Del felice progresso* have, as already noted, their own unity. In the first place, a more traditional and exterior unity is provided by the ethical and above all pedagogical tone possessed by all the various meanings of history. But there is also a more profound and structural unity because all the coordinates of time—past, present, and future—are expressed. *Historia* as memory and narration of past events looks to the past. Panegyric praises (and also involves criticism of) aspects of the present situation. Indications about the future emerge in considerations about how to maintain power and especially in the emphasis placed on the education of young princes. The figure of the prince himself provides the unifying articulation of these three coordinates of time, because he is the center round which they revolve. He is the outcome of those past events and must imitate them; he is the recipient of criticism and praise; and he is responsible for the young and the future of the city. And in his turn the prince is the reflection—in a real sense, the mirror—and in a metaphorical sense the father of the city.[59] Time and history therefore root themselves in and revolve around Borso, the ideal prince, and around Ferrara, "a paradise on earth."

Much more could be said about this typically humanistic point of view, sustained by a physician, but I want now to emphasize some particularly significant points. First, up to now I have used not *historia* but history (in the modern sense of the term) to interpret Savonarola's images of history. But it needs to be specified that, in *Del felice progresso*, he uses the word *historia*. Moreover, he uses this term both to indicate events in the past that are exemplary and should be read at table,[60] and to designate an anecdote, a little story or example: for instance, an *exemplum* taken from his memory, from his experience (what he has seen), or from texts that he has read.[61]

This leads to the second point: that all the meanings of history that Michele presents here one after another—*magistra vitae*, chronicle, encomium, eloquence—are continually interspersed with and supported by *narrationes* and *exempla* as well as *dicta*, proverbs, and metaphors taken from common experience and everyday life.

Third, the use of the same term for great history and for little anecdotes might seem to indicate an as yet undifferentiated concept. But it seems to me that what is shown here is in reality the weight and significance that Michele attached in every case to accurate description and lively presentation of evidence. If history to be educational must be true, truthful anecdotes (based on experience) and proverbs (which everyone understands because they refer to shared experience) reinforce the reliability of the author. They impress the historical events narrated more firmly on the memory of the reader; further, they clarify the theories that are being expounded. Moreover it is precisely through history that we succeed—Michele points out—in understanding present customs and controversies. For example, the origin of the dispute over the relative preeminence of arms and letters has also a historical cause, in that we learn (from history) that at the beginning of civilization in cities, military defense was of the greatest importance.[62] Likewise, we learn that the prohibition against women drinking wine derives from the sagacity and prudence of the Romans,[63] and is therefore naturally a custom that should be imitated and continued.

Finally, the treatise that I have been discussing up to now exists, as noted previously, in both Latin and vernacular versions. Evidently the intended audience of the vernacular text (which I have been following), in this case the prince and the court, seems to have influenced the writing. In fact, *dicta, exempla*, anecdotes, and metaphors abound in this version and are much rarer in the Latin one.[64]

The foregoing considerations should be kept in mind as we turn to Savonarola's scientific and medical output. However brief, they are sufficient to permit us to consider Savonarola's notions, from both a conceptual and a terminological point of view, of *historia, narratio*, and *exemplum* as still very fluid and as expressing a transitional phase.

Stories, Experience, and Medicine

In the context of late medieval and early modern practical medicine, two interrelated phenomena, which appear to require interpreting together, stand out. The first concerns the *practicae*, the texts used by both students and practitioners to acquire command of medicine as an *ars operativa*. Between the fifteenth and the sixteenth century they show two tendencies. On the one hand, there is a methodical and speculative element in the treatment of the subject matter, with attention to problems of method and, above all, to the classical works that had either been recovered or made available in supposedly more faithful (and more elegant) versions.[65] This bookish and textual tendency in part continues that of earlier *practicae*, which had already in the fourteenth century presented themselves as primarily teaching tools because they were commentaries or compendia on the *Liber nonus Almansoris*[66] or sections of the *Canon* of Avicenna.[67] In these texts, too, the development is in the direction of textual references enriched from the layers of previous commentaries. On the other hand, some *practicae* display instead a marked attention to the role of *experientia*, to the weight of the knowledge acquired in the course of professional practice.[68] Especially in the latter group of texts, the authors consider it necessary to refer to, and frequently introduce, particular cases,[69] instances of successfully treated pathological conditions, the health practices of different peoples and regions, the practical knowledge of *matrones* and *empirici*,[70] and situations and information from daily life.[71]

The second important phenomenon is the notable development of the *regimina sanitatis*,[72] a literary genre long present in medieval medicine. These texts, which are increasingly numerous by the fifteenth century, are the symptom of a more broadly diffused attention to the problem of health and in general to the strength and well-being of the body. They also reflect the enhanced interest in the individual and the particular that is typical of certain aspects of late medieval medicine;[73] indeed, many of the late *regimina* are personalized. Texts of this type provide one of the principal contexts in which practical medicine develops, and they constitute a strong connection between learned study of medicine and professional practice. They are writings of learned doctors of medicine and therefore come from the university milieu; but they are intended not for teaching purposes, but for the use of a public that is in general not learned but distinguished. They are addressed for the most part to monarchs, lords, or rich and prestigious elites of merchants and officials, and for this reason they are often written in the vernacular.

As is well known, a *regimen* is organized around correct conduct with regard to the six *res non naturales* (air, food and drink, physical exercise, sleep, evacuation, and the emotions) in order to acquire a generally well-balanced lifestyle, capable of guaranteeing both health and longevity.[74] In these later *regimina* a progressive tendency toward increased emphasis on food is notable, together with a superimposition of the concepts of food that is *medicinalis* and medicine that is *cibalis*. There is also a great expansion of the section devoted to diet and cooking, a characteristic that is strongly correlated with the elevated social destination of these texts and of the importance—including ritual importance—attached in the signorial milieu to eating. Because of this emphasis, the *regimina* also contain information about learned synonyms and vernacular terms for foodstuffs, regional recipes, and ways of producing and preparing food. They too might be considered—although no such study has as yet been done—under the aspect of collections of natural and technical facts, comparable in function and epistemological importance to contemporary developments relative to the classification of simples and in general of *materia medica*. Moreover, this literary genre fits very well with the new figure of the physician-adviser and the ethical and pedagogical purposes of his new function. Indeed, in the medical division of the *res—naturales*, constituents of the human body; *contra naturales*, diseases; and *non naturales*—only the *non naturales* depend on deliberation, and it is only with regard to them that the individual can exercise choice. Therefore, they are the aspects of health on which the physician/pedagogue can give directives and advice that are not just technical but often also strongly weighted with moral values relative to temperance, prudence, and general sobriety of behavior. From this point of view, some *regimina* could be considered as a sort of lay sermons to be compared to the religious preaching of the same period.[75]

The medical output in both Latin and the vernacular of Savonarola—who in fact taught practical medicine at Padua—involves both of these genres, *practica* and *regimen*. As he declares in the prologue of the

Practica major written in Padua and dedicated to his colleague and friend Sigismondo Polcastro (a younger professor of medical theory), the duty of the author of a practical text is to assemble "practicabilia." He must keep utility in mind, and not linger (as Michele often repeats in the body of the text) in excessively theoretical and deep *speculationes* that stray much too far from the operative purpose and are therefore useless for its realization. Notwithstanding these pronouncements, the *Practica major* exhibits a complex and carefully planned structure built round the exposition of *signa*, *causae*, and *curae*, all of which Savonarola explains and discusses for every pathological condition. The treatise is organized according to the scheme "from head to toe," but above all according to the complexional typology of the organs and patients. Moreover, despite his emphasis on practical utility, at the end of every section Michele does not hesitate to expound *dubia* and *quaestiones* that are more properly theoretical. Obviously the text is not lacking in references to the *auctores*, but these include also recent or contemporary writers. Regional schools and currents of thought and their opponents are cited, and local therapeutic practices are noted.[76]

Savonarola gives particular attention to the actual results of professional practice, attested to by the insertion of *exempla* and small case histories, which are often very specific. For example, he reports (the term used is "narrabo," 13r) the unfortunate case of the wife of an illustrious fellow citizen, Biagio da Meliara, who was visited and treated by no fewer than two *solemnes medici*; the politically worrying case of the repeated miscarriages of the Marchesa del Monferrato, finally satisfactorily resolved with an heir (266v); the case, which could be verified in the city, of an elderly woman giving birth (262r); the blundering mistaken diagnosis of a case of uterine mole by both Marsilio di Santasofia and Pietro da Tossignano (265v); and the case of a macrocephalic child whom Savonarola attended until his death at the age of ten (267r). If these accounts of cases are particularly numerous in Savonarola's *Practica*, they are not exceptional in the genre of *practicae*. Indeed, the use of an *exemplum* of a case satisfactorily treated, following an earlier mistaken diagnosis by others, appears in earlier *practicae*, for example in the *Summa curationis* of Guglielmo da Saliceto.[77]

But if we turn to the later vernacular medical writings of Savonarola in Ferrara, we find a more multifaceted situation. The facts narrated are both much more numerous and different in kind. I will sketch here their typology, drawing on those texts, already mentioned, whose doctrinal content is the same as that of sections of the Latin *Practica*. It should be noted first, however, that even if the content is identical, at times corresponding exactly to the sequence in which the same material is treated in the *Practica*, the form of the text is different. The later texts are all *regimina sanitatis*, written in the vernacular and addressed to different recipients. The dedicatee of *De gotta* is the prince, Borso (in fact, in its third part, it contains a true *consilium* on the prince's disease). "Ferrarese women" are the declared recipients of the *De regimine pregnantium*, which is thus a specialized *regimen* (although it is not concerned exclusively with pregnancy).[78] The *regimen* for the plague, also obviously specialized, is dedicated to all the citizens of Ferrara, both rich and poor. The dietary *Libretto* is dedicated to Leonello: it concerns the appropriate diet for elites in general, but sometimes also indicates whether a particular food is or is not right according to the individual complexion of the prince, Leonello.

But let us come to the facts that Michele expounds here. In the first place, the clinical cases are here too, with the same *exempla* that appear in the *Practica* but also with the addition of others. Thus, the Marchesa of Monferrato reappears in *De regimine* (104), but at her side is the case of the woman who sold Michele fruit and vegetables in the piazza (52) and that of the woman who was his baker (168) and who successfully treated her own little boy with strange remedies (ashes). Also reported (95) are the reactions to vomiting of Michele's own daughter-in-law, now in her third pregnancy. He also noted (91) the case of a medical colleague and fellow citizen who went mad from excess of melancholic fumes, believed himself a goose, and was therefore nicknamed Johngoose. There was also the case of the postpartum complications of another citizen of Ferrara (130), which Michele successfully cured, and the incident of the noble girl who, concealing her illegitimate pregnancy during a masquerade at carnival time, was taken with labor pains and obliged to reveal her shame (143). Two hermaphrodites whom Michele had seen (140), and the report about the conjoined twins born in Padua in 1320 (113),[79] may serve to close this incomplete summary. From these instances it is immediately clear that the cases introduced in the vernacular works are at the same time more familiar and stranger; moreover, they are tied to urban life. At times they are not completely pertinent from a scientific point of view, but they are appropriate and adapted to propound moral exhortations or to satisfy curiosity and to divert, making the reader more attentive and readier to memorize what is being taught.

Michele's personal memories are still more suitable from this point of view, because they are marked by an emotionally involved, first-person experience.[80] These recollections concern events occurring during his experience as a court physician; for instance, the memories of his treatment of a gouty and suffering Nicolò are particularly vivid. The duke uses Christ's words on the Mount of Olives, and Michele appears to him (if the duke follows instructions) as a true Savior.[81] Another vivid memory (in this case salacious) regards Leonello's use of and curiosity about the aphrodisiac qualities of oysters; Michele's response is sarcastic and his medical advice is facetious.[82] But there are also memories of his personal life, his childhood,[83] stories his mother told him, events Michele once saw and heard that he has understood only now.

Experiential data introduced by formulas such as "it can be seen," "as I have seen," and "we can see by experience that it is true" are properly aimed at rendering Michele's instructions more comprehensible and truthful. In this set of facts Michele includes malformations typical of certain regions and local customs.[84] The Friuli region produces a large number of lame people, because of midwives' mistakes (*De regimine*, 142). In Puglia (143) little monsters are almost always born with the fetus; this is a well-known phenomenon, and is there quite normal. The strange habits of eels during May make them particularly indigestible in that period (70). The addiction of the Ferrarese to "boiled milk" makes them fall easily into the disease of the stone (77). The experience (in this case already structured as an art) of cattle breeders, shepherds, and farmers is a much-appreciated source of many true facts. From these professions Michele draws comparisons, confirmation, and also practical indications concerning human physiology.[85]

But proverbs and sayings (mostly of popular origin) are given pride of place due to their overall efficacy.[86] They are efficacious because they summarize, one might say, the authority connected with case histories, the emotional tone of memories, and the evidence deriving from experience. While *casus*, *exempla*, memories, and anecdotes all concern a singular fact or (better) an individual, proverbs can express general statements. In any case, proverbs concern well-known facts and are based on everyday experience that is shared by the narrator and the listener;[87] this is added to and confirmed by continuous factual proofs. Such experiences are set in an expository formula that is also rooted in collective memory. Furthermore, proverbs are in themselves verbal formulas and conceptual grids that, in spite of their low level of elaboration, are suitable to gathering experiential data. They create, in other words, the epistemological conditions for collecting and organizing future experiences. Proverbs are thus ideal formulas for reinforcing recommendations or prohibitions, according to their indisputable obviousness.[88] Often, the proverbs used by Savonarola have no medical connotations, but they generally refer to concrete, well-known facts, so that he could easily adapt them to exemplify dietetic events. Thus, he applied the saying "The eye of the master fattens the horse" to the need to watch over the health of a wet nurse and the food she eats.[89] Other proverbs derive more directly from specific medical fields—pregnancy, diet, gout, plague—that he, as a physician, deals with, but these situations (generation, food, common diseases) are frequent, if not normal, occurrences in everyone's experience. They have thus generated sayings that condense facts, signs, risks, and precautions connected with the conditions and situations in question. Space does not permit further examination of these proverbs and *dicta* here, although they are of great interest not only from the point of view of historical linguistics, but also and specifically from that of epistemology. Suffice it to say that they recur very frequently in the four *regimina* but are almost totally absent from the *Practica*, where, if anything, they are replaced by the sentences of authors.

The experiential concreteness (supported by details, facts, memories, and proverbs) with which Savonarola evidently wished to endow his *regimina* is also assured more subtly and pervasively through the elaboration of what may be termed concrete metaphors. Such metaphors are employed in at least two cases, in the *De regimine* and *De gotta*. These metaphors summarize the whole physiological process or course of disease with which Michele is dealing via an assembly of similes all drawn from everyday experience (this is why I call them concrete metaphors) that are intended to describe every phase of generation or of the course of gout. The concrete metaphor of generation and birth, from courtship to delivery, is perfect in every detail; it runs through the whole of the *De regimine,* unifying it. Here I shall attempt to give a brief outline, in order to bring out its consistent unity. After comparing, from a Christian perspective, the whole process of generation with that of the Creation (from the original chaos to the final ordered result, thus adopting a metaphor that is anything but concrete), Michele proceeds to describe the details and problems inherent in conception, pregnancy, and birth. His similes, which taken together define a metaphorical birth,[90] do not appear continuously but are gradually inserted in his treatise, as follows.

Menstrual blood (which according to medieval embryological theory was in pregnancy retained to form the matter of the fetus) is *flores,*[91] because women know that a beautiful fruit can derive from it. Courtship (*De regimine,* 22) is the "nightingale that sings the sweet song taught by nature" (this is better done during the cold season, as Avicenna confirms). Erotic foreplay (which is necessary for the emission of women's semen) is whetting a knife and starting a skirmish (28, 23, 41); a sympathetic uterus during successful reproductive intercourse behaves like a stomach that absorbs food greedily when it has been enticed by savory appetizers (40–42). A sort of "big lump of dough" is created in this way, as when bread is made (26); this humid stuff, which has been mixed in the matrix by natural heat, "boils and foams" as "water in a cauldron" (31); in this way three clots are formed (heart, brain, liver). During this first period it is wise "not to knock on the door lest the lord of the manor should come out"; this means abstention from sexual intercourse (84–85). In the meantime, the features of the child start to be formed, as when painters draw the lineaments before filling in the colors (34). The fetus is a beautiful fruit (35, 42, 76, 82), hanging by but a thin stalk from the branch that nourishes it, and it is exposed to the strength of the winds. The pregnant woman is obviously a vessel; less obviously, she is a bricklayer who carefully balances the weight she transports when she moves (60); she is also a barrel that could gush wine (85). Especially toward the end, she is a ship, bound with rotten bands, to be governed with firm helm (consisting of the wisdom of the midwife) not toward an unfortunate harbor but toward an easy delivery (57, 108, 110, 116). At last, the fruit appears, in the position boys assume when they dive into the river (39, 112), and rolling like a barrel on its guides (122); and, if all has gone well, the superfluous blood also comes out flowing as easily as metal when bells are cast (129). To conclude: "The most noble and perfect fruit in the world is the newborn human baby" (but, if it is illegitimate, it is a false coin, 197).

In his treatise on gout, the concrete metaphor is more of a piece. It is presented by Savonarola right from the beginning and is clearly intended to support the whole of the first part, which is devoted to the description of the causes, symptoms, and course of the illness. In this instance, the metaphor is military.[92] Gout has an army of knights and infantry, guided by captains and chamberlains, and trained by counselors. She occupies pinnacles and castles and, when she triumphs, she celebrates with her brigade of aches and pains (each type of which has its own name and precise characteristics and makes its incursion during a particular stage of the illness).

It would be superfluous, I think, to point out that no metaphorical form is used in the lengthy sections of the *Practica* devoted to these two subjects. It is thus obvious that the type of exposition here described has been adopted bearing in mind the intended audience of the vernacular treatises. As regards *De regimine*, this consists of "the women of Ferrara," midwives, and fathers of families. I shall not dwell now on the problems of controlled divulgation and stratification of the intended audience (and thus of literary genres) raised by this treatise.[93] Rather, I shall confine myself to noting that the concrete metaphor for conception and pregnancy is built up with facts that are evident to, and have been experienced by, all sections of the audience. They are facts drawn from the kitchen, the vegetable garden, from games and trades familiar to all. The gout metaphor, on the other hand, was not intended for everyone, nor was it comprehensible to all. The military similes, the weapons listed, the court titles, the parties and the pleasures described specifically indicate that this elitist metaphor revolves around the court of the lord (to whom, indeed, the treatise is dedicated). This court is a reflection of that of Madame Gout, being composed of the brigade of the closest friends, counselors, functionaries, and captains of the prince. It is a gouty brigade of ill people who have experienced what Michele describes, and can immediately understand it through the translation of what they feel into the court and military language in which the diagnostic and therapeutic content is expressed.

Texts such those of Savonarola are excellent examples of the process that has been defined as an "incursion of everyday life" into Scholastic medicine[94] and into late medieval and early modern scientific texts more generally. Moreover, Savonarola's and other medical texts of this type were produced during a period of transition, from both the linguistic and the epistemological point of view. Indeed, referring to everyday life in any form—*exemplum*, memories, proverbs, facts of experience—meant enhancing the value of a less logically elaborated level of experience as a fruitful heuristic tool.[95] This was something of which the heavily textual and doctrinal medicine of the Quattrocento stood in just as much need as it did of new translations and new perspectives on method.

The facts that Michele does not collect but simply uses are usually called *exempla*. The *exemplum* has been investigated primarily as regards its use in preaching;[96] its experiential concreteness[97] together with its evolution toward *historia* (conceived here as a broader *narratio*) have been emphasized.[98] Obviously this analysis also needs to be carried out in other contexts such as that of medicine,[99] but it can be noted that in this instance too we are dealing with a period of transition. Indeed, Savonarola uses the term *historia* to introduce an anecdote/*exemplum* in *Del felice progresso*; he uses *exemplo* to introduce an anecdote only three times.[100] As far as I know, he never uses *historia* in the Latin medical texts, where the term is replaced by the (rarely used) term *casus*. In the vernacular medical texts, Michele does not name these facts, data, and stories (neither *historia* nor *caso* appears); he simply uses them. It appears that even the vocabulary related to the facts narrated is still fluid; it is in the process of creation (as regards the vernacular) and of transformation (as regards the Latin).

What is certain is that these facts—however they may be defined and narrated—are drawn from experience. The accentuation of this feature and the abundance with which these narratives are spread in Savonarola's vernacular medical works might lead one to think that the reference to experience is merely a rhetorical device. In my opinion this is not so. The various audiences for whom the vernacular texts were written needed not only to be attracted and persuaded, but to understand the truth of what Michele was explaining. The use of the vernacular, which conveys insistent and multifaceted reference to experiential facts, also implies an empirical epistemology, which tends to highlight sense-derived data accessible even to nonspecialists. I therefore hold that the studies currently under way on the vernacularization of scientific texts in early modern times should bear in mind this epistemological-philosophical aspect (which calls for cooperation between historical linguists and historians of science).

Finally, if the emphatic reference to sense experience is not merely a rhetorical device, this does not make it neutral. Indeed, it should not be forgotten that this reference plays a vital role in the *regimina*, texts designed to advise, exhort, and convince. At bottom, these texts aimed to change not only the condition of the body but the ethics of the audience's lifestyle. This remark holds good for all the *regimina* but is true particularly for those of Savonarola, who was so strongly aware that his duty as a physician was to be also a wise confidant, counselor, and pedagogue. An epistemology of persuasion, based not only on rhetorical devices but also on evidence and truth, can therefore connect (by means of references to historical or physiological true facts) moral advice and therapeutic recommendations within an ethical context. The goal sought is to achieve the well-being of the prince and the citizens within the framework of the well-being of the city as a whole. It is no coincidence that Michele frequently compares the role of the prince with that of the physician, the art of curing and salubrious living with the art of governing well.[101]

Notes

I would like to thank Nancy Siraisi for her generous help in translating this essay. Sources cited in boldface type appear in the Primary Sources of the bibliography.

1. Tachau 1988; Murdoch 1989, especially 3–9; Reina 2002; Spinosa 2002, especially 178–183.

2. Murdoch 1982 and Parodi 1987; McVaugh 1975, especially 106–107, 112–113.

3. Agrimi and Crisciani 1990; Hamesse 2002; Spinosa 2002.

4. Agrimi and Crisciani 1988, chapter 1; Agrimi and Crisciani 1990. Nicholas of Poland can be considered as an extreme case of appreciation of experiential evidence, which (he claimed) we must trust because it is essentially substantiated by divine illumination (see Eamon and Keil 1987). It is noteworthy that Savonarola does not criticize this extreme evaluation of experience: in fact he refers to Nicholas twice in the *Practica*, appreciating some of his statements because he is an *experimentator* (see Pesenti 1977, 96).

5. Agrimi and Crisciani 1988; Crisciani 1990; Bynum and Porter 1993. Baud 1983 correctly affirms that *medicina operativa* is the threshold of tolerance of the university didactic system.

6. The role of sense perception is surely and consciously stressed in many alchemical medieval texts (Crisciani 1998, 2003a, and forthcoming). I have not yet examined mystical works from this point of view; see, however, the fruitful hypothesis proposed by Pereira 1996.

7. I have written some tentative essays in this direction: Crisciani, 1998, 2001, 2003a, and forthcoming.

8. BLAIR in this volume, especially her concluding considerations and bibliography. See also Crisciani 2001, 695–696.

9. See in this volume BLAIR and the considerations on *sensata experientia* and *historia* developed by POMATA.

10. See specially POMATA and SIRAISI in this volume. Regarding the relationship between history and medicine see Momigliano 1987 and Rechenauer 1991; for the early modern period see POMATA in this volume and Pomata 1996. Moreover, see the following judgment on Savonarola's humanistic historical attitude issued by his friend Sicco Polenton: "Diligentiam laudo tuam quod non, utique plerique solent, Galieno et Avicene solum, sed, cum datur otium, antiquitati, eloquentiae, ac omni virtuti studes. Legisti enim apud Ciceronem, puto, eum qui nesciat historias, puerum semper esse" (quoted in Samaritani 1976, 70–71).

11. For instance, the rules concerning the necessity to go down from generalities to particulars, and to clarify theories with *exempla*, are fundamental in practical medicine (see here, below) and are also evoked and followed in Savonarola's *Del felice progresso* (**Savonarola** 1996, 154, 78).

12. See below in this article.

13. O'Neill 1975.

14. Bernard de Gordon and Henri de Mondeville gave precise pedagogical instructions and suggestions in their medical and surgical works (Nagel 1983). More exceptional was the case of Arnald of Villanova, who, as physician and counselor of some kings, vigorously suggested political and religious instructions, prophetic injunctions, and programs of reform (for an overview see Perarnau 1995).

15. Agrimi and Crisciani 1988 (appendix with the edition of two *orationes* of Gentile da Foligno and Jacopo da Forlì) and Schlam 1978.

16. See, among others, Wack 1990 and Jacquart 1987.

17. **Arnald of Villanova** 1998.

18. Benzoni 1907.

19. See Defennu 1955; De' Reguardati 1977; Nicoud 2000; Pesenti 2003; Crisciani 2003c.

20. Naso 2000.

21. Park 1985.

22. Rotzoll 2000.

23. See also Pesenti 1997; Palmer 1981; Crisciani 2004; and, in general, Biow 2002.

24. The basic information on Savonarola is to be found in Segarizzi 1900 and Samaritani 1976; see also Thorndike 1959, 4:183–214; Pesenti 1977; Pesenti 1984, 187–196; Bertoni 1921.

25. On the Paduan period of Savonarola's life see Pesenti 1977 and 1983; more generally on the cultural context of the University of Padua see Poppi 1993 and Bylebyl 1979.

26. Quoted in Pesenti 1977, 88: "suo ingenio singulari, sua curandis humanis corporibus providentia et arte, suisque voluminibus et libris, quos plures condidit, medicine disciplinam maxime illustravit."

27. Pesenti 1977, 98 (this definition is provided in the *Summa de pulsibus*).

28. For manuscripts and editions of these works and for some lost texts of Savonarola see Pesenti 1984, 187–196.

29. On medical humanism in Ferrara see Münster 1963 and Nutton 1997.

30. See the last part of GRAFTON's essay in this volume; see also Bertoni 1903, 1921; Garin 1961; Gundersheimer 1988; Papagno and Quondam 1982; Quondam 1978; Bertozzi 1994; Cappelli 1889; Castelli 1991, 1992; Del Nero 1996; Tissoni Benvenuti 1991, 1987a, 1987b, 1994.

31. On the characteristics of "Padanian humanism" see the previous note, especially the essays of Tissoni Benvenuti; Anselmi, Avellini, and Raimondi 1988; Anselmi 1992.

32. Agrimi 1984. For some documentary data about the presence of the *Secretum secretorum* in the d'Este library, see Crisciani 2003b. Pier Andrea de' Bassi (in his commentary on *Teseida*) had already used the relationship between Aristotle and Alexander to describe that between Guarino, the teacher, and Leonello, the prince (Tissoni Benvenuti, 1994, 392, 402). Savonarola quotes and uses the *Secretum* in his *Speculum physionomie* (**Savonarola** 1997), and he says in *Del felice progresso* (**Savonarola** 1996, 84): "E vogly, pregotte, signuor mio, a la memoria rivocare quanto splendore e quanta gloria ricevuto ha e tutavia riceve Alexandro Magno per haver havuto sempre appresso di se Aristotile e per havere la doctrina di quello con gran diligentia observata" (it should, however, be emphasized that the most important source followed by Savonarola in this work, sometimes literally, is the *De regimine principum* of Egidio Romano). On the diffusion of the *Secretum secretorum* see Rapisarda 2001 and Williams 2003; on the presence of this text in the European court libraries see Williams 2004.

33. On this genre see, among others, Quaglioni 1987.

34. This theme is strongly expressed in *De nuptiis* (**Savonarola** 1992), where Savonarola harshly condemns the adulation, lies, slanders, and gossip so frequent in court life.

35. **Savonarola** 1954. Savonarola also wrote two (unedited, and both in the vernacular) texts of instructions for a good confession, one for the clergy and the other for courtiers.

36. On Savonarola's bilingualism see Gualdo 1996, 1999; Crisciani 2003b; also the introductions of Mastronardi (**Savonarola** 1996) and Biamini (**Savonarola** 1992).

37. Thomann 1997; Agrimi 2002.

38. Park 1999, Nicoud 2002, and Gualdo forthcoming.

39. A hypothesis about these texts, which seem to be neither translations nor original treatises but the result of a process of editing in the vernacular, is tentatively proposed in Crisciani 2003b.

40. See, among others, Dionisotti 1999; Altieri Biagi 1984; Lusignan 1987; Slack 1979; Copeland 1991; Badia 1996; Cifuentes 2001; Gualdo 2001; Voigts 1990, 1996; Jacquart 2001; *Vernacularization of Science* 1998.

41. Pier Candido Decembrio, *De laudibus Mediolanensium urbis panegyricus*; Leonardo Bruni, *Laudatio florentinae urbis*; Aeneas Sylvius Piccolomini, *Descriptio urbis Viennensis* (cf. **Savonarola** 1996, Introduzione, 33–35). Before them, Bonvesin Dalla Riva had described the *mirabilia* of Milan and Giovanni Garzoni those of Bologna. Numerous examples of this genre can be found for Rome, either in the form of an account of a journey to the city or in the form of description of monuments.

42. For a summary of the contents, see the Introduzione of Segarizzi (**Savonarola** 1902, 4–5).

43. See Regoliosi 1991, 1994, and 1995a.

44. "Poesiam vero historia plus magnifacio, cum velata et moralis sit philosophia. Hec enim morali naturalique philosophia veluti filia supponitur, nam de moribus hominum deque naturali historia, ut de ventis, aquis, plantis ceterisque vegetalibus pertractat, de cursu siderum earumque proprietatibus se intromittit, hecque scientias omnes non mediocriter discurrit et summatim omnium scientiarum agnitionem habet; hec autem veluti divina habetur, nam, cum plurimi, quibus scientiarum omnium cognitio non lateat, ad hanc capessendam caperetur ostendere, sed quoddam celeste munus se esse fateretur. . . . Historia namque, preterita que sunt, tantum cum eloquentia narrat, poesis vero preterita, presentia et futura non absque eloquentia profert et inventione gaudet, qua historia caret" (**Savonarola** 1902, 23, 29).

45. See in general Biondi 1984 and Sapegno 1984.

46. On Borso cf. Chiappini 1971.

47. **Savonarola** 1996, Introduzione, 16; see also Mastronardi 1993–94.

48. Similar, but much more propagandistic, is the work of Lodrisio Crivelli on the seizure of power by Francesco Sforza (Ianziti 1988, 35–47). We can also compare the more sympathetic account of Savonarola with the description of the same events written by Fra Giovanni da Ferrara.

49. See Cochrane 1981; Guenée 1973.

50. **Savonarola** 1996, Introduzione.

51. On this topos of humanistic historiography see Koselleck 1984; Nadel 1964; Landfenster 1972; Hampton 1990; Seifert 1976; Regoliosi 1991, 1994; and the classic essay of Cantimori 1938. See in general *Storiografia umanistica* 1994.

52. **Savonarola** 1996, 81, 83, 207.

53. **Savonarola** 1996, 83.

54. This suggestion appears also in **Savonarola** 1992, 123. On one occasion, however, Savonarola expressed the idea that reading or hearing about history was not always the most essential thing. See the criticism in his plague treatise (**Savonarola** 1953, 4) of anyone who "reads the histories of Livy and the fables of Ovid without getting bored, but who might instead weary of this quite long treatise, which however, is very useful."

55. **Savonarola** 1996, 196.

56. **Savonarola** 1996, 67.

57. **Savonarola** 1996, 77: "Io forzato mi ho di scrivere il vero, il perche' so ad te cussi piu' piacere che se scrivesse il falso"; 160: ". . . che biem son vero testimonio. . . ."

58. On the debate about these *orationes fictae* see GRAFTON in this volume, especially the third section: "The Historian's Speeches: Rhetorical Decorum as a Hermeneutical Tool."

59. The definitions of the prince as a good father of a family and as a mirror of right behavior are constantly used throughout the text.

60. **Savonarola** 1996, 83, 196, 207, 241.

61. About these "short" or "simple" forms of narration (*exemplum*, proverb, etc.) see Jauss 1985 and, from another point of view, Bremond, Le Goff, and Schmitt 1982.

62. **Savonarola** 1996, 90.

63. **Savonarola** 1953, 76.

64. Mastronardi's hypothesis is that the Latin version was intended to circulate among learned humanists in northern Italian courts. See also Phillips 1979.

65. On the structure of the *practicae* see Demaitre 1975 and, for the Renaissance period, Wear 1985; see in general Bylebyl 1979, 1985.

66. See the *practicae*-commentaries of Gerardo da Solo, Giovanni Arcolani, and Giovanni Matteo Ferrari da Grado (Agrimi and Crisciani 1988, chapter IV).

67. A *practica* arranged in this way is the *Introductorium sive ianua ad opus practicum* by Cristoforo Barzizza. Concerning the use of the *Canon* in sections, and its use as an ordered structure for the material of *practicae*, see Siraisi 1987. In his *Practica*, commenting on the ninth-century Arab physician Razi, Ferrari da Grado affirms that it is necessary to "particularizare," i.e., to go to the specific individual case when Razi's exposition is too general. The general trend, however, is always toward more textual stratification of commentaries.

68. Cf. the practical works of Valesco de Taranta, Antonio Guainerio, and Savonarola himself: see Jacquart 1990, 1998; Pesenti 1978.

69. See Murray Jones 1991 and Nutton 1991a. For an explicit appreciation of the didactic, indispensable role of the *exemplum*, see **Arnald of Villanova** 1988, 135: "Quia tamen proponimus doctrinam istarum consideracionum patefacere in exemplo particularis operacionis more sapencium docencium res

universales per exempla rerum particularium, ideo operacionem flebotomie de qua in pluribus plures dubitant ad exemplum introducemus." The *exemplum* can be either a narration of an actual successful treatment or a construction of a fictitious, but always singular and particular, possible clinical situation that the physician could face in his profession.

70. Antonio Guainerio lays much emphasis on this type of material in his *Opus preclarum ad praxim*, but so do others.

71. The last (which can be called a sort of practical anthropology) is not only a tendency in the *practicae* but also in some commentaries of the late Middle Ages. Buridan himself, in commenting on Aristotle's *Physica*, introduces a lot of facts and *exempla* drawn from daily life; in medicine, the same is true of the commentary on Avicenna's *Canon* by Jacques Despars. Examples can also be found in Pomponazzi, for instance in his unedited commentary lectures on the Aristotelian *Meteora* (see the forthcoming edition of these lectures by Stefano Perfetti).

72. Nicoud 1998 is fundamental on this subject; see also Gil Sotres 1993, 1996; Mikkeli 1999.

73. This is also true of the genre of *consilia* (see Agrimi and Crisciani 1994a, Crisciani 1996, and Crisciani 2004), but only in part; in fact, *consilia* become primarily texts for teaching when they are gathered together into collections.

74. See Garcia Ballester 1993.

75. Therefore, they might be studied in the context of the connections between medicine and religion/theology so deeply explored by Ziegler 1998 and others recently.

76. **Savonarola** 1559b, 15r about "mos Turcorum"; 13v about the different use of a syrup by Venetians and Paduans; 30r about differences in treatment between Padua and Bologna; 23r about "mos Schiavonum"; on habits of "Furlani" see 35r, 42r (these are only a few references among the numerous ones spread all through the *Practica major*).

77. They are particularly stressed in the *Chirurgiae* from the time of their origins, in order to legitimate the author's doctrine through the enumeration and description of his successful treatments; see Agrimi and Crisciani 1994b and Siraisi 1994a.

78. The third part of *De regimine* (**Savonarola** 1952) is devoted to physical and moral treatment of children (from birth to the age of seven), while this subject is totally absent from the *Practica*.

79. See O'Neill 1974.

80. On "emotive features" in a scientific texts see Taavitsainen 1994. It is noteworthy that in Savonarola's vernacular *regimina* verbs in the second person abound, which—according to Taavitsainen—is one of the typical emotive features of vernacular and practical texts.

81. See *Trattato sulla Peste*, Introduzione of Belloni (**Savonarola** 1953, xxxiv).

82. **Savonarola** 1952, 71.

83. **Savonarola** 1952, 112 (about a strange nickname, incomprehensible in his childhood); 172 (about a skull full of quicksilver that he saw when he was a boy of ten).

84. Not by chance, Jacquart has noted, some of the physicians of the fifteenth century sometimes act as ethnologists (Jacquart 1990).

85. **Savonarola** 1952, 17, 20, 55, and passim.

86. As is well known, collections of proverbs were quite widespread in the fifteenth century; some of them are very famous. I would stress that Savonarola's proverbs are only very rarely in Latin and quoted from literary sources; they are not collected, but scattered throughout the text; and most of them concern concrete and humble subjects of everyday life.

87. On proverbs as "short form" see Jauss 1985. His considerations on the peculiar medieval set of these forms (which differ from those of Jolle) and on their chronological evolution are particularly noteworthy and can usefully be transferred to the context of scientific texts. Jauss further stresses the simplicity, but also the complexity, modularity, and openness in such forms as the proverb and the *exemplum*, which he defines as "a world of lived experience based on what has happened" ("mondo dell'esperienza attraverso l'accaduto," 60). Cf. also Davis 1975; Mieder 1994; Mieder and Dundes 1994.

88. For an analysis of one of these proverbs from the point of view of the history of disease see Nicoud 1994.

89. **Savonarola** 1952, 198; see also the narration of an anecdote (about a man of Forlì's way of drinking wine), which evolves into a proverb (131), and is finally adapted as a rule about the appropriate diet for new mothers, whose food must be augmented only gradually.

90. See Zuccolin 2003 on this metaphor. On the medieval theories of embryology that underlie this description see Martorelli Vico 2002; Dunstan 1990; and especially Goodman 1990.

91. This is the only simile on this subject that we can find in the *Practica major* (251r), where menstruation is also defined as *monstrum*.

92. Crisciani 2003b.

93. The three genres interacting here are a specialized regimen for pregnancy, a specialized didactic text for midwives, and a sort of *libro della famiglia*.

94. See Nicoud 1998 (concluding considerations); Jacquart 1990.

95. Savonarola repeats many times in *De regimine* his firm intention to avoid subtle questions and speculations. In the *Trattato sulla peste* (**Savonarola** 1953, 40) he refers to some beautiful speculations and questions that the interested reader must study in Savonarola's "trattato litterale" (that is, in the *Practica canonica de febribus*). He also states that writing in the vernacular does not admit theoretical speculations ("Ho pur fatto questo cussì in volgare non per stare su le speculatione, le quale non cape il volgo . . ."). On the contrary, in the *Practica major*, although speculative subtleties and theoretical doubts must be avoided (as we have seen), Savonarola sometimes presents one strange effect "ut ad

speculationem te moveat" (16r). If Savonarola worried about inserting theoretical speculations in his vernacular texts, he also worried about the dangers that divulgation to a broader audience could involve. In fact, at the end of *De regimine*, he expresses his fear that this treatise could cause death because many people could begin "altrui medicare senza alcuno timore, fondandose solamente su le recepte, non su le regole che ensegnano come quelle applicare se debono" (166). These are the obvious problems involved in his general enterprise, which can be defined as a controlled divulgation of scientific notions.

96. Among the numerous references which can be quoted here, see at least Bremond, Le Goff, and Schmitt 1982; Zorzetti 1980; Vitale Brovarone 1980; Berlioz 1980; Del Corno 1989 (especially on the evolution of *exempla* in preaching); Von Moos 1988; Menzel 1998. For medicine, see Murray Jones 1991. A recent discussion on *exempla* is provided in Berlioz and Polo de Beaulieu 1998; cf. also Ricklin, forthcoming.

97. It is perhaps useful to remember some medieval definitions of *exemplum*. For Jacques de Vitry *exempla* must express "corporalia et palpabilia, et talia que per experienciam norunt"; for Etienne de Bourbonne *exemplum* is the same as *sermo corporeus* (both sentences are quoted in Del Corno 1989, 10).

98. The historical evolution of the *exempla* literature has up to now been explored by linguists in order to evaluate the process of laicization of the *exemplum*, and to determine the origins of the genre of the novella (see Albanese et al. 2000; Auzzas et al. 2003). It is evident that a deep change is to be seen also in medicine, as some long *historiae* in the *Consilia* of Giovanni Battista da Monte suggest (they have not yet been studied from this point of view).

99. Until recently, we could refer only to few hints suggested in Murray Jones 1991; Agrimi and Crisciani 1994a; Agrimi 1998. Now the essay of POMATA in this volume provides a solid basis to carry out a deeper research about the development of these genres (*exempla* and *historia*) even in late medieval medicine.

100. **Savonarola** 1996, 78, 92, 203.

101. See, for instance, **Savonarola** 1996, 87: "... che certo ne' il principo, ne' il medico, biem che seppano le regole di l'arte, non puoteno conseguire di sua opera degna laude senza exercitio et experientia"; and 213: "... che molto vale la experientia nel rezere e governare i stati, come quella vale nel medicare i corpi."

10

Historiae, Natural History, Roman Antiquity, and Some Roman Physicians

Nancy G. Siraisi

Throughout the second half of the sixteenth century Rome was the scene of a battle of the books over the potability of Tiber water. On the one side, Giovanni Battista Modio denounced the water as undrinkable and any physician who pronounced it wholesome as incompetent. On the other, in a series of publications spread over some thirty years, Alessandro Traiano Petroni and Andrea Bacci energetically defended the river.[1] The issue was one of public health, and the context was (or soon became) the efforts of sixteenth-century popes to provide alternative sources of water for the city. All three of these authors were physicians and all three appealed, as one might expect, to medical knowledge and to their own medical experience. But in addition, two of the three also incorporated into their arguments both the natural history of rivers and waters and substantial historical and antiquarian information about water supply and water management in ancient Rome.

The episode illustrates the ease with which some Renaissance learned physicians integrated medicine with natural historical, historical, and antiquarian learning, and as such it is only one of many similar examples. Indeed, the relations of medicine to both history and natural history are well known, though perhaps seldom considered together. As Arnaldo Momigliano once noted, in antiquity and the Renaissance, medicine and history had much in common—including attention to examples drawn from records of the past and narrative of particulars.[2] Similarly, physicians played a very large part in the sixteenth-century development of botany and natural history, which to a significant extent grew out of medical interest in *materia medica*.[3] Moreover medicine, natural history, and antiquarianism as practiced in the sixteenth century were all disciplines that habituated their practitioners to integrate the analysis of texts with consideration of material evidence. Yet against this broader background, the quarrel about the potability of the Tiber also belongs to a very specific milieu, one in which references to the ancient Roman past were both part of the universal currency of European humanistic culture and had immediate local resonance. Indeed, the particulars of medicine, history, antiquities, and natural history were often necessarily local, whether they involved patients, past events, monuments, or natural features. Thus the quarrel over the Tiber serves to introduce the range of uses of *historia* in the medical world of late sixteenth-century Rome. Roman physicians concerned themselves with *historia* in many forms: with medical case histories and other *historiae* about individuals, with natural history, and with the written record and material remains of the past. In so doing they responded to an environment shaped not only by local health conditions and the experience of medical study, professional practice, or teaching in Rome, but also by the religious life and institutions of the papal city and by flourishing local cultures of natural history and antiquarianism.

Even the briefest summary reveals the place of Rome in the Renaissance and early modern revival and expansion of the study of plants, animals, and minerals. These developments culminated in the interests and activities of some of the seventeenth-century Lincei, but they began much earlier.[4] Previous contributions included Nicholas V's mid-fifteenth-century sponsorship of new translations of the Aristotelian works on animals and of Theophrastus on plants and the institution of a chair in medicinal simples at the *studium urbis* in 1513 (teaching of the subject seems to have occurred consistently from 1539). A botanical garden was established at the Vatican in about 1570.[5] In mid- and late sixteenth-century Rome, natural history engaged the attention of such men as Ippolito Salviani, professor of practical medicine, author of a large illustrated work on fish, and correspondent of Aldrovandi, and Michele Mercati, student of Cesalpino, prefect of the Vatican botanic garden, and mineralogist.[6] Another was Andrea Bacci, one of the physicians on whom this essay will focus.

The flourishing state of Roman antiquarianism in the fifteenth and sixteenth centuries scarcely requires emphasis. The interest of Roman humanists in antiquities took many forms, not all of which were necessarily exclusively connected with the past of the city itself.[7] But both the customs and the physical remains of the ancient city were, obviously, subjects of special interest. Attempts to perpetuate or recreate ancient Roman traditions reveal a strong (if almost entirely imaginary) sense of identity with the city's ancient past. The best known example is doubtless the celebration of ancient Roman festivals in the circle of Pomponio Leto, which may have played a part in bringing disaster on the Roman Academy.[8] However, the notion of identity or continuity with ancient Roman forebears also found expression in the claims of some noble Roman families to improbable antique genealogies as well as in Altieri's identification of local wedding customs as survivals of ancient Roman tradition.[9] More pragmatically, from the time of Alberti and Biondo in the fifteenth century, the mapping, description, illustration, and preservation of the physical remains, monuments, and landscape of the ancient city were the focus of intense scholarly and artistic interest and endeavor.[10] In the mid-sixteenth century, the household of Cardinal Alessandro Farnese, known for his collection of antiquities and patronage of architecture and the arts, was a major center of antiquarian activity. There the cardinal's physician, Girolamo Mercuriale, profited from his association with antiquaries to compose a lengthy work on ancient athletics.[11] In some areas practical considerations combined with antiquarian interest. For example, the needs of the growing city directed attention to the ancient water supply and made the restoration of some of the ancient aqueducts (attempted as early as the reign of Nicholas V and carried on with more success in the next century) important projects of papal government. Moreover, fascination with ancient art and architecture, interest in nature, and the need for copious water came together in the taste for gardens that was a prominent feature of the villa culture of aristocrats and prelates in sixteenth-century Rome and its environs.[12]

Furthermore, although in the later sixteenth century some aspects of the study of the remains of pagan antiquity became increasingly problematic in the religious climate of Counter-Reformation Rome, that same climate fostered yet another specifically Roman form of antiquarianism. A new focus of intensive investigation and description was the physical remains of early Christian Rome, the *vitae* of early Roman saints and especially martyrs, and the narrative of early Roman ecclesiastical history.[13]

Brief and inadequate as the foregoing sketch of natural history and antiquarianism in sixteenth century Rome necessarily is, it may nonetheless serve to illustrate the cultural environment inhabited by the physicians from whom I shall draw most of my examples. Andrea Bacci (1524–1600),[14] Marsilio Cagnati (1543–1612),[15] and Angelo Vittori (1547–after 1632)[16] form a group loosely linked by some common experiences and by personal connections with Alessandro Petroni (d. 1585), a renowned physician in Rome in the middle years of the century.[17] Bacci, Cagnati, and Vittori all came to Rome as young men, soon after completing studies elsewhere. Rome was a magnet for the ambitious in the sixteenth century, and they, like many others, were no doubt drawn by the opportunities for professional advancement in one of Europe's greatest cities, the religious capital of the Catholic world, and a major center of learning.[18] In a highly competitive medical world that offered no single secure path to professional authority, social status, or recognition, all three achieved substantial success. Each of them acquired some combination of influential patrons and/or patients, academic status, authorial reputation, or responsible appointment. Bacci was born and completed his early education in small towns in Le Marche, studied medicine at the University of Siena, and moved to Rome around 1550. There, he obtained an appointment to the chair of medicinal simples at the university (1567), a grant of Roman citizenship (1581), the patronage of Cardinal Ascanio Colonna, and ultimately, it would appear, the position of physician to Sixtus V.[19] Bacci's successful climb seems to have been owing chiefly to his copious written works on natural history and medicine, printed during his lifetime in both Rome and other cities, rather than to any great renown as a practitioner.[20]

In Rome, both Cagnati and Vittori became disciples of Petroni. Simultaneously attending physician at the hospital of Santo Spirito and papal physician, Petroni provided one model of a successful Roman medical career, just as his personal connection with St. Ignatius Loyola suggests close involvement with the religious life of the city.[21] Cagnati, originally from Verona, graduated in medicine from the University of Padua, where his teachers included Jacopo Zabarella. Cagnati's Roman career combined prestigious academic and institutional responsibilities with practice among houses of religious orders and senior clergy. He became professor of theoretical and, subsequently, practical medicine at the Sapienza, succeeded Petroni as the attending physician at Santo Spirito, was several times appointed the city's protomedico, and was physician to several cardinals. His written works range from a philological miscellany to treatises on such practical subjects as the flooding of the Tiber and the health effects of Roman air. His contributions to the two last topics may well have been connected with his position as protomedico, since that office usually involved responsibilities having to do with public health.[22] Both Bacci and Cagnati clearly belonged to a circle of learned physicians, naturalists, and antiquarians in Rome. Bacci's personal contacts included, in addition to his friend Petroni, Mercuriale (another physician intensely concerned with Roman antiquities), the physician and polymath Girolamo Cardano during the latter's last years in Rome, and the antiquarians Fulvio Orsini (1529–1600) and Pedro Chacon (1526–1581). Bacci also maintained significant scientific contacts outside Rome, corresponding with Aldrovandi and with a naturalist in Peru.[23] In his youth, Cagnati seems to have known Onofrio Panvinio (1529–1568), whom he praised as "the man to whom we owe more than any of the others who undertook the work of unearthing the antiquities of the city from darkness."[24] Another friend was Michele Mercati, in Cagnati's words "most expert in the medical art and the *res herbaria* and outstanding in the study of *metallica* at Rome." Cagnati visited Mercati's collection of minerals and was the recipient of a copy of his unpublished *Metallotheca*.[25]

Vittori, who graduated from Perugia before coming to Rome, seems—unlike Bacci and Cagnati—never to have held a professorship of medicine, but to have been primarily a practitioner. Certainly his published output was much less than either of theirs. If his posthumously printed *Consultationes* are in any way representative, most of his patients were members of the Roman nobility and about one-third were women. He also seems occasionally to have served as an expert witness in secular court cases.[26] But it is clear that his most important patrons were among the clergy. From about 1585, he was physician to the Oratorian community; his patients included Filippo Neri, whose autopsy he subsequently recorded both in testimony during Neri's canonization process and in a separately published pamphlet.[27] He became a papal physician (to Gregory XIII) and, presumably because of his close relations with the Curia, is said to have been an expert witness in several other canonization processes.

Drawing on the writings of these men, I shall focus on three topics: history and antiquarianism in the service of contemporary Roman public health; *historiae* of natural springs and Roman baths; and case histories, sacred and profane. The first two topics by no means exhaust either the natural historical or the antiquarian interests of these physicians. Bacci also wrote on poisons, on precious stones, and on the unicorn. The last of these treatises makes a noteworthy, if not very successful, attempt to establish grounds on which reports of phenomena outside the realm of one's own experience could be believed (Bacci concluded that unicorns existed, but that most of the supposed specimens of the horn in Europe were fakes).[28] Cagnati's *Observationes variae* is really a humanist philological miscellany, though one chiefly concerned with medical topics. Most of the entries consist of emendations or interpretations of passages in ancient medical and natural philosophical authors, but also include essays on, for example, the history of silk, ancient Greek coinage, ancient maps, and a correction of Patrizi's dating of Strabo.[29] But these physicians' writings in the three areas named above perhaps reveal most clearly the uses of *historia* in the medical world of late sixteenth-century Rome.

History, Antiquarianism, and Public Health

The importance of environment and lifestyle was an enduring feature of Hippocratic-Galenic medical teaching, reflected in numerous medieval and Renaissance expositions of regimen (CRISCIANI, in this volume). In the sixteenth century, however, attention to the influence of place as an aspect of medicine seems to have intensified, a development illustrated—as it was probably encouraged—by a new interest in the Hippocratic *Airs, Waters, Places*.[30] In the case of Rome moreover, the period was one of massive rebuilding during which the papal government also displayed much concern for the city's health and water supply.[31] The physicians who are the focus of this paper wrote not only on water supply but also on diet, the disease environment, and exercise in local context. In regard to water, the Tiber presented the major health concerns, the water of the restored aqueducts being generally regarded as good. One issue was the debate over the quality of the water, summarized above; another was the river's propensity to flood. Bacci and Cagnati both wrote on the flooding of the Tiber as a contemporary urban problem affecting the health of the city. Their contributions differed as to goals and methodology, but each made extensive use of historical examples and argument.

Bacci's interests in natural history were broad, but water seems to have been central to his concerns. His treatment of the Tiber was certainly more expansive than that of Modio, Petroni, or Cagnati. The Tiber comes in for a good deal of attention in Bacci's principal work, *De thermis*, a massive survey of all kinds of waters, rivers, baths, and springs, to which I shall return. There, he expatiated on the Tiber in the course of discussing rivers in general. For example, he claimed that even though such esteemed medical authorities as Hippocrates and Celsus considered river water less wholesome than either rainwater or water from springs, this did not apply to rivers of superior length and grandeur, such as the Nile and the Tiber.[32]

But Bacci reserved his fullest treatment of the Tiber for a vernacular work devoted exclusively to that river. His first version of *Del Tevere*, published in 1558, was an immediate response to Modio's pamphlet. He expanded the second edition (1576) into a substantial treatise in three books, of which the first presented arguments about the nature and goodness of the water, the second gave medical advice (topics included storage and purification, health effects, and a diatribe against artificially chilled drinks), and the third addressed flood control.[33] Many of the central arguments and examples refer not only to the natural history of water but also to the human history of ancient Roman practices relating to water supply. Moreover, in his dedicatory letter to the Senate and people of Rome, Bacci also invoked a kind of metahistory linking nature and human affairs, asserting that "if the Tiber had had the empire of the world and given laws and true religion to all people, so it could be the rule and example of good qualities of all other natural waters."[34] Modio too had argued from Roman history. He insisted that the ancient Romans had only used the Tiber for drinking early in their history, before they grew wealthy, powerful, and technologically advanced enough to construct the aqueducts, of which he gave a capsule account, presumably drawn from the first century treatise on the waters of Rome by Frontinus.[35] But Bacci's counterarguments involved much more thorough and widely based examination of the history of changes in Roman water supply and use. He drew on multiple sources to argue that the impulse behind the construction of the aqueducts was the result not of any inherent defect of the Tiber but of urban development. He culled Livy and Virgil for remarks about the original small size and subsequent increase of the early population. He drew attention particularly to the need for water caused by the introduction of public baths and the problems caused by the runoff of the baths flowing into the Tiber. From Pliny he drew information about the ancient Roman propensity for frequent washing.[36]

But much his most extensive use of historical arguments is found in Book 3 on flood prevention, where he seems to have tried to use historical research as a practical tool for modern reform. After lengthy discussion of the causes of floods, he provided a list of ancient and modern Tiber floods and a lively *historia* of that of 1557, which "in very few hours made most of Rome navigable."[37] He then embarked on a critique of contemporary measures of control. His central contention was that flooding was getting progressively more severe because modern administrators and engineers neglected the methods of the ancients. Accordingly, Bacci combed Tacitus and Suetonius for information about the efforts of Augustus and Tiberius to moderate floods and examined inscriptions recording later imperial flood control projects.[38]

Bacci's interest in this topic was the direct result of practical involvement. He recorded that in 1566 an edict of Pius V had invited "ingenious spirits" to suggest remedies for the problem of flooding. As a result, Bacci had participated in consultations on the subject with Cardinals Sforza and Montepulciano, the "mastri di strada," and "many noble architects and engineers."[39] Bacci thought very little of the "secrets" proposed by some of the technical experts: "I believe and hold for certain that every time they depart from the method that the ancients regularly used in this business, they will always deceive themselves."[40] He went on to describe and evaluate a series of schemes that had supposedly been tried under successive Roman emperors. In this way, ancient engineering could be the model for modern; thus his account of "the fourth remedy, to make an embankment along the Tiber in imitation of the emperor Aurelian" was accompanied by an approving reference to Cosimo I's embankment of the Arno.[41] Bacci drew his information about ancient engineering projects primarily from textual sources: the histories of Tacitus and Dio Cassius, Pliny, Plutarch, and Cicero's correspondence. But he also appealed to material evidence, in the shape of ruins that he thought were the remains of ancient locks or barriers intended to control or divert tributaries, and to his own reading of the landscape itself. Thus, he described at length the effects on the terrain of the ancient construction and subsequent blockage of the canal known as the Curiana between the Velino (a tributary of the Tiber) and Nera rivers. The account is introduced with the remark: "I will relate a *historia* that cannot be read elsewhere, but that is drawn out of the true sense of ancient memoirs"—in this instance, Cicero's letters to Atticus—"and from the site and nature of that lake."[42]

Bacci drew an explicit parallel between his use of historical investigation in the interests of improved flood control and procedure in medicine: "We conclude and hold for a maxim that it is impossible completely to eliminate [floods]. . . . However, it is necessary to proceed in the manner of a good physician who uses great care to find out all the causes of the disease and the appropriate remedies. And with all this, in the cure the most important thing is experience." In the vocabulary of Renaissance medicine, "experience" could, of course, still frequently refer to experiences reported in ancient texts. In this instance, although Bacci also acknowledged the excellence of modern architects (naming Bramante and Michelangelo), his preferred "experience" was that of the ancients.[43] Bacci advocated not only careful selection among ancient engineering solutions and but also the recreation of ancient administrative offices. He recommended the appointment of experienced men with good knowledge of mathematics and surveying as *curatores* of the Tiber, officials who would be responsible for dredging, keeping the banks in order, and prohibiting people from throwing refuse into the river.[44]

Very different in approach is Cagnati's brief *Medical Disputation on the Flooding of the Tiber*, published almost twenty-five years after the enlarged edition of *Del Tevere*. The *Medical Disputation* is one of a pair of treatises, the other being Cagnati's *On the Roman Air*, that cannot easily be separated, as the essential subject of both is the Roman disease environment.[45] Writing shortly after a severe flood in 1598, Cagnati set out to demonstrate that floods in general and floods of the Tiber in particular were neither a portent nor a cause of either epidemics or individual disease. He asserted that floods had natural causes, usually heavy rainfall. They might sometimes be preceded by notable astrological conjunctions, but this was not always the case. Furthermore, many floods occurred without subsequent epidemics or other calamities. He based all these claims on historical examples drawn from both ancient Roman authors and the recent past. Thus, the flood of 1530 had caused great damage to Roman citizens, but 1530, far from being calamitous, was a fortunate year, in which Charles V was crowned at Bologna by Clement VII and there was peace between France and the Empire. The idea that floods preceded epidemics was contradicted by instances when they did not—for example in Rome in the 1490s, 1530, and 1557. In other years, when a flood did precede an unhealthy summer and an epidemic, as was the case in Rome in 1589–1590, the cause was not the flood but the "constitution of the year [*constitutio anni*]."

The interest in the history of natural events and in epidemiological history evident in Cagnati's use of examples in the *disputatio* on the flooding of the Tiber is even more apparent in his work on Roman air. His stated purpose was to dispel the belief that the air of Rome was unhealthy. Given the reputation for unwholesomeness attached to Roman air from antiquity to the early twentieth century, his endeavor seems more suggestive of civic pride than of empiricism. Yet Cagnati defended Roman air with arguments based on empirical and historical considerations. As a counter to ancient medical opinion about its insalubriousness, he pointed to changes since antiquity in the pattern of urban habitation. In order to demonstrate "that there are no more nor more pernicious diseases in the city than elsewhere," he provided a year by year summary of health conditions in Rome from 1568 to 1580, a compilation that clearly implies that he was keeping some kind of epidemiological record or diary.[46] Moreover, it seems likely that his chronology of Roman floods and epidemics was intended to be part of a larger project. He noted elsewhere that he was working on "a chronology of natural things," which included attempts to correct Pliny's correlation of the Olympiad system with Roman dating with regard to both astronomical events and events in human history.[47]

In the 1580s and 1590s, Petroni, Cagnati, and Bacci all wrote at length on diet: Petroni and Bacci with specific reference to Rome and Cagnati clearly in the context of the contemporary Roman religious environment.

Petroni's work was a regimen for the inhabitants of modern Rome, who were "of many and various nations, but principally Italians, Spanish, French, and Germans" and many different walks of life.[48] Though Petroni claimed that no one before him had written a regimen specifically for Rome, much of his advice was fairly standard. Nevertheless, the work is localized in various ways. For example, it discusses the waters of both the Tiber and the new fountains and aqueducts as well as Roman wines and wines from places near Rome. Petroni also made the possibly ironic claim that since Roman food was less nourishing than food abroad, foreigners found they needed to eat more in Rome than they did at home. The remark may be an indirect allusion to the food shortages that were the principal dietary problem for the majority of Rome's population in the late sixteenth century.[49] Moral, religious, or historical aspects of diet played little part in Petroni's work, although he made a few references to Lent and the practice of some religious of abstaining entirely from meat. By contrast, Cagnati's *De continentia* centers on moral and religious issues and has substantial antiquarian content. It sets out to define true continence in the light of ancient Roman practice, Christian teaching, and the requirements of human physiology.[50] Cagnati denounced dietary excess and praised religious fasting. But his insistence on the importance for health of eating sufficient food and more than once a day (chapters 9–10) may also reflect awareness of current conditions in Rome.

Bacci's contribution to the literature of diet was to bring the full weight of his antiquarian and historical, as well as his medical, learning to bear on the subject of wine and the dining arrangements of the ancients. In *De naturali vinorum historia, de vinis Italiae et de conviviis antiquorum*, as in *De thermis* and *Del Tevere* much earlier in his career, Bacci combined natural historical, medical, and antiquarian exposition. *De naturali vinorum historia*, published in 1596 and dedicated to Bacci's patron Cardinal Colonna, is divided into seven books, of which two are primarily antiquarian, one medical, and four natural historical. The books mainly devoted to natural history (2, 5, 6, and 7) treat the properties of vines, grapes, and wines in general and the characteristics of wines from different regions of Europe, with special attention to Italian wines imported to Rome. Book 3, on the use of wine, addresses medical issues, among them the temperament of wine, whether it nourishes, its physiological effects (in moderation and in excess), wine in the regimen of health, according to season, sex, age, and temperament, and its use in sickness.[51] Of the two antiquarian books (1 and 4), the first, devoted to wine in antiquity, consists largely of efforts to identify or explain various terms found in ancient authors to describe wines or wine products and mixtures; the second describes "the dinner parties of the ancients [*De conviviis antiquorum*]." As this description suggests, antiquity and natural history fill much more of the volume than medicine (although extensive use of medical authorities and discussion of medical considerations is by no means confined to book 3). Yet in this work, just as much as in *Del Tevere*, Bacci seems to have been intent on evaluating both antiquarian and natural historical knowledge in terms of current practical—and often medical—usefulness. But whereas in *Del Tevere* he had called for the revival of ancient practices, in *De vinis* he was often highly critical of them.

Thus, although Bacci drew heavily on ancient medical authorities for Book 3, his exposition emphasized historical relativity. He maintained that changes in diet and lifestyle since antiquity had been so great that ancient teachings about the nature and appropriate use of wine could no longer all be regarded as valid.[52] His discussion of ancient methods of viticulture, wine making, and wine additives often involved comparisons with practices current in his own day, by no means always to the advantage of the ancients. Hence his remark that Pliny's description of ancient additives made the resulting concoction sound more like medicine than wine.[53] His own accounts of viticulture and wine production, ancient and modern, often seem as much concerned with the evaluation of methods of agriculture and technology as with natural history.[54]

In this work, Bacci's antiquarianism finds its fullest expression in book 4, "on the dinner parties of the ancients." The goal of Bacci's exposition was, as he said, to show "how from the common custom of life of those times and with what actions the Roman man spent the hours of the day and arranged his meals."[55] Numerous subsections describe or explain every imaginable detail of ancient Roman dining.[56] Most of the evidence comes from Bacci's reading of what he described as an immense forest of ancient texts, although he also referred to "the authority . . . of various marble tombs."[57] But he reserved special attention for the one aspect of ancient dining habits that seemed to have religious significance, namely whether Jesus and the apostles had reclined on couches or sat on chairs at the Last Supper.[58] The topic had aroused interest among Roman antiquaries in Bacci's circle of acquaintances since Girolamo Mercuriale had proposed a recumbent Last Supper in an early edition of his *De arte gymnastica*.[59] Subsequently the antiquaries Pedro Chacon and Fulvio Orsini both endorsed Mercuriale's hypothesis (indeed, according to Mercuriale they plagiarized his work).[60]

By contrast, Bacci—taking what seems to have been a minority position—vigorously denied that Christ and the apostles could have adopted a practice that he associated with the luxury, vices, and excess of pagan Roman aristocrats.[61] In a move that simultaneously reflected both Bacci's own relation to his patron Ascanio Colonna and the ideals for the households of cardinals expressed by sixteenth-century popes and Catholic reformists, he contrasted this pagan Roman luxury with the present lifestyle of the cardinals. "In the moderation of our own times," he declared, "our princes have very decent tables, and especially here in the Roman curia we see that they are frugal, and of praiseworthy moderation in splendor, yet very well appointed."[62] The following description of the "honest sobriety [*hac honesta sobrietate*]" of Cardinal Ascanio Colonna's "elegant daily dinner [*elegantem . . . quotidianam mensam*]" does not sound very frugal, with its lavish table settings, numerous attendants, song for "a chorus of grateful guests," and recipe for *olla podrida* (to compliment the cardinal's Spanish tastes), calling for, inter alia, beef, pork, partridges, and doves.[63] But Bacci's description gives a lively picture of the guests and conversation at the table of this cultivated, Hispanophile, and wealthy prelate.[64]

Historiae of Natural Springs and Roman Baths

I return now to Bacci's *De thermis*, a magnum opus worthy of fuller analysis than is possible here. Bacci's vast survey of the properties, uses, and locations of all kinds of springs and baths throughout Europe and beyond begins with a general natural historical overview of varieties of water—the seas, rivers, rainfall, and springs. There follow two books on the medicinal uses of different waters, both externally in bathing and taken internally. Three following books describe, respectively, sulfurous and hot springs, "mineral waters," and "metallic waters." A final book is devoted to the baths of ancient Rome. The topic of springs, baths, and waters was hardly a new one for either medicine or natural history, and much of the content of this work is highly derivative. For his accounts of the properties of different waters Bacci chiefly relied, as one might expect, on ancient natural philosophical and medical authorities. For many of his *historiae* about particular springs and baths in Italy, as he acknowledged, he drew extensively on the considerable body of late medieval and early Renaissance writing on springs and spa waters by such authors as Bartolomeo Montagnana, Antonio Guainerio, Michele Savonarola, and Ugolino Montecatino.[65] Yet Bacci's treatment went beyond his predecessors in both encyclopedic scope and organization in a way that seems characteristic of late Renaissance and early modern natural history (see OGILVIE and PINON in this volume). The sections of the work on natural hot springs and mineral and metallic waters are informed by Bacci's conviction that it was insufficient either to generalize about the properties of minerals and metals or merely to discuss them in terms of their uses in alchemy or medicine. Nor was it enough simply to describe the springs, spas, or baths of a particular region. Instead, what was necessary was to consider the qualities and effects of minerals and metals when mixed with different kinds of waters.[66] Consequently, each of these sections of the work begins with a substantial general account—termed in the table of contents and elsewhere the *historia*—of the properties of the type of water in question followed by descriptions (again, *historiae*) of individual baths, springs, and spas.[67] Thus, the descriptions of individual sites are arranged, as Bacci remarked, not according to region but to genus: "We follow the order of their genera, which is *scientificus*, and thus suitable for a *historia*" (see MACLEAN in this volume).[68] He evidently took considerable pains to gather descriptions of baths, springs, and spas in regions outside Italy and indeed outside Europe from recent travel writers and, in some instances, personal informants. For example, his information about Turkish baths in Hungary and salt mines in Transylvania came from chorographical works by Georg Werner and Georg Reychersdorff; for waters in India and Africa he drew on the travel accounts of Ludovico Varthema and Alvise Cadamusto. He plumbed Barthélemy de Chasseneux's vast compilation on human and natural hierarchies for a few remarks about hot springs in Burgundy and a brief pamphlet by a local physician for an account of the baths at Aachen.[69] From time to time, his *historiae* of individual spas or baths included passages of human history. Edward Carne, Queen Mary Tudor's ambassador to the papacy, told him about the many mineral springs in Britain, but he drew from Polydore Vergil the information that Bath had been founded by King Badaudus who had constructed the baths there and rejected an explanation that ascribed them to Julius Caesar.[70]

His decision to include a historical and antiquarian account of ancient Roman baths in a work addressed to "physicians and students of nature" is also characteristic of its time, place, and author. Bacci, like others in Rome, had access to Mercuriale's *De arte gymnastica*, perhaps the first notable product of the intersection of medicine with Roman antiquarianism, before its publication. Indeed, it seems likely that, as Bacci more or less explicitly acknowledged, he was inspired to include this section by Mercuriale's work.[71]

Mercuriale had set out to recreate the lost world of athletics in the ancient baths and palaestra, but was reserved and cautious about modern applications, at any rate in civil life. By contrast, Bacci believed that knowledge of ancient Roman bathing practices, like knowledge of ancient Roman flood control, could offer useful practical guidance. He remarked that even the "very expert antiquaries [*peritissimi antiquarii*] of our age" had not completely succeeded in explaining the structure and uses of the baths, particularly as regards washing practices; he proposed to remedy the last defect and "especially to expound those things that are useful in medicine."[72] But although he stressed that his interest was purely in the medical or sanitary aspects of bathing he nevertheless provided his readers with a list of Roman baths and the emperors responsible for their construction, the texts of related inscriptions, an account of the ancient aqueducts supplying the water for the baths (largely drawn from Frontinus), a plan of the Baths of Diocletian, and his own conclusions about the heating system and the uses of different sections of the baths.[73] Renaissance and early modern accounts of Roman baths normally include denunciations of luxury and immorality, and Bacci was no exception. But he was unambiguous in his conviction that frequent washing and exercise at the baths contributed importantly to the "elegance of life ... and health and strength of body" of the ancient Romans.[74] In this instance, too, he sought to draw practical utility from ancient example hallowed by Roman imperial dignity and magnificence. His description of the medical usefulness of bathing draws, as one might expect, on ancient medical authorities, but emphasizes its value for the treatment or prevention of modern diseases, notably the *morbus gallicus*.[75] Moreover, he insisted that the ancients had private as well as public baths and that at least this custom could and should easily be imitated.[76]

If Bacci's account of the Roman baths was in some sense a supplement to Mercuriale, Cagnati seems to have intended a critique of some of Mercuriale's basic definitions. According to Mercuriale, gymnastica was divided into three categories: legitimate *gymnastica medica*, that is exercise for the sake of health; partially legitimate *gymnastica bellica*, or exercises preparatory for war; and vicious and degenerate athletics for the sake of competition or entertainment.[77] Writing at a time when *De arte gymnastica* had been reissued in several successively enlarged editions, Cagnati inserted into a work on regimen a short dialogue on gymnastic and its relation to medicine in which he set forth a very different view. The initial setting of the dialogue is a visit to the Thermae Antoninae by a visitor from Verona—presumably Cagnati himself—and his Roman host, during which they discuss the Greek and Roman history of gymnasia and athletics, and consider whether ancient gymnasia contributed more to vice or to health. But the scene rapidly shifts to a dialogue within a dialogue in which the most famous physicians of Verona in Cagnati's midcentury youth, namely Girolamo Fracastoro (d. 1553) and Antonio Fumanelli, are made to discuss the history and function of gymnastic, with special reference to the views of Plato and Galen. The thrust of the argument is to insist that any and all exercise that contributes to health is part of medicine, and of medicine alone. But this does not include exercise purely in order to build "athletic habitus," which belongs to gymnastic, is not part of medicine, and is not necessary for health. Finally, the dialogue reverts to the original two speakers and the Veronese guest and his Roman host conclude that gymnastic has nothing to do with medicine and that no part of gymnastic is anything other than *vitiosa*.[78]

Case Histories, Sacred and Profane

"Just as history is very useful for the good government of public affairs because it is said to be the mistress of life, so also is it extremely useful for curing illness. We learn this sufficiently from the example of Hippocrates, which teaches that written descriptions of epidemics [*historias Epidemicas*] should not be taken lightly." With these words, Cagnati introduced his account of two epidemics in Rome in 1591 and 1593.[79] His parallel between civil history and medical narrative neatly exemplifies both the responsiveness of medical thought to the enthusiasm for history in the broader culture and the role of the Hippocratic *Epidemics* in fostering the proliferation of narratives about specific outbreaks or individual cases of disease in sixteenth-century medical writing. Many signs of an increasing presence of narrative in medicine had already emerged in the fourteenth and fifteenth centuries. Among them may be noted *consilia* for individual patients (though these are primarily recommendations for treatment rather than narratives of disease), advice on interrogating the patient, anecdotes embedded in works on surgery or *practica*, descriptions of autopsies, and accounts of remarkable cures (see CRISCIANI in this volume). What is striking in the medical literature of the sixteenth century is therefore not the introduction but the multiplication and expansion of narrative or descriptive elements (see POMATA in this volume). This development reflected primarily growing medical interest in particulars and in observation, but also the availability of the complete Hippocratic *Epidemics*, which contains some of the principal examples of ancient medical narrative.[80] Similarly, Renaissance editions of Galen made it easier to appreciate Galen's analyses of his own cases in *Methodus medendi* and elsewhere. Thus sixteenth-century physicians had not one but two powerful models for case narratives. The Hippocratic case histories recorded the course of disease from onset to (indifferently) favorable or unfavorable outcome, with minimum analysis. Some of Galen's accounts of his cases provided copious analysis but usually illustrated his success. At the same time, much medical writing about individual cases continued to take the traditional form of the *consilium*.

All three of the Roman physicians seem to have regarded *historiae* about individual cases as essential tools of medical teaching and practice. Of the three, Bacci was probably the least concerned with medical practice. Nevertheless, in his work on poisons he used a narrative of one of his own cases—the illness and death of Jacopo, a weaver from Trastevere who had been bitten by a mad dog—as a peg on which to hang an elaborate analysis of causes, signs, accidents, and treatment. In introducing the case, Bacci invoked the Hippocratic model with the remark that "examples and particular histories . . . please the audience and the histories in the *Epidemics* of Hippocrates amplify information about disease no less than the teaching of the *Aphorisms*." The narrative is Hippocratic in its attention to events from day to day and its record of unfavorable outcome in the shape of Jacopo's agonizing death. Yet his decision to choose just one striking case from among "the many examples of this condition that I have seen (and I cured some of them)" and use it essentially as a rhetorical device to introduce his own exposition is more Galenic than Hippocratic.[81]

By contrast, Cagnati, who (as already noted) drew on records of epidemics to prove his contentions about floods and the Roman air, clearly did perceive the systematic compilation of records of disease as providing useful empirical evidence. Yet another example of this attitude appears in his treatment of the concept of critical days of illness. Commenting on a Hippocratic aphorism that stated that the fourth day of illness was predictive of the patient's condition on the seventh, he tested the claim against a series of seven detailed case histories of patients of his own and concluded that it was not valid.[82] Moreover, he seems to have regarded the compilation of such records as an ongoing task of modern medicine to which he had consciously dedicated himself: "Many authors of our century have already recognized the usefulness of *historiae*. I indeed first imitated their studies of diseases for my own sake. [Diseases] have by now offered themselves for my observation during more than thirty-two years of medical practice, and I have spent no little labor and time in describing and examining them. But then it seemed that if I published a specimen of that study and labor, this could bring something of value to people studious of the science of medicine."[83] Yet as this remark shows, Cagnati, who had already published a large collection of his philological *observationes*, chose to present only selected examples of what may have been an equally extensive collection of personally observed medical *historiae*. For Cagnati, the term *historia* in a medical context evidently embraced both types of record that occur in the Hippocratic *Epidemics*: disease patterns and individual case histories. Indeed, he drew explicit parallels between cases described by the Hippocratic author and ones he had attended himself. For example, Olivier, a noble Frenchman had "exactly the same [*ad unguem*]" symptoms as Heraclides who lay sick in the house of Aristocydes.[84]

But at least to the extent that it is possible to judge from their published writings, Angelo Vittori seems to have spent far more effort in assembling *historiae* of individual cases than either Bacci or Cagnati. Moreover, although Vittori was certainly conscious of ancient models for medical narrative and had learned the importance of recording and analyzing *historiae* from Petroni, he also drew extensively on quite different traditions: *consilia*, accounts of marvelous cures, and hagiography. In his lifetime, Vittori seems to have published only one item, his *Medica disputatio* on the autopsy of St. Filippo Neri. The history of this work in itself exemplifies one kind of relation between *historiae* of disease and the elaboration of theoretical argument. Vittori was present at Neri's autopsy and on the spot wrote up a brief factual report (*historia*) of the findings. Subsequently, in addition to testifying orally during the canonization process, he wrote three lengthy interpretive versions of his original report. In these expositions, each more elaborate than the last, he surrounded his original *historia* with arguments assembled from ancient and recent medical authorities in order to demonstrate that Filippo Neri's condition could not have been of natural (and therefore must have been of supernatural) origin. By 1613, when the third version appeared, the work had expanded into a formal medical treatise replete with *quaestiones*, discussion of causes, and citations of authorities.[85]

But the main evidence for Vittori's practice of collecting and preserving information about individual cases comes from the large volume of his *consultationes* published some time after his death by Vincenzo Mannucci, a professor of medicine and mathematics in Vittori's native Perugia. The work is in reality a miscellany. Some three-quarters of the *consultationes* in the volume correspond to *consilia* of the type traditional since the thirteenth century and frequently assembled into collections in the fourteenth and fifteenth; with them are interspersed a few brief treatises or *quaestiones* on random medical topics.[86] The *consultationes/consilia* are letters of advice for individual patients, often in response to a written request, probably more likely to have come from the attending physician than from the patient. Some of them include a brief descriptive *historia* of the patient's illness, but the emphasis is on recommendations for treatment. The presence of lengthy theoretical discussions suggests rewriting for publication, whether by Vittori himself or by his posthumous editor.[87] The careful identification of the elite social standing of many patients (by such terms as *nobilissimus, princeps, illustris*, or *patritius Romanus*) serves as a reminder that one purpose served by the compilation of *consilia* was presumably to emphasize a practitioner's distinction, as measured by the status of his clientele as well as by his skill.

But the volume also contains evidence of a different approach to the record of information about individual cases. "The *historia* of a young man who died of variolae noted by the author when he was practicing medicine under the supervision of Alessandro Petroni . . . proposed as an example of accurately observing the progress of diseases" is a case history in the full sense of the term.[88] It follows the course of the disease in a nineteen-year-old *princeps* day by day from onset on Monday, January 24, to the patient's death on Wednesday, February 9, 1575. The model for the day-by-day narrative of symptoms and treatment is Hippocratic, but the account is enriched with theoretical discussion and consideration of causes and concludes with autopsy findings in a way quite unlike anything in the Hippocratic case histories. Petroni's systematic teaching on case history was presumably associated with his own collection of cases, the publication of which was eagerly expected by Schenck von Grafenberg in 1584.[89] In reality, Petroni's collection seems never to have been published, presumably because of his death in the following year.

More advice about case histories is also to be found in the most substantial single item in Vittori's *Consultationes*, a treatise addressed to a recent graduate "in which it is fully explained how a young man instructed in the art of medicine should proceed to its actual practice."[90] Vittori began by informing the new graduate, somewhat discouragingly, that the knowledge of the *scientia* of medicine that he had acquired at the university merely put him in the same position as someone who owned a flute without being able to play it.[91] The detailed set of recommendations that follow, many of which stress prudence and recognition of medicine's limitations, lead the young practitioner through proper procedure before, during, and after a visit to a patient. Appropriate preparation beforehand consists of finding out something about the patient's health history and looking up likely diseases; during daily visits the patient must be carefully interrogated and observed.[92] Such standard recommendations could no doubt be paralleled in many earlier deontological works. But a separate section advising the neophyte what to do on returning home reminds him that Hippocrates "observed the events of diseases in many places, and described them rather accurately, as one can see in the histories in the *Epidemics*." Hence, "the physician who, in imitation of this exemplar, wishes to learn the art from things themselves will find it very useful if he himself has, even from his earliest years in practice, compiled and written down day by day a *historia* of every patient containing the symptoms, causes, occasions, and cures of that disease as each occurred."[93] Only such a history—studied in conjunction with appropriate consultation of medical books—would provide the right basis for thoughtful consideration and interpretation of the signs.

Finally in Vittori's case, *historiae* and evidence are connected in another context. As already noted, the *Consultationes* contains two items that suggest his involvement as an expert witness in secular court cases. In both, his role was to interpret physiological information contained in what he termed the *historia ex processu*, the story told in the criminal charge.[94] But both Vittori himself and the editor of the volume evidently attached far more importance to his testimony in ecclesiastical processes in the great age of Counter-Reformation saint making. In addition to a reissue of Vittori's treatise on Filippo Neri's autopsy, the volume includes his interpretations of a series of miracles of healing attributed to Sts. Ignatius Loyola, Francis Xavier, Filippo Neri, and the fifteenth-century Franciscan Diego of Alcalá, for each of whom a canonization process was instituted in Vittori's lifetime.[95] In his preface, Mannucci drew attention to the fact that Vittori had been called to testify by the Rota Romana not only in Neri's but also in other canonization processes, so that presumably some of the discussions of miracles are based on his testimony on those occasions.[96] In both criminal and canonization testimony Vittori followed essentially the same procedure, briefly recounting the *historia*, then adducing medical authorities to support his interpretation. In a case of suspected poisoning, he marshaled his authorities to show that death occurred naturally. In the case of the miracles he followed the same methodology to show, just as he had done with Neri's autopsy, that the events in question could only have happened supernaturally.[97] In both civil and ecclesiastical cases (apart from that of Neri), Vittori's testimony rested on reported *historiae*, not his own experience. Whereas the original *historia* that he wrote of Neri's autopsy and in all probability at least some of his *consultationes* were records of personal experience, the same was clearly not true of his testimony as an expert witness in criminal trials. There is, for example, no suggestion that he had attended the supposed victim of poisoning himself. Still further from his own experience were many of the miracles about which he testified, since they had supposedly taken place far from Rome and in some cases many years earlier.[98] But in both civil and ecclesiastical instances, the *historiae* in question were, of course, sworn evidence given in a court of law. Vittori's insistence on a careful, personally observed and recorded history of each patient as a prerequisite for good medical practice was not necessarily incompatible with other criteria for judging the validity of medical *historiae* in other contexts.

Conclusion

Historia had many uses and many meanings for learned physicians in late sixteenth-century Rome. Bacci, Cagnati, and Vittori all gained their initial training in medicine elsewhere, but in Rome they applied their medical learning in ways shaped by the Roman context of teaching, professional activity, and intellectual environment. Thus, Bacci and Cagnati were part of a Roman intellectual milieu that attached central importance to the study and discussion of the city's classical and Christian antiquities. From their own medical standpoint, they perceived antiquarian knowledge as fraught with current, practical implications for Roman public health concerning such mundane matters as the proper use of exercise, the best means of flood control, and the right way to make wine. At the same time, they participated fully in the contemporary expansion of *historia naturalis*. Of the physicians discussed in this paper, only Bacci made substantial contributions to the subject, but Cagnati's relation with Michele Mercati is surely indicative of strong shared interest. Moreover Bacci appears to have seen *historia naturalis* as yet another field of intellectual inquiry that had implications for human health and well-being. He seems indeed to have set out systematically to bring *historia naturalis*, the human past, and medicine together, choosing repeatedly to write on subjects that involved all three, as he did in *De thermis*, the treatise on the Tiber, and the book on wine. Methodologically, moreover, Bacci seems to have approached *historia naturalis*, the study of antiquities, fragments of civil history, and medicine in much the same way. He largely compiled from texts, including the texts of recent authors, but added some items from personal informants, noted material evidence, and made significant use of his own experience.

Taken together, the works of all three physicians provide numerous examples, both medical and natural historical, of the use of the term and concept *historia* in the sense of a brief factual narrative or description. But Cagnati and Vittori were especially concerned with *historia* in a very specific sense, that is the record of human disease in Rome. Both proclaimed their allegiance to the Hippocratic *Epidemics*, but they put the Hippocratic model into practice so differently as to preclude easy generalizations about the influence of the *Epidemics* on empiricism in early modern medicine. Cagnati's endeavors to use compilation as a basis for systematic analysis and his interest in historical chronology have very little in common with Vittori's analyses of *historiae* of individual patients, criminal cases, or miracles. *Historiae* were, increasingly evidence, but distinctions between firsthand and reported evidence had by no means been established, and empiricism and learning might still go hand in hand.

Notes

My thanks to participants in the Historia workshop at the Max Planck Institut für Wissenschaftsgeschichte, June 2003, and to Silvia de Renzi for helpful comments on an earlier version of this paper.

Sources cited in boldface type appear in the Primary Sources of the bibliography.

1. **Modio** 1556; **Petroni** 1552; **Bacci** 1558, subsequently reissued in an enlarged edition in three books as **Bacci** 1576 (in 1599 Bacci published an additional fourth book); **Petroni** 1581, translated into Italian as **Petroni** 1592, book 2, chapter 5; D'Onofrio 1986, 33–42.

2. Momigliano 1985.

3. See, for example, on the interaction between attention to Dioscorides and the expansion of botanical knowledge, Ferri 1997 and Findlen 2000.

4. I make no attempt to list the growing bibliography on the Lincei and life sciences, but mention may at least be made of Freedberg 2002; Clericuzio and De Renzi 1995; and De Renzi 2000.

5. For a summary with bibliographical references of the sponsorship of translations of ancient works on the life sciences by Nicholas V and Sixtus IV, see Siraisi 1993. For the history of the chair of simples at the university and the Vatican botanic garden, see Coffin 1991, 210–214 and bibliography there cited.

6. On Salviani (1514–1572), see Renazzi 1804; Castellani 1975; and, for his exchanges with Aldrovandi, Pinon 2002. On Mercati, see Premuda 1974; Accordi 1980.

7. Among many studies of antiquarianism, Momigliano 1950 is foundational. Mention should also be made of Barkan 1999, Haskell 1993, and Weiss 1973. Specifically on antiquarianism in Rome, see Ferrary 1996; Gaston 1988; Grafton 1995a; Herklotz, 1999; Mandowsky and Mitchell 1963; Rowland 1998, chapters 1 and 2; Schreurs 2000.

8. On Leto (1427–1498) and the suppression of the Roman Academy by Paul II in 1468, see Rowland 1998, 10–17, and D'Amico 1983, 91–97. The accusation of paganism was only one among several charges laid to the academicians.

9. See Bizocchi 1995, 9–13, and, regarding an inventive forger of such genealogies, Petrucci 1979. On Marco Antonio Altieri and his *Li nuptiali* (ca. 1500), see Klapisch-Zuber 1985, 247–260, and Kolsky 1987.

10. See Grafton 1995a, 87–104; Curran and Grafton, 1995. For the extensive antiquarian interests of Alberti, see Grafton 2000, chapter 7.

11. On Alessandro Farnese, see Robertson 1992. For his collections of antiquities and the antiquarians in his household, Robertson 1992, 220–224; Herklotz 1999, 214–226. On Mercuriale in this context, see Siraisi 2003.

12. Delumeau 1957, 1:327–349; D'Onofrio 1986; Rinne 2000; Rinne 2001–02; Coffin 1991; Coffin 1979, especially chapters 8 and 9; Lazzaro 1990, chapter 9. For an example of antiquities and antiquarianism in a garden setting, see Cellauro 1995.

13. I make no attempt to supply a full bibliography on Christian antiquarianism and ecclesiastical history in sixteenth-century Rome, but see Grafton 1995a, 112–117; Fiorentino 1982; and Ditchfield 1998 and 2000. Regarding studies of ancient paganism see MULSOW in this volume.

14. On Bacci, see Crespi 1963; Caetani 1924, cols. 40–43; D'Onofrio 1986, 39–42; **Carafa** 1751, 2:358; **Marini** 1784, 1:464, and in the supplement by P. Mandosius, 13–16; Renazzi 1804, 2:195; Conte 1991, 2:861; Saffrey 1994; Thorndike 1959, 5:484–485 and 6:315–316; Simili 1970, which includes an edition of several of Bacci's letters to Aldrovandi; and Stefanutti 1979. I have yet to see Filippo Pio Massi, *Andrea Bacci, memorie sparse* (Rome, 1883) and L. Münster, "Studi e ricerche sull'opera scientifica di Andrea Bacci da S. Elpidio,"

in *Atti del III Convegno della Marca per la storia della medicina* (Fermo, 1959), 99–103.

15. On Cagnati, see Stabile 1973; Renazzi 1804, 3:41, 92; Conte 1991, 2:947; **Carafa** 1751, 2:361.

16. On Vittori, see the editor's preface to **Vittori** 1640, A3r-v; and **Marini** 1784, 1:458–459. For his birth and approximate death dates, see his testimony of August 16, 1610, in **Incisa della Rocchetta et al.** 1957–63, 4:35, which states that he was then sixty-three years old, and the preface to **Vittori** 1640, which states that he continued to practice medicine until the eighty-sixth year of his age.

17. On Petroni, see **Marini** 1784, 1:422–423 and 454–455; and D'Onofrio 1986, 38–39.

18. On the expansion of Rome's population in the sixteenth century and the large numbers of immigrants, see Delumeau 1957, 1:188–220.

19. Bacci's chair is described as in *botanica* by Crespi 1963, 29, and Renazzi 1804, 2:195, but the university *rotuli* edited in Conte 1991, 2:861, name his chair as in *simplicia medicamenta*. In **Bacci** 1596, book 4, part 1, p. 131, Bacci noted that he had been personal physician to Sixtus for eighteen years before his elevation to the papacy. Bacci's subsequent appointment as papal physician is noted in Crespi 1963, 29, but questioned by D'Onofrio 1986, 39.

20. D'Onofrio 1986, 39, quotes a negative opinion of Bacci's skills as a practitioner from 1645, but given the conditions of premodern medicine, it is difficult to know how to evaluate such judgments.

21. For Cagnati's association with Petroni as pupil and subsequently as colleague at Santo Spirito, see Stabile 1973, 301. Lectures on medicine, surgery, and anatomy were given at Santo Spirito in the early seventeenth century and probably also before. See De Angelis 1948, 23–28; De Angelis 1952, 92–94; De Renzi 1999, 108–109; Grégoire 1979, 238–239. Vittori early in his career learned to practice medicine under Petroni's supervision, possibly at the hospital (see further here 344 and note 88). Although Petroni thus evidently taught medicine, or the practice of medicine, and was both learned and Latinate, he does not appear to have taught at the university. His name does not occur in the *rotuli*, but they are not complete for the sixteenth century (Conte 1991, 1:viii–x). For Petroni's connection with Ignatius Loyola, who visited him when he (Petroni) was ill, see *Acta Sanctorum Julii*, 1731, 617, and *Relatio* 1644, 34. The story of this visit is also found in **Bardi** 1644, 10.4, p. 217.

22. For an account of the office of protomedico, who in Rome was the chief member of the Roman College of Physicians, see Pertile 1902 (reprint 1966), vol. 2, part 2, pp. 212–214; Garofalo 1950 (with reference to Cagnati as protomedico in 1595 [25, 31] and 1602 [15]), Gentilcore 1994, and Carlino 1999, 70–75. My thanks to Gianna Pomata for drawing my attention to the works of Pertile and Garofalo. As in other cities, the College of Physicians was composed of a small number of elite physicians practicing in the city; some of them might also teach in the studium, but this was not necessarily the case. The protomedico and

the College were responsible for the licensing and supervision of all medical practitioners and apothecaries in Rome. Pomata 1998, 138–139, points out that in the late sixteenth and seventeenth centuries the protomedici of Bologna increasingly assumed authority over diet.

23. **Bacci** 1596, 3.3, p. 93: "familiares nostri Petronius, et Cardanus"; 3.4, p. 94: "Hieronymum Mercurialem, peritissimum hac aetate virum, et familiarem meum"; 4.1, p. 127: "venerandae memoriae hic in urbe Roma Petrus, Ciacconus Canonicus Toletanus, atque his Fulvius Ursinus ex Patritiis Romanis summae doctrinae, ac rerum antiquarum studiosissimus." See Simili 1969 for Bacci's correspondence with Aldrovandi and for Bacci's correspondent in Peru (433).

24. **Cagnati** 1605, 95.

25. **Cagnati** 1587, 3.17, p. 236, and 4.4, p. 265. Mercati's *Metallotheca* was first published in 1717.

26. **Vittori** 1640, no. 32, pp. 103–107

27. For Vittori's connections with the Oratorians, attendance on St. Filippo Neri, and presence at and report on Neri's autopsy, see **Incisa della Rocchetta et al.** 1957–63, no. 40, 1:151–154; no. 229, 2:235–236; no. 241, 2:259–267; no. 334, 3:303–305; 4:35–37. Vittori's treatise on Neri's autopsy is **Vittori** 1613. For discussion of his analysis of Neri's postmortem, see Siraisi 1998, 2001.

28. **Bacci** 1586; **Bacci** 1603. Bacci's work on the unicorn first appeared in Italian (**Bacci** 1573); I have consulted the Latin translation, **Bacci** 1598. For his opinions on the reality of the unicorn and the prevalence of fake horns, see **Bacci** 1598, proemium.

29. **Cagnati** 1587, 3.20, pp. 243–259 (Patrizi); 4.5, pp. 268–270 (maps); 4.6, pp. 270–260 [error for 276] (coins); 4.11, pp. 294–309 (silk).

30. I make no attempt to provide a full bibliography of Renaissance Hippocratism. Some valuable studies are Mercati 1917; Lonie 1985; Smith 1979; Nutton 1989a. For Renaissance editions and translations of the Hippocratic corpus, see Maloney and Savoie 1982. *Airs, Waters, Places* had been translated into Latin but not commented upon during the Middle Ages (Kibre 1985, 25–28). Sixteenth-century commentaries on *Airs, Waters, Places* include **L'Alemant** 1557; **Cardano** 1663 (originally published Basel, 1570), 8:1–212; **Baldini** 1586; **Settala** 1590.

31. See Delumeau 1957, 1:223–339; D'Onofrio 1986; Sansa 2002; San Juan 2001, chapter 4; Coffin 1991, chapter 3; Pecchiai 1944.

32. **Bacci** 1571, 1.9, pp. 22–25.

33. See Bacci references in note 1, above. The theme of the dangers of cold drinks goes back to antiquity.

34. **Bacci** 1576, dedicatory letter.

35. **Modio** 1556, 4v–6v. Frontinus's (1st century CE) *De aquis urbis Romae* became known to humanists in the fifteenth century and was printed in several late fifteenth- and early sixteenth-century editions.

36. **Bacci** 1576, book 2, pp. 91–96.

37. **Bacci** 1576, book 3, pp. 251–255, with phrase quoted at 254. The list of historic floods continues until 259.

38. **Bacci** 1576, book 3, pp. 259–269, with references to inscriptions at 268, 269. Floods in the sixteenth century seem indeed to have become more frequent and more devastating; see Di Martino and Belati 1986, 19 and 55–81; Delumeau 1957, 1:339–353. A number of other treatises were also written on the subject, for example **Castiglione** 1599, which draws on Bacci (I owe this reference to Laurent Pinon).

39. **Bacci** 1576, book 3, p. 269. On proposed remedies, including Bacci's, see D'Onofrio 1980, 311 ff.

40. **Bacci** 1576, book 3, p. 276.

41. **Bacci** 1576, book 3, p. 283–285: "Quarto rimedio, di far un'argine al Tevere, ad imitatione di Aureliano Imperatore," with reference to Cosimo at 284–285.

42. **Bacci** 1576, book 3, p. 302–308, with the remark quoted at 302. See Cicero, *Ad Atticum* 4.15.

43. **Bacci** 1576, book 3, pp. 288–289.

44. **Bacci** 1576, book 3, pp. 290–296.

45. **Cagnati** 1599b. (*De Tiberis inundatione* occupies pp. 1–22 of this collection) and **Cagnati** 1599a.

46. **Cagnati** 1599a, 21: "aegritudines nec frequentius nec perniciosius in urbe, quam alibi grassantur." The list of epidemics occupies 21–24.

47. **Cagnati** 1587, 4.8, p. 264 [bis; misprint for 280]. The chapter continues until p. 268 [bis; misprint for 284]. On sixteenth-century efforts to reconstruct the chronology of ancient history, see McCuaig 1991.

48. See note 1, above. On the diversity of peoples and occupations in the city, see **Petroni** 1581, 3.25, p. 164 (**Petroni** 1592, 196) and Delumeau 1957, 1:188–220. For the substantial Spanish presence and political and cultural influence in sixteenth-century Rome, see Dandelet 2001.

49. On dearth in Rome and the attempts of the papacy and city magistrates to ensure the supply of grain, see Delumeau 1957, 1:592–625, and Reinhardt 1991. **Petroni** 1581, book 2, pp. 31–76 (**Petroni** 1592, 35–90), is on water and wine; for the eating habits of foreigners in Rome, **Petroni** 1581, 3.25, p. 164 (**Petroni** 1592, 196). According to Carafa 1751, 2:359, Sallustio Salviani, son of Ippolito and professor of medical theory at the Sapienza, strongly disapproved of Petroni's treatise and planned to refute it.

50. **Cagnati** 1591a.

51. **Cagnati** 1601, 40–44, also addressed the medicinal uses of wine, vigorously defending the practice of Roman physicians who permitted or prescribed it for fever patients.

52. **Bacci** 1596, book 3, preface, pp. 88–90.

53. **Bacci** 1596, 1.8, p. 12.

54. See, for example, **Bacci** 1596, 1:31–32, pp. 47–51, on containers for wine.

55. Bacci 1596, book 4, preface, p. 129.

56. For example, **Bacci** 1596, book 4, pp. 130–143, 150–164, 187–195.

57. Bacci 1596, book 4, p. 144. However, he seems to have drawn some of his information from Fulvio Orsini's very learned appendix to **Chacon** 1588, despite his disagreement with Chacon's conclusions (see below); compare **Chacon** 1588, 116, and **Bacci** 1596, book 4, part 1, p. 141, on *coenatoria vestis*.

58. Bacci 1596, book 4, pp. 144–147.

59. **Mercuriale** 1573, 1.11, pp. 52–55 [misprint for 57], with reference to the Last Supper on 54 [misprint for 56]. Mercuriale had previously discussed the Roman practice of reclining in the first edition of this work, namely **Mercuriale** 1569, 1.11, fols. 20r–22r, but in a purely secular context.

60. **Chacon** 1588. **Mercuriale** 1672, 1.11, appendix (added in the edition of 1601), pp. 77–78: "Petrus Ciacconius, et Fulvius Ursinus rerum antiquitarum peritissimi, quique multis annis post meam gymnasticam de triclinio scripserunt"; Blunt 1938–39. On Chacon, see Ruiz 1976 (I owe this reference to William Stenhouse).

61. **Bacci** 1596, book 4, p. 148: "Quid magis detestabile putandum Christianae pietati, ut quemadmodum rite tanti mysterii iubemur imitari solemnia, sic in throno accumbentes sanctissimam celebraremus coenam?"

62. **Bacci** 1596, On cardinals' households in the sixteenth century, see Fragnito 1993 and Antonovics 1972.

63. **Bacci** 1596, book 4, pp. 148–149; the recipe for "Oglia poderida in mensis Hispanicis inclyta" and the "Chorus gratiarum convivialium" are on 165–167.

64. On Ascanio Colonna, see Petrucci 1982. A contemporary quoted by Fragnito 1993 (41) commented approvingly that Cardinal Ascanio "lived with much splendor," although Antonovics 1972, 324–325, notes that he had to scale back in 1590.

65. See Park 1999.

66. **Bacci** 1571, 4.1, p. 190: "Imo si ipsae rerum facultates absolute sumantur, et non considerentur in temperamento cum aquis; non possunt de eis fieri verae demonstrationes ... quam ob rem tantopere necessarium est singulares earum historias scribere."

67. **Bacci** 1571, table of contents: "In IIII [libro] qui Thermeusis inscribitur conscribuntur Historiae Thermarum sulphurearum ... Balnearum Historiae singulares CXXII. ... In V libro qui est Mineralis, describuntur Historiae aquarum Mineralium. ... Historiae singulares ICX. ... In VI libro, qui est Metallicus, describuntur Historiae ex Metallis. ... Historiae singulares Balnearum et Fontium CLXIII."

68. Bacci 1571, 4.3, p. 196.

69. **Bacci** 1571, 4.11, p. 262: "ut scribit hodie Georgius Vernherus"; I have consulted **Werner** 1557 (Werner fl. 1505), but there are other early editions. For Reychersdorff, see **Bacci** 1571, 5.4, p. 282, and **Reychersdorff** 1550, 7r, 18r.

References to Varthema (**Varthema** 1991; originally published Rome, 1510) and Cadamusto, that is Alvise Ca da Mosto (1432–1488) (**Ca da Mosto** 1966), are found **Bacci** 1571, 4.3, p. 194, and 5.4, p. 280. **Bacci** 1571, 4.7, p. 227, cites Bartholomaeus Burgundius: see **Chasseneux** 1571, 12.18, 257v; on Aachen, **Bacci** 1571, 4.10, p. 247, cites **Ruremundanus** 1564.

70. Bacci 1571, 256.

71. Bacci 1571, 7.7, p. 448. Onofrio Panvinio also knew Mercuriale's as yet unpublished work on gymnastic in manuscript (Ferrary 1996, 130).

72. Bacci 1571, 7.5.

73. Bacci 1571, 7.3–6, pp. 432–446

74. Bacci 1571, 7.12, p. 459. The denunciation of decadence at the baths is found at 7.14, pp. 464–466.

75. Bacci 1571, 7.17, pp. 472–474.

76. Bacci 1571, 7.18, p. 474. The entire chapter up to page 477 is a discussion of private, domestic baths.

77. Mercuriale 1573, 1.5, pp. 13–17.

78. Marsilio Cagnati, *Fumanellus, seu dialogus de gymnastica*, in **Cagnati** 1605 (an enlarged edition of **Cagnati** 1591a), fols. 94r–133r.

79. Cagnati 1599c, 25–73. The work opens: "Ut ad rempublicam bene gerendam Historia conducit plurimum; quae idcirco sapienter magistra vitae dicta est; sic quoque ad valetudinem curandam utilissima est. Quod Hippocratis exemplo satis docemur, a quo, non temere historias Epidemicas descriptas, existimare, consentaneum est."

80. Regarding Renaissance Hippocratism, see note 30 above. The complete *Epidemics* was one of the most important Hippocratic texts to become newly available in the early sixteenth century (partial versions in Latin translation were available in the Middle Ages). For a summary and bibliography regarding narratives about cases in late medieval and Renaissance medicine before the sixteenth century, see Siraisi 1997, 201–204, and bibliography there cited. On the development of case history, collections of *observationes*, and similar productions in the sixteenth and seventeenth centuries see Pomata 1996 and Nance 2001, especially chapter 2.

81. Bacci 1586, 69. The narrative of the case occupies 68–71, the analysis 71–81. See note 20 above.

82. Cagnati 1591b, 22–23; the case histories are dated between November 1584 and July 1590.

83. Cagnati 1599c, 25–26.

84. Cagnati 1599c, 31: "velut Heraclides, qui decumbebat apud Aristocydem: nam huic et ex naribus sanguis erupit, et alvus turbata est, et per vesicam purgatus est. Haec autem, de illo Heraclide dicta, Oliverio cuidam nobili Gallo ad unguem evenisse, ipse testari potest, qui vivit, et bene valet." Cf Hippocrates *Epidemics* 1.15.

85. See note 27 above.

86. For example, **Vittori** 1640, no. 11, pp. 36–37 (seasickness); no. 24, pp. 85–86, "An idem homo possit esse in quibusdam sapiens, in aliis vero insipiens"; no. 81, pp. 287–299 (Hippocrates and pulse).

87. For example, **Vittori** 1640, no. 15, pp. 48–52.

88. **Vittori** 1640, no. 70, pp. 240–254, "Historia iuvenis extincti variolis notata ab auctore dum medicinam faceret sub disciplina Alexander Petronii, ubi et describitur, ratio curandi variolas, et exemplar observandi progressum morborum accurate proponitur."

89. **Schenck** 1584, Proemium; see POMATA in this volume, 134. I owe this reference to Gianna Pomata.

90. **Vittori** 1640, no. 82, pp. 299–379, "Ad medicum, qui assumpto gradu doctoratus, erat in patriam reversurus. Ubi diffuse ostenditur via qua iuveni arte medica instructo ad illam exercendam progrediendum sit."

91. **Vittori** 1640, 299: "Scientiae in nostra mente sint velut tibia in manu tibicinis: quare sicut nesciens modulari . . . inutiliter tibiam possideret."

92. **Vittori** 1640, 316–327.

93. **Vittori** 1640, 342, "Quid agendum cum domum redieris."

94. **Vittori** 1640, no. 32, p. 103, and no. 74, p. 264.

95. The section on miracles is placed at the end of the volume, with its own brief introduction (**Vittori** 1640, 380–381). *Consultationes* nos. 83–97, pp. 381–449, and 98, pp. 443–449, concern miracles of healing attributed to intercession of saints (83–87, Loyola; 88–94, Xavier; 95–97, Neri; 98, Diego). The treatise on Neri's autopsy, not numbered among the *consultationes*, occupies 415–443.

96. Two of the miracles attributed to Loyola that are discussed by Vittori correspond to those included in the canonization process. Compare **Vittori** 1640, 387–388, no. 85, with *Acta Sanctorum Julii* 1731, 617, no. 1086 (Magdalena Talavera/Valavera) and ***Relatio*** 1644, 39, no. 4, cure dated 1601; and **Vittori** 1640, 390–391, no. 87, with *Acta Sanctorum Julii* 1731, 617, no. 1083 (Isabella Rebelles) and ***Relatio*** 1644, 35, no. 1, cure dated 1564. These cures presumably occurred in Barcelona and Gandia (Valencia), the places with which the women are identified.

97. His procedure seems to have been very similar to that of Paolo Zacchia, who also gave testimony both in secular legal cases involving accusations of murder and in canonization processes. See the *consilia* and *responsa* included in **Zacchia** 1651, book 9, nos. 1–10, pp. 659–680 (miracles), and nos. 12–16, pp. 682–692 (criminal cases).

98. In addition to those from Spain mentioned in note 96, Vittori also discussed miracles reported to have taken place at Malacca at the time when the body of St. Francis Xavier (d. 1552) briefly rested there and at Goa, where he is entombed.

11

Description Terminable and Interminable: Looking at the Past, Nature, and Peoples in Peiresc's Archive

Peter N. Miller

The pages of Peiresc's archive that are not copied out from documents, nor sent to or received from other people, take the form of descriptions. And description can be very tedious. Nevertheless, much early modern historical scholarship takes the form of description. On our side of the divide, we know *and feel* that just telling what happened, or just describing what has been seen or discovered, is inadequate, ever exposed to the devastating "So what?" question. "Historians" have come to think of themselves as storytellers, with a beginning, a middle, an end, and, above all, a point to their stories. "Antiquaries," seemingly, did not. And yet, at the beginning of the seventeenth century, this mutual repulsion of history and description did not exist; in certain contexts *historia* actually meant description.

Historians of medicine, looking into the teaching and writing that emerged from the University of Padua at the beginning of the seventeenth century—Fabricius of Acquapendente, but also, as POMATA has shown in this volume, Aselli and Harvey—have shown that *historia* took the shape of a "description" of the parts of the whole (Galen) or of a particular person's illness (Hippocrates). SIRAISI, in this volume, has brilliantly followed up this insight, showing how the antiquarian revolution of the sixteenth century was taken up by doctors. The next step in this inquiry might be to suggest an impact of the medical revolution on antiquaries and antiquarian forms of historical scholarship—via description.

Nicolas-Claude Fabri de Peiresc (1580–1637) was a man who studied many, many things,[1] but he belongs to this particular story, too. For though he came to Padua to study law—in a long tradition of legal antiquarianism[2]—we know from Gassendi's *vita* that he attended the lectures of Acquapendente and after returning to Provence continued to seek out his publications. Indeed, Gassendi reports that when presented with Harvey's *De motu cordis*, Peiresc replied that he had heard bits of it from Acquapendente and that, anyway, it was Sarpi who had discovered the existence of valves.[3] As Andrew Cunningham and Gianna Pomata have argued, here and elsewhere, Acquapendente was a crucial figure in promoting what we can consider a historical approach to medicine, reforming Aristotle in the light of Galenic and Hippocratic approaches to the observational and individuating character of medical cases.[4] In Pomata's contribution to this volume, we see that the crucial taking-up of Fabricius's demarche was by Aselli and Harvey; Peiresc was devoted to both of them, repeating the former's anatomy on a convict at Aix in 1634 and so becoming the first to observe the lymphatics in human beings. This was one of the few accomplishments of his that was recognized by contemporaries or near contemporaries.[5] Indeed, Peiresc's combination of admiration for Bacon and Harvey precisely aligns him with the intellectual prosopography of the early Royal Society.

His archive preserves working papers in fields we call anatomy, anthropology, archaeology, art history, astronomy, botany, epigraphy, glyptics, history, numismatics, paleontology, and zoology. Of course, neither he nor any of his colleagues knew of these terms. Their world of learning had different divisions. But if we look at these studies, some of which are finished (or almost finished) texts, others mere notes recorded in the midst of some activity, and which number in the thousands of pages, we find that they take the form of descriptions. These, supplemented by complementary passages drawn from his equally voluminous correspondence, serve as the material foundation for the present study. Peiresc's intellectual practice, and the fate of his work in the history of scholarship, is an exemplary case in the as yet unwritten history of learned description.

The most serious treatment of the meaning of description in early modern Europe has been written by an art historian, Svetlana Alpers. Though concerned to rehabilitate Dutch art, her observations can be profitably extended to the history of historiography. For the contrast she draws between an Italian art that appealed because it was both narrative and narrated the emotions of individuals (she even refers to Alberti's definition of *istoria*) and a Dutch art of mapping "places, not actions or events" effectively refracts Momigliano's famous distinction between the diachronic ancient historian and the synchronic, protostructuralist antiquarian through the prism of description.[6] Momigliano himself recognized that early modern antiquarianism had something in common with the New Science. In a very early essay of 1935 he used the term "empiricism" to refer to the antiquarian scholarship of G. B. Heyne, one of the pioneers of modern history at Göttingen. But by the 1960s his emphasis was on the antiskeptical character of close observation.[7]

By suggesting a cultural-historical connection between an artisanal culture and the New Science on the one hand, and a specific presentational format on the other, Alpers allows for the possibility that *ekphrasis*—verbal description of the visual—could be a mode common to both the natural and human sciences.[8] If we look into the Peiresc archive, we find just that: "description" links the various continents of Peiresc's world of learning. It also reminds us how important words still were if one wanted to be as precise and detailed as possible.[9] I would suggest that just as Martin Kemp has shown how Leonardo's intensely detailed but unfocused and then unknown anatomical drawings can be used to explore the antinomies of visual description in the Renaissance, so, too, an exploration of Peiresc's intensely detailed but unfocused and unknown ekphrastic technologies takes us into an antiquarian's study in the early seventeenth century.[10]

Carlo Ginzburg's swashbuckling "Ekphrasis and Quotation" argues that ancient orators and writers harnessed the authoritativeness of direct experience (autopsy) through a rhetorical use of vivid descriptions ("*enargheia* was the aim of *ekphrasis*") that aimed to convince their audiences of the truth of their account. Detail mattered because of the working presumption that the only way to know it was from actually being present. He even suggests that the rhetorical context—*evidentia* rather than evidence, in his terms—explains something of the gulf between Momigliano's historians and his antiquarians.

But even Ginzburg misses the power of "description." This is evident in his quotation of a striking passage from Manuel Chrysoloras's letter to the Emperor John VIII Palaeologus in 1411 (since quoted by others as well).[11] After describing the reliefs on the Arch of Constantine, Chrysoloras explained that "Herodotus and some other writers of history are thought to have done something of great value when they describe these things; but in these sculptures one can see all that existed in those days among the different races, so that it is a complete and accurate history [*historian*]—or rather not a history so much [as] a direct experience [*autopsian*], so to speak, and presence [*parousian*] of everything that existed anywhere at that time."[12] In a note, Ginzburg explains that he took the translation from Michael Baxandall, changing only the rendering of *autopsian*—from "exhibition" to "direct experience"— and *parousian*—from "manifestation" to "presence." So far, so good. But, following Seifert and, more recently, POMATA in this volume, we would also want to change *historian*, too: to "description." In which case, the full import of Chrysoloras is an even more striking statement of an antiquarian approach still alive and well in the age of Peiresc: history as description and direct examination of objects in order to make the past present.

Perhaps the most striking way to elucidate the centrality both of Peiresc's practice of description and of description's place in the New Science is to compare—briefly—Peiresc with Bacon.[13] If we take the hint and look more closely into the third part of the *Instauratio magna*, which Bacon called "Natural and Experimental History," we find much that Peiresc would have found immediately appealing. Not only do the topics of those "Particular Histories" converge closely with those of the descriptions found in Peiresc's archive (up to 35 of Bacon's 128 subjects are achieved by Peiresc), but the way Bacon tells the handful of histories he managed to write is followed closely by Peiresc.

When Bacon uses "history" in these experimental and natural histories it includes, for instance, names of winds, what people have said about them, when they blow, where they come from, and what they do to trees and plants.[14] The connotations of "history" in the "History of Life and Death" are still more varied. They range from an account of the life span of plants and animals, to how animals live, to an account of how fire works to dry things (a process is described rather than an object), to particular instances of the process (in Germany, for example). The "history" of the longevity of humans begins with a collection of textual evidence. The history of "the operation of the spirits" is a description of how different kinds of spirits work in people and what they do to people. The history of the circulation of the blood not only tells how blood might move, but also facts relevant to circulation and heating. The history of the "operation of the juices of the body" describes the structure, function, maintenance, and then the lived example of the thing.[15] In the history of "Dense and Rare," the history of "contraction and expansion" is expressed first in a table, or list, and then as a narrative of different experiments that produced expansion or contraction. Accounts of expansion in plants, liquids, gems, and trees—sometimes quite bizarre accounts, at that—are also offered as "history."[16]

In these different instances the section on "history" is typically followed by one called "Major Observations" (*Observatio major/Observationes majores*). In other words, these "histories" were presented by Bacon as a preliminary stage of work in which the raw facts were made available for subsequent elaboration—just as *historia* was acknowledged a preliminary in anatomical exposition, at Padua and elsewhere. If we combine the emphasis on a simple factual style of description with the view that history's purview includes the structure and workings of nature but not its purpose, then Peiresc might well have read the third part of the *Instauratio magna* as a manifesto for the Paduan natural history he was already familiar with. Indeed, Pecquet begins the sixth chapter of his book (following sixteen pages of discussion of Aselli and his immediate followers), in which he acknowledges Peiresc's role as the discoverer of the lymph system in humans, by declaring, "Thus far, my reader, you have an exact history of the lacteal veins [*Ita (mi lector) habes exactam Lactearum Venarum historiam*]."[17]

The historiography of the New Science has of course been written in terms of "observation" rather than "description." But observations lived—and live—only in their description. Description is how most early moderns learned of observations that were conducted elsewhere.[18] And description may also have helped even those who were present remember what it was that they saw: there is at least a vestigial link between Peiresc's organized preservation of labeled working notes and the (albeit less flexible) humanist commonplace book.[19]

Just as "observation" is usually studied by historians who connect it to the New Science, rather than to antiquities or anthropology, "experiment," which is itself so closely linked to observation, typically excludes the human sciences—though it seems a good way of describing Peiresc's testing the measurements of his different vases, to take just one example.

"Description" has, however, come to be very closely associated with the practice of experiment in the seventeenth century, especially in the work of Peter Dear.[20] As indicated by the specificity of a given description, experiment "was a single, historical occurrence, not a generalized statement. These things, we seem to be told, had happened by the action of or in the presence of a particular person, at a particular time and place." Dear wants to reserve this approach to the English.[21]

From this perspective, Peiresc would have to have been English. Even more: Dear's description of English, as opposed to Continental, Catholic natural philosophizing also makes sense of Peiresc's antiquarianism. "Boyle and his allies lacked such a framework, which is why they so frequently characterized their work as a Baconian collecting of facts—there was no clear way forward to making universal knowledge about the structure of nature."[22]

The contrast between a Peiresc and a Descartes—to keep to Dear's dichotomy—can be captured in an anecdote. Peiresc's archive contains an attestation from several residents of Aix, including one "Peyron Isnard called Charet, the son of Chaillon," that they saw three suns in the sky the previous Lent (of 1629) though they could not remember the exact date or hour. Charet claimed to have often seen similar things, even to the number of three.[23] It was the same appearance of parhelia in Italy in 1629 that led Descartes's friends to ask for an explanation; he gave it in the form of his *Meteors*. It is worth also noting the difference in their approaches: Peiresc narrates the event with an emphasis on who saw what and when, while Descartes felt he had "to examine methodically [*par ordre*] all the Meteors."[24]

It remains difficult to talk about description without doing it—to say that scholars described carefully, like saying that they compared many things together, seems meaningless. But the alternative, immersion in the details of a description, runs into the obstacle of a modern historical sensibility educated to be impatient with anything but interpretation. Peiresc and his friends had different priorities. After he first looked through Cornelius Drebbel's describing machine, the camera obscura, Constantijn Huyghens wrote that painting was dead, by comparison, "because this is life itself, or something even more elevated, if the words were not lacking [*car c'est icy la vie mesme, ou quelque chose de plus relevé, si la parole n'y manquoit*]."[25]

The Past

As we are well aware, travel and description were intimately linked, at the level of theory as well as practice.[26] Peiresc himself, confronted with the fables of Vincent Leblanc, was forced—unusually—into a theoretical statement of his own. He insisted that Leblanc leave out all the far-fetched stuff, "ordering him to select and express after his own manner, what ever he found meerly historicall, and containing a credible narration of things ... that it should be left to Philosophers to dispute those questions, and did not become a Relater to play the Dogmatist, especially contrary to the common opinion" and "that he should reap praise enough, sound and without spot, from the naked History [*nuda historia*] of his Travells."[27] Was description what Peiresc understood by "nuda historia"? Let us turn to two examples, drawn from his first two trips: to Italy and to northern France, the Low Countries, and England.

At the very beginning of his Italian trip of 1599, at the Camposanto in Pisa, Peiresc copied out a one-line inscription. It was followed by a 15-line *ekphrasis*:

In the middle of the picture of that tomb was seated a figure wearing a belt slung low, the rest being broken off. Above that figure there was another, very little, dressed in Greek style with a *pallium*. Behind this one was a woman *stolata*, with her hands outstretched between two trees that are probably laurels, on each of which there is a bird, which seems almost to be a *picus martias*. At the woman's feet is a sheep. One of those trees is on a mountain, opposite which is a naval anchor, with three rams above. Behind the tree on that same left side there is a peasant with a hat, who holds a pan in one hand and a fishing rod in the other, and pulls a fish from the water. Above the water there is a beardless head of Serapis, crowned with rays and a hat. To the right of the tomb there is the figure of a bearded man, seated with an instrument in his left hand, pointing with his right toward the tree, on which there is another bird like the others, and at the foot of the tree a ram who climbs up to this deity. Opposite the ram there is another figure—maybe of a shepherd, with a bare head, carrying another sheep on his shoulders.[28]

There are drawings in Peiresc's own hand interspersed throughout the archive, and also two surviving volumes of drawings executed by artists at his behest. These contain bowls, vases, and sarcophagi, among other things. Almost immediately after his death scholars began picking through these materials looking for an ensemble of rarities (like

Montfaucon for gnostic gems, for example) or for pieces that filled out knowledge collected from elsewhere, or, more recently, for drawings of monuments that no longer exist.[29] And yet, in Peiresc's archive it is the word—and the verbal description—that dominates.

Travel itineraries, of which there are many different sorts in the archive, often contained descriptions, and some were even structured as a series of descriptions. Such was the memoir prepared by Peiresc for his sometime factotum Denis Guillemin, the prieur de Roumoulles, traveling to Angers in June 1609. Peiresc's precise instructions are telling: Guillemin was to discover and describe in writing whatever tombs could be located of the counts of Provence and dukes of Anjou. Second, he was to draw them in color or in pencil ("de faire faire en couleur, ou en crayon sur du papier de mesme grandeur que ceste feuille")—except for those that were already familiar. Third, "de marquer bien les lieux ou il en trouverà."[30]

Peiresc's requests for investigation of the abbey church of St. Aubin of Angers were, if possible, even more particular. First, Guillemin was to do whatever possible (literally: "employer toute la faveur qu'il pourra avoir") to see the charters of Charlemagne containing the abbey's original privileges. Second, to note "the wax seals that are attached to the parchment of those charters, and try to recognize the letters written on them, and to see if the image is bearded or not, and if the crown is with flowers or not." Third, "to take care, very exactly, if among the wax seals there is any hair of a head or beard, as Jan du Boardigne wrote in his *Annals of Anjou*, ch. ix, and see if it is possible if any mention is made there that he had put the hairs of a beard in the wax or not."[31] Fourth, to try to obtain permission to take an impression of the seal, as he had obtained from elsewhere. Fifth, to see if any other Carolingian charters existed—without wasting any time on the Capetians "car nous les avons tous." Sixth, to note the names of the keeper of the charters and the superior of the abbey, so that if they provided help, the favor could be returned at some point in the future.[32] Finally, Peiresc reminded his agent just how these impressions were made: "Remember that these impressions are made by throwing molten sulfur onto an oily clay, which one had pressed onto the seal in question."

Peiresc was among the first to take the Middle Ages seriously. He collected Merovingian coins, explored churches, copied tomb inscriptions, and did not judge worth on the standard of classical antiquity. We tend to think about Mabillon, Montfaucon, and Muratori as the pioneers of national medieval history, but early in the seventeenth century, and very much on his own, Peiresc began to assemble materials for a history of the monuments of the French monarchy. Hundreds of these notes and sketches survive, and they fill an entire register.[33] According to Gassendi, it was the project to defend France against the Habsburgs that put Peiresc in mind "from that time forwards, to think of an Edition of all Authors, especially those of that age, who had written the Antiquities and History of France."[34] But as early as 1609, in that memoir to Guillemin, we find Peiresc thinking about reproducing the visual evidence found on coins, seals, tombstones, and glass to produce an iconographic history of the kings of France.[35] Indeed, Peiresc has been placed in a line of early modern French scholars who created the study of the "Monuments de la Monarchie Française," though their ambition was only to be realized in the nineteenth century by the "Monumenta Germaniae Historica."[36]

Peiresc would have been exposed to ecclesiastical antiquities in Italy, but his approach was his own. For example, from his reading in the history of religion he had come to think about the physical orientation of churches. First, he asked Selden whether the English churches faced more toward the equinox or the solstice. Then—though in fact we cannot be entirely sure of the chronology—he examined the churches of Paris. It turned out that the most ancient ones pointed at the sector of the circle between the equinox and the winter sunrise, St. Victor and St. Benoît excepted.[37] Peiresc did what in his own day might have been called an "experience": he had the royal mathematician, Jacques Alleaume, draw for him a compass, with Notre Dame at the center, and mark the points where the sun rose at the solstice and equinox. And, lo and behold, almost all the churches of Paris did, as he suspected, fall into this quadrant. The map survives and we can see that the exceptions were indeed St. Victor and St. Benoît "le betourné"[38] (fig. 11.1).

Peiresc visited and described the contents of many churches.[39] But he was not only looking and writing; he was also drawing both inscriptions and blazons.[40] Indeed, these rough drawings are usually embedded in the verbal descriptions in order to provide what Peiresc must have viewed as necessary clarity.[41] Sometimes the images to be drawn were so

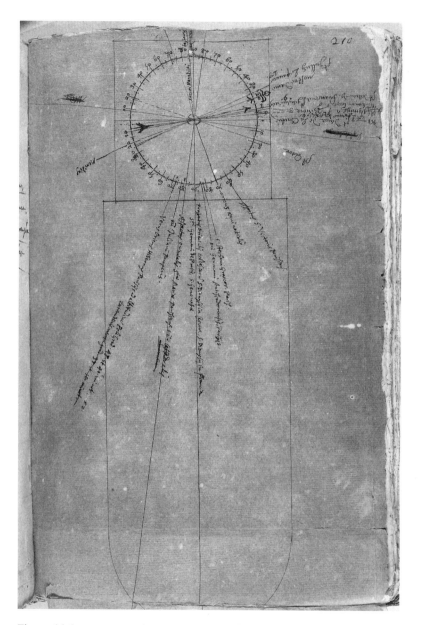

Figure 11.1
Bibliothèque Inguimbertine, Carpentras, MS 1971, fol. 210: orientation of the churches of Paris relative to the position of the sun's rising between the equinox and winter solstice.

important—or so complicated—that Peiresc seems to have turned to "professional" draftsmen.[42]

Paradoxically, the images that were most important for Peiresc, and the ones that were most frequently copied, were not of paintings or sculptures but rather the more "linguistic" ones associated with heraldry. These are treated as historical documents, and their description, whether in words or lines, is always detailed.[43] There are many notes that Peiresc made, always on site, that involved copying a tomb sculpture, describing it, and drawing its coat of arms.[44] When Peiresc moved beyond genealogical reconstruction he was less interested in an individual's perceptual universe than his cultural community—the enracination of heraldry in history.[45] This explains the attention that Peiresc, a student of medieval France, paid to these images. The precision—if not beauty—of his own sketches follows from this evidentiary role, just like his *ekphrasis* of the tapestry series in the "Salles des Gardes" (antechamber of the King) depicting the battle of Formigni won by the French over the English on 15 April 1451.[46] The tapestries described in "FIGURES DES ROYS EN LA SALE DU PALAIS" show Peiresc using the content of an image to guess at the use of its physical setting.[47] Similarly, an overfireplace painting becomes evidence for the shift from ambulatory to fixed sessions of government.[48]

Men like Peiresc, or William Camden, Clarenceux king-of-arms, composed armories in the same way that they turned their classical learning inside out and composed poems in Greek or Latin. For, understanding how heraldic coats could be deconstructed to yield historical evidence, Peiresc was also in position to advise on the construction of new coats of arms. And, indeed, a document entitled *1624. sigillum equestris ordinis Provinciae* puts Peiresc in the position of advisor to the Parlement of Provence, then seeking to draw up a great seal and a second for the syndics of the notables.[49] Much more interesting are the pages of notes in which Peiresc worked out the rationale for the different choices he had made in the process of heraldry. It is an instance of turning historical knowledge into an explicit and theoretically coherent visual code.[50]

Church-as-museum was another key function. Peiresc carefully studied artifacts in the basilica of St. Denis and the Sainte Chapelle, both in Paris—in fact, as we have noted, his comments are deemed to be of such importance that they are incorporated into the standard history of the basilican treasures.[51] Some of Peiresc's most intense descriptions of jewels and cameos come from these churches. Alongside the verbal descriptions we find drawings in his own hand, as well as by artists, and also lists of jewels.[52]

Peiresc was fascinated by Charlemagne in particular. In Peiresc's archives there are a series of wonderful drawings taken from the church of Notre Dame at Aix-la-Chapelle of Charlemagne, of his sarcophagus, and of the cupola mosaic that have been described as of "the highest worth."[53] There are inscriptions,[54] and there are discussions of charters that bear on the chronology of his reign.[55] There are even reports of conversations about his relationship to later rulers of France, like those Peiresc had with Bignon and Du Chesne about Hugh Capet.[56] But one of the most detailed descriptions focuses on a portrait head he viewed in the Louvre in 1621. "A marble head, about 800 years old, with the beard completely shaved . . . the hairs seem long on the head, and all the same in curls down the front and all around the scalp, making large bubbles of hair." Peiresc now compared it with other visual evidence, including seals and mosaics. What emerges is a comparative study of the image of Charlemagne, with especial attention to facial hair.[57]

Now, why was Peiresc so especially interested in how Charlemagne looked? To answer this question, and to make sense of the impetus behind the document of 1621, we need to go backward and return to one of the most important epistolary relationships of his apprenticeship, with the Roman antiquary Lelio Pasqualini.[58] In these discussions of gems, with their extremely close attention to workmanship and depiction, we watch them moving from antiquarianism to the history of style.[59]

It is in Peiresc's letter of 5 September 1605 that the appearance of Charlemagne is first discussed. He had found a coin of Louis the Pious that also portrayed a beardless Charlemagne, "just like yours," referred to in an earlier letter.[60] But it was not until November 1608 that the subject of Charlemagne assumed central importance in their correspondence. Peiresc's brother, Vallavez, was in Paris, but having some time free, Peiresc sent him to Aix-la-Chapelle, where he procured the drawings that are known to scholars.[61] Peiresc, who had already raised the question of Carolingian coins as evidence, now asked Pasqualini for everything he or his friends had that was "minted at the time of our French kings . . . of the first or second family."[62]

Peiresc then turned directly to the question of Charlemagne. Interestingly, however, this extensive discussion was not actually sent to Pasqualini, but survives only in Peiresc's own draft of the letter.[63] Peiresc reported that Pasqualini's assertion of Charlemagne's beardlessness was challenged in other authors and especially in a special privilege that Charlemagne had conceded to the abbey of Angers in which his facial hair was included in the wax of his seal, for authenticity. Peiresc had not seen the hairs himself, but he had received word of this from a reliable person and had read it in the annals of the church, "in the second part, ninth chapter." This explains the careful questions about Charlemagne's charter that Peiresc had addressed to Guillemin, the prior of Roumoulles, in preparation for his trip to Angers in 1609, discussed above. "And I know this for certain," Peiresc added, "having myself seen similar hairs in some of his seals in the abbey of St. Denis." This had always seemed to him an annoyance, as it prevented him from making a good copy of the seal, "but now I no longer am of this opinion." He was certain, however, that no mention of the hair was to be found in the document "that you have observed so carefully."[64]

From hairs and charters, Peiresc turned to the visual evidence. From the "legitimate," by which Peiresc no doubt meant "authentic," images of Charlemagne it was clear ("Basta si") that he had a beard, "not very long, really, but such that one could not say that he was beardless." Peiresc went beyond the evidence of coins. He was sending Pasqualini impressions of three seals he had taken from those in St. Denis. He compared these with mosaics made by Pope Leo III in the church of Santa Susanna and in the Sala Leonina at San Giovanni, and with a cameo preserved in St. Denis that, he boasted, "no one had noted before me." Moreover, official documents, capable of being dated very close to Charlemagne's own time—he noted that seals were then affixed directly to the parchment and did not depend from it, an innovation that came later—corresponded to the other images.[65]

Even the fact that Pasqualini's coin of Charlemagne was of gold gave pause. Out of politeness or persuasion, his response steered clear of the question of its authenticity. But, ventriloquizing, he explained "by not having seen a gold coin of those princes, he believed Sr. Petau of the

Parlement of Paris, that that no coins were struck in that metal, and no gold money used in those centuries, but those of the Saracens, with Arab characters, basing his opinion on an author of that time, who describes payment in gold Arab money." This claim, which became a pillar of one of the twentieth century's great historical chestnuts, the "Pirenne thesis," is yet further evidence of the inventiveness of seventeenth-century antiquarians: using cross-cultural material evidence for establishing a medieval history. In this case, however, Peiresc was not prepared to go along, because he believed that all absolute rulers minted gold money whenever possible—the argument from principle—and because he possessed a gold coin very like Pasqualini's—the argument from collections.[66]

Nature

Peiresc's Italian trip was the beginning of his serious investigation of nature, as it was of his serious study of the past. He met Galileo in Padua and the great luminaries of natural history, Ulisse Aldrovandi in Bologna and Giovanni Vincenzo Della Porta in Naples. In the north, he met Carolus Clusius. Peiresc corresponded with all these men.[67] Natural history and antiquarianism, coins and flowers, were talked about with the same people, in the same way. Clusius is an interesting example. The terms in which he approaches and discusses the history of flowers are exactly like those used by Peiresc for his antiquarian researches.[68] In fact, botany was an early interest of Peiresc's, stimulated by contact with Prospero Alpino at Padua, Richer de Berval at Montpellier, and Jean Robin in Paris.[69] In addition to Clusius,[70] Peiresc shared this passion with Jérôme de Winghe, with whom he also exchanged seeds.[71]

It is as a botanist and naturalist that Peiresc might appear most like his contemporaries.[72] But once we start to look more closely at his verbal, rather than visual, descriptions, we begin to understand that language could be as precise a tool as the eye itself. Take the four-page essay on copulating slugs. "It was Friday 24 August 1635," Peiresc began, "that while walking to Trebeillane, I was invited to turn aside a bit from the path to see the austerity of the abbot and the situation of the hermitage of *St. Honoré de Roque Fauour*, in the territory of Ventabreu. We left the carriage at the passage across the River Arc and mounted on horseback, with Father Théophile Minuti of the Minims, Mr. Lombard and Sr. Balthasar Grange, and Perrot, my man, along with the guides." The hermit turned out to have been away, and did not leave the key to the church in any obvious place. Fortunately, there was an ill-secured wall and they were able to enter through it. Under the overhanging boughs of the large tree at the entry to the church there was a little cabana, and in it Mr. Lombard noticed two very large slugs spiraled together as if pear-shaped, attached to it by some sticky white substance. "After Mr. Lombard advised me of his discovery, I approached, and because the branch of that tree was not too high off the ground, to consider this marvel at my ease, I put a knee on the ground, and remained there a good half-hour, always more ravished in admiration, and more hard pressed to guess what it could be." Lombard wanted to cut them down and take them home, "which I absolutely forbade and similarly did not suffer anyone to touch them, so that with patience we might discover something more, without turning these animals away from their natural instinct, and without doing them any violence, which might disrupt their activity." So they all watched, carefully, the slow movements of the slugs, and the infinite number of smaller animals that crawled between them. "After, therefore, having for a long while considered this marvel, having brought my finger close to this 'pear' without however touching it, I saw leaving from the bottom, the two little horns of the slug." These were followed by two others, and then the animals began to uncoil themselves. Peiresc described their bodily motions and the possible sexual use of an organ. After this exciting description of slugs making love ("faire l'amour" was actually Peiresc's term) he turned to the more banal measuring of their uncoiled length and a description of color.[73] Seeing through Peiresc's words seems to offer incontrovertible proof for Lorraine Daston's attempt to put "attentiveness," or *Aufmerksamkeit*, at the center of the new scientific persona.[74]

Peiresc, the master of looking closely, was also a student of looking. Of course, he would have been interested in optical effects because of his early exposure to the telescope. But in addition, as David Freedberg has noted, Peiresc was also at the forefront of microscopic research, obtaining microscopes for himself in 1622 and for Cesi in 1623.[75]

It was only a decade later, however, that Peiresc put the eye at the center of his activities. In a letter written to the Dupuy brothers in Paris at the end of 1633 he sought their help in obtaining a new, unspecified and still unknown recent book about optics.[76] In the spring of 1634, Peiresc was paid a visit by the nuncio, Giorgio Bolognetti, who found him at work dissecting the eyes of animals.[77] According to Gassendi, the ancients held that vision was in the "crystalline" humor, the moderns that it was in the retina, and Peiresc that it was in the vitreous humor.[78]

Over the next year Peiresc pursued an experimental program in which extreme, unusual visual effects in a human (himself) were noted, and animal dissections were then performed to try to explain them.[79] Peiresc began with mirrors, of the sort used in microscopes and telescopes. These were the subject of a series of observations in the spring of 1634. "From Wednesday 19 April 1634 I observed and then showed M. Gassendi," begins one, which was also labeled for filing purposes "EFFECTS OF MIRRORS / and concave and convex glasses on the conversion of species of images."[80] He devoted a whole memoir to his observations of 21 April, labeled "EFFECTS OF MIRRORS in convex and concave DIAPHANOUS BODIES for the reception, reflection, amplification, diminution and reversal or reconversion of species of images."[81]

Peiresc was constantly attentive to the optical effects that he himself experienced. We possess a diary-like document, from the middle of the following month, May 1634, in which he describes the relative darkening or lightening of the window frame in his room depending on the background lighting and the position of his head relative to his body. These are notes taken as the effects were experienced. They are minutely descriptive. Friday 19 May: "After returning from church in the morning, awakening in my chair after a good quarter-hour's nap, and after looking at the window frame in my room, by chance and having refixed my view on the green portfolio that was on my knees, but so situated as to be a little in the shade of that window frame, I saw very clearly the image of that window frame get brighter, with its natural appearance of clear and dark." The next entry is for Sunday 21 May. The same "accident" occurred. The object seemed closer and there was left-right inversion. The image seemed to move as he moved his head. Saturday 27 May: "lying on my bed after dinner, for my colic," looking out toward his window, "I saw the same appearance of the frame, in both my eyes, each pitched up toward the nose so that if it were on paper in would be like this, of the sort that one must examine and research the causes if possible." He then sketched on his paper the optical effect he was describing. Sunday 28 May: the same, this time after returning from mass and sitting in his usual chair after having taken "a little nap." And so on.

Peiresc conducted his exploration of these visual effects deliberately, opening and closing his eye repeatedly and noting the differences in what he saw. He also tried to imagine what it could all mean. "Nota," Peiresc added at the end: "It is necessary to examine the effects of vision when one places the head between the legs, and while one looks at a landscape from below, because when one looks upside down, it could serve as an explanation for the inversion of images which occurs at the back of our eye, and which seems (to our imagination) completely contrary to the natural situation of objects."[82]

This "It is necessary" must have stayed with Peiresc. At the same time that he was conducting these experiments on what he saw, he was beginning to dissect the eyes of animals. His assumption was that the cause of an optical effect lay in some physiological fact. He did not turn to theory, but to experimental biology.

Gassendi described this project as naive—a rare public dissent from his friend's approach—and did not believe that anything substantial could be extracted from necessarily idiosyncratic personal experiences. It is worthwhile, at this point, to again recur to the Paduan medical tradition and to its English outgrowth. For Gassendi also remained skeptical of Harvey's assertion that repeated direct description—*historiae* in the anatomical sense—could ever add up to knowledge. Did his reaction to Peiresc's dissections reflect this same dissent? Even though Peiresc, unlike Harvey, never did attempt to offer a retrospective epistemological theory of description, could Gassendi's criticism of the limits of Peiresc's experimentalism reflect a discomfort with a similar sensibility?[83]

Gassendi did, nevertheless, note that Peiresc's research program discovered many new, discrete facts about the eye.[84] Much of this would have come from the series of animal autopsies that began in earnest in August 1634. Indeed, we know that others in this circle thought much more highly of them. In the postscript of a letter to Gassendi in September 1634, Ismael Bouilliau wrote, "I saw at Mr. de Thou's, in the hands of the Dupuy brothers, some pretty *mémoires* on anatomies of the eyes of fish and animals. I hope that you will discover for us some beautiful secrets of optics."[85] This refers to an extraordinary series of memoirs and notes, beginning from August, that records the handiwork of Sr. Cayre, "master surgeon and anatomist of this university," under the instruction of Peiresc. Eyes of cats, whales, owls, eagles, and various fish were cut open and examined.[86] Inside the owl's eye, Peiresc described what he thought he saw in minute detail—including the palace in Aix, right across from his window, but inverted, of course.[87]

At the end of the month, Peiresc organized his thoughts on this matter and outlined a possible future course of research in a long memoir entitled "EXPERIMENTS ON THE EYES, both of natural mirrors and the effects of their reflection as of the comparison of the effects of LENSES and glasses convex, concave, and flat, and of phials filled with water, and of the doubling of images." For labeling purposes he was more terse: "1634. August 29 & 31./ NATURAL MIRRORS IN THE EYES." The main claim is presented right at the start: "We have seen from experience, first in the eye of a *lamia* and then in that of a dolphin, tuna, beef, sheep, and even that of a screech-owl, that at the back of the concavity, all clear of vitreous material, the *burning candle is painted* and represented *reversed, as in a concave mirror.*"[88]

Humbert has noted that this was the same discovery pointed out by Descartes in his *Dioptrique* (1637, but the work was done in 1629), though there is no indication that Peiresc knew of it, despite his close ties with Mersenne.[89] He also dismissed Peiresc's approach as "puerile"— a much less polite echo of Gassendi's acknowledgment that he had failed to persuade Peiresc that vision rested in the retina (the best he could do was to get Peiresc to agree that it was in no single part of the eye). Yet Kepler was himself "tortured" by the problem of the righting of inverted images on the retina.[90] And Peiresc was not the sort of person who could have been satisfied by Kepler's solution—declaring victory and leaving the problem of inversion unresolved.

Over the next months Peiresc was absorbed in animal dissection.[91] Among the most detailed descriptions was one of a "monster" fished up off Marseilles on 9 May 1635, which none recognized and which he thought might belong to Rondelet's first type of whale.[92] It had been caught, so Peiresc tells us, on Wednesday 9 May 1635 at 9 in the morning, and its eye arrived at his home on the evening of Monday the 14th, brought by Monsieurs Fort and Sabolini in a glass vial filled with "eau nittre"—they obviously had had dealings enough with local naturalists. But, Peiresc noted, by the time it had got to them there was already substantial decay. On the 16th Peiresc assembled his dissection team, led by Cayre the anatomist from Aix and observed by Gaultier, the Prieur de la Valette. They worked after dinner. There follows a description of how and where Cayre made his cuts and what the eye looked like and did at every stage. Peiresc must have been standing and writing while this was happening. The famous contemporary Dutch anatomies suggest something of the mixture of show and concentration that must have been happening in Peiresc's house in Aix.[93]

At the end of January 1635 (24th–25th) Peiresc drew up a provisional balance of his research on "INVERSION OF IMAGES painted in our two eyes."[94] He returns to the reversal and/or multiplication of images that he had experienced himself in May of the previous year and then explored in the animal anatomies of the previous autumn. This part of the essay had been copied over in a fair hand; yet, as it typically decays into unfinished, imperfect observations, Peiresc adds yet another autobiographical fact to the existing heap. He describes lying in bed before dawn, balancing his portfolio on his lap and writing a letter on a folded piece of quarto paper on which he had left a substantial margin, and yet seeing writing in the margin when he experienced that same darkening of the window frame.[95]

The way he proposed to work with this puzzle was, as ever, through further experimentation: "Dont il fault faire quelques experiences...." Humbert, as we have noted above, was uncomfortable with Peiresc's long descriptions of particular, personal optical phenomena. But of course this misses the much more fundamental point: because Peiresc lived in a generation that turned so many received opinions on their head, he could never be sure which of these "puerile," quotidian events might turn out to be a decisive proof for something new. It comes as no surprise that in these papers Peiresc's refrain is always: "Il en faut reiterer l'experience."[96]

Peiresc's close looking and close describing come together in his study of astronomy. His observation notes are extremely detailed and, especially for the study of the Jovian moons, are among the largest surviving treasure troves for the early seventeenth century. His place in the history of astronomy was once more prominent than it since has become.[97] Appropriately enough for someone whose approach so seamlessly blends Bacon and Harvey, Peiresc's interest in astronomy began in Padua in 1600 when he met Galileo. Although their direct epistolary contact was sporadic in the years that followed, they seem to have stayed more or less in touch through intermediaries, especially Paolo Gualdo, in the intervening years.[98]

We can date the beginning of Peiresc's own astronomical work to his contact with Galileo's *Sidereus nuncius*.[99] Over the next two years Peiresc amassed the largest surviving early modern archive devoted to Jupiter and its satellites. In its density of information and diversity of approaches it far surpasses the remains of Galileo's papers, which are mostly from a later date (probably late 1611 or early 1612). Those few who have worked on the surviving register of this material have been especially interested in documenting what Peiresc did and when he did it.[100] Some have concluded that Peiresc was a "better" astronomer than Galileo, others that he was a much worse one. But both miss the point, for the two men were seeking different things. And the difference between them turns, in fact, on the role of description.

Galileo was not interested in description—or, rather, he was interested in it only insofar as it was necessary to support his theoretical inquiry. For instance, the argument of the *Sidereus nuncius* is built on three months of observation, but the supporting material that survives is fairly scanty. Galileo gives the date and time, a visual display of the disposition of Jupiter and its satellites, and an indication of their distances in planetary diameters. Very rarely we find a note explaining some visual effect or unusual appearance of the objects in question. Overall the presentation is very similar to what appears in the *Sidereus nuncius* (1610): position, orientation, relative distance and luminosity of the Jovian system on a night-by-night basis.

The comparison with Peiresc's daily observation log is striking. Running for about forty folio pages, it records what Peiresc saw and what he thought of it, from 24 November 1610 to 17 April 1612 (fols. 189–227). The core, each night, is a visual representation of the disposition of Jupiter and its satellites—sometimes three and even four times in a night— but it is accompanied by a verbal description of the observational conditions, a verbal description of what is seen, and sometimes also drawings and comments on other celestial phenomena.[101] A comparison with what survives of Galileo's log for the same period covered by Peiresc's (roughly, 19 December 1610 through 14 June 1611, and again, though much more sporadically, from 20 November 1611 through 26 March 1612) makes plain the thickness of Peiresc's descriptions. Also, while Galileo keeps to what is narrowly relevant for his inquiry, Peiresc seems always to be keeping his eyes open for other things—there are descriptions of the appearance of the Moon, Venus (fol. 194r), Mars, Saturn (with its "cinctum" or belt, 196r, 210r), the conjunction of Mars, Jupiter, and Mercury, the motion and retrograde motion of Jupiter past the heart of Leo, and the nebula in Orion, which he saw during his first week of observing.[102] But, above all, Peiresc's commitment to verbal description as a necessary *accompaniment* to visual description stands out on every page, starkly contrasting with Galileo's purely visual, or diagrammatic, presentation. Because of this density of description, which in fact only increased over time, where Galileo could get a whole month on a page of his log, Peiresc was barely getting six days on his (see, e.g., fol. 215v). Finally, ever aware of the technical limitations of the tools he used, Peiresc kept trying to acquire additional telescopes, and by the autumn of 1611 he had four of them. He recorded his observations using each of them in turn, all labeled, so as to indicate the range of possible distortions (from fol. 208; a discussion of the telescopes at 235r).

This information was recorded in real time. But Peiresc also went back and redescribed other, older observations. He was, in particular, either fascinated by or fixated on the first weeks of Galileo's observations as recorded in the *Sidereus nuncius*, and so we possess a whole series of drawings of Jupiter and its satellites. He began by recording, on a page titled *Maioris Planetae Medicei septem Absolutae circa Iovem circumvolutiones ex Galileo*, all the positions published by Galileo (66r). Then, working with Galileo's assumption that the angular diameter of Jupiter was equivalent to one minute of arc, Peiresc ruled "graph paper" and designated one box as equal to one diameter of Jupiter or one minute of arc. He arrayed the moons on either side of Jupiter in their precisely designated places and then mapped out the positions given by Galileo (66r–68r).

This makes for a much more precise picture of the Jovian system than Galileo had given. But then Peiresc went one step further: he began to draw in the orbits of the individual satellites. Galileo, it will be recalled, simply called the moons I-II-III-IV, and did not distinguish among them until some time at the end of 1610 or beginning of 1611. Tracing their positions meant understanding them as discrete bodies. Peiresc gave them names (Cosmus Maior, Cosmus Minor, Marie, and Catherine) and even prepared "commentaries" discussing their properties.[103] On the graphed paper he traced their orbits around Jupiter; sometimes only one, sometimes two, and sometimes as many as three at the same time (68r). Also, by doing this Peiresc was able to give the position of Jupiter's moons even for nights when Galileo did not—and all for a time when Peiresc had not yet begun his own observations (fig. 11.2). How did he do this?

To solve this puzzle will take us to the heart of the different approaches of the two natural philosophers. Peiresc prepared computational tables for the four moons. Working backward from his calculation of the period of the moons, he calculated the "anomaly," or the position of the satellites in their rotation around Jupiter, at many times per night for each night of an entire year. This table records not observation but reconstruction; not prognostication, but description. It is, as Bacon might have called it, *historia*.

Peiresc takes Galileo's information about the time he did his observation and gives the equivalent time for Aix-en-Provence (i.e., the hour after sunset, given the different sunset times). Computation then allows him to give the position of the satellites in sexagesimal degrees, integer degrees, minutes, and seconds. He also gives the distance of the moon from the planet in units, and its direction (direct or retrograde, east of apogee, west of apogee, east of perigee, west of perigee). This amount of information allows for an extraordinary visualization.

Thus, not only does Peiresc now have the ability to see where Galileo had been blinded—say on 9 or 14 January—but to describe with such precision as to actually make it possible to determine the shape of the Jovian system at any point in the past, even when it had not been observed by human beings—as well as to project its orientation at some future time.

Working with his own observational notes, Peiresc proceeded to compute the position and direction of the four moons of Jupiter for every night between November 1610 and October 1611—sometimes at several

Looking at the Past, Nature, and Peoples in Peiresc's Archive 377

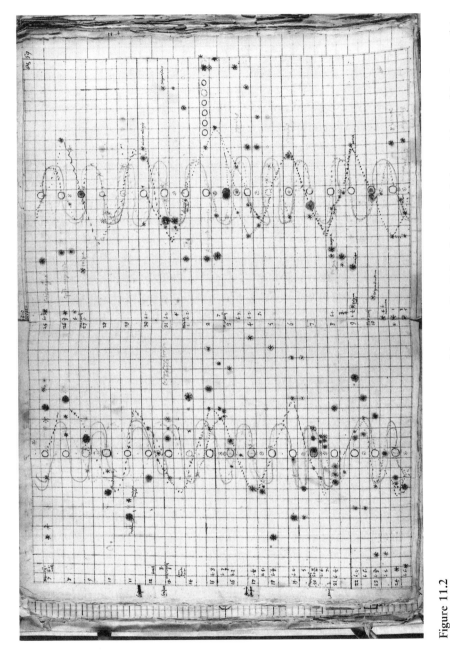

Figure 11.2
Bibliothèque Inguimbertine, Carpentras, MS 1803, fols. 67v–68r: Peiresc's sketch of the positions of Jupiter and its satellites, and their motions, for January–February 1610, the period of Galileo's first observations.

different moments during the night (fols. 7–18). This is an absolutely remarkable amount of information. This description of the positions of Jupiter and its moons is no less an accomplishment than a historian reconstructing every moment of the past year's history of a particular subject. The "table" is, then, at one and the same time both a description *and* a history, for the competent user could "read" it and "see" the positions of the system as they changed over time. Moreover, one suspects that its termination in the fall of 1611 suggests a general terminus ad quem of Peiresc's project, even though observations continued through the first few months of 1612.

Galileo sought to discover certain laws of planetary motion and cosmology and with them to predict the future. This required a modicum of reconstruction in order to check the accuracy of his observations. Peiresc, by contrast, devoted much more attention to the retrospective aspect than the predictive, as if reconstructing the *past* life of Jupiter and its moons mattered more to him. We might think of this as Peiresc's "antiquarian astronomy." But at the same time, it needs to be distinguished from the textual recovery of antique astronomical authors and arguments that Kepler had termed, in his letter to Maestlin of February 1601, "philological."[104]

When Peiresc returned to astronomy in the 1630s his eyesight was much weaker, but his descriptive powers even greater.[105] He now had the aid of Pierre Gassendi, a truly excellent astronomer. In fact, in the history of astronomy, the highlight of November 1631 was Gassendi's observation of the transit of Mercury. Peiresc in Provence and Gassendi in Paris had prepared for this event. It was another triumph of description over theory, since Mercury was far smaller than Kepler had predicted (because the solar system was actually much bigger).[106] But the more substantial evidence for his activities at the time is found in the extended log, devoted to sunspot observation, that began that same month and continued through January of 1632. It offers another opportunity to study his descriptive practice as an astronomer twenty years after those breakthrough observations of Jupiter's moons. The "log" takes the form of the day and date down the left side, the description of what was seen in the center, with the time of day at the right.[107] For each of the daily entries he also wrote out a paragraph or more of detail. A comparison again makes clear the scientific suppleness of verbal *ekphrasis*.

The powerful visual momentum of Peiresc's descriptions was fully realized in his project to map the moon. It is first mentioned, casually, in the postscript to a letter to Schikard of 4 September 1634. Peiresc noted that Gassendi was now working with two good painters in different locations to paint the phases of the moon, with all its specificity, using a telescope.[108]

From Gassendi's astronomical diary we know that Mathieu Frédeau, a local artist who had worked with Peiresc on some zoological drawing, had executed a pastel of the full moon of 9–10 July. This was unsatisfactory, and another, even more obscure painter, Claude Saulvat, was brought in for the first time on 26 August 1634.[109] This was also a failure. It rained, and the next attempt was on 2 September. It rained on the 3rd, but on the 4th Saulvat painted with help from Gassendi and Gaultier. On 7 September there was a full moon but the skies were cloudy. They worked together on 8–12 September. They resumed work on 24 September, after the new moon, but the rain came, so no observing was done until the 30th. But the first week in October was cloudy again.

Then nothing. The project is not mentioned again until 3 March 1635, when Saulvat turned up with Gassendi at Peiresc's observatory for an eclipse and provided expert opinion on the moments when the shadow crossed certain parts of the moon. Only in August 1636, with the arrival of Claude Mellan on his way back from Rome, was the project of a selenography revived. Gassendi appeared on 16 September with Galileo's newly sent telescope. Mellan painted, in color, on 23, 24, and 25 September, but instead of putting his eye in the scope, he painted off the telescopic image projected onto paper. From 2 through 10 October Mellan worked with Gassendi each night save the 4th. We know that Mellan was also working on 13 and 14–16 October (around midnight), then on the 21st and 22nd (in the morning). Bad weather in November limited observation to the 2nd, 8th, 11th, 19th, and 22nd. After the new moon, observation picked up again on 30 November at twilight, then 1, 2, 5, and 7 December 1636. Then no more.[110] Mellan was (back?) in Aix in April when he executed the charcoal sketch of Peiresc now in the Hermitage. Printing of the lunar atlas was interrupted—or truncated? —by the death of Peiresc in June.

The reception history of this project reflects on the fate of much of Peiresc's work.[111] In a letter to Cassiano Dal Pozzo of 2 June 1637 Peiresc explained that Mellan had done only two phases because he was so disappointed in the low quality of the local printers; he was waiting to return to Paris to engrave ("scolpare") four or five other phases, in order to print them all together, since "the one without the other [is] not able to preserve the reputation of the work."[112] This now appears to have had the power of prediction. For if we turn to the most famous early mapping of the moon, executed by one of Gassendi's friends, Hevelius, only a decade after Peiresc's death, both the scale and detail of Peiresc's activities have been lost. Even though Gassendi gave him engravings of two phases, Hevelius writes as if there had been only one.[113] Moreover, though one would have expected that Gassendi would have given a full oral history of the project, Hevelius writes that he learned of Peiresc's very small step forward through the *Vita Peireskii* alone. How far had Peiresc dropped out of the story? Hevelius notes that though it was not his preference, others had suggested to him to name lunar locations after modern astronomers: "Oceanum Coperniceum, Oceanum Tychonicum, Mare Kepplerianum, Lacum Galileaei, Paludem Maestlini, Insulam Scheinerianam, Peninsulam Gassendi, Montem Mersenni, Vallem Bullialdi, Sinum Wendelini, Promontorium Crugerianum, Fretum Eichstandiuanum, Desertum Lennemanni, & sic deinceps."[114] Only ten years in the ground, Peiresc had already been written out of astronomy's triumphalist narrative of observation and discovery.[115]

Peoples

Peiresc's study of living people and their cultural forms is also based on observation and description. Margaret Hodgson's pathbreaking work on early anthropology remains relevant today, but it has been greatly amplified, expanded, and amended in these last years, primarily by scholars of early modern travel. In what follows, many of these lines of development will be evident. But it was, especially, Peiresc's interest in ritual, understood as historical evidence lived as practice, that differentiates his approach, as a student of the past, from that of even the most sophisticated traveler.

As in his historical and natural philosophical descriptions, Peiresc's memoirs of peoples living elsewhere are fixed in a particular time and place. Sometimes these are drawn from reports passed along to him by others. For example, a note on the Jewish inhabitants of Cairo ("Juifs. Samaritains. Juifs de la Columbe au Cayre"), begins with just such a contemporary approach: "Dans le Cayre touts les Juifs sont constraints d'habiter en une mesme contree qui n'est pas esloignee de celle des Françoys." The body of the memoir describes the different numbers of Jews, Samaritans, and Caraites in the city. In the margin, Peiresc noted that "le P. Gilles dict qu'il n'y a pas 12. familles de Samaritains en tout le Levant"—an eyewitness account that was wrong. The Caraites were said to have more than 60 synagogues—again a marginal note possibly attributable to de Losches. Peiresc noted that in addition the Jews had another quarter in Old Cairo named Bezeyin, where they buried their dead. From talking about Jews and mention of the Jewish cemetery, Peiresc came to note down a description of the Turkish cemetery.[116]

In the report of M. de Monts about Canada, Peiresc took special note of his comments about the weapons (bows and arrows) and boats (canoes) covered in painted tree bark.[117] He noted that M. de Monts had prepared drawings of various animals. In keeping with the thinking of the seventeenth century, among these was one of a native. His painted body, clothes, and weapons were all described.[118]

From Tunis, Peiresc had received a memo from d'Arcos on Moorish foodways. After noting the size and shape of their cups, he observed that while the Moors drank only water at home, they did frequent taverns run by Christians, where they drank wine and spirits to excess. Despite running contrary to religious law this vice was tolerated, even publicly. Tableware was typically of wood since silver was banned. No tablecloth or napkin was used but rather people sat on the ground and ate off tables and plates made of wood; the rich used leather. Salt was not used and a salt shaker not present because of the heavy use of salt in preparing their meats. "Ordinarily," d'Arcos concluded, "all is boiled and little or nothing roasted."[119]

Despite these marvelous memoirs that dot Peiresc's collection, it was not so much that the jarring encounter with the strange opened his eyes to the familiar, as that the awareness of difference generated by a sensitized historical sensibility enabled him to see everyday practices as products of history. Indeed, from a heraldic perspective, pageants, processions, receptions, and, of course, births, marriages, and deaths were forms of living history. This perception went back at least to Peiresc's visit to Rome. Gassendi tells us that "he was present at the Performance of Religious Ceremonies, as much as he thought he might with safety. For, being but of a weakly complexion, he was loath to thrust himself into a tumultuous Crowd of People."[120] In London with the French ambassador in 1606, Peiresc was trapped in a drinking bout with several Englishmen. He could not keep up, or the liquor down, but played along all the same. The humor of the spectacle appealed to his hosts, and as his behavior became the talk of the court he was eventually summoned by James I to tell the story in person.[121]

The most interesting of Peiresc's eyewitness accounts of ritual focus on those involving the king and the royal family: entries, funerals, marriages, and the Estates-General.[122] Peiresc's interests follow closely those of an exact contemporary and correspondent, Théodore Godefroy, who published his *Le ceremonial de France, ou Description des ceremonies, rangs, & seances observées aux couronnemens, entrées, & enterremens des roys & roynes de France, & autres actes et assemblées solemneles* in Paris in 1619, during Peiresc's residence there as the influential private secretary to their mutual friend, Guillaume Du Vair.

Peiresc's archive preserved, for example, "La cérémonie du sacre du roy Louis XIII" observed during his entry to Reims, along with documentation of the inscriptions borne on the city's arches.[123] Much more substantial is a long document of royal entries in Provence and Languedoc in November 1622. Some are in Peiresc's own hand and some in that of his brother.[124] Peiresc's activities in this regard are representative; interesting, however, is his use of the word "dessein" in this context to refer not to the visual depiction of the painted arches and displays but to his verbal description.

In 1625 the royal match between Charles I of England and Henrietta Maria was observed by Vallavez, then in Paris (Peiresc having returned to Provence in 1623). He wrote a "Relation de ce qui c'est faict tant aux Fiançailles de Madame Henriette Marie de France soeur du Roy, avec Charles, premier Roy de la Grand Bretagne, lesquelles furent faictes au Louvre dans la Chambre du Roy le ieudy 8.me jour de May, iour de l'ascension 1625, Qu'au mariage de ladit Dame qui fut faict a l'Eglise Notre Dame, le dimache unzieme iour dudit moys et an."[125] The physical space of the king's antechamber was described first—including its furniture —then the movements, appearance, arrangement, and actions of the principals. The celebrations received equally detailed attention, with Vallavez noting that the parties spilled over the next few days, with Cardinal Richelieu in particular having prepared "une superbe collation de confitures"—the equivalent of an English banqueting course of sweetmeats—accompanied by a concert of voices and instruments, itself followed by a fireworks display in the garden, which Vallavez described as "the most superb and beautiful invention that was seen in a long while."[126]

But it was the church service that generated another kind of representation. Vallavez drew the "theater" created in Notre Dame for the marriage ceremony. There are two sketches, the first of the arrangement of the principals around the altar table, which is drawn in some detail head-on. Other seats are presented in a bird's-eye view.[127] The second is really a map, presenting the whole church and marking the locations where the principals were seated in the king's chamber for the "engagement."[128]

The death of Henri de Gondi, cardinal de Retz, in 1622 was the occasion for an extraordinary demonstration. Accounts of royal funerals were of course published, as were those of great nobles.[129] Peiresc's own archive preserves many of these.[130] His own narrative starts in a fair hand but soon decays into a draft filled with crossing-out, insertion, and additions. Yet it ends with the word "Fin," suggesting that Peiresc at some point envisioned the work as a whole and as a literary product before abandoning it—a bit like the projected commentary on the Jovian moons.

Like other documents of this sort, it describes the decoration of the church and the catafalque, the positions of the marchers and mourners, and the content of the funeral oration.[131] But it diverges from the norm in its focus less on the person of the deceased than on the concrete corporate structure that the ritualized expression of grief recalled into being. Gondi's death, in Peiresc's account, is but the occasion for the crystallization of ancient custom, preserving, still, ancient history. This survival, rather than the particular person's passing, is what fires his description.

He begins with the story of the cardinal's life and death, but then turns immediately to the funeral. Organization comes first—the procedures and personalities who decided its timing and format. Whether the curés would wear their stoles or the dean his distinguishing black velvet bonnet required a meeting of the dean and chapter to decide, and Peiresc relates it to us, with both sides of the argument. There were decisions to be made about which churches would march on which side of the street—and here the decision was taken against "la plus ancienne coustume" but rather in accord with what the late cardinal had himself decided for the entry of the queen. There was the inevitably political dimension to the question of whether the Parlement wanted to hold its own memorial ceremony or participate in the church's and the question of whether the court *ever* went to the funerals of cardinals—i.e., the inevitable disputes about priority between civil and ecclesiastical temporalities and the equally inevitable resort to the ancient registers of the Parlement for advice from precedent. The narrative of the corporate bodies' jostling for prominence and control of the proceedings is full of personality and pique. It *is* a literary document. Thus, Peiresc noted that while the *parlementaires* formed up as a body, the domestics present failed to say anything about their former employer, "which the men of that chamber found very strange."

"Coustume" frequently rears its head. Peiresc, as always, was keenly attentive to anything that could possibly cast a glance backward onto earlier decisions, actions, or attitudes. Custom in ritual, like oral history, was one such valuable avenue to the past. And, again, many maintained that it was not done ("n'estoit pas tenu") to proceed as a body to a funeral, that "the register contained no example of a similar ceremony in the funerals of cardinals" (325v). And nothing was more dramatic than the account of how the "pesle mesle" sitting of the great ladies of the court disturbed the carefully planned—and counted—seating arrangement in the church, such that perceived lessers took seats reserved for their betters, setting in motion exactly the sort of disputes that could be expected. The president of the Cour des Comptes "murmured that the chamber could not suffer and was not accustomed to be preceded by them" (335v), while the *chantre* and archdeacon, "fearing some violence," sought to work out a compromise, which turned into a debate about the priority of Church to Parlement (336).

The same interest in custom is played out in the description of the place and preeminence of Paris's churches and civic corporations in procession. Peiresc shows convincingly how a ritual act could be read as a historical document. For instance, Peiresc read the location of criers as a hint of the survival of the antique. In addition to the palace, Notre Dame, and the university, criers stood in front of St. Denis; at the Pallus, or Pailleux market, which was at the "marché neuf"; at the gate of Paris; at the beginning of the bridge of Notre Dame; at the Place de Grève; at the Baudoyer gate joining the *barrière* of St. Gervais; at the Petit Pont; and at the Place Maubert. Peiresc noted that there were many other celebrated crossroads, but the ceremony remained from those times when there were no habitations beyond the circuit of the criers. Peiresc is here coaxing urban history out of living ritual; those churches were precisely the ones given primacy "which much exceed 300 years of age, around the time that the second circuit [of walls] was made, which enclosed St. Germain de l'Auxerroys, St. Eustache, St. Magloire, St. Accoy, and the little St. Antoine" (325r).

The order of marching was also an order of seniority, with the oldest churches marching last. Nor was this taken lightly. When the priests of St. Hippolyte got ahead of those of St. Sauveur and spread across the street, they were retreated and placed behind St. Sauveur. Precedence always mattered. Nor was it beyond imagining that one church group would respond to the encroachment of another with blows, using batons and even their processional crosses to ward them off (328v). And, occasionally, changes were made for aesthetic reasons: when the little but ancient and highly privileged churches began to march, instead of walking side by side across the street they went single file, so as to lengthen their few numbers and give an impression of robustness (329r). To make sure this complex narrative could be easily visualized, at least in part, Peiresc drew up a table of which curés walked on which side of the road (330r). There was hidden history to be excavated here as well. That St. Bartholemy marched with St. Hilaire du Mont and St. Estienne du Mont, despite its being located in the Cité, the oldest part of Paris, showed that the order was "made according to the order of antiquity; and in fact, St. Bartholemy was not built until much later, upon the ruins of the Abbey of St. Magloire, which was burned only during the reign of King Henri I, in 1034" (328v).

For this extraordinary story and document, many of Peiresc's working papers survive. There are his notes on the outline summary of events and chronology (357). We have a similar set of summary references to the legal texts referred to in the debates about seating and precedence (384). Remarkably, and perhaps uniquely for Peiresc's oeuvre, we possess the outline of the essay, with the different parts crossed off as if to indicate their having been accomplished. This focuses exclusively on the ceremony in the church but descends to the detail of sentence-by-sentence (385). We also have a few paragraphs, worked out in prose and then struck through (386r). We even have the overview of the entire essay, in outline form, with strikethroughs (386v). A broadside of the *mandement*, in his collection, ordering participation in the funeral would have provided Peiresc with much of the basic information he used in the essay (389). Finally, the work had a title—and if not exactly a title page, it is a title that could fill a page: "A very precise relation, of all the order and all the ceremonies that one observed at the obsequies and funeral of the late Cardinal de Rhetz, as much for the transport of his body from the place of his death, up to Paris, and for the procession of his interment, as well as the meeting of the companies that assisted at his service, as well as the deliberations and expedients that were taken in the diverse conflicts and difficulties created by reason of rank and precedence, as much among the clergy as among other persons of quality who had been invited. Following which all the clergy wanted to arrange themselves so that each would be in an honored place, without the mutual recriminations, instead of which all was confusion and tumult, as usual. Together with the rolls and commands delivered to this effect, and the acts preserved in the registers of different places, in the chapter of the Cathedral of Paris and in the sovereign companies, the Parlement, the Courts of *Comptes* and *Aydes*, the Hôtel de Ville, and the University of Paris" (358).

We also know who helped Peiresc with this. The key figure was Herbert, archpriest of St. Marie Madeleine. We have the document signed by him (351r) and dated 16 November. Peiresc acknowledged this help in a letter, a draft of which was kept in this file. In it, we can also see Peiresc directing specific questions at Herbert, exploring matters of further interest or of abiding unclarity. The questions were specific: how many? where? when? who? But they also reveal the truth behind the whole: it was done "for the contentment of Monsieur de Lomenie, who gave the subject for all this research." Lomenie was the chancellor and one of Peiresc's correspondents (405r). Why he might have been interested in this subject is not stated. Herbert's answers in the form of documentation came later (411). Someone else (M. de Montmaur) provided Peiresc with the detailed contents of the funeral oration (413). Peiresc was also assisted by another memoir, this from Blanc, vicar-general of Notre Dame (392), which in turn provoked more questions and answers (402, 403, 404). Peiresc collected also the procès-verbaux from the Parlement and different courts (367, 373, 381, 384). The detail in Peiresc's essay is extraordinary; in these working materials, and in the incredible precision of the questions he wanted answered, we see how committed his work was to the reconstruction of the past.

From marching orders Peiresc worked backward to corporate structure and the history that survived latent, and for all intents and purposes lost, in that structure. A memoir in Peiresc's hand, dated October 1622, "DESNOMBREMENT DU CLERGÉ DE NOTRE DAME," belongs to the inquiry sparked by the funeral essay. It is an extraordinary representation of the human wealth of the church, but also of its many institutional dependencies and the people who ran them. The various church offices are enumerated and their occupants named, from the dean all the way down to the clerks and chaplains. Peiresc also gives occasional indication of their costs and revenues.[132] What escaped him gives some direction to what he was looking for. "One wants to know [*On desire sçavoir*]," another page begins, "the number and standing of the servants who came to pray at the obsequies of the defunct." "In what place lodged the nine last criers who brought the final churches," begins another line of inquiry. But the motivation behind some questions still seems opaque.[133] A small side of paper is covered with questions about rank and precedence in processing: "En quel rang marcherent."[134] Peiresc also wanted to know about the number of presbyters at different churches (407). Some of these questions were formulated by Peiresc and then answered by someone who knew the answers: Herbert, "Archiprebstre de la Madeleine." We know this because of the draft letter to him that accompanies these notes. It too is full of questions about marching order and precedence.[135]

But this is not the only extraordinary document of Peiresc's political anthropology. In fact, others date from his trip to England in 1606. There, amidst new acquaintances who were to accompany his learned adventures for the next decades—men like Camden, Cotton, Spelman, and Selden—he also showed an already refined taste for political institutions and their long historical development. In London at the end of May and beginning of June, Peiresc was able to witness firsthand three important acts of state: the royal audience of a new ambassador—the French, in whose suite he had traveled—the investiture of new members into the Order of the Garter at Windsor, and the meeting of a session of Parliament. These experiences evoked careful descriptions from Peiresc.

That meeting of Parliament provided Peiresc with another occasion for a startling piece of description. The "Parlement General" of England met on Tuesday 6 June (N.S.) in the Great Chamber of Westminster, whose details of appearance Peiresc described. It was another memory space.[136] He then described the clothing worn by those present, with greatest care devoted to the king's attire. But what is special about this description is that Peiresc sat and sketched its disposition, providing us with one of the earliest surviving depictions of a meeting of the English parliament—a wonderful acknowledgment of the power of mapping for the description of human culture (fig. 11.3).

There are two other such "maps" of the ancient constitution in action. Like that of the English Parliament, they capture the French representative body, the Estates-General meeting in 1614. They support texts written by Vallavez, and annotated by Peiresc, who was present in his capacity as Du Vair's secretary. The first, copied out of the *registres* of the Parlement of Paris in Peiresc's hand, is the "Proces verbaux des propositions et deliberations faictes pour les rangs de la Procession generale que le Roy Louis XIIIe fait a Paris le Dimanche 26.me Octobre *1614* avant l'ouverture des Estats Generaux. Ensemble de l'ordre qui y fut observé fort exactement descript et inseré dans le Reg.^{re} du Parlement." It is labeled "*1614* 23–26 Octobre. PROCESSION GENERALE pour LES ESTATS." It gives, for each of the enumerated days, the individuals present and some narrative of the events.[137] Peiresc would have been interested in the account and order of the procession: clerics first, from lesser to greatest, then the royal suite, then the great nobles, the parlement, and the corporations of Paris, continuing with their arrival at the church of Notre Dame and their seating arrangement. The verbal description was followed by a seating chart.[138]

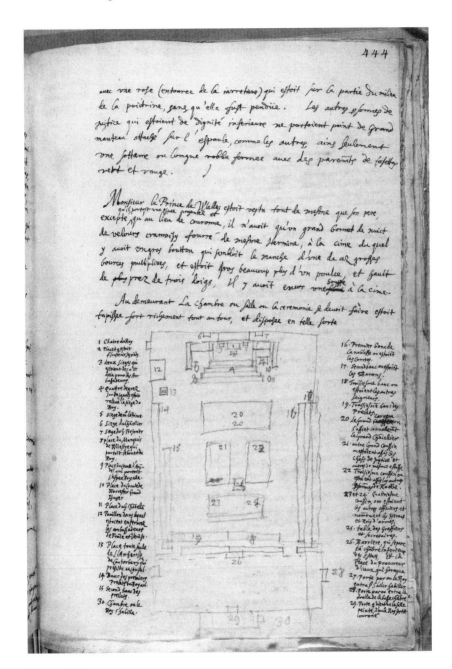

Figure 11.3
Bibliothèque Inguimbertine, Carpentras, MS 1794, fol. 444r: Peiresc's drawing of the House of Commons in session, with key.

Even more impressive is a document in the hand of Vallavez, "Memoires par Monsieur de Valavez de l'ouverture des estats faicte par le Roy Louys 13 en la Grand Sale de Boubon le Lundy 27 Oct. [1614]," which Peiresc labeled "L'OUVERTURE DES *ESTATS*."[139] It described the seating arrangement, then the room itself, and finally the principals and their attire. But, here, too, it is what Peiresc the mapper does with information that is so fascinating. He prepared an extraordinarily detailed drawing of the scene. It captures everything in Vallavez's account save the colors and the clothing, but it goes beyond it in precision and in scope—adding information about the seating of the second and third estates (469).

These detailed accounts of human actions, mostly recorded as they unfolded in time, reflect Peiresc's concern with preserving information for its possible later usefulness, even if the significance of any given detail at any given moment was hard to discern. In this sense, his records of funerals, parliamentary meetings, or royal marriages are like his experiments on optics or pebble formation or astronomical observation: as detailed as possible and as comprehensive as possible. This breadth, in turn, reflects a rejection of inherited criteria of relevance and a constructive sort of ground clearing. We are comfortable with this in the context of the New Science, as a Baconian declaration, or as Boyle's practice; less so, typically, with suspension of judgment as a rule for the historical sciences. But Peiresc's practice of description takes us back to a time when these boundaries were not fixed and the future shape of so many intellectual inquiries was still to be determined. And, of course, Peiresc made the right decision: thanks to it historians at the beginning of the twenty-first century can use his archive to reconstruct medieval monuments as well as the movements of priests and planets.

Notes

I am very grateful to the editors for months of attention, comments, and careful reading, and to the collective assistance of the entire *Historia* group over four happy weeks in Berlin. I also wish especially to thank Noel Swerdlow for patiently discussing Peiresc's astronomical work with me and explaining its relationship to Galileo's. Any misunderstandings and mistakes are mine.

Sources cited in bold face type appear in the Primary Sources of the bibliography.

1. See Miller 2000.
2. Kelley 1970; Shapiro 2000. For Peiresc see Miller 2001a, 68–70.
3. **Gassendi** 1657, year 1628, 28.
4. Cunningham 1985; Pazzini 1957; French 1994, esp. ch. 11.
5. Peiresc describes this event in many contemporary letters; in the postscript of one to Schikard of 4 September 1634, Peiresc noted that "Il [Gassendi] vid cez jous passez les veines lactees d'Asellius, sur un corps humain qui avoit esté pendu et estranglé, une heure et demy aprez sa mort." Peiresc to Schikard, 4 September 1634, letter 4, unpaginated, Württembergische Landesbibliothek 563, fol. [2]r; **Gassendi** 1657, year 1634, 104–105. It is to this that reference is made in **Pecquet** 1654, 18. I thank Gianna Pomata for pointing me in this direction.
6. Alpers 1983, xxi.
7. Momigliano 1955, 169; Momigliano 1990, 56–57.
8. Alpers 1960, 197, noted that "*ekphrasis* originated in late antiquity as a rhetorical mode of praising and describing people, places, buildings, and works of art" but then focused exclusively on art. More recent work on *ekphrasis* remains exclusively focused on art (painting and literature); for example, Carrier 1987, 31. Even Ruth Webb—who explains that "not only is ekphrasis not conceived as a form of writing dedicated to the art object" but it is not even restricted to objects and could include anything or anyone—does not escape from the same trap (Webb 1999, 7–18).
9. Winkler and Van Helden (1992, 212, 216) argue in the case of Galileo that he eschewed images for reasons of class rather than precision. For the explanatory power of images in the New Science see Lefèvre, Ren, and Schoepflin 2003 and Freedberg 2002.
10. Kemp 1993, 88–90. But Kemp notes that at least the later Leonardo thought that "illustrations reigned supreme for description, while the text remained best adapted to explaining how something worked" (94).
11. See Brown 1996, 76–77; Haskell 1993, 90–92; Burke 2003, 276.
12. Ginzburg 1988, 18.
13. I have done this, in part, in Miller 2000, 22, 28.
14. F. **Bacon** 1858–74, 19:409–412.
15. F. **Bacon** 1858–74, 10:24–114.
16. F. **Bacon** 1858–74, 10:198–227.
17. **Pecquet** 1654, 17.
18. See for example Ogilvie 1997, 308.
19. Blair 1992.
20. Dear 1991, 137.
21. Dear 1985, 154. The English kind "takes the form of historical reportage of events—accounts of what the author witnessed as a result of chance observation or, more typically, as a result of deliberate contrivance, often including

place, date, and even names of witnesses" (Dear 1990, 663). Dear contrasts this with a Jesuit, or Catholic, science, devoted to proving a universal statement from particulars (Dear 1995, chapter 2).

22. Dear 1991, 162.

23. Bibliothèque Inguimbertine, MS. 9531, fol. 189; **Gassendi** 1657, year 1629, 36.

24. Quoted in Sabra 1981, 61.

25. Alpers 1983, 12.

26. Blair 1992, 542n3; Rubiés 1996.

27. **Gassendi** 1657, year 1619, 191.

28. Bibliothèque Nationale, MS. Latin 8958, fol. 276r contains the inscriptions and dated drawing of a griffin; fol. 276v the long description.

29. Bibliothèque Nationale, Cabinet des Estampes Res. Aa53–54. For example, Stern 1956.

30. Bibliothèque Nationale, MS. N.a.f. 5171, Peiresc to Roumoulles, 1 June 1609, fol. 708.

31. For this practice, see Bedos-Rezak 2000, 1527.

32. Bibliothèque Nationale, MS. N.a.f. 5171, Peiresc to Roumoulles, 1 June 1609, fol. 708v.

33. Bibliothèque Inguimbertine, MS. 1791.

34. **Gassendi** 1657, year 1618, 184.

35. Leclerq 1934, 2717; Bibliothèque Nationale, MS N.a.f. 5171, fol. 709.

36. Leclerq 1934, 2708–2747; on Peiresc, 2710–2723.

37. **Gassendi** 1657, year 1622, 207.

38. Bibliothèque Inguimbertine, MS. 1791, fol. 210.

39. Bibliothèque Inguimbertine, MS. 1791, fols. 211–255, 522.

40. Bibliothèque Inguimbertine, MS. 1791, fols. 52, 492.

41. For example Bibliothèque Inguimbertine, MS. 1791, fols. 81, 83.

42. For example, for an overdoor painting, Bibliothèque Inguimbertine, 1791, fol. 102, and a tomb at fol. 516.

43. Compare with the work of Pastoureau 1982a, 106.

44. Bibliothèque Inguimbertine, MS. 1791, fol. 33.

45. This is the "héraldique érudit" referred to by Pastoureau 1982c, 337.

46. Bibliothèque Nationale, MS. N.a.f. 5174, fol. 329. Alongside the Constable is his standard, which is also meticulously described and is also drawn on a separate page (fol. 339). Bibliothèque Inguimbertine, MS. 1791, fols. 74–79, fol. 134. On the importance of the boar, see Pastoureau 1982b.

47. Bibliothèque Inguimbertine, MS. 1791, fol. 103.

48. Bibliothèque Inguimbertine, MS. 1864, fols. 232–233.

49. Bibliothèque Inguimbertine, MS. 1864, fol. 303.

50. Bibliothèque Inguimbertine, MS. 1864, fols. 307–308.

51. Peiresc's graphic realism—vertical copying of vertical inscriptions (Bibliothèque Nationale, MS. N.a.f. 5174, fol. 34/121) is unmatched even in modern studies like Montesquiou-Fezensac and Gaborit-Chopin 1973-77, 2:341.

52. Bibliothèque Inguimbertine, MS. 1791, fols. 79, 124, 130, 131, 511, 511bis; Bibliothèque Nationale, MS. N.a.f. 5174, fols. 128–129. Marjon van der Meulen published some of these (Meulen 1997, 223–226).

53. Bibliothèque Inguimbertine, MS. 1791, fols. 479–481. See Stephany 1957.

54. Bibliothèque Inguimbertine, MS. 1791, fol. 85.

55. Bibliothèque Inguimbertine, MS. 1791, fol. 523v.

56. Bibliothèque Inguimbertine, MS. 1791, fol. 198.

57. Bibliothèque Inguimbertine, MS. 1791, fol. 132.

58. **Gassendi** 1657, year 1607, 121. For their relationship see Jaffé 1993. Three sets of copies of letters exchanged by Peiresc and Pasqualini do survive: a complete set in the Bibliothèque Méjanes at Aix-en-Provence (209 (1027)), another at the Bibliothèque Inguimbertine at Carpentras (MS. 1809), and a set in the Bibliothèque Nationale (MS. N.a.f. 5172) that seem to have been stolen from this Carpentras register by Libri.

59. Peiresc to Pasqualini, 4 December 1602, Bibliothèque Inguimbertine, MS. 1809, fol. 252v; Bibliothèque Méjanes, MS 209 (1027), p. 38.

60. Peiresc to Pasqualini, 5 September 1605, Bibliothèque Inguimbertine, MS. 1809, fol. 351v; Bibliothèque Méjanes, MS. 209 (1027), p. 50.

61. See **Gassendi** 1657, year 1608, 127. Palamède de Fabri was the sieur de Vallavez.

62. Peiresc to Pasqualini, 2 November 1608, Bibliothèque de l'Ecole de Médecine, MS. H.271, fol. 6r; Bibliothèque Méjanes, MS. 209 (1027), p. 80.

63. The section on Charlemagne is in Bibliothèque Nationale, MS. N.a.f. 5172, fol. 316v; the copies in Aix are made from this; Bibliothèque Inguimbertine MS. 1809 is an inferior copy.

64. Bibliothèque Méjanes, MS. 209 (1027), 87–88; Bibliothèque Inguimbertine, MS. 1809, fol. 300v.

65. Peiresc to Pasqualini, Bibliothèque Méjanes, MS. 209 (1027), 88–93.

66. Peiresc to Pasqualini, Bibliothèque de l'Ecole de Medecine, MS. H.271, fol. 12.

67. Peiresc to Aldrovandi, 30 October 1601, Bibliothèque Inguimbertine, MS. 1809, fol. 378; Peiresc to Aldrovandi, 28 November, 1601, Bibliothèque Inguimbertine, MS. 1809, fol. 381; Peiresc to Della Porta, 25 January 1602, Bibliothèque Inguimbertine, MS. 1809, fol. 382; Bibliothèque Nationale, MS. N.a.f. 5172, fol. 107. For Clusius, see Bibliothèque Inguimbertine, MS. 1809, fol. 402, and **Peiresc** 1889-1898, 7:941–960. Aldrovandi is by far the most studied, though in addition to the works of Olmi and Findlen see more narrowly Carrara 1998. I am grateful to Riccardo Di Donato for giving me the volume containing Carrara's article.

68. Ogilvie 1997, 275, 391–393.

69. Nardi 1980, 312; Legré 1899–1904.

70. Bibliothèque Inguimbertine, MS. 1809, fol. 402; Joret 1893–94, 437–442.

71. Bibliothèque Inguimbertine, MS. 1821, fol. 218.

72. There are many drawings of animals and animal parts; there are accounts of the flora and fauna of extra-European lands; there are descriptions of exotic animals to hand; and there are real animals, kept at home for study.

73. Bibliothèque Inguimbertine, MS. 1821, fols. 82–83.

74. Daston 2001a.

75. Bibliothèque Inguimbertine, MS. 1774, 407–409, in Humbert 1951b.

76. Peiresc to Dupuy, [November] 1633, **Peiresc** 1888–98, 2:645–646.

77. **Gassendi** 1657, year 1634, 95; Peiresc to Barberini, 5 May 1634, Bibliotheca Apostolica Vaticana, MS. Barb.Lat. 6503, fol. 88.

78. **Gassendi** 1657, year 1634, 95. For an overview of this history see Lindberg 1976.

79. This body of work has not found its way into the history of anatomy of the eye. See Koebling 1967, Hirschberg 1899–1918, Sudhoff 1907. For an early version of the "high road" in the history of optics see Wilde 1838. For parts of the experimental history that have been written see Koebling 1968; Lux 1989, 40, 123.

80. Bibliothèque Inguimbertine, MS. 1774, 446.

81. Bibliothèque Inguimbertine, MS. 1774, 444.

82. Bibliothèque Inguimbertine, MS. 1774, 487–488.

83. See French 1994, 315–317. But on Gassendi's indubitable sympathy for experimentalism, see 331–333.

84. **Gassendi** 1657, year 1634, 99.

85. Boulliaut [sic] to Gassendi, 7 September 1634, Bibliothèque Inguimbertine, MS. 1810, 48–50.

86. Bibliothèque Inguimbertine, MS. 1774, 434–435, 436. The first documented dissection is of the eye of a *lamia*, "gros poisson du poids de 4. quintaulx," on 11 May, with additional notes from the 11th, 13th, 14th, 15th, and 16th (469–470).

87. Bibliothèque Inguimbertine, MS. 1774, 436.

88. Bibliothèque Inguimbertine, MS. 1774, 403–406. This memo is interpolated, more or less directly, into Peiresc to Schikard, 29 August 1634, Württembergische Landesbibliothek, MS. 563, unfoliated. At the same time that he was composing these documents, Peiresc drew up another essay on much the same material, but phrased in terms of lunettes and mirrors: "Experiances des LUNETTES VERDES/ de Mr Gassend & autres tant CONCAVES et CONSERVES que CONVEXES," dated 30 August 1634. Bibliothèque Inguimbertine, MS. 1774, 457–458.

89. Humbert 1951a. It is this grudging work that is listed in the bibliography prepared by Turner 1969, 59.

90. Lindberg 1976, 203.

91. Bibliothèque Inguimbertine, MS. 1774, 426–427.

92. Bibliothèque Inguimbertine, MS. 1774, 428–431; **Gassendi** 1657, year 1634, 101.

93. Bibliothèque Inguimbertine, MS. 1774, 438–440.

94. Bibliothèque Inguimbertine, MS. 1774, 477, 478–482.

95. Initially, Peiresc wrote, he attributed this particular phenomenon "aux lunettes ordinaires que j'avoys suspendües devant mes ieulx"—an indication that Peiresc wore eyeglasses, at least later in his life—but further experimentation revealed it had nothing to do with the glasses themselves.

96. Bibliothèque Inguimbertine, MS. 1774, 478–482.

97. Peiresc and his circle are discussed in Bigourdan 1918, 13–69; Humbert 1948.

98. For example, Gualdo to Peiresc, quoted in Rizza 1961, 438.

99. He received news of its publication from Pignoria in the spring of 1610 and probably received his copy of the *Sidereus nuncius* from his friend and teacher Giulio Pace with a letter of 23 August 1610; "de ce live de Galileus que vous avez demandé" was what Pacius wrote (quoted in Rizza 1961, 437).

100. See Chapin 1958; Le Paige 1891–92; Costabel 1983; Bernhardt 1981.

101. This material cries out for careful scrutiny by a historian of astronomy.

102. Despite looking directly at the belt and sword of Orion, and despite commenting explicitly on the meaning of "Nebulosa"—or perhaps because of it—Galileo did not spot the nebula in Orion (**Galileo** 1993, 123–131).

103. The use of the names of these French queens—of Florentine derivation—for satellites III and IV appears in the 1611 title page Peiresc commissioned for his never-finished project; until that time they were named for the Medici grand dukes Ferdinandus and Franciscus.

104. Jardine 1984, 27.

105. There are of course other documents from the intervening period.

106. **Gassendi** 1657, year 1631, 62. For Peiresc's failure and Gassendi's success see Humbert 1950, 30, quoting selectively from Peiresc to Gassendi, 22 December 1631. Peiresc and his observer-friends, the doctors Le Febre and Nöel, saw that the day dawned cloudy and went back to bed; when the sky cleared they found themselves at mass where an unusually long sermon kept them until 11 o'clock—Gassendi had observed the transit at 10:30. Interestingly, Peiresc showed no disappointment whatsoever. For an overview, see Van Helden 1976.

107. The material is divided up among three manuscripts. The log begins on 3 November 1631 and runs through February 1632. The beginning and end are found in Bibliothèque Inguimbertine, MS. 1832. However, material stolen by Libri and later recuperated is divided between Bibliothèque Nationale, MS. N.a.f.

5856 and N.a.f. 5174, with fols. 380–383 and 388–390 in the former and 384–387 in the latter.

108. Peiresc to Schikard, 4 September 1634, letter 4, unpaginated, Württembergische Landesbibliothek, MS. 563 [2]r.

109. It was Frédeau who sketched Peiresc's "Alzaron." See Rizza 1961, 100 for Peiresc's letter of recommendation to the Barberini for Frédeau. From correspondence with Borilly, we know that Peiresc was using him to draw as early as 1630 (**Peiresc** 1888–98, 5:23). Gassendi, *Opera omnia*, vol. 4, quoted in Humbert 1936, 16; Humbert 1931.

110. Humbert 1931. Jaffé (1990, 175) has drawn on this chronology to explain the dating of Mellan's printed engravings.

111. Kopal and Carder 1974, 9; Van de Vyver 1971; Ashworth 1993, 323–324; Whitaker 1989; Whitaker 1999, 17–35. But it is not mentioned in Winkler and Van Helden 1993.

112. Peiresc to Cassiano, 4 June 1637, **Peiresc** 1989, 269.

113. **Hevelius** 1647, 206–207. Humbert 1931, 199.

114. **Hevelius** 1647, 224.

115. Gassendi did better with J. Caramuel Lobkowitz, to whom he sent a map, and who seems to have used it in his lunar observation of 14 August 1642. Unlike Hevelius, he proposed to name lunar features after modern scholars: "Tous nos amis y seront," he wrote to Gassendi, "toi-même, et Peiresc, et Mersenne, et Naudé." Riccioli adopted this scheme in 1651 and finally translated Peiresc to the moon. Humbert 1931, 200.

116. Bibliothèque Inguimbertine, MS. 1864, 261.

117. Bibliothèque Inguimbertine, MS. 1821, 125v.

118. Bibliothèque Inguimbertine, MS. 1821, 126.

119. "Relation des mesures et des Vases dont on use à Thunis en Barbarie faicte par le S. d'Arcos" [Peiresc's title], Bibliothèque Nationale, MS. Dupuy 688, fol. 63. The last and longest paragraph deals with foodways.

120. **Gassendi** 1657, year 1600, 35. This same combination of historical desire and concern for physical well-being characterized Aby Warburg's experience of Rome in June 1929. For Peiresc, as for Warburg, being part of the ritual trumped worries about health, especially the worries of others.

121. **Gassendi** 1657, year 1606, 99.

122. The literature on this subject matter is now substantial. See for example, Watanabe-O'Kelly and Simon 2000; Wisch and Scott Munshower 1990.

123. Bibliothèque Inguimbertine, MS. 1791,102, 128–132.

124. Bibliothèque Inguimbertine, MS. 1794, 186–91; 195–201. I have not yet been able to consult, for comparison, *La voye de laict, ou, Le chemin des heros au palais de la gloire: ouvert a l'entrée triomphante de Louys XIII. Roy de France & de Nauarre en la cité d' Auignon le 16. de Nouembre 1622* (Avignon, 1623).

125. Bibliothèque Inguimbertine, MS. 1795, 43–81.

126. Bibliothèque Inguimbertine, MS 1795, 55.
127. Bibliothèque Inguimbertine, MS. 1795, 90.
128. Bibliothèque Inguimbertine, MS. 1795, 91.
129. See Tate 1771, 1:204.
130. Bibliothèque Inguimbertine, MS. 1795.
131. Bibliothèque Inguimbertine, MS. 1795, fol. 322.
132. Bibliothèque Inguimbertine, MS. 1795, fol. 394.
133. Bibliothèque Inguimbertine, MS. 1795, fol. 402.
134. Bibliothèque Inguimbertine, MS. 1795, fol. 404v.
135. Bibliothèque Inguimbertine, MS. 1795, fol. 405, undated autograph draft.
136. Bibliothèque Inguimbertine, MS. 1795, fols. 443–444.
137. Bibliothèque Inguimbertine, MS. 1794, 314–321.
138. Bibliothèque Inguimbertine, MS. 1794, 319–320r; 320v–321v, summary on 322r.
139. Bibliothèque Inguimbertine, MS. 1794, 465–468.

Bibliography

Primary Sources

Manuscripts

Bayerische Staatsbibliothek, Munich: MS.CLM 12021. Michele Savonarola, "Directorium ad actum practicum."

Biblioteca Ambrosiana, Milan: B Ambr. N 26 Sup.: "Selecta ex Aristotel. Hist. De Animalibus et partibus" (here referred to as MS Ambrosiana).

Biblioteca Braidense, Milan: MS. AC.X.32. Lodrisio Crivelli, "Series triumphi Francisci Sfortiae."

Biblioteca Corsiniana, Rome: MS Linceo VII. Antonio Persio, "De natura ignis et caloris."

Biblioteca Estense, Modena: MS. Lat. 114: alpha.W.6.6. Michele Savonarola, "De vera re publica et digna seculari militia."

Biblioteca Estense, Modena: MS. VIII.B.20; MS.VIII.B.30. Michele Savonarola, "Confessionali."

Biblioteca Universitaria, Bologna: MS 936. "Marcelli Malpighii Opera."

Bibliotheca Apostolica Vaticana, Vatican City: MS. Barberini-Latini 6503.

Bibliothèque de l'École de Médecine, Montpellier: no. H.271.

Bibliothèque Inguimbertine, Carpentras: nos. 1774, 1791, 1794, 1795, 1797, 1803, 1809, 1810, 1821, 1831, 1832, 1864, 9531.

Bibliothèque Méjanes, Aix-en-Provence: no. 209 (1027).

Bibliothèque Nationale de France, Paris: Collection Dupuy, 663, 688, 746.

Bibliothèque Nationale de France, Paris: Fonds français, 9530, 9531, 9532.

Bibliothèque Nationale de France, Paris: MS latin 8958.

Bibliothèque Nationale de France, Paris: Nouvelles acquisitions françaises, 5171, 5172, 5174.

British Library, London: Add. MS 6789: Harriot manuscripts.

British Library, London: Sloane MS 3315, MSS of Francis Glisson, n. 381: "De historia, sive de plena enumeratione experimentorum."

Universitätsbibliothek, Basel: MS. K III 42. Conrad Weigand, "Quadripartitum de quercu."

Württembergische Landesbibliothek, Stuttgart: MS. 563.

Printed

Aconcio, Jacopo. 1944. *De methodo e opuscoli religiosi e filosofici*. Ed. Giorgio Radetti. Florence: Vallecchi.

Acta Sanctorum Julii. 1731. Vol. 7. Antwerp: J. du Moulin.

Aelian. 1731. *Kl. Ailianou sophistou poikile historia, Cl. Aeliani sophistae varia historia, cum notis integris Conradi Gesneri, Johannis Schefferi, Tanaquilli Fabri, Joachimi Kuhnii, Jacobi Perizonii et interpretatione Latina Justi Vulpeii*. Ed. Abraham Gronovius. Leiden, Amsterdam, Rotterdam, Utrecht, and The Hague: Luchtmans & Langerak, Wetstein & Smith, Boom & Waasberge, Beman, Poolsum, and Scheurleer.

Agricola, Rudolph. 1967. *De inventione dialectica*. (Facsimile reprint of edition of Cologne, 1523.) Niewkoop: B. de Graaf.

Albertus Magnus. 1495. *De animalibus libri vigintisex novissime impressi*. Venice: per Joannem & Gregorium de Gregoriis.

Albertus Magnus. 1916–21. *De animalibus libri xxvi nach der Cölner Urschrift*. Ed. Hermann Stadler. 2 vols. Münster i. W.: Verlag der Aschendorffschen Verlagsbuchhandlung.

Alciati, Andrea. 1530. Prefatory letter to Giacomo Bracelli, *Libri quinque de bello Hispaniensi*. Haguenau: per Johannem Secerium.

Aldrovandi, Ulisse. 1599. *Ornithologiae hoc est de avibus historiae libri XII. Cum indice septemdecim linguarum copiosissimo*. Bologna: apud Franciscum de Franciscis, fol.

Aldrovandi, Ulisse. 1637. *De quadrupedib. digitatis viviparis libri tres, et de quadrupedib. digitatis oviparis libri duo*. Ed. Bartholomaeus Ambrosinus. Bologna: apud Nicolaum Tebaldinum.

Aleandro, Girolamo. 1617. *Antiquae tabulae marmoreae Solis effigie, symbolisque exculptae, accurata explicatio, qua priscae quaedam mythologiae, ac nonnulla praeterea vetera monumenta marmorum, gemmarum, nomismatum illustrantur*. Paris: Officina Nivelliana.

Alsted, Johann. 1615. *Philosophia digne restituta libros complectens quatuor*. Herborn: G. Corvini.

Allen, P. S., et al., eds. 1906–58. *Opus epistolarum Des. Erasmi Roterodami*. 12 vols. Oxford: Clarendon.

Amatus Lusitanus. 1570. *Curationum medicinalium centuria septima*. Lyon: apud G. Rouillium.

Amatus Lusitanus. 1628. *Curationum medicinalium centuriae septem*. Barcelona: S. and J. Mathevat.

Argenterio, Giovanni. 1610. *Opera*. Fol. Hanau: apud haeredes Claudii Murnii.

Aristotle. 1498. *De natura animalium libri nouem; De partibus animalium libri quattuor; De generatione animalium libri quinq*. Venice: impressum mandato & expensis nobilis uiri Domini Octauiani Scoti ciuis Modoetiesis per Bartholameum de Zanis de Portesio.

Aristotle. 1513. *De natura animalium*. Trans. Theodore Gaza. Venice: in ædibus Aldi et Andreæ Asulani.

Aristotle. 1550. *Libri omnes cum Averrois commentariis*. Fol. Venice: apud Iuntas.

Aristotle. 1584. *Liber qui decimus Historiarum inscribitur, nunc primum latinus factus a Iulio Caesare Scaligero et commentariis illustratus*. Ed. Silvius Caesar Scaliger. 8vo. Lyon: apud Antonium de Harsy.

Aristotle. 1597. *Organum*. Ed. Julius Pacius. Editio secunda. 4to. Frankfurt: apud heredes Andreae Wecheli, Claudii Marnii et Ioannis Aubrii.

Aristotle. 1619. *Opera omnia*. Ed. Guillaume Du Val. 2 vols. Fol. Paris: typis regiis.

Arnald of Villanova. 1988. *Tractatus de consideracionibus operis medicine sive de flebotomia*. Ed. Luke Demaitre and Pedro Gil Sotres. (*Arnaldi de Villanova opera medica omnia*, vol. IV.) Barcelona: Publicacions de la Universitat de Barcelona.

Arnald of Villanova. 1998. *Regimen Almarie (Regimen ad castra sequentium)*. Ed. Luis Garcia Ballester, Juan Antonio Paniagua, and Michael McVaugh. (*Arnaldi de Villanova opera medica omnia*, vol. X.2.) Barcelona: Publicacions de la Universitat de Barcelona.

Aselli, Gaspare. 1627. *De lactibus sive lacteis venis, quarto vasorum mesaraicorum genere novo invento . . . dissertatio*. Milan: apud Io. Baptistam Bidellium.

Athenäum. 1798–1800. *Athenäum: eine Zeitschrift von August Wilhelm Schlegel und Friedrich Schlegel*. Berlin: F. Vieweg.

Bacci, Andrea. 1558. *Del Teuere, della natura et bontà dell'acque & delle inondationi. Libri II*. Rome: V. Luchino.

Bacci, Andrea. 1571. *De thermis . . . libri septem opus locupletissimum non solum medicis necessarium, verumetiam studiosis variarum rerum naturae perutile*. Venice: apud Vincentium Valgrisium.

Bacci, Andrea. 1573. *L'Alicorno, discorso . . . nel quale si tratta della natura dell'Alicorno, & delle sue virtù eccellentissime*. Florence: Giorgio Marescotti.

Bacci, Andrea. 1576. *Del Tevere . . . libri tre, ne' quali si tratta della natura, & bontà dell'acque, & specialmente del Tevere, & dell'acque antiche di Roma, del Nilo, del Pò, dell'Arno, & d'altri fonti, & fiumi del mondo. Dell'uso dell'acque, & del bevere in fresco, con nevi, con ghiaccio, & con salnitro. . . .* Venice: Aldo Manuzio.

Bacci, Andrea. 1586. *De venenis, et antidotis prolegomena, seu communia praecepta ad humanam vitam tuendam saluberrima*. Rome: apud Vincentium Accoltum.

Bacci, Andrea. 1596. *De naturalis historia de vinis Italiae et de conviviis antiquorum libri septem* . . . Rome: ex officina N. Mutij.

Bacci, Andrea. 1598. *De monocerote seu unicornu, eiusque admirandis viribus et usu tractatus.* Stuttgart: imprimebat Marcus Fürsterus.

Bacci, Andrea. 1603. *De gemmis et lapidibus pretiosis, eorumq[ue] viribus & usu tractatus italica lingua conscriptus: nunc vero non solum in Latinum sermonem conuersus verum etiam utilissimis annotationibus & observationibus auctior redditus.* Frankfurt: ex officina Matthiae Beckeri, impensis Nicolai Steinii.

Bacon, Francis. 1620. "Preparative towards a Natural and Experimental History [= Parasceve]." In *The Philosophical Works of Francis Bacon*, vol. 4. Ed. James Spedding, Robert L. Ellis, and Douglas D. Heath. London: Longman, 1857–1858.

Bacon, Francis. 1858–74. *Works.* Ed. James Spedding, Robert L. Ellis, and Douglas D. Heath. 14 vols. London: Longman.

Bacon, Francis. 1878. *Novum organum.* Ed. Thomas Fowler. Oxford: Clarendon Press.

Bacon, Francis. 1973. *The Advancement of Learning.* Ed. G. W. Kitchin. London: J. M. Dent & Sons.

Bacon, Francis. 1996. *Philosophical Studies, ca. 1611–1619.* Ed. Graham Rees. Oxford: Oxford University Press.

Bacon, Francis. 2000a. *The Advancement of Learning.* Ed. Michael Kiernan. Oxford: Oxford University Press.

Bacon, Francis. 2000b. *The Instauratio magna: Last Writings.* Ed. Graham Rees. Oxford: Oxford University Press.

Bacon, Roger. 1920. *Secretum secretorum cum glossis et notulis et tractatus brevis et utilis ad declarandum quedam obscure dicta.* Ed. Robert Steele. (*Opera hactenus inedita Rogeri Baconi*, V.) Oxford: Clarendon Press.

Baillet, Adrien. 1685. *Jugemens des sçavans sur les principaux ouvrages des auteurs.* Paris: Antoine Dezallier.

Bailly, J. S. 1969. "Histoire de l'astronomie ancienne." In M. Péchaux and M. Fichant, eds., *Sur l'histoire des sciences.* Paris: F. Maspero.

Baldini, Baccio. 1586. *In librum Hyppocratis De aquis, aere, et locis commentaria.* Florence: ex officina Bartholomaei Sermartellii.

Barbaro, Ermolao. 1943. *Epistolae, orationes et carmina.* Ed. Vittore Branca. Florence: Bibliopolis.

Barbaro, Ermolao. 1993. *Castigationes plinianae et in Pomponium Melam.* Ed. G. Pozzi. 4 vols. Milan: Antenore.

Bardi, Girolamo. 1644. *Medicus politico-catholicus.* Genoa: Farroni.

Baronio, Cesare. 1597. *Annales ecclesiastici.* Vol. 2. Rome: Ascanius et Donangelus.

Bartholin, Thomas. 1654. *Historiae anatomicae rariores.* Copenhagen: Typis Academicis Martzani.

Bartholomaeus Anglicus. 1485. *Hier beghinnen de titelen d[er] sijn de namen der boeken daer men af spreke[n] sal en[de] oec die capittelen der eerwaerdighen mans bartolome[us] engelsman en[de] een gheoerdent broeder van sinte franciscus oerde. Ende heest xix. boeke[n] die sprekende sijn vande eygenscappen der dingen.* Haarlem: Jakob Bellaert.

Bartholomaeus Anglicus. 1488. *Proprietates rerum domini bartholomei anglici.* Heidelberg: Printer of Lindelbach (Heinrich Knoblochtzer?).

Baudouin, François. 1561. *De institutione historiae universae et ejus cum jurisprudentia conjunctione prolegomenon: libri II.* Paris: A. Wechel.

Bauhin, Caspar. 1609. *Institutiones anatomicae corporis virilis et muliebris historiam exhibentes . . . Hippocrat. Aristot. Galeni auctoritat. illustratae & novis inventis plurimis.* Basel: apud Joann. Schroeter.

Bauhin, Caspar. 1623. *Pinax theatri botanici, sive index in Theophrasti, Dioscoridis, Plinii et botanicorum qui a seculo scripserunt, opera.* 4to. Basel: sumptibus et typis Ludovici Regis.

Bauhin, Jean. 1593. *Traicté des animauls aians aisles, qui nuisent par leurs piqueures ou morsures, avec les remèdes; oultre plus une histoire de quelques mousches ou papillons non vulgaires apparues en l'an 1590, qu'on a estimé fort venimeuses.* Montbéliard: J. Foillet.

Becichemo, Marino. 1504. *Lucuplentissima oratio qua Brixiano senatui gratias agit. Aurea praelectio in C. Plinium Secundum.* Brescia: Angelo Britannico.

Belon, Pierre. 1551. *L'histoire naturelle des estranges poissons marins, avec la vraie peincture et description du daulphin et de plusieurs autres de son espece.* Paris: R. Chaudière.

Belon, Pierre. 1553. *De aquatilibus libri duo cum iconibus ad vivam ipsorum effigiem quoad ejus fieri potuit expressis.* Paris: Ch. Estienne.

Belon, Pierre. 1555. *L'histoire de la nature des oyseaux avec leurs descriptions et naïfs portraicts retirez du naturel.* Paris: G. Cavellat & G. Corrozet.

Benedetti, Alessandro. 1507. *Plinii Secundi Caii Historia naturalis libri XXXVII ab Alexandro Benedicto Veronensi physico emendatiores redditi.* Venice: impressum per Joannem Rubeum & Bernardinum fratresque Vercellenses.

Benedetti, Alessandro. 1967. *Diaria de bello Carolino* (1496). Ed., trans., and intro. D. M. Schullian. New York: Renaissance Society of America.

Benedetti, Alessandro. 1998. *Historia corporis humani sive anatomice* (1502). Ed., trans., and intro. Giovanna Ferrari. Florence: Olschki.

Benivieni, Antonio. 1994. *De abditis nonnullis ac mirandis morborum et sanationum causis.* Ed. Giorgio Weber. Florence: Olschki.

Beurer, Johann Jakob. 1594. *Synopsis historiarum et methodus nova.* Hanau: Antonius.

Beyerlinck, Laurentius. 1666. *Magnum theatrum vitae humanae.* Lyon: Huguetan and Ravaud.

Blanke, Horst Walter, and Dirk Fleischer, eds. 1990. *Theoretiker der deutschen Aufklärungshistorie.* Stuttgart: Frommann-Holzboog.

Boaistuau, Pierre. 1560. *Histoires prodigieuses les plus memorables.* Paris: V. Sertenas.

Boccaccio, Giovanni. 1998. *Genealogia deorum gentilium. De montibus, silvis, fontibus, lacubus, fluminibus, stagnis seu paludibus, de diversis nominibus maris.* Ed. Vittore Branca. (*Tutte le opere di Giovanni Boccaccio*, vol. 7/8, t. 1.) Milan: Mondadori.

Boccadiferro, Lodovico. 1566. *Lectiones super tres libros de anima Aristotelis.* Venice: ex officina Ioan. Baptistae Somaschi & fratres.

Bock, Hieronymus. 1539. *New Kreütter Buch von Underscheydt, Würckung und Namen der Kreütter so in teütschen Landen wachsen: Auch der selbigen eygentlichem und wolgegrundtem Gebrauch in der Artznei, zu behalten und zu fürdern Leibs Gesuntheyt fast nutz und tröstlichen, vorab gemeynem Verstand.* Strasburg: Wendel Rihel.

Bodin, Jean. 1566. *Methodus, ad facilem historiarum cognitionem.* Paris: apud Martinum Iuvenem, sub insigni D. Christophori e regione gymnasii Cameracensium.

Bodin, Jean. 1572. *Methodus, ad facilem historiarum cognitionem.* 2d ed. Paris: apud Martinum Iuvenem, via S. Io. Lateranensis, ad insigne Serpentis.

Bonet, Théophile. 1679. *Observations et histoires chirurgiques tirées des œuvres de quatre excellens médecins.* Geneva: J. A. Chouët.

Bonifacio, Baldassarre. 1656. *Historia ludicra.* 2d ed. Brussels: Mommartius.

Bossuet, François. 1558. *De natura aquatilium carmen, in universam Gulielmi Rondelitii . . . quam de piscibus marinis scripsit historiam; cum vivis eorum imaginibus, opusculum nunc primum in lucem emissum.* Lyon: M. Bonhomme.

Boyle, Robert. 1772. *The Works of the Honourable Robert Boyle.* Ed. Thomas Birch. 6 vols. London: J. and F. Rivington.

Bracelli, Giacomo. 1530. *Jacobi Bracelli Genuensis . . . libri quinque de bello Hispaniensis.* Haguenau: per Johannem Secerium.

Brasavola, Antonio Musa. 1556. *Index refertissimus in omnes Galeni libros qui ex Iuntarum tertia editione extant.* Fol. Venice: apud Iuntas.

Browne, Thomas. 1981. *Sir Thomas Browne's Pseudodoxia epidemica.* Ed. Robin Robbins. 2 vols. Oxford: Oxford University Press.

Brucker, J. J. 1742–44. *Historia critica philosophiae a mundi incunabulis ad nostram usque aetatem deducta.* Leipzig: Christopher Breitkopf.

Brunfels, Otto. 1532. *Herbarum vivae eicones: Ad naturae imitationem, summa cum diligentia et artificio effigiatae, una cum effectibus earundem, in gratiam veteris illius & iamiam renascentis herbariae medicinae.* Strasburg: apud Ioannem Schottum.

Bruni, Leonardo. 1987. *The Humanism of Leonardo Bruni: Selected Texts.* Trans. and intro. Gordon Griffiths, James Hankins, and David Thompson. Binghamton, N.Y.: Center for Medieval and Early Renaissance Studies.

Bruni, Leonardo. 2002. "De studiis et litteris liber ad Baptistam de Malatestis." Trans. Craig Kallendorf. In Craig Kallendorf, ed., *Humanist Educational Treatises*. Cambridge, Mass.: Harvard University Press, 92–125.

Burton, Robert. 1893. *The Anatomy of Melancholy*. Ed. A. R. Shilleto. 3 vols. London: George Bell & Sons.

Ca da Mosto, Alvise. 1966. *Le navigazioni atlantiche del veneziano Alvise Da Mosto*. Ed. Tullia Gasparrini Leporace. Rome: Istituto poligrafico dello Stato.

Cagnati, Marsilio. 1587. *Variarum observationum libri quatuor*. Rome: apud Bernardinum Donangelum.

Cagnati, Marsilio. 1591a. *De continentia vel de sanitate tuenda liber primus*. Rome: apud Ascanium et Hieronymum Donangelos.

Cagnati, Marsilio. 1591b. *In Hippocratis Aphorismum secundae sectionis vigesimum quartum commentarius*. Rome: apud Ascanium & Hieronymum Donangelos.

Cagnati, Marsilio. 1599a. *De Romani aeris salubritate commentarius*. Rome: apud Aloysium Zannettum.

Cagnati, Marsilio. 1599b. *De Tiberis inundatione medica disputatio... Epidemia Romana, disputatio, scilicet de illa populari aegritudine, quae anno 1591, et de altera, quae anno 1593, in urbem Romam invasit*. Rome: apud Aloysium Zannettum.

Cagnati, Marsilio. 1599c. *Romana epidemia. Descriptio scilicet, et examen vulgaris aegritudinis, quae in Urbe anno 1591, et alterius, qua anno 1593, orta est*. Printed with his *De Tiberis inundatione* (1599), with separate internal title page but continuous pagination, pp. 25–73.

Cagnati, Marsilio. 1601. *De urbana febres curande ratione commentarius*. Rome: apud Aloysium Zannettum.

Cagnati, Marsilio. 1605. *De sanitate tuenda libri duo. Primus de continentia, alter de arte gymnastica*. Padua: apud Franciscum Bolzettum.

Caius [Kay], John. 1570. *De canibus britannicis, liber unus. De rariorum animalium et stirpium historia, liber unus. De libris propriis, liber unus*. London: W. Seres.

Camerarius, Joachim. 1590. *Symbolorum et emblematum ex re herbaria desumptorum centuria una collecta a Joachimo Camerario... in quibus rariores stirpium proprietates historiae ac sententiae memorabiles non paucae breviter exponuntur*. Nuremberg: impensis J. Hofmanni et H. Camoxii.

Camerarius, Joachim. 1595. *Symbolorum et emblematum ex animalibus quadrupedibus desumptorum centuria altera, collecta a Ioachimo Camerario. Exponuntur in hoc libro rariores tum animalium proprietates tum historiae ac sententiae memorabiles*. Nuremberg: P. Kauffmann.

Camerarius, Joachim. 1596. *Symbolorum & emblematum ex volatilibus et insectis desumptorum centuria tertia collecta a Joachimo Camerario... in qua multae rariores proprietates ac historiae et sententiae memorabiles exponuntur*. Nuremberg: P. Kauffmann.

Camerarius, Joachim. 1702. *Symbolorum ac emblematum ethicopoliticorum: Centuriae quatuor: prima, arborum & plantarum; secunda, animalium quadrupedium; tertia, avium & volatilium; quarta, piscium & reptilium.* Mainz: sumptibus L. Bourgeat.

Camerarius, Joachim. 1986. *Symbola et emblemata (Nürnberg 1590 bis 1604).* Ed. Wolfgang Harms and Ulla-Britta Kuechen. 2 vols. Graz, Austria: Akademische Druck- und Verlagsanstalt.

Campanella, Tommaso. 1954. "Rationalis philosophiae pars quinta, videlicet: Historiographiae liber unus, iuxta propria principia." In *Tutte le opere di Tommaso Campanella*, ed. Luigi Firpo. Milan: Arnaldo Mondadori Editore, 1: 1222–1255.

Campofulgosus. *See* Fulgosus.

Cano, Melchior. 1776. *Opera.* Ed. Hyacinth Serry. Bassano (but Venice: Remondini).

Cano, Melchior. 1973. *L'autorità della storia profana (De humanae historiae auctoritate).* Ed. and tr. Albano Biondi, preface by Luigi Firpo. Turin: Giappichelli.

Capivaccius, Hieronymus. 1603. *Opera omnia.* Ed. Johannes-Hartmannus Beyerus. Fol. Frankfurt: e Paltheniana curante Jona Rhodio.

Carafa, Giuseppe Maria. 1751. *De gymnasio Romano et de eius professoribus.* Rome. (Rpt., Bologna: Forni, 1971.)

Cardano, Girolamo. 1643. *De propria vita liber.* Paris: apud Iacobum Villery.

Cardano, Girolamo. 1663a. *De libris propriis.* In Cardano 1663c.

Cardano, Girolamo. 1663b. *In librum Hippocratis De aere, aquis, et locis commentarii.* In Cardano 1663c, 8: 1–212.

Cardano, Girolamo. 1663c. *Opera omnia.* Ed. C. Spon. 10 vols. Fol. Lyon: Jean-Antoine Huguetan and Marc-Antoine Ravaud. (Facsimile, New York and London: Johnson Reprint, 1967.)

Cardano, Girolamo. 2004. *De libris propriis. The Editions of 1544, 1550, 1557, 1562, with Supplementary Material.* Ed. Ian Maclean. Milan: FrancoAngeli.

Cartari, Vincenzo. 1647. *Imagini delli dei de gl'antichi.* Venice. (Reprint ed. by Walter Koschatzky. Graz: Akademische Druck- und Verlagsanstalt, 1963.)

Casali, Giovanni Battista. 1646. *Sacrae prophanaeque religionis vetustiora monumenta, hoc est, symbolicus et hieroglyphicus aegyptiorum cultus, sacra ab antiquioribus diis colendis superstitiose adinventa, et veriores sacratioresque primorum christianorum ritus.* Rome: et vaeneunt Parisiis.

Castellani, Luigi Francesco. 1788. *Vita del celebre medico mantovano Marcello Donati.* Mantua: Pazzoni Alberto erede.

Castelli, Bartolomeo. 1642. *Lexicon medicum graeco-latinum.* Venice: apud Io. Baptistam Cestari, et Franciscum Bolzettam.

Castelli, Bartolomeo. 1700. *Amaltheum Castello-Brunonianum, sive Lexicon medicum graeco-latinum.* Padua: sumptibus Jacobi de Cadorinis.

Castiglione, Iacomo. 1599. *Trattato dell'inondazione del Tevere*. Rome: appresso Guglielmo Facciotto, ad istantia di Giouanni Martinelli.

Cesalpino, Andrea. 1593. *Quaestionum peripateticorum libri v; Daemonum investigatio peripatetica; Quaestionum medicarum libri ii; De medicamentorum facultatibus libri ii*. 4to. Venice: apud Iuntas.

Chacon, Pedro. 1588. *De triclinio Romano, Fulvi Ursini appendix*. Rome: in ædibus S.P.Q.R., apud Georgium Ferrarium.

Champier, Symphorien. 1503. *La nef des dames vertueuses*. Lyon: Jacques Arnoullet.

Champier, Symphorien. 1510. *Le recueil ou croniques des hystoires des royaulmes d'Austrasie ou France orientale que maintenant on dit Lorraine*. Lyon: pour V. de Portunaris.

Champier, Symphorien. 1516a. *Les grans croniques des gestes et vertueux faictz des très-excellens catholiques illustres et victorieux ducz et princes des pays de Savoye et Piémont*.... Paris: pour Jehan de la Garde.

Champier, Symphorien. 1516b. *Medicinale bellum inter Galenum et Aristotelem*. Lyon: Simon Vincent.

Champier, Symphorien. 1516c. *Symphonia Platonis cum Aristotele et Galeni cum Hippocrate... Platonica medicina de duplici mundo*.... Paris: impressum est hoc opus apud Badium.

Champier, Symphorien. 1532. *Claudii Galeni Pergameni historiales campi, per D. Symphorianum Campegium... in quatuor libros congesti*. Basel: apud A. Cratandrum et J. Bebelium.

Champier, Symphorien. 1884. *L'antiquité de la cité de Lyon*. Lyon: H. Georg.

Chasseneux, Barthélemy de. 1571. *Catalogus gloria mundi... Opus in libros XII divisum*. Venice: apud Vincentium Valgrisium.

Cicero, Marcus Tullius. 1933. *De natura deorum*. Trans. H. Rackham. In Cicero, *De natura deorum; Academica*. Cambridge, Mass.: Harvard University Press; London: William Heinemann.

Cicero, Marcus Tullius. 1988. *De oratore*. Trans. H. Rackham. Cambridge, Mass.: Harvard University Press.

Clusius [L'Ecluse], Carolus. 1576. *Rariorum aliquot stirpium per Hispanias observatarum historia, libris duobus expressa*. Antwerp: ex officina Christophori Plantini, architypographi regii.

Clusius, Carolus. 1583. *Rariorum aliquot stirpium, per Pannoniam, Austriam, & vicinas quasdam provincias observatarum historia, quatuor libris expressa*. Antwerp: ex officina Christophori Plantini.

Clusius, Carolus. 1601. *Rariorum plantarum historia*. Antwerp: ex officina Plantiniana apud Joannem Moretum.

Coelius Rhodiginus, Ludovicus. 1542. *Lectionum antiquarum libri XXX*. Basel: Frobenius et Episcopius.

Comte, Auguste. 1949. *Cours de philosophie positive*. Ed. Charles Le Verrier. Paris: Garnier.

Condillac, Etienne Bonnot de. 1798. *Traité des systèmes.* In *Oeuvres de Condillac*, vol. 2. Paris: C. Houel.

Cordus, Euricius. 1534. *Botanologicon.* Cologne: apud Ioannem Gymnicum.

Cornarius, Diomedes. 1599. *Tractatus consiliorum medicinalium.* Leipzig: M. Lantzenberger.

Costeo, Giovanni. 1565. *De venarum mesaraicarum usu liber.* Venice: [eredi di Luca Antonio Giunta il vecchio].

Cramerus, Daniel. 1601. *Isagoge in Metaphysicam Aristotelis, editio secunda, cum nova ad Tychonem Brahe dedicatione et ad Rudolphum II acclamatione.* 8vo. Wittenberg: impensis Bechtholdi Raben.

Crato von Crafftheim, Johann. 1560. *Ad artem medicam isagoge, additae sunt in libros Galeni de elementis, de natura humana, de atrabile, et de temperamentis, et facultatibus naturalibus periochae Johannis Baptistae Montani, cum epistola Jo. Cratonis, qua recte Galenum legendi ratio breviter ostenditur. . . .* Venice: ex officina Valgrisiana.

Crato von Crafftheim, Johann. 1591–1611. *Johannis Cratonis et aliorum medicorum consilia et epistolae medicinales, studio Laurentii Scholzii.* 7 vols. Frankfurt and Hanau: Wechel.

Crippa, Bernardinus. 1566. *In Aristotelis librum de animalium motu latine redditum ecphrasis.* 8vo. Venice: per Gratiosum Perchacinum.

Da Monte, Giovanni Battista. 1554. *In tertiam primi epidemiorum sectionem explanationes.* Venice: apud Balthassarem Constantinum.

Da Monte, Giovanni Battista. 1558. *Consultationum medicinalium centuria secunda, nunc primum opera, et studio Io. Cratonis . . . edita.* Venice: in officina Erasmiana, apud Vincentium Valgrisium.

Da Monte, Giovanni Battista. 1559. *Consilia medica omnia.* Nuremberg: [apud Joannem Montanum, & Ulricum Neuberum] (*praefatio* Girolamo Donzellini).

Da Monte, Giovanni Battista. 1583. *Consultationes medicae.* Basel: H. Petri & P. Perna.

Dandini, Girolamo. 1610. *De corpore animato.* Paris: apud Claudium Chappeletum.

Decembrio, Angelo. 2002. *De politia litteraria.* Ed. Norbert Witten. Munich and Leipzig: Saur.

Descartes, René. 1910. *Le monde, ou Traité de la lumière.* In Descartes, *Oeuvres*, ed. Charles Adam and Paul Tannery, vol. 11. Paris: Cerf.

Descartes, René. 1977. *Règles utiles et claires pour la direction de l'esprit en la recherches de la verité.* The Hague: Nijhof.

Descartes, René. 1985. *Discourse on the Method.* In Descartes, *The Philosophical Writings*, trans. John Cottingham et al. Cambridge: Cambridge University Press.

Diogenes Laertius. 1702. *The Lives of the Ancient Philosophers, Containing an Account of their Several Sects, Doctrines, Actions, and Remarkable Sayings.*

Extracted from Diogenes Laertius, Causabon [sic], *Menagius, Stanley, Gassendus, Charleton, and others.... With an Appendix Containing the Lives of Several Later Philosophers... Taken from Eunapius. And an Account of Women Philosophers, Written Originally in Latin by Aeg. Menagius to Madame Dacier.* London: Tho. Newborough.

Diogenes Laertius. 1995. *Lives of Eminent Philosophers.* Trans. R. D. Hicks. 2 vols. Cambridge, Mass.: Harvard University Press.

Dionysius of Halicarnassus. 1895. *De arte rhetorica.* Leipzig: B. Teubner.

Dioscorides. 1518. *De medica materia libri sex.* Ed. Marcellus Virgilius. Fol. Florence: per haeredes Philippi Iuntae.

Dodoens, Rembert. 1557. *Histoire des plantes, en laquelle est contenue la description entiere des herbes, c'est à dire, leures especes, forme, noms, temperament, vertus & operations: non seulement de celles qui croissent en ce païs, mais aussi des autres estrangeres qui viennent en usage de Medecine.* Trans. Carolus Clusius. Anvers: de l'Imprimerie de Jean Loë.

Dodoens, Rembert. 1568. *Florum, et coronariarum odoratarumque nonnullarum herbarum historia.* Antwerp: ex officina Christophori Plantini.

Dodoens, Rembert. 1583. *Stirpium historiae pemptades vi sive libri xxx.* Fol. Antwerp: ex officina Christophori Plantini.

Dodoens, Rembert. 1581. *Medicinalium observationum exempla rara.* Leiden: ex officina Christophori Plantini.

Donati, Marcello. 1586. *De medica historia mirabili libri sex.* Mantua: per Franciscum Osanam.

Donati, Marcello. 1604. *Scholia sive dilucidationes eruditissimae in Latinos plerosque historiae Romanae scriptores.* Venice: apud Juntas.

Doni, Anton Francesco. 1972. *La libraria.* Ed. V. Bramanti. Milan: Longanesi.

Draud, Georg. 1625. *Bibliotheca classica.* Frankfurt am Main: Ostern.

Dresser, Matthaeus. 1606. *Orationum libri tres.* 3 vols. Leipzig: Apelius.

Du Cange, Charles du Fresne. 1938. *Glossarium mediae et infimae latinitatis.* Paris: Libraire des sciences et des arts.

Ducci, Lorenzo. 1604. *Ars historica.* Ferrara: apud Victorium Baldinum typographum cameralem.

Du Laurens, André. 1599. *Historia anatomica humani corporis.* Frankfurt: apud Matthaeum Beckerum impensis Theodorici de Bry viduae et duorum filiorum.

Du Laurens, André. 1600. *Opera anatomica.* Fol. Frankfurt: apud Matthaeum Beckerum, impensis Theodorici de Bry viduae et duorum filiorum.

Du Laurens, André. 1602. *Historia anatomica humani corporis partes singulas uberrime enodans.* Frankfurt: Zacharias Palthenius sumptibus Jonae Rodii.

Dullaert, Joannes. 1521. *Questiones in librum Predicabilium Porphyrii.* Fol. Paris: venundantur a Bernardo Aubry.

Eliot, George. 1963. *Essays.* Ed. Thomas Pinney. New York: Columbia University Press.

Emiliani, Giovanni. 1584. *Naturalis de ruminantibus historia, vario doctrinae genere referta*. Venice: F. Ziletti.

Encyclopédie. 1751–80. *Encyclopédie, ou Dictionnaire raisonné des sciences, des arts et des métiers*. Ed. Denis Diderot. Paris and Neuchâtel.

Erasmus, Desiderius. 1518. Ep. ded. to Ernest of Bavaria, in *Quintus Curtius de rebus gestis Alexandri Magni regis Macedonum cum annotationibus Des. Erasmi Roterodami*. Strasbourg.

Erasmus, Desiderius. 1703. *Opera omnia . . . in decem tomos distincta*. Ed. Jean Le Clerc. Leiden: cura & impensis Petri Vander Aa.

Estienne, Charles. 1545. *De dissectione partium corporis humani*. Paris: apud Simonem Colinaeum.

Eustachius [Eustachi], Bartholomaeus. 1564. *Opuscula anatomica*. 4to. Venice: Vincentius Luchinus excudebat.

Fabricius, Johann Andreas. 1752–54. *Abriss einer allgemeinen Historie der Gelehrsamkeit*. 3 vols. Leipzig: in der Weidmannischen Buchhandlung.

Fabricius Montanus, Joannes. 1555. *Differentiae animalium quadrupedum secundum locos communes, opus ad animalium cognitionem apprime conducibile*. 8vo. Zurich: per Andream Gessnerum et Iacobum Gessnerum.

Falloppia, Gabriele. 1606. *Observationes anatomicae*. Ed. Johann Crato von Crafftheim. In Falloppia, *Opera omnia*. Venice: apud Jo. Antonium, & Jacobum de Franciscis.

Ferdinando, Epifanio. 1621. *Centum historiae, seu Observationes et Casus medici*. Venice: Tommaso Baglioni.

Fernández de Oviedo y Valdes, Gonzalo. 1526. *Oviedo dela natural hystoria delas Indias*. Toledo.

Fonseca, Petrus. 1604. *Commentaria in Metaphysicorum Aristotelis libros*. 3 vols. 4to. Cologne: impensis Lazari Zetzneri.

Fontanus, Jacobus. 1611. *Phisiognomia Aristotelis ordine compositorio edita ad facilitatem doctrinae*. 8vo. Paris: apud Joannem Paquet.

Foreest [Forestus], Pieter van. 1584. *Observationes et curationes*. Antwerp: Christophorus Plantinus.

Fox Morcillo, Sebastián. 1557. *De historiae institutione liber*. In Wolf 1579.

Franckenberger, Andreas. 1586. *Institutionum antiquitatis et historiarum, pars prima, in libros sex distributa*. Wittenberg: Crato.

Frischlin, Nicodemus, ed. 1588. *Selectae orationes e Q. Curtio, T. Livio, C. Salustio, C. Caesare, M. Cicerone in usum Scholae Martinianae apud Brunsvicenses*. Wolfenbüttel: Corneus.

Fuchs, Leonhart. 1530. *Errata recentiorum medicorum lx numero adiectis eorundem confutationibus*. 4to. Haguenau: in aedibus Johannis Secerii.

Fuchs, Leonhart. 1542. *De historia stirpium commentarii insignes, maximis impensis et vigiliis elaborati, adjectis earundem vivis plusquam quingentisi imaginibus*. Basel: in officina Isingriniana.

Fuchs, Leonhart. 1543. *New kreüterbuch in welchem nit allein die gantz histori das ist namen gestalt statt und zeit der wachsung natur krafft und würckung des meysten theyls der kreüter so in teütschen vnnd andern landen wachsen mit dem besten vleiss beschriben sonder auch aller derselben wurtzel stengel bletter blumen samen frücht und in summa die gantze gestalt allso artlich vnd kunstlich abgebildet vnd contrafayt ist das dessgleichen vormals nie gesehen noch an tag komen.* Basel: Durch M. Isingrin.

Fulgosus [Fregoso], Baptista. 1578. *Bap. Fulgosii factorum dictorumque memorabilium libri IX.* Paris: Cavellat.

Furlanus, Daniel. 1574. *In libros Aristotelis de partibus animalium commentarius primus.* 8vo. Venice: apud Joan[nem] Baptistam Somaschum.

Gaillard, Pierre Droit de. 1576. *Méthode qu'on doit tenir en la lecture de l'histoire.* Paris: P. Cavellat.

Galen. 1821–33. *Opera omnia.* Ed. C. G. Kühn. 20 vols. Leipzig: Cnoblauch. (Facsimile, Hildesheim: Olms, 1997.)

Galen. 1944. *On Medical Experience.* First edition of the Arabic version with English translation and notes by Richard Walzer. London: Oxford University Press.

Galen. 1979. *On Prognosis.* Ed. Vivian Nutton. Berlin: Akad. Verlag.

Galen. 1985. *Three Treatises on the Nature of Science. On the Sects for Beginners, An Outline of Empiricism, On Medical Experience.* Trans. Richard Walzer and Michael Frede, intro. Michael Frede. Indianapolis: Hackett.

Galen. 1991. *On the Therapeutic Method Books I and II.* Ed. R. J. Hankinson. Oxford: Oxford University Press.

Galileo Galilei. 1993. *Sidereus nuncius.* Ed. Andrea Battistini. Venice: Marsilio.

Gassendi, Pierre. 1641. *Viri illustris Nicolai Claudii Fabricii de Peiresc Senatoris Aquisextentis vita.* Paris. (English translation, Gassendi 1657.)

Gassendi, Pierre. 1657. *Mirrour of True Nobility and Gentility: Being the Life of the Renowned Nicolaus Claudius Fabricius, Lord of Pieresk.* London.

Gassendi, Pierre. 1658. *Vita Peireskij.* In Gassendi, *Opera omnia.* Lyon: sumptibus Laurentii Anisson. (Facsimile, Stuttgart: Frommann Verlag, 1964.)

Gassendi, Pierre. 1727. *Opera omnia.* 6 vols. Florence: Tartini & Franchi.

Gellius, Aulus. 1927. *Attic Nights.* Trans. J. C. Rolfe. 3 vols. Cambridge, Mass.: Harvard University Press.

Gerard, John. 1633. *The herball; or, Generall historie of plants.* Ed. Thomas Johnson. London: printed by Adam Islip, Joice Norton, and Richard Whitakers.

Gerard, John. 1636. *The Herbal.* Ed. and intro. Thomas Johnson. Fol. London: Adam Islip, Joice Norton, and Richard Whitaker.

Gessner, Conrad. 1541. *Historia plantarum et vires ex Dioscoride, Paulo Aegineta, Theophrasto, Plinio & recentioribus Graecis iuxta elementorum ordinem.* Paris: apud Odinum Petit.

Gessner, Conrad. 1548. *Pandectarum sive partitionum universalium libri XXI.* Zurich: Ch. Froschauer.

Gessner, Conrad. 1549. *Partitiones theologicae, pandectarum universalium liber ultimus.* Zurich: Ch. Froschauer.

Gessner, Conrad. 1551. *Historiae animalium liber I, de quadrupedibus viviparis. Opus philosophis, medicis, grammaticis, philologis, poetis et omnibus rerum linguarumque variarum studiosis, utilissimum simul iucundissimumque futurum.* Zurich: Ch. Froschauer.

Gessner, Conrad. 1553. *Icones animalium quadrupedum viviparorum et oviparorum, quae in historia animalium describuntur, cum nomenclaturis singulorum Latinis, Italicis, Gallicis et Germanicis plerunque, per certos ordines digestae.* Zurich: Ch. Froschauer.

Gessner, Conrad. 1554. *Historiae animalium liber II, de quadrupedibus oviparis. Adiecta sunt etiam novae aliquot quadrupedum figurae. . . .* Zurich: Ch. Froschauer.

Gessner, Conrad. 1555a. *Historiae animalium liber III, qui est de avium natura.* Zurich: Ch. Froschauer.

Gessner, Conrad. 1555b. *Icones avium omnium, quae in Historia avium Conradi Gesneri describuntur, cum nomenclaturis singulorum latinis, italicis, gallicis et germanicis plerunque, per certos ordines digestae.* Zurich: Ch. Froschauer.

Gessner, Conrad. 1556. *De raris et admirandis herbis, quae sive quod noctu luceant, sive alias ob causas, Lunariae nominantur, commentariolus. . . .* Zurich: apud Andream Gesnerum F. & Iacobum Gesnerum, fratres.

Gessner, Conrad. 1558. *Historiae animalium liber IIII, qui est de piscium et aquatilium animantium natura. Cum iconibus singulorum ad vivum expressi fere omnibus DCXVI. Continentur in hoc volumine G. Rondeletii et P. Bellonii de aquatilium singulis scripta.* Zurich: Ch. Froschauer.

Gessner, Conrad. 1560a. *Icones animalium quadrupedum viviparorum et oviparorum, quae in historiae animalium libro I. et II. describuntur, . . . Editio secunda, novis eiconibus non paucis, et passim nomenclaturis ac descriptionibus auctior.* Zurich: Ch. Froschauer.

Gessner, Conrad. 1560b. *Nomenclator aquatilium animantium. Icones animalium aquatilium in maris et dulcibus aquis degentium . . . et nominum confirmandorum causa descriptiones quorumdam.* Zurich: Ch. Froschauer.

Gessner, Conrad. 1562. *De libris à se editis epistola ad Guilielmum Turnerum, theologum et medicum excellentissimum in Anglia.* Zurich: Ch. Froschauer.

Gessner, Conrad. 1577. *Epistolarum medicinalium Conradi Gesneri, philosophi et medici Tigurini, libri III.* Zurich: Ch. Froschauer.

Gilles, Pierre. 1533. *Ex Aeliani historia per Petrum Gyllium latini facti, itemque ex Porphyrio, Heliodoro, Oppiano, tum eodem Gyllio luculentis accessionibus aucti libri XVI. De vi & natura animalium.* Lyon: S. Gryphe.

Giovanni da Ferrara, Fra. 1936. *Ex annalium libris marchionum Estensium excerpta.* Ed. Luigi Simeoni. In *Rerum italicarum scriptores*, 20.2. Bologna: Zanichelli.

Giraldi, Lilio Gregorio. 1580. *Historia deorum gentilium.* In *Opera.* Basel: Thomas Guarinus.

Goclenius, Rodolphus. 1613. *Lexicon philosophicum*. Frankfurt: M. Becker.

Goodman, Godfrey. 1622. *The creatures praysing God: or, The religion of dumbe creatures, an example and argument, for the stirring up of our devotion, and for the confusion of atheisme*. London: printed by Felix Kingston.

Gregorio de Valencia. 1580. *De idololatria contra sectariorum contumelias disputatio*. Ingolstadt: D. Sartorius.

Gross, Johann Georg. 1625. *Urbis Basil. Epitaphia et inscriptiones omnium templorum*. . . . Basel: J. J. Genathi.

Guarino, Battista. 2002. "De ordine docendi et studendi." Trans. Craig Kallendorf. In Kallendorf, ed., *Humanist Educational Treatises*. Cambridge, Mass.: Harvard University Press, 260–309.

Guarinoni, Cristoforo. 1601. *Commentaria in primum librum Aristotelis de Historia animalium, in quibus animantium differentiae conspicuae fiunt, et sedulo ad certam divisionis normam, ordinantur*. 4to. Frankfurt: imprimebat Ioannes Saurius, sumptibus Nicholai Steinii.

Hartlib Papers Project. 1995. *The Hartlib Papers*. Ed. Judith Crawford et al. CD-ROM edition. Ann Arbor: UMI.

Harvey, William. 1674. *Observationes et historiae omnes et singulae è Guiljelmi Harvaei libello de generatione animalium excerptae, & in accuratissimo ordine redactae. Studio Justi Schraderi*. Amsterdam: Abraham Wolfgang.

Heineccius, Johann Gottlieb. 1756. *Elementa philosophiae rationalis: ex principiis admodum evidentibus justo ordine adornata*. Edinburgh: apud G. Hamilton & J. Balfour.

Helvicus, Christophorus [Christoph Helwig]. 1618. *Theatrum historicum: sive chronologiae systema novum*. Giessen: Hamphelius.

Herbarius. 1484. *Herbarius, seu De virtutibus herbarum*. Magonza: Peter Schoeffer.

Herberstein, Sigmund von. 1556. *Rerum moscoviticarum commentarii*. . . . First ed. Basel: J. Oporinus, 1549.

Herbert, Edward, of Cherbury. 1663. *De religione gentilium errorumque apud eos causis*. Amsterdam. (Reprint, Stuttgart: Frommann-Holzboog, 1967.)

Herold, Johannes. 1555. *Exempla virtutum et vitiorum*. Basel: Henricpetri.

Hevelius, Johannes. 1647. *Selenographia: sive, Lunae descriptio; atque accurata . . . delineatio*. Gdansk: typis Hunefeldianis.

Historia generalis plantarum. 1587–88. *Historia generalis plantarum in libros XVIII per certas classes artificiose digesta*. Fol., 2 vols. Lyon: apud Gulielmum Rovillium.

Holbach, Paul Henri Thiry, Baron d'. 1994a. *Système de la nature*. Hildesheim: Olms.

Holbach, Paul Henri Thiry, Baron d'. 1994b. *Système social*. Paris: Fayard.

Hooke, Robert. 1705. "Lectures Concerning Navigation and Astronomy" [1683]. In *Posthumous Works of Robert Hooke*. London. (Facsimile, New York, 1969.)

Hortus sanitatis. 1954. *An Early English Version of Hortus sanitatis: A Recent Bibliographic Discovery by Noel Hudson*. London: Bernard Quaritch.

Huet, Pierre-Daniel. 1679. *Demonstratio evangelica*. Paris: S. Michallet.

Huet, Pierre-Daniel. 1810. *Memoirs of the Life of Peter Daniel Huet*. Trans. John Aikin. London: Longman et al.

Imperato, Ferrante. 1599. *Dell'historia naturale . . . libri XXVIII, nella qvale ordinatamente si tratta della diuersa condition di miniere, e pietre; con alcune historie di piante, & animali; sin'hora non date in luce*. Naples: C. Vitale.

Incisa della Rocchetta, Giovanni, et al. 1957–63. *Il primo processo per San Filippo Neri nel codice vaticano latino 3798 e in altri esemplari dell'Archivio dell'Oratorio di Roma*. 4 vols. Studi e Testi, nos. 191, 196, 205, and 224. Vatican City: Biblioteca Apostolica Vaticana.

Isidore of Seville. 1962. *Etymologiarum sive originum libri XX*. Ed. W. M. Lindsay. Oxford: Clarendon Press.

Jonstonus, Joannes. 1657a. *Historiae naturalis de avibus libri VI*. Amsterdam: apud Ioannem Iacobi fil. Schipper.

Jonstonus, Joannes. 1657b. *Historiae naturalis de insectis libri III; de serpentibus et draconibus libri II*. Amsterdam: apud Ioannem Iacobi fil. Schipper.

Jonstonus, Joannes. 1657c. *Historiae naturalis de piscibus et cetis libri V*. Amsterdam: apud Ioannem Iacobi fil. Schipper.

Jonstonus, Joannes. 1657d. *Historiae naturalis de quadrupedibus libri*. Amsterdam: apud Ioannem Iacobi fil. Schipper.

Keckermann, Bartholomäus. 1614. *Operum omnium quae extant tomus primus [—secundus]*. 2 vols. Geneva: Aubert.

Kennedy, George A., trans. 2003. *Progymnasmata. Greek Textbooks of Prose Composition and Rhetoric*. Leiden: Brill.

Kepler, Johannes. 1954. *Gesammelte Werke*. Vol. 16. Ed. Max Caspar. Munich: Beck.

Kessler, Eckhard, ed. 1971. *Theoretiker humanistischer Geschichtsschreibung*. Munich: Fink Verlag.

L'Alemant, Adrien. 1557. *Hippocratis medicorum omnium principis De aere, aquis, et locis . . . commentariis quatuor illustratus*. Paris: apud Aegidium Gorbinum.

Landi, Bassiano. 1605. *Anatomia corporis humani seu de humana historia*. Frankfurt: Spiessius, Porisius.

La Popelinière, Henri Lancelot Voisin, sieur de. 1599. *L'histoire des histoires, avec l'Idée de l'histoire accomplie*. Paris: Marc Orry.

Le Clerc, Daniel, and Jean-Jacques Manget, eds. 1699. *Bibliotheca anatomica, sive recens in anatomica inventorum thesaurus locupletissimus*. 2d ed. 2 vols. Geneva: sumptibus Johan. Anthon. Chouët & Davidis Ritter.

Le Clerc, Jean. 1699. *Parrhasiana ou pensées diverses sur des matiéres de critique, d'histoire, de morale et de politique. Avec la défense de divers ouvrages de Mr. L.C. par Theodore Parrhase*. Amsterdam: chez les héritiers d'Antoine Schelte.

Le Clerc, Jean. 1712a. *Ars critica*. 4th ed. 3 vols. Amsterdam: apud Janssonio-Waesbergios.

Le Clerc, Jean. 1712b. *Oratio inauguralis, de praestantia et utilitate historiae ecclesiasticae*. Amsterdam: Schelte.

Le Clerc, Jean. 1715. "Praefatio Theodori Goralli [Joannis Clerici], in qua consilium eius aperitur et ratio interpretandi veteres traditur." In *C. Pedonis Albinovani Elegiae III et fragmenta, cum interpretatione et notis Jos. Scaligeri, Frid. Lindenbruchii, Nic. Heinsii, Theod. Goralli et aliorum*. Amsterdam: Schelte.

Le Clerc, Jean. 1991. *Epistolario*. Vol. 2, 1690–1705. Ed. Maria Grazia and Mario Sina. Florence: Olschki.

Leibniz, G. W. 1670. Preface to Mario Nizolio, *De veris principis et vera ratione philosophandi contra pseudophilosophos libri IV*. Frankfurt: Hermann a Sande.

Leoniceno, Niccolò. 1492. [*De Plinii et plurium aliorum medicorum in medicina erroribus liber*]. Ferrara: per magistrum Laurentium de Valentia et Andream de Castronovo.

Leonico Tomeo, Niccolò. 1540. *Conversio in latinum atque explanatio primi libri Aristotelis de partibus animalium*. 8vo. Venice: apud Ioannem de Farno fratres.

Libellus de natura animalium. 1958. *Libellus de natura animalium: A Fifteenth-Century Bestiary*. Ed. J. I. Davies. London: Dawson's.

Liberius, Christianus. 1681. *Bibliophilia sive de scribendis, legendis et aestimandis libris exercitatio paraenetica*. Utrecht: apud Franciscus Halma.

Linden, Jan Antonides van der. 1651. *De scriptis medicis*. Amsterdam: Blaeu.

Lipenius [Lipen], Martinus. 1679. *Bibliotheca realis medica*. Frankfurt: cura et sumptibus Johannis Friderici.

L'Obel, Mathieu de. 1576. *Plantarum seu stirpium historia . . . cui annexum est Adversariorum volumen*. Antwerp: ex officina Christophori Plantini.

Lommius, Jodocus. 1560. *Medicinalium observationum libri tres*. Antwerp: ex officina Gulielmi Sylvii.

Lucinge, René de. 1993. *La manière de lire l'histoire*. Ed. Michael Heath. Geneva: Droz.

Maccius, Sebastian. 1593. *De historia libri tres*. Venice: Dei.

Macer, Floridus Odo. 1510. *Herbarum varias qui vis cognoscere vires*. Paris: Baquelier.

Maimonides, Moses. 1579. *Aphorismi Rabi Moysis medici ex Galeno medicorum principe collecti*. Basel: ex officina Henricpetrina.

Maimonides, Moses. 1641. *De idololatria*. Amsterdam: Blaeu.

Malpighi, Marcello. 1967. "Risposta a *De recentiorum medicorum studio dissertatio epistolaris ad amicum* di Sbaraglia." In Malpighi, *Opere scelte*. Ed. Luigi Belloni. Turin: UTET.

Marchesi, Giorgio Viviano. 1726. *Vitae virorum illustrium foroliviensium*. Forlì: ex typographia P. Syluae.

Marini, Luigi Gaetano. 1784. *Degli archiatri pontifici*. 2 vols. Rome: Pagliarini.

Marschalk [Marescalcus], Nicolaus. 1517–20. *Historia aquatilium latine ac grece cum figures*. Rostock: Thurius.

Martini, Cornelius. 1605. *Metaphysica commentatio*. 12mo. Strasbourg: Ioannes Carolus.

Mascardi, Agostino. 1662. *Dell'arte historica*. Venice: per il Baba.

Mascardi, Agostino. 1859. *Dell'arte historica*. Florence: Le Monnier, 1859.

Melanchthon, Philipp. 1834–60. *Opera quae supersunt omnia*. Ed. Carl Gottlieb Bretschneider and Heinrich Ernst Bindseil. 28 vols. Halle: Schwetschke.

Menabeni, Apollonio. 1581. *Tractatus de magno animali, sive bestia et de ipsius partium in re medica facultatibus; cui adjungitur historia cervi rangiferi et gulonis filfros vocati*. Vienna: L. Nassinger.

Mercklin, Georg Abraham. 1686. *Lindenius renovatus*. Nuremberg: impensis Johannis Georgii Endteri.

Mercuriale, Girolamo. 1569. *Artis gymnasticae apud antiquos celeberrimae, nostris temporibus ignoratae, libri sex*. Venice: apud Juntas.

Mercuriale, Girolamo. 1573. *De arte gymnastica libri sex*. Venice: apud Juntas.

Mercuriale, Girolamo. 1672. *De arte gymnastica libri sex*. Amsterdam: sumptibus Andreae Frisii. (Facsimile, Ilkley: Scolar Press, n.d.)

Mexía, Pedro. 1989. *Silva de varia lección*. Ed. Antonio Castro. 2 vols. Madrid: Catedra.

Meyer, Frederick G., Emily Emmart Trueblood, and John L. Heller. 1999. *The Great Herbal of Leonhart Fuchs: De historia stirpium commentarii insignes, 1542 (Notable Commentaries on the History of Plants)*. 2 vols. Stanford: Stanford University Press.

Michael of Ephesus. 1541. *Scholia in Aristotelis libros aliquot*. 8vo. Basel: apud Bartholomeum Westhemerum.

Michael of Ephesus. 1559. *Scholia, id est brevis sed erudita atque utilis interpretatio in iv libros Aristotelis de partibus animalium, trad. Dominicus Monthesaurus*. 8vo. Basel: per Petrum Pernam.

Michelet, Jules. 1959. "Journal des idées." In Michelet, *Ecrits de jeunesse*. Paris: Gallimard.

Modio, Giovanni Battista. 1556. *Il Tevere. . . . Dove si ragiona in generale della natura di tutte le acque, e in particolare di quella del fiume di Roma*. Rome: apresso à Vincenzo Luchini.

Monceaux, François. 1605. *Aaron purgatus: sive de vitulo aureo libri duo, simul cheruborum Mosis, vitulorum Jeroboami, theraphorum Michae formam & historiam*. Arras: Riverius.

Mondella, Luigi. 1543. *Epistolae medicinales*. (Also contains *Annotationes in Antonii Musae Brasavolae simplicium medicamentorum examen*.) Basel: apud Michelem Isingrinium.

Mondella, Luigi. 1551. *Dialogi medicinales decem, in quibus multa et varia in artis theoremata, tum historiae et experimenta doctissime explicantur.* Zurich: apud Froschoverum.

Mondella, Luigi. 1568. *Theatrum Galeni, hoc est universale medicinae a medicorum principe Galeno diffuse traditae promptuarium.* Basel: per Eusebium Episcopium, et Nicolai fratris haeredes.

Mondella, Luigi. 1587. *Loci communes medicinae universae, e medicorum principe Galeno diffuse, sparsimque traditae.* Cologne: apud Ioannem Gymnicum.

Montaigne, Michel de. 1965. *Essays.* Trans. Donald Frame. Stanford: Stanford University Press.

Montaigne, Michel de. 1988. *Essais.* Ed. Pierre Villey. 3 vols. Paris: Presses Universitaires de France.

Monteux, Sébastien de. 1537. *Dialexeon medicinalium libri duo.* 4to. Lyon: apud Michaelem Parmenterium.

Morgagni, Giovanni Battista. 1719. *Adversaria anatomica omnia.* Padua: excudebat Josephus Cominus, Vulpiorum aere.

Morhof, Daniel Georg. 1732. *Polyhistor, literarius, philosophicus et practicus.* Lübeck: P. Roeckmann.

Münster, Sebastian. 1544. *Cosmographia, Beschreibung aller Lender...* Basel: H. Petri.

Nanni Giovanni. 1498. *Commentaria super opera diversorum auctorum de antiquitatibus loquentium.* Rome: Silber.

Nifo, Agostino. 1546. *Expositiones in omnes Aristotelis libros de Historia animalium libros xi. De partibus animalium et earum causis libros iv. ac de Generatione animalium libros v.* Fol. Venice: apud Hieronymum Scotum.

Nifo, Agostino. 1547. *Expositiones in Aristotelis libros Metaphysices.* Fol. Venice: apud Hieronymum Scotum.

Novalis (Friedrich von Hardenberg). 1907. "Fragmenta." In Novalis, *Schriften.* Ed. J. Minor. Vol. 3. Jena: Diderichs.

Odonus, Caesar. 1563. *Aristotelis sparsae de animalibus sententiae in continuatam seriem ad propria capita revocatae.* 4to. Bologna: apud Alexandrum Benaccium.

Origen. 1984. *Origen, Spirit and Fire: A Thematic Anthology of His Writings.* Ed. and trans. Hans Urs von Balthasar. Washington: Catholic University of America.

Ortega y Gasset, José. 1941. *Toward a Philosophy of History.* Urbana: University of Illinois Press.

Pacius, Julius. 1597. *In Porphyrii Isagogen et Aristotelis Organum commentarius analyticus.* 4to. Frankfurt: apud heredes Andreae Wecheli, Claudii Marnii et Ioannis Aubrii.

Pascal, Blaise. 2003. *Pensées: Edition établie d'après l' "ordre" pascalien.* Ed. Philippe Sellier. Paris: Pocket.

Patin, Charles. 1684. *Quod medicus debeat esse polymathes. Oratio habita in Archi-Lyceo Patavino die 3 nov. 1684*. Venice: Giovanni Francesco Valvasense.

Patrizi, Francesco. 1560. *Della historia diece dialoghi*. Venice: appresso Andrea Arrivabene.

Patrizi, Francesco. 1583. *La militia romana di Polibio, di Tito Livio e di Dionigi Alicarnasseo, da Francesco Patricii dichiarata e con varie figure illustrata*. Ferrara: per Domenico Mamarelli.

Patrizi, Francesco. 1594. *Paralleli militari*. Rome: appresso Luigi Zannetti.

Pausanias. 1935. *Description of Greece*. Trans. W. H. S. Jones. Vol. 4. Cambridge, Mass.: Harvard University Press.

Pecquet, Jean. 1654. *Experimenta nova anatomica*. 2d ed. Paris: ex officina Cramosiana.

Peiresc, Nicolas Claude Fabri de. 1888–98. *Lettres de Peiresc*. Ed. Philippe Tamizey de Larroque. 7 vols. Paris: Imprimerie Nationale.

Peiresc, Nicolas Claude Fabri de. 1891. *Correspondance de Peiresc avec plusieurs Missionaires et Religieux de l'ordre des Capucins 1631–1637*. Ed. P. Apollinaire de Valence. Paris: Picard.

Peiresc, Nicolas Claude Fabri de. 1989. *Lettres à Cassiano dal Pozzo*. Ed. Jean-François Lhote and Danielle Joyal. [Clermont-Ferrand]: Adosa.

Perizonius, Jacob. 1685. *Animadversiones historicae*. Amsterdam: H. & Vidua T. Boom.

Perizonius, Jacob. 1703. *Q. Curtius in integrum restitutus et vindicatus*. Leiden: apud Henricum Teering.

Perizonius, Jacob. 1740a. "Oratio de fide historiarum contra Pyrrhonismum historicum, dicta Lugd. Bat. postr. Non. Febr. 1702." In Perizonius, *Orationes*, vol. 12. Leiden: apud J. A. Langerak, 103–154.

Perizonius, Jacob. 1740b. *Dissertationes septem*. Leiden: Langerak.

Perlini, Gerolamo. 1610. *Historia medica physiologica, pathologica et therapeutica de curatione mulieris, cui accedit methodus compendiaria scribendi hujusmodi historias*. Rome: apud G. Facciottum.

Perlini, Gerolamo. 1613a. *Binae historiae, seu instructiones medicae ... quibus loco praefationis adiuncta est & praeposita methodus compendiarie scribendi huiusmodi historias seu instructiones*. Hanau: Typis Wechelianis, apud haeredes Joannis Aubrii.

Perlini, Gerolamo. 1613b. *Praelectiones urbanae ... super variis locis Prorrheticorum Hippocratis*. Hanau: Typ. Wechelianis imp. haeredum J. Aubrii.

Petrarca, Francesco. 1948. "On His Own Ignorance and That of Many Others." Trans. Hans Nachod. In Ernst Cassirer, Paul Oskar Kristeller, and John Herman Randall, Jr., eds., *The Renaissance Philosophy of Man*. Chicago: University of Chicago Press, 47–133.

Petroni, Alessandro Traiano. 1552. *De aqua Tiberina: opus quidem novum, sed ut omnibus qui hac aqua utuntur utile, ita & necessarium*. Rome: apud Valerium & Aloisium Doricos fratres.

Petroni, Alessandro Traiano. 1581. *De victu Romanorum et de sanitate tuenda libri quinque*. . . . Rome: in aedibus Populi Romani.

Petroni, Alessandro Traiano. 1592. *Del viver delli Romani, e di conservar la sanità . . . libri cinque dove si tratta del sito di Roma, dell'aria, de' venti, delle stagioni, dell'acque, de' vini, delle carni, de' pesci, de' frutti, delle herbe, e di tutte l'altre cose pertinenti al governo de gli huomini, e delle donne d'ogni età e conditione*. Rome: Domenico Basa.

Physiologus. 1588. *Sancti Patris Nostri Epiphanii, Episcopi Constantiæ Cypri, ad Physiologum. Eiusdem in die festo Palmarum sermo*. Ed. Gonzalo Ponce de Leon. Antwerp: ex officina Christophori Plantini, architypographi regii.

Physiologus. 1979. *Physiologus*. Trans. Michael J. Curley. Austin: University of Texas Press.

Piccolomini, Aeneas Sylvius. 2002. "De liberorum educatione." Trans. Craig Kallendorf. In Craig Kallendorf, ed., *Humanist Educational Treatises*. Cambridge, Mass.: Harvard University Press, 126–259.

Pietro d'Abano. 1482. *Expositio problematum Aristotelis*. Venice: Johannes Herbort.

Pignoria, Lorenzo. 1628. *Symbolarum epistolicarum liber primus*, Padua: I. B. de Martinis.

Pignoria, Lorenzo. 1647. *Annotationi . . . al libro delle imagini del Cartari*. In Cartari 1647.

Pignoria, Lorenzo. 1669. *Magnae Deum matris Idaeae et Attidis initia. Ex vetustis monumentis nuper Tornaci Nerviorum erutis*. Amsterdam: Frisius.

Pignoria, Lorenzo. 1670. *Mensa Isiaca, qua sacrorum apud Aegyptos ratio & simulacra subjectis tabulis aeneis simul exhibentur & explicantur. Accessit ejusdem authoris de Magna Deum Matre discursus, et sigillorum, gemmarum, amuletorum aliquot figurae & earumdem . . . interpretatio*. Amsterdam: Andreas Frisius.

Pliny the Elder. 1549. *C. Plinii Secundi historiæ mundi libri XXXVII*. Ed. Sigmund Gelen. Basel: Froben.

Pliny the Younger. 1969. *Letters and Panegyricus*. Trans. Betty Radice. 2 vols. Cambridge, Mass.: Harvard University Press.

Poliziano, Angelo. 1971. *Opera omnia*. Ed. Ida Maier. 3 vols. Turin: Bottega d'Erasmo. (Facsimile of 1573 Basel ed.)

Pomey, François Antoine. 1659. *Pantheon mythicum, seu fabulosa deorum historia*. Lyon: Molin.

Porcacchi, Thomaso. 1565. *Il primo volume delle cagioni delle guerre antiche*. Venice: Gabriel Giolito.

Quintilian. 1989. *Institutio oratoria*. Trans. H. E. Butler. 4 vols. Cambridge: Harvard University Press.

Rabelais, François. 1532. *Les horribles et espoventables faictz et prouesses du très renommé Pantagruel, roy des Dipsodes, filz du grand géant Gargantua, composez nouvellement par Maistre Alcofrybas Nasier*. Lyon: C. Nourry.

Rabelais, François. 1990. *Gargantua and Pantagruel*. Trans. Burton Raffel. New York: W. W. Norton.

Raimundus Sibiuda [Raymond Sebonde]. 1540. *Liber creaturarum: Theologia naturalis, sive liber creaturarum specialiter de homine*. . . . Leiden.

Raimundus Sibiuda. 1648. *Theologia naturalis, sive Liber creaturarum*. Leiden: sumptibus Petri Compagnon, via Mercatoria, sub signo Cordis boni.

Ramus, Petrus. 1583. *Scholarum metaphysicarum, libri xiv in totidem Metaphysices libros Aristotelis*. Ed. Joannes Piscator. 8vo. Frankfurt: apud haeredes Andreae Wecheli.

Ray, John. 1686–1704. *Historia plantarum; species hactenus editas aliasque insuper multas noviter inventas & descriptas complectens: In qua agitur primo de plantis in genere, earumque partibus, accidentibus & differentiis*. 3 vols. London: typis M. Clark, prostant apud H. Faithorne.

Ray, John. 1691. *The Wisdom of God Manifested in the Works of the Creation*. London: printed for Samuel Smith, at the Princes Arms in St. Pauls Church-Yard.

Ray, John. 1710. *Historia insectorum: . . . opus posthumum jussu Regiae Societatis Londinensis editum; cui subjungitur appendix De scarabeis Britannicis, autore M. Lister*. London: impensis A. & J. Churchill.

Reineck, Reiner. 1583. *Ad Reinerum Reineccium liber epistolarum historicarum, seu de editionibus et operis eius historicis per ann. xvi. scriptarum*. Helmstedt: Lucius.

Relatio. 1644. *Relatio facta in Consistorio Secreto coram Sanctissimo D. N. Gregorio Papa XV a Francisco Maria Episcopo Portuensi Sanctae Romanae Ecclesiae Cardinalis a Monte die XIX Ianuarii MDCXXII super vita, sanctitate, actis canonizatione, et miraculis Beati Ignatii, fundatoris Societatis Iesu*. Paris: Typographia Regia.

Reusner, Nicolaus. 1589. *Icones aliquot clarorum virorum Germaniae, Angliae, Galliae, Ungariae*. Basel: Conr. Valdkirch.

Reychersdorff, Georgius a. 1550. *Chorographiae Transylvaniae . . . descriptio*. Vienna: Aquila.

Reyes Franco, Gaspar de los. 1661. *Elysius jucundarum quaestionum campus*. Brussels: F. Vivien.

Rhaedus, Thomas. 1609. *De accidente proprio theoremata philosophica, quibus essentia breviter declarata quaestiones etiam dubiae enodantur, et rationes philosophicae contra propriorum communicationem a Bellarmino, Keckermanno, Goclenio, Timplero aliisque excogitatae diluuntur*. 4to. Rostock: literis Reusnerianis.

Riccobono, Antonio. 1995. *De poetica*. In *Aristoteles Latine interpretibus variis*. Ed. E. Kessler. Berlin: Wilhelm Finck.

Roberti, Gaudenzio. 1691–92. *Miscellanea italica erudita*. 4 vols. Parma: typis Iosephi ab Oleo, & Hippolyti Rosati.

Robortello, Francesco. 1968. *In librum Aristotelis de arte poetica explicationes. Paraphrasis in librum Horatii, qui vulgo de arte poetica ad Pisones inscribitur*. Munich: W. Fink. (Rpt. of Florence: Torrentinus, 1548.)

Robortello, Francesco. 1975. *De arte sive ratione corrigendi antiquorum libros disputatio.* Ed. and trans. G. Pompella. Naples: L. Loffredo.

Rondelet, Guillaume. 1554–55. *Liber de piscibus marinis.* 2 vols. Fol. Lyon: apud Matthiam Bonhomme.

Rondelet, Guillaume. 1558. *La première partie [La seconde partie] de l'histoire entière des poissons, composée premièrement en Latin . . . , maintenant traduite en François.* Lyon: M. Bonhomme.

Rubeus, Ioannes Baptista. 1618. *Commentaria et quaestiones in universam Aristotelis Metaphysicam.* 4to. Venice: apud Ioannem Guerilium.

Ruland, Martin, the Elder. 1556. *De lingua graeca.* Zurich: apud Andream Gesnerum f. et Jacobum Gesnerum.

Ruland, Martin, the Elder. 1567. *Synonyma graeca: copia Graecorum verborum omnium.* Augsburg: Francus.

Ruland, Martin, the Elder. 1580. *Curationum empiricarum et historicarum in certis locis et notis hominibus optimè, riteque probatarum & expertarum centuria.* Basel: Henricpetri.

Ruland, Martin, the Elder. 1612. *Lexicon Alchymiae.* Frankfurt: Palthen.

Ruremundanus, Franciscus Fabricius. 1564. *De balneorum naturalium, praecipue eorum que sunt aquis grani et porceti, natura et facultatibus, et qua rationi illis utendum sit, libellus perutilis.* Cologne: Cholinus.

Saccardini, Costantino. 1621. *Libro nomato la verità di diverse cose, quale minutamente tratta di molte salutifere operationi spagiriche, et chimiche.* Bologna: Gio. Paolo Moscatelli.

Salviani, Ippolito. 1557. *Aquatilium animalium historiae, liber primus, cum eorumdem formis, aere excusis.* Rome: I. Salviani.

Sardi, Alessandro. 1577. *De moribus ac ritibus gentium libri III.* Mainz: per Franciscus Behem. Typis Godefridi Kempensis.

Savonarola, Michele. 1505. *De gotta la preservatione e cura per lo preclaro medico maestro Michel Savonarola ordinata. . . .* Pavia: [Jacob dal Borgofrancho].

Savonarola, Michele. 1559a. *De balneis et thermis naturalibus omnibus Italiae.* In Tommaso Giunta, ed., *De balneis omnia quae extant. . . .* Venice: apud Juntas.

Savonarola, Michele. 1559b. *Practica major.* Venice: apud Juntas.

Savonarola, Michele. 1563. *Practica canonica de febribus.* Venice: apud Juntas. (With *De vermibus, De pulsibus, urinis et egestionibus, De balneis.*)

Savonarola, Michele. 1902. *Libellus de magnificis ornamentis regie civitatis Paduae.* Ed. Arnaldo Segarizzi. In *Rerum Italicarum Scriptores*, vol. 24, parte 15. Città di Castello: S. Lapi.

Savonarola, Michele. 1952. *De regimine pregnantium et noviter natorum usque ad septennium.* In *Il trattato ginecologico-pediatrico in volgare 'Ad mulieres ferrarienses de regimine pregnantium' di Michele Savonarola.* Ed. Luigi Belloni. Milan: XLII Congresso della Società Italiana di Ostetricia e Ginecologia.

Savonarola, Michele. 1953. *Libellus de aqua ardenti, de preservatione a peste et eius cura*, ed. In *I trattati in volgare della peste e dell'acqua ardente*. Ed. Luigi Belloni. Milan: LIV Congresso nazionale della Società di Medicina Interna.

Savonarola, Michele. 1954. *De cura languoris animi ex morbo venientis*. Ed. Cesare Menini. Ferrara: Istituto di Storia della Medicina dell'Università di Ferrara.

Savonarola, Michele. 1982. *Libreto de tute le cosse che se manzano, un libro di dietetica di Michele Savonarola, medico padovano del secolo XV*. Ed. Jane Nystedt. Tesi di dottorato di ricerca, Istituto di lingue romanze. Stockholm: University of Stockholm, Gotab.

Savonarola, Michele. 1988. *Libreto de tutte le cosse che se magnano, un'opera dietetica del secolo XV*. Ed. Jane Nystedt. Stockholm: Alqvist & Wiksell International.

Savonarola, Michele. 1992. *De nuptiis Batibecho et Serabocha*. Ed. Paola Biamini. In Biamini, "Peccati di lingua alla corte estense. Il 'De Nuptiis Batibecho et Serabocha' di Michele Savonarola." *Schifanoia* 11:101–179.

Savonarola, Michele. 1996. *Del felice progresso di Borso d'Este*. Ed. Maria Aurelia Mastronardi. Bari: Palomar Ed.

Scaliger, Julius Caesar. 1557. *Exotericarum exercitationum liber XV. De subtilitate, ad Hieronymum Cardanum*. Paris: Vascosanus.

Scaliger, Julius Caesar. 1592. *Exotericarum exercitationum liber XV. De subtilitate, ad Hieronymum Cardanum*. 8vo. Frankfurt: apud Andreae Wecheli haeredes, Claudium Marnium et Joannem Aubrium.

Scaynus, Antonius. 1599. *Paraphrasis cum adnotationibus in libros Aristotelis de anima . . . de motione animalium*. Fol. Venice, [Padua: Laurentius Pasquatus].

Schaller, Georg. 1579. *Thierbuch, sehr künstliche und wol gerissene Figuren, von allerley Thieren, durch die Weitberhümpten Jost Amman und Hans Boksperger, sampt einer Beschreibung ihrer Art, Natur und Eigenschafft, auch kurtzweiliger Historien*. Frankfurt: S. Feyerabend.

Schegk, Jakob. 1556. *Perfecta et absoluta definiendi ars ab Aristotele tractata et exposita sexto topicorum, qui est de definitione, et septimo, qui est de eodem et uno, libris*. 4to. Tübingen: apud viduae Ulrici Morhardi.

Schegk, Jakob. 1584. *Commentaria in viii libros Topicorum Aristotelis*. 8vo. [Heidelberg]: apud Ioannem Mareschallum.

Schenck von Grafenberg, Johann. 1584. *Observationes medicae de capite humano.* . . . Basel: ex officina Frobeniana.

Schenck von Grafenberg, Johann. 1584–97. *Observationes medicae, rarae, novae, admirabiles et monstrosae*. Basel: Froben; Freiburg: ex officina Martini Beckleri.

Schenck von Grafenberg, Johann. 1609. *Paratereseon, sive Observationum medicarum rararum, novarum, admirabilium et monstrosarum volumen in tomis septem*. Frankfurt: e typographeo Nicolai Hoffmanni, impensa Jonae Rhodii.

Selden, John. 1668. *De diis Syris*. Leipzig: Laurentius S. Cörner.

Selvatico, Giovan Battista. 1605. *Galeni historiae medicinales.* Hanau: Typis Wechelianis, apud Claudium Marnium & haeredes Johannis Aubrii.

Settala, Luigi. 1590. *In librum Hippocratis Coi De aeribus, aquis, locis commentarii V.* Cologne: Joan. Baptistae Ciotti aere.

Smetius, Henricus a Leda. 1611. *Miscellanea medica.* Frankfurt: Rhodius.

Speroni, Sperone. 1740. *Opere.* 5 vols. Venice: Domenico Occhi.

Tanner, Adam. 1621. *Dissertatio peripatetico-theologica de coelis.* Ingolstadt: ex typographeo Gregorii Haenlin.

Tate, Francis. 1771. "Of the Antiquity, Variety and Ceremonies of Funerals in England." In T. Hearne, ed., *A Collection of Curious Discourses by Eminent Antiquarians upon Several Heads in Our English Antiquities.* 2 vols. London: W. and J. Richardson.

Taurellus, Nicolaus. 1596. *Synopsis Aristotelis Metaphysices: ad normam Christianae religionis explicatae et completae.* 8vo. Hanau: apud Guilelmum Antonium.

Taurellus, Nicolaus. 1597. *Alpes caesae, hoc est, Andreae Caesalpini Itali, monstrosa et superba dogmata, discussa et excussa.* 8vo. Frankfurt: apud Zachariam Palthenium.

Tenison, Thomas. 1678. *Of Idolatry: A Discourse in which is Endeavoured a Declaration of, its Distinction from Superstition; its Notion, Cause, Commencement, and Progress.* London: Tyton.

Textor, Benedictus. 1534. *Stirpium differentiae ex Dioscoride secundum locos communes, opus ad ipsarum plantarum cognitionem admodum conducibile.* 12 mo. Paris: apud Simonem Colinaeum.

Theophrastus. 1483. *[Historia plantarum; De causis plantarum.]* Ed. Giorgio Merula, trans. Theodore Gaza. Treviso: per Bartholomaeum Confalonerium de Salodio.

Thomas à Kempis. 1941. *The Imitation of Christ, from the First Edition of an English Translation Made c. 1530.* Ed. Edward J. Klein, trans. Richard Whitford. New York: Harper & Brothers.

Thomas Aquinas. 1587. *In tres libros Aristotelis de anima praeclarissima expositio.* Venice: apud haeredem Hieronymi Scoti.

Toland, John. 1696. *Christianity Not Mysterious: or, A treatise Shewing, that There Is Nothing in the Gospel Contrary to Reason, nor above it: and that no Christian Doctrine Can Be Properly Call'd a Mystery.* London: Sam. Buckley.

Tommasini, Giacomo Filippo. 1669. *De vita, bibliotheca et museo Laurentii Pignorii . . . dissertatio.* In Pignoria 1669, 61–94.

Tonjola, Johann. 1661. *Basilea sepulta retecta continuata.* Basel: E. König.

Topsell, Edward. 1607. *The Historie of Foure-Footed Beastes: Describing the True and Lively Figure of Every Beast, with a Discourse of Their Severall Names, Conditions, Kindes, Vertues (Both Naturall and Medicinall) Countries of Their Breed, Their Love and Hate to Mankind, and the Wonderfull Worke of God in*

Their Creation, Preservation, and Destruction. London: printed by William Iaggard.

Turgot, Anne-Robert-Jacques. 1973. *Turgot on Progress, Sociology, and Economics.* Trans. Ronald L. Meek. Cambridge: Cambridge University Press.

Turner, William. 1544. *Avium praecipuarum, quarum apud Plinium et Aristotelem mentio est, brevis et succinta historia.* Cologne: J. Gymnich.

Universal History. 1779. *An Universal History from the Earliest Accounts to the Present Times.* 23 vols. London: C. Bathurst.

Valencia, Gregor de. 1580. *De idolatria contra sectariorum contumelias disputatio.* Ingolstadt: D. Sartorius.

Valla, Giorgio. 1501. *De expetendis et fugiendis rebus opus.* Venice: in aedibus Aldi Romani, impensa, ac studio Ioannis Petri Vallae filii.

Valla, Lorenzo. 1962. *De rebus a Ferdinando Hispaniorum rege et maioribus eius gestis.* In Valla, *Opera omnia.* Ed. E. Garin. 2 vols. Turin: Bottega d'Erasmo.

Valla, Lorenzo. 1973. *Gesta Ferdinandi regis Aragonum.* Ed. Ottavio Besomi. Padua: Antenore.

Valleriola, François. 1577. *Commentarii in librum Galeni De constitutione artis medicae.* 8vo. Lyon: apud Carolum Pesnot.

Valleriola, Francois. 1605. *Observationum medicinalium libri vi.* Lyon: apud Franciscum Fabrum.

Vallés, Francisco. 1589. *In libros Hippocratis de morbis popularibus commentaria.* Turin: apud haeredem Nicolai Bevilaquae.

Valverde, Juan de. 1556. *Historia de la composicion del cuerpo humano.* Rome: Antonio Salamanca y Antonio Lofrerij.

Varthema, Ludovico. 1991. *Itinerario dallo Egypto alla India.* Ed. Enrico Musacchio. Bologna: Leopoldo Fusconi editore.

Velasco de Tarenta. 1490. *Practica quae alias Philonium dicitur.* Lyon: apud haeredem Nicolai Bevilaquae.

Vergerio, Pier Paolo. 2002. "De ingenuis moribus et liberalibus adulescentiae studiis liber." Trans. Craig Kallendorf. In Kallendorf, ed., *Humanist Educational Treatises.* Cambridge, Mass.: Harvard University Press, 2–91.

Vergil, Polydore. 2002. *On Discovery.* Ed. and trans. Brian P. Copenhaver. Cambridge, Mass.: Harvard University Press.

Verville, Béroalde de. 1600. *L'histoire des vers qui filent la soye.* Tours: M. Sifleau.

Vesalius, Andreas. 1555. *De humani corporis fabrica libri septem.* Basel: Oporinus.

Vico, Giambattista. 1948. *The New Science.* Trans. T. G. Bergin and M. H. Fisch. Ithaca: Cornell University Press.

Vico, Giambattista. 1968. *The New Science of Giambattista Vico.* Rev. trans. Thomas Goddard Bergin and Max Harold Fisch. Ithaca: Cornell University Press.

Vincent of Beauvais. 1965. *Speculum quadruplex, sive speculum majus: naturale, doctrinale, morale, historiale*. Graz: Akademische Druck-und-Verlagsanstalt. (Facsimile edition of Douai: ex officina typographica Balthazaris Belleri, 1624.) *Generalis prologus*, ch. 16, vol. 1.

Viperano, Giovanni Antonio. *De scribenda historia*. 1569. In Wolf 1579.

Vittori, Angelo. 1613. *Medica disputatio. De palpitatione cordis, fractura costarum, aliisque affectionibus B. Philippi Nerii Congregationis Oratorii Romani fundatoris. Qua ostenditur praedictas affectiones fuisse super naturam*. Rome: ex typographia Camerae Apostolicae.

Vittori, Angelo. 1640. *Consultationes post obitum auctoris in lucem editae a Vincentio Mannuccio Perusino Practicae medicinae, et mathematicarum professore, primario, et iubilato in Perusino Gymnasio*. Rome: ex officina typographica Caballina.

Vossius, Gerhard Johannes. 1623. *Ars historica sive de historiae et historices natura*. Leiden: Joannes Maire.

Vossius, Gerhard Johannes. 1668. *De theologia gentili et physiologia christiana: sive de origine et progressu idololatriae*. Amsterdam: Blaeu.

Vossius, Gerhard Johannes. 1699. "Ars historica." In *Opera*. Amsterdam, 4:1–48.

La voye de laict, ou, Le chemin des heros au palais de la gloire: ouvert a l'entrée triomphante de Louys XIII. Roy de France & de Nauarre en la cité d' Auignon le 16. de Nouembre 1622. 1623. Avignon.

Werdmüller, Otto. 1555. *Similitudinum ab omni animalium genere desumptarum libri vii*. 8vo. Zurich: per Andream Gessnerum et Iacobum Gessnerum.

Werner, Georg. 1557. *De admirandis Hvngariæ aqvis, hypomnemation: Georgio Vuernhero authore*. Fols. 179–198 in *Rerum Moscoviticarvm commentarij, Sigismundo Libero authore. Rvssiæ breuissima descriptio, & de religione eorum varia inserta sunt. Chorographia totius imperij Moscici, & vicinorum quorundam mentio*. Antwerp: In ædibus Ioannis Steelsij [Typis Ioannis Latij].

Wheare, Degory. 1684. *Relectiones hyemales de ratione et methodo legendi utrasque historias, civiles et ecclesiasticas*. Cambridge: ex officina Joan. Hayes.

Wilkins, John. 1649. *A Discourse Concerning the Beauty of Providence in All the Rugged Passages of It Very Seasonable to Quiet and Support the Heart in These Times of Publick Confusion*. London: printed for Sa. Gellibrand.

Wilkins, John. 1675. *Of the Principles and Duties of Natural Religion: Two Books*. London: printed by A. Maxwell for T. Basset, H. Brome, R. Chiswell.

Willughby, Francis. 1686. *De historia piscium libri quatuor*. Ed. John Ray. Oxford: e Theatro Sheldoniano.

[Wolf, Johannes, ed.] 1576. *Io. Bodini Methodus historica: duodecim eiusdem argumenti scriptorum, tam veterum quam recentiorum, commentariis adaucta*. Basel: Petrus Perna.

Wolf, Johannes, ed. 1579. *Artis historicae penus octodecim scriptorum tam veterum tam recentiorum monumentis*. 3 vols. Basel: Petrus Perna.

Wotton, Edward. 1552. *De differentiis animalium libri decem.* Fol. Paris: apud [Michaelem] Vascosan.

Zabarella, Jacopo. 1586. *Liber de naturalis scientiae constitutione.* Venice: Meietus.

Zabarella, Jacopo. 1586–87. *Opera logica.* Ed. Julius Pacius. Fol. [Heidelberg]: apud Joannem Mareschallum [excudebat Joannes Wechelus].

Zacchia, Paolo. 1651. *Quaestiones medico-legales in quibus eae materiae medicae, quae ad legales facultates videntur pertinere, propununtur, pertractantur, resolvuntur. . . . Editio tertia, correctior, auctiorque, non solum variis passim locis, rerum et subjunctis quae nunc recens prodeunt, partibus octava et nona.* Amsterdam: ex typographo Joannis Blaeu. Book 9, *Consilia et responsa XXXIV ad materias medico-legales pertinentia,* 657–731.

Zacutus Lusitanus, Abraham. 1636–42. *De medicorum principuum historia, . . . ubi medicinales omnes historiae, de morbis internis, quae passim apud principes medicos occurrunt, concinno ordine disponuntur, paraphrasi et commentariis illustrantur.* 2 vols. Amsterdam: Laurentius.

Zwinger, Theodor. 1565. *Theatrum vitae humanae . . . cum gemino indice.* Basel: J. Oporinus and Froben fratres.

Zwinger, Theodor, ed. 1566. *Aristotelis De moribus ad Nicomachum.* Basel: J. Oporinus and Eusebius Episcopius.

Zwinger, Theodor, ed. 1582. *Ethicorum Nicomachiorum libri decem.* Basel: Episcopius.

Zwinger, Theodor. 1586. *Theatrum humanae vitae . . . cum tergemino elencho.* Basel: Eusebius Episcopius.

Secondary Literature Cited

Accordi, Bruno. 1980. "Michele Mercati (1541–1593) e la Metallotheca." *Geologica romana* 19:1–50.

Ago, Renata. 2004. "'Così si volta questa ruota di parole': biblioteche e lettori nella Roma del Seicento." *Quaderni storici* 39.115:119–138.

Agrimi, Jole. 1984. Review of Jane Nystedt, ed., *Libreto de tute le cosse che se manzano, un libro di dietetica di Michele Savonarola, medico padovano del secolo XV. Aevum* 58:358–365.

Agrimi, Jole. 1998. "Aforismi, parabole, esempi. Forme di scrittura della medicina operativa: il modello di Arnaldo da Villanova." In Massimo Galuzzi et al., eds., *Le forme della comunicazione scientifica.* Milan: Franco Angeli, 361–392.

Agrimi, Jole. 2002. *'Ingeniosa scientia nature.' Studi sulla fisiognomica medievale.* Florence: Edizioni del Galluzzo-SISMEL.

Agrimi, Jole, and Chiara Crisciani. 1988. *Edocere medicos: medicina scolastica nei secoli XIII–XV.* Milan: Guerini.

Agrimi, Jole, and Chiara Crisciani. 1990. "Per una ricerca su experimentum-experimenta: riflessione epistemologica e tradizione medica (secc XIII–XV)." In Pietro Janni and Innocenzo Mazzini, eds., *Presenza del lessico greco e latino nelle lingue contemporanee*. Macerata: Pubblicazioni della Facoltà di Lettere e Filosofia dell'Università degli Studi di Macerata, vol. 55, 9–49.

Agrimi, Jole, and Chiara Crisciani. 1994a. *Les 'consilia' médicaux*. Typologie des sources du Moyen Age Occidental, vol. 69. Turnhout: Brepols.

Agrimi, Jole, and Chiara Crisciani. 1994b. "The Science and Practice of Medicine in the Thirteeenth Century According to Guglielmo da Saliceto, Italian Surgeon." In Luis Garcia Ballester et al., eds., *Practical Medicine from Salerno to the Black Death*. Cambridge: Cambridge University Press, 60–87.

Albanese, Gabriella, Lucia Battaglia Ricci, and Rosella Bessi, eds. 2000. *Favole, parabole, istorie. Le forme della scrittura novellistica dal medioevo al Rinascimento*. Rome: Salerno Editrice.

Albright, William F. 1949. *Von der Steinzeit zum Christentum*. Bern: Dalp.

Allen, Don Cameron. 1970. *Mysteriously Meant: The Rediscovery of Pagan Symbolism and Allegorical Interpretation in the Renaissance*. Baltimore: Johns Hopkins University Press.

Allut, P. 1859. *Etude biographique et bibliographique sur Symphorien Champier*. Lyon: Scheuring.

Alpers, Svetlana. 1960. "Ekphrasis and Aesthetic Attitudes in Vasari's *Lives*." *Journal of the Warburg and Courtauld Institutes* 23:190–215.

Alpers, Svetlana. 1983. *The Art of Describing*. Chicago: University of Chicago Press.

Altieri Biagi, Maria Luisa. 1984. "Forme della comunicazione scientifica." In Alberto Asor Rosa, ed., *La letteratura italiana. III, Le forme del testo. II, La prosa*. Turin: Einaudi, 891–947.

Das Amerbach Kabinett. 1991. 5 vols. Basel: Öffentliche Kunstsammlung Basel.

Anselmi, Gian Mario. 1992. *Il tempo ritrovato. Padania e Umanesimo tra erudizione e storiografia*. Modena: Mucchi.

Anselmi, Gian Mario, Luisa Avellini, and Ezio Raimondi. 1988. "Il Rinascimento padano." In Alberto Asor Rosa, ed., *La letteratura italiana. Storia e geografia. II, L'età moderna*. Turin: Einaudi, 521–591.

Antoni, Carlo. 1951. *Der Kampf wider die Vernunft. Zur Entstehungsgeschichte des deutschen Freiheitsgedankens*. Trans. Walter Goetz. Stuttgart: Koehler.

Antoni, Carlo. 1968. *La lotta contro la ragione*. Florence: Sansoni. (Originally published 1942.)

Antonovics, A. V. 1972. "Counter-Reformation Cardinals: 1534–90." *European Studies Review* 2:301–327.

Arbusow, Leonid. 1951. *Liturgie und Geschichtsschreibung im Mittelalter; in ihren Beziehungen erläutert an den Schriften Ottos von Freising 1158, Heinrichs Livlandchronik, 1227, und den anderen Missionsgeschichten des bremischen Erzsprengels: Rimberts, Adams von Bremen, Helmolds*. Bonn: L. Röhrscheid.

Arnaldi, Girolamo. 1998. "Cronache con documenti, cronache 'autentiche' e pubblica storiografia." In Giuliana Albini, ed., *La scrittura del Comune: amministrazione e memoria nelle città dei secoli XII e XIII.* Turin: Scriptorium, 121–140.

Ashworth, William B., Jr. 1990. "Natural History and the Emblematic World View." In David C. Lindberg and Robert S. Westman, eds., *Reappraisals of the Scientific Revolution.* Cambridge: Cambridge University Press, 303–332.

Ashworth, W. B., Jr. 1993. "The Map of the Moon of Gassendi, Peiresc and Mellan." *Annales de Haute-Provence* 113:323–324.

Ashworth, William B., Jr. 1996. "Emblematic Natural History of the Renaissance." In Jardine, Secord, and Spary 1996, 17–37.

Assmann, Jan. 1998. *Moses der Ägypter.* Munich: Hanser.

Athanassiadi, Polymnia. 1992. *Julian: An Intellectual Biography.* Oxford: Routledge.

Atkinson, John. 2000. "Originality and Its Limits in the Alexander Sources of the Early Empire." In Bosworth and Baynham 2000, 307–325.

Atran, Scott. 1990. *Cognitive Foundations of Natural History: Towards an Anthropology of Science.* Cambridge: Cambridge University Press.

Aufrère, Sydney H. 1990. *La momie et la tempête. Nicolas-Claude Fabri de Peiresc et la "curiosité egyptienne" en Provence au début du XVIIe siècle.* Avignon: Barthélemy.

Auzzas, Ginetta, Giovanni Baffetti, and Carlo Delcorno, eds. 2003. *Letteratura in forma di sermone: i rapporti tra predicazione e letteratura nei secoli XIII–XVI.* Florence: Olschki.

Backus, Irena. 2003. *Historical Method and Confessional Identity in the Era of the Reformation (1378–1615).* Leiden: Brill.

Badia, Lola. 1996. *Textos catalans tardomedieval i 'ciencia de natures,'* Barcelona: Reial Academia de Bonas Lletras de Barcelona.

Balme, D. M. 1987a. "Aristotle's Biology Was Not Essentialist." In Allan Gotthelf and James G. Lennox, eds., *Philosophical Issues in Aristotle's Biology.* Cambridge: Cambridge University Press, 291–312.

Balme, D. M. 1987b. "Aristotle's Use of Division and Differentiae." In Allan Gotthelf and James G. Lennox, eds., *Philosophical Issues in Aristotle's Biology.* Cambridge: Cambridge University Press, 69–89.

Barkan, Leonard. 1999. *Unearthing the Past: Archeology and Aesthetics in the Making of Renaissance Culture.* New Haven: Yale University Press.

Barnes, Annie. 1938. *Jean Le Clerc (1657–1736) et la république des lettres.* Paris: Droz.

Baroncini, Gabriele. 1992. *Forme di esperienza e rivoluzione scientifica.* Florence: Olschki.

Barret-Kriegel, Blandine. 1988. *Les historiens et la monarchie.* 4 vols. Paris: Presses Universitaires de France.

Barrow, John D., and Frank J. Tipler. 1986. *The Anthropic Cosmological Principle*. Oxford: Clarendon Press.

Battafarano, Italo Michele. 1991. *Tommaso Garzoni: Polyhistorismus und Interkulturalität in der frühen Neuzeit*. Bern: Lang.

Baud, Jean-Pierre. 1983. *Le procès de l'alchimie. Introduction à la légalité scientifique*. Strasbourg: Cerdic.

Bauer, Barbara. 2000. "Der Polyhistor physicus, ein Spiegel der naturwissenschaftlichen und wissenschaftstheoretischen Diskussionen auf der Schwell zum Zeitalter Newtons." In Waquet 2000, 179–220.

Bäumer, Änne. 1990a. "Edward Wotton: aristotelische Zoologie aristotelischer als bei Aristoteles." In Klaus Döring and Georg Wöhrle, eds., *Vorträge des ersten Symposions des Bamberger Arbeitskreises "Antike Naturwissenschaft und ihre Rezeption."* Wiesbaden: Harrassowitz, 187–211.

Baümer, Änne. 1990b. "Das erste zoologische Kompendium in der Zeit der Renaissance: Edward Wottons schrift Über die Differenzen der Tiere." *Berichte zur Wissenschaftsgeschichte* 13:13–29.

Baxandall, Michael. 1963. "A Dialogue on Art from the Court of Leonello d'Este." *Journal of the Warburg and Courtauld Institutes* 26:304–326.

Baxandall, Michael. 1988. *Painting and Experience in Fifteenth-Century Italy*. Oxford: Oxford University Press.

Beagon, Mary. 1992. *Roman Nature: The Thought of Pliny the Elder*. Oxford: Clarendon Press.

Bedos-Rezak, Brigitte. 2000. "Medieval Identity: A Sign and a Concept." *American Historical Review* 105:1489–1533.

Bellini, Eraldo. 2002. *Agostino Mascardi tra ars poetica e ars historica*. Milan: Vita e pensiero.

Bellomo, Manlio. 2000. *I fatti e il diritto, tra le certezze e i dubbi dei giuristi medievali (secoli XIII–XIV)*. Rome: Il cigno Galileo Galilei.

Bentley, Jerry. 1978. "Erasmus, Jean le Clerc, and the Principle of the Harder Reading." *Renaissance Quarterly* 31:309–321.

Benzoni, Andrea. 1907. "Un giudizio di Pietro Tomasi" (with edition of the *Consilium*). *L'Ateneo veneto* 30:24–40.

Berchtold, Alfred. 1990. *Bâle et l'Europe*. 2 vols. Lausanne: Payot.

Berlioz, Jacques. 1980. "Le récit efficace: l'exemplum au service de la prédication (XIIIe–XVe siècles)." *Mélanges de l'Ecole française de Rome. Moyen Age* 92:113–146.

Berlioz, Jacques, and Marie Anne Polo de Beaulieu, eds. 1998. *Les exempla médiévaux: nouvelles perspectives*. Paris: Honoré Champion.

Bernhardt, Jean. 1981. "Les activités scientifiques de Fabri de Peiresc." *Nouvelles de la république des lettres* 1:165–184.

Bertelli, Sergio. 1973. *Ribelli, libertini e ortodossi nella storiografia barocca*. Florence: La Nuova Italia.

Bertoni, Giulio. 1903. *La Biblioteca Estense e la coltura ferrarese ai tempi di Ercole I*. Turin: Loescher.

Bertoni, Giulio. 1921. *Guarino da Verona fra letterati e cortigiani a Ferrara*. Geneva and Florence: Olschki.

Bertozzi, Marco, ed. 1994. *Alla corte degli Estensi. Filosofia, arte e cultura a Ferrara nei secoli XV e XVI*. Ferrara: Università degli Studi.

Bettini, Maurizio. 1998. *Nascere: storie di donne, donnole, madri ed eroi*. Turin: Einaudi.

Bezold, Friedrich von. 1918. "Zur Entstehungsgeschichte der historischen Methodik." In Bezold, *Aus Mittelalter und Renaissance*. Munich: Oldenbourg.

Bietenholz, Peter G. 1994. *Historia and Fabula: Myths and Legends in Historical Thought from Antiquity to the Modern Age*. Leiden: Brill.

Bigourdan, G. 1918. *Histoire de l'astronomie d'observation et des observatoires en France. Première partie. De l'origine à la fondation de l'observatoire de Paris*. Paris: Gauthier-Villars.

Biographie médicale. 1855. *Biographie médicale par ordre chronologique: d'après Daniel Leclerc, Eloy, etc*. Ed. and rev. A. L. J. Bayle and Auguste Thillaye. Paris: A. Delahays.

Biondi, Albano. 1984. "Tempi e forme della storiografia." In Alberto Asor Rosa, ed., *La letteratura italiana*. III, *Le forme del testo*. II, *La prosa*. Turin: Einaudi, 1075–1099.

Biow, Douglas. 2002. *Doctors, Ambassadors, Secretaries: Humanism and Professions in Renaissance Italy*. Chicago: University of Chicago Press.

Bizzocchi, Roberto. 1995. *Genealogie incredibili: scritti di storia nella Europa moderna*. Bologna: Società editrice il Mulino.

Black, Robert. 1987. "The New Laws of History." *Renaissance Studies* 1:126–156.

Blair, Ann. 1992. "Humanist Methods in Natural Philosophy: The Commonplace Book." *Journal of the History of Ideas* 53:541–551.

Blair, Ann. 1996. "Bibliothèques portables: les recueils des lieux communs à la Renaissance tardive." In M. Baratin and Ch. Jacob, eds., *Le pouvoir des bibliothèques. La memoire des livres en Occident*. Paris: A. Michel, 84–106.

Blair, Ann. 1997. *The Theater of Nature: Jean Bodin and Renaissance Science*. Princeton: Princeton University Press.

Blair, Ann. 1999. "The *Problemata* as a Natural Philosophical Genre." In Grafton and Siraisi 1999, 171–204.

Blair, Ann. 2003. "Reading Strategies for Coping with Information Overload ca. 1550–1700." *Journal of the History of Ideas* 64:11–28.

Blair, Ann, and Anthony Grafton. 1992. "Reassessing Humanism and Science." *Journal of the History of Ideas* 53:535–540.

Blanke, Horst Walther. 1991. *Historiographiegeschichte als Historik*. Fundamenta Historica, vol. 3. Stuttgart: Frommann-Holzboog.

Blunt, Anthony. 1938–39. "The Triclinium in Religious Art." *Journal of the Warburg Institute* 2:271–276.

Bödeker, Hans Erich, et al., eds. 1986. *Aufklärung und Geschichte*. Göttingen: Vandenhoeck & Ruprecht.

Bodson, L. 1986. "Aspects of Pliny's Zoology." In French and Greenaway 1986.

Boehm, Laetitia. 1965. "Der wissenschaftstheoretische Ort der historia im frühen Mittelater." In Clemens Bauer, Letitia Boehm, and Max Müller, eds., *Speculum historiale. Geschichte im Spiegel von Geschichtsschreibung und Geschichtsdeutung: Johannes Spörl aus Anlass seines 60. Geburtstag*. Freiburg: Karl Alber, 663–693.

Boockmann, Hartmut, and Hermann Wellenreuther, eds. 1987. *Geschichtswissenschaft in Göttingen: eine Vorlesungsreihe*. Göttingen: Vandenhoeck & Ruprecht.

Borghero, Carlo. 1983. *La certezza e la storia: cartesianismo, pirronismo e conoscenza storica*. Milan: Franco Angeli.

Borst, Arno. 1994. *Das Buch der Naturgeschichte. Plinius und seine Leser im Zeitalter des Pergaments*. Heidelberg: C. Winter.

Borst, Arno. 1995. *Das Buch der Naturgeschichte: Plinius und seine Leser im Zeitalter des Pergaments*. 2d ed. Heidelberg: Universitätsverlag C. Winter.

Bosworth, A. B. 2000. Introduction to Bosworth and Baynham 2000, 1–22.

Bosworth, A. B., and E. J. Baynham, eds. 2000. *Alexander the Great in Fact and Fiction*. Oxford: Oxford University Press.

Bots, Hans, and Françoise Waquet, eds. 1994. *Commercium litterarium, 1600–1750: la communication dans la République des lettres = Commercium litterarium, 1600–1750: Forms of Communication in the Republic of Letters*. Amsterdam: APA-Holland University Press.

Bots, Hans, and Françoise Waquet. 1997. *La République des lettres*. Paris: Belin.

Branca, Vittore, ed. 1981. *Giorgio Valla tra scienza e sapienza*. Florence: Olschki.

Braun, Lucien. 1990. *Conrad Gessner*. Geneva: Slatkine.

Bremond, Claude, Jacques Le Goff, and Jean-Claude Schmitt. 1982. *L'exemplum*. Typologie des sources du moyen âge occidental, vol. 40. Turnhout: Brepols.

Brown, John Lackey. 1939. *The Methodus ad Facilem Historiarum Cognitionem of Jean Bodin: A Critical Study*. Washington: Catholic University of America Press.

Brown, Patricia Fortini. 1996. *Venice and Antiquity: The Venetian Sense of the Past*. New Haven: Yale University Press.

Bruyère, Nelly. 1984. *Méthode et dialectique dans l'oeuvre de La Ramée*. Paris: Vrin.

Buckley, Michael J., S.J. 1987. *At the Origins of Modern Atheism*. New Haven: Yale University Press.

Buffière, Félix. 1956. *Les mythes d'Homère et la pensée greque*. Paris: Les Belles Lettres.

Buonora, Paolo. 2003. "Cartografia e idraulica del Tevere (secoli XVI–XVII)." In Alessandra Fiocca, Daniela Lamberini, and Cesare Maffioli, eds., *Arte e scienza delle acque nel Rinascimento*. Venice: Marsilio, 169–193.

Burckhardt, Andreas. 1967. *Johannes Basilius Herold. Kaiser und Reich im protestantischen Schrifttum des Basler Buchdrucks um die Mitte des 16. Jahrhunderts*. Basel and Stuttgart: Helbing & Lichtenhahn.

Burke, Peter. 1969. *The Renaissance Sense of the Past*. London: Edward Arnold.

Burke, Peter. 2003. "Images as Evidence in Seventeenth-Century Europe." *Journal of the History of Ideas* 64:273-296.

Burkert, Walter. 1983. *Homo Necans: The Anthropology of Ancient Greek Sacrificial Ritual and Myth*. Berkeley: University of California Press.

Burnyeat, Myles F., ed. 1983. *The Skeptical Tradition*. Berkeley: University of California Press.

Burnyeat, Myles F., and Michael Frede, eds. 1997. *The Original Sceptics: A Controversy*. Indianapolis: Hackett.

Butterfield, Herbert. 1955. *Man on His Past: The Study of the History of Historical Scholarship*. Cambridge: Cambridge University Press.

Bylebyl, Jerome J. 1979. "The School of Padua: Humanistic Medicine in the 16th Century." In Charles Webster, ed., *Health, Medicine and Mortality in the 16th Century*. Cambridge: Cambridge University Press, 335–370.

Bylebyl, Jerome J. 1985. "Medicine, Philosophy and Humanism in Renaissance Italy." In John W. Shirley and F. David Hoeniger, eds., *Science and the Arts in the Renaissance*. London: Associated University Presses, 27–49.

Bylebyl, Jerome J. 1991. "Teaching 'Methodus Medendi' in the Renaissance." In Fridolf Kudlien and Richard J. Durling, eds., *Galen's Method of Healing*. Leiden: Brill, 157–189.

Bylebyl, Jerome J. 1993. "The Manifest and the Hidden in the Renaissance Clinic." In Bynum and Porter 1993, 40–68.

Bynum, W. F., and Roy Porter, eds. 1993. *Medicine and the Five Senses*. Cambridge: Cambridge University Press.

Caetani, Leone. 1924. *Saggio di un dizionario bio-bibliografico italiano*. Rome: R. Accademia Nazionale dei Lincei.

Cantimori, Delio. 1938. "Rhetoric and Politics in Italian Humanism." *Journal of the Warburg Institute* 1:83–102.

Cappelli, Adriano. 1889. "La biblioteca estense nella prima metà del secolo XV." *Atti e memorie della Deputazione ferrarese di Storia Patria*, 1st ser., 9:1–30.

Cardini, Roberto, and Mariangela Regoliosi, eds. 1996. *Umanesimo e medicina. Il problema dell' 'individuale.'* Rome: Bulzoni Editore.

Carlino, Andrea. 1994. *La fabbrica del corpo. Libri e dissezione nel Rinascimento*. Turin: Einaudi. Trans. as Carlino 1999a.

Carlino, Andrea. 1999a. *Books of the Body: Anatomical Ritual and Renaissance Learning*. Chicago: University of Chicago Press.

Carlino, Andrea. 1999b. *Paper Bodies: A Catalogue of Anatomical Fugitive Sheets, 1538–1687*. London: Wellcome Institute for the History of Medicine.

Carney, Elizabeth. 2000. "Artifice and Alexander History." In Bosworth and Baynham 2000, 263–285.

Carrara, Eliana. 1998. "La nascita della descrizione antiquaria." In Elena Vaiani, ed., *Dell'antiquaria e dei suoi metodi. Atti delle giornate di studio*. Annali della Scuola Normale Superiore di Pisa, 4th ser. Quaderni, 2. Pisa, 32–49.

Carrier, David. 1987. "Ekphrasis and Interpretation: Two Modes of Art History Writing." *British Journal of Aesthetics* 27:20–31.

Cary, George. 1956. *The Medieval Alexander*. Ed. D. J. A. Ross. Cambridge: Cambridge University Press.

Castellani, Carlo. 1975. "Salviani, Ippolito." In *Dictionary of Scientific Biography* 12:89–90.

Castelli, Patrizia, ed. 1991. *La rinascita del sapere. Libri e maestri dello Studio ferrarese*. Venezia: Marsilio.

Castelli, Patrizia, ed. 1992. *Ferrara e il Concilio, 1438–1439*. Ferrara: Università degli Studi di Ferrara.

Céard, Jean. 1977. *La nature et les prodiges: l'insolite au XVIe siècle en France*. Geneva: Droz.

Cellauro, L. 1995. "The Casino of Pius IV in the Vatican." *Papers of the British School at Rome* 50:183–214.

Cerutti, Simona. 2003. *Giustizia sommaria. Pratiche e ideali di giustizia in una società di Ancien Régime (Torino XVIII secolo)*. Milan: Feltrinelli.

Cerutti, Simona, and Gianna Pomata. 1999. "Fatti: una proposta per un numero di Quaderni storici." *Quaderni storici* 34.100:199–206.

Cerutti, Simona, and Gianna Pomata, eds. 2001a. *Fatti: storie dell'evidenza empirica. Quaderni storici* 36.108.

Cerutti, Simona, and Gianna Pomata. 2001b. "Premessa." In Cerutti and Pomata 2001a, 647–663.

Chapin, Seymour L. 1954. "An Early Bureau of Longitude: Peiresc in Provence." *Navigation: Journal of the Institute of Navigation* 4:59–66.

Chapin, Seymour L. 1958. "The Astronomical Activities of Nicolas Claude Fabri de Peiresc." *Isis* 47:13–29.

Chatelain, Jean-Marc. 1997. "Encyclopédisme et forme rhapsodique au XVIe siècle." In Jean Bouffartigue and François Mélonio, eds., *L'entreprise encyclopédique. Littérales*. Paris: Centre des Sciences de la Littérature, Université de Paris X-Nanterre, 21:97–111.

Cherchi, Paolo. 1997. "La Piazza Universale: somma di altre somme." In Paola Barocchi and Lina Bolzoni, eds., *Repertori di parole e immagini. Esperienze cinquecentesche e moderni data bases*. Pisa: Centro di Ricerche Informatiche per i Beni Culturali, Accademia della Crusca, 119–157.

Chiappini, Luciano. 1971. "Borso d'Este." In *Dizionario biografico degli italiani*. Rome: Istituto della Enciclopedia Italiana, 13:134–143.

Cifuentes, Luis. 2001. *La ciencia en català a l'etat mitjana i el renaixement*. Barcelona and Palma de Mallorca: Universitat de Barcelona and Universitat de Illes Baleares.

Clark, Willene B., and Meradith T. McMunn, eds. 1989. *Beasts and Birds of the Middle Ages: The Bestiary and Its Legacy*. Philadelphia: University of Pennsylvania Press.

Clericuzio, Antonio, and Silvia De Renzi. 1995. "Medicine, Alchemy and Natural Philosophy in the Early Accademia dei Lincei." In D. S. Chambers and François Quiviger, eds., *Italian Academies of the Sixteenth Century*. London: Warburg Institute, 175–194.

Clucas, Stephen. 2001. "Corpuscular Matter Theory in the Northumberland Circle." In Christoph Lüthy, John E. Murdoch, and William R. Newman, eds., *Late Medieval and Early Modern Corpuscular Matter Theories*. Leiden: Brill, 181–207.

Cochrane, Eric. 1976. "Science and Humanism in the Italian Renaissance." *American Historical Review* 81:1037–1059.

Cochrane, Eric. 1981. *Historians and Historiography in the Italian Renaissance*. Chicago: University of Chicago Press.

Coffin, David R. 1979. *The Villa in the Life of Renaissance Rome*. Princeton: Princeton University Press.

Coffin, David R. 1991. *Gardens and Gardening in Papal Rome*. Princeton: Princeton University Press.

Conte, Emanuele, ed. 1991. *I maestri della Sapienza di Roma dal 1514 al 1787: i rotuli e altre fonti*. Rome: Istituto Storico Italiano.

Copeland, Rita. 1991. *Rhetoric, Hermeneutics, and Translation in the Middle Ages: Academic Traditions and Vernacular Texts*. Cambridge: Cambridge University Press.

Copenhaver, Brian. 1978. *Symphorien Champier and the Reception of the Occultist Tradition in the Renaissance*. The Hague: Mouton.

Costabel, Pierre. 1983. "Peiresc et Wendelin. Les satellites de Jupiter de Galilée à Newton." In A. Rhembold, ed., *Peiresc et la passion de connaître*. Paris: Vrin.

Cotroneo, Girolamo. 1966. *Jean Bodin teorico della storia*. Naples: Giannini.

Cotroneo, Girolamo. 1971. *I trattatisti dell'ars historica*. Naples: Giannini.

Couzinet, Marie-Dominique. 1996. *Histoire et méthode à la Renaissance: une lecture de la Methodus ad facilem historiarum cognitionem de Jean Bodin*. Paris: Vrin.

Couzinet, Marie-Dominique. 2000. "L'inspiration historique chez Francesco Patrizi." *Epistemon*, 19 January. <http://www.cesr.univ-tours.fr./Epistemon/trivium/couz-ent.asp>.

Cranz, F. Edward, and Charles B. Schmitt. 1984. *A Bibliography of Aristotle Editions 1501–1600*. Baden-Baden: Koerner.

Crawford, Michael H., and C. R. Ligota, eds. 1995. *Ancient History and the Antiquarian: Essays in Honor of Arnaldo Momigliano.* London: Warburg Institute-University of London.

Crespi, M. 1963. "Bacci, Andrea." In *Dizionario biografico degli Italiani.* Rome: Istituto della Enciclopedia Italiana, 5:29–30.

Crisciani, Chiara. 1990. "History, Novelty, and Progress in Scholastic Medicine." In Michael McVaugh and Nancy G. Siraisi, eds., *Renaissance Medical Learning: The Evolution of a Tradition. Osiris* 6:118–139.

Crisciani, Chiara. 1996. "L'individuale nella medicina tra Medioevo e Umanesimo: i Consilia." In Cardini and Regoliosi 1996, 1–20.

Crisciani, Chiara. 1998. "Esperienza, comunicazione e scrittura in alchimia (secoli XIII–XIV)." In Massimo Galuzzi et al., eds., *Le forme della comunicazione scientifica.* Milan: Franco Angeli, 85–110.

Crisciani, Chiara. 2001. "Fatti, teorie, 'narratio' e i malati a corte. Note su empirismo in medicina nel tardo medioevo." In Cerutti and Pomata 2001a, 695–718.

Crisciani, Chiara. 2003a. "Artefici sensati: experientia e sensi in alchimia e chirurgia (secc. XIII–XIV)." In Chiara Crisciani and Agostino Paravicini Bagliani, eds., *Alchimia e medicina nel Medioevo.* Florence: Edizioni del Galluzzo-SISMEL, 135–153.

Crisciani, Chiara. 2003b. "Michele Savonarola, medico: tra Università e corte, tra latino e volgare." In Nadia Bray and Loris Sturlese, eds., *La filosofia in volgare. Atti del XII convegno della SISPM.* Turnhout: Brepols, 433–449.

Crisciani, Chiara. 2003c. "Tra università, corte e città: note su alcuni medici 'pavesi' del sec. XV." *Annali di storia delle università italiane* 7:55–69.

Crisciani, Chiara. 2004. "'Consilia,' responsi, consulti: i pareri del medico tra insegnamento e professione." In Carla Casagrande et al., eds., *"Consilium." Teorie e pratiche del consigliare nella cultura medievale.* Florence: Edizioni del Galluzzo-SISMEL, 259–278.

Crisciani, Chiara (forthcoming). "Experience and Sense Perception in Alchemical Knowledge: Some Notes." In *Intellect and Imagination, Proceedings of the XI International Conference of SIEPM (2002).* Porto.

Cropper, Elizabeth, et al., eds. 1992. *Documentary Culture: Florence and Rome from Grand-Duke Ferdinand I to Pope Alexander VII.* Bologna: Nuova Alfa.

Cunningham, Andrew. 1985. "Fabricius and the 'Aristotle Project' in Anatomical Teaching and Research at Padua." In Andrew Wear, Roger K. French, and Iain M. Lonie, eds., *The Medical Renaissance of the Sixteenth Century.* Cambridge: Cambridge University Press, 195–222.

Cunningham, Andrew. 1997. *The Anatomical Renaissance: The Resurrection of the Anatomical Projects of the Ancients.* Aldershot: Scolar Press.

Curran, Brian, and Anthony Grafton. 1995. "A Fifteenth-Century Site Report on the Vatican Obelisk." *Journal of the Warburg and Courtauld Institutes* 58:234–248.

Curtius, Ernst Robert. 1953. *European Literature and the Latin Middle Ages.* Trans. Willard R. Trask. Princeton: Princeton University Press.

Da Costa Kaufmann, Thomas. 2001. "Antiquarianism, the History of Objects, and the History of Art before Winckelmann." *Journal of the History of Ideas* 62:523–541.

D'Amico, John. 1983. *Renaissance Humanism in Papal Rome.* Baltimore: Johns Hopkins University Press.

Dandelet, Thomas J. 2001. *Spanish Rome 1500–1700.* New Haven: Yale University Press.

Daston, Lorraine. 1991a. "Baconian Facts, Academic Civility and the Prehistory of Objectivity." *Annals of Scholarship* 8:337–364.

Daston, Lorraine. 1991b. "Marvelous Facts and Miraculous Evidence in Early Modern Europe." *Critical Inquiry* 18:93–124.

Daston, Lorraine. 1996. "Strange Facts, Plain Facts and the Texture of Scientific Experience." In Suzanne Marchand and Elizabeth Lunbeck, eds., *Proof and Persuasion: Essays on Authority, Objectivity and Evidence.* Turnout: Brepols, 42–59.

Daston, Lorraine. 1997. "The Cold Light of Facts and the Facts of Cold Light: Luminescence and the Transformation of the Scientific Fact, 1600–1750." *EMF: Studies in Early Modern France* 3:17–44.

Daston, Lorraine. 1998. "The Nature of Nature in Early Modern Europe." *Configurations: A Journal of Literature, Science, and Technology* 6:149–172.

Daston, Lorraine. 2001a. *Eine kurze Geschichte der wissenschaftlichen Aufmerksamkeit.* Reihe "Themen," 71. Munich: Carl Friedrich von Seimans Stiftung.

Daston, Lorraine. 2001b. "Perché i fatti sono brevi?" In Cerutti and Pomata 2001a, 745–770.

Daston, Lorraine. 2003. "Die Akademien und die Neuerfindung der Erfahrung im 17. Jahrhundert." *Nova Acta Leopoldina,* n.s. 87, no. 325, 15–33.

Daston, Lorraine. 2004. "Attention and the Values of Nature in the Enlightenment." In Lorraine Daston and Fernando Vidal, eds., *The Moral Authority of Nature.* Chicago: University of Chicago Press, 100–126.

Daston, Lorraine, and Katharine Park. 1998. *Wonders and the Order of Nature, 1150–1750,* New York: Zone Books.

David, Jean-Michel. 1998. *Valeurs et mémoire à Rome: Valère Maxime ou la vertu recomposée.* Paris: de Boccard.

Davis, Natalie Zemon. 1975. "Proverbial Wisdom and Popular Error." In Davis, *Society and Culture in Early Modern France.* Stanford: Stanford University Press.

De Angelis, Pietro. 1948. *L'insegnamento della medicina negli ospedali di Roma: cenni storici.* Rome: Scuola Medica Ospitaliera di Roma.

De Angelis, Pietro. 1952. *L'arciospedale di Santo Spirito in Saxia nel passato e nel presente.* Rome: [Tip. Pasquino]. Collana di studi storici sull'ospedale di Santo Spirito in Saxia e sugli ospedali romani.

Dear, Peter. 1985. "Totius in Verba: Rhetoric and Authority in the Early Royal Society." *Isis* 76:144–161.

Dear, Peter. 1987. "Jesuit Mathematical Science and the Reconstitution of Experience in the Early Seventeenth Century." *Studies in the History and Philosophy of Science* 18:133–175.

Dear, Peter. 1990. "Miracles, Experiments, and the Ordinary Course of Nature." *Isis* 81:663–683.

Dear, Peter. 1991. "Narratives, Anecdotes, and Experiments: Turning Experience into Science in the Seventeenth Century." In Dear, ed., *The Literary Structure of Scientific Argument: Historical Studies*. Philadelphia: University of Pennsylvania Press.

Dear, Peter. 1995. *Discipline and Experience: The Mathematical Way in the Scientific Revolution*. Chicago: University of Chicago Press.

Deffenu, Gonario. 1955. *Benedetto Reguardati, medico e diplomatico di Francesco Sforza*. Milan: Hoepli.

Deichgräber, Karl. 1965. *Die griechische Empirikerschule*. Berlin and Zurich: Weidmannsche Verlagsbuchhandlung.

Deitz, Luc. 1999. "Space, Light, and Soul in Francesco Patrizi's *Nova de universis philosophia*." In Grafton and Siraisi 1999, 139–170.

Del Corno, Carlo. 1989. *Exemplum e letteratura: tra medioevo e Rinascimento*. Bologna: Il Mulino.

Del Nero, Domenico. 1996. *La corte e l'università. Umanisti e teologi nel Quattrocento ferrarese*. Florence: Titania.

Delumeau, Jean. 1957. *Vie économique et sociale de Rome dans la seconde moitié du XVIe siècle*. 2 vols. Paris: E. de Boccard.

Demaitre, Luke. 1975. "Theory and Practice in Medical Education at the University of Montpellier in the 13th and 14th Centuries." *Journal of the History of Medicine and Allied Sciences* 30:103–123.

Demaitre, Luke. 1976. "Scholasticism in Compendia of Practical Medicine 1250–1450." *Manuscripta* 20:81–95.

De' Reguardati, Fausto M. 1977. *Benedetto de' Reguardati da Norcia, 'medicus tota Italia celeberrimus.'* Trieste: Lindt.

De Renzi, Silvia. 1999. "'A Fountain for the Thirsty' and a Bank for the Pope: Charity, Conflicts, and Medical Careers at the Hospital of Santo Spirito in Seventeenth Century Rome." In Ole Peter Grell and Andrew Cunningham with Jon Arrizabalaga, eds., *Health Care and Poor Relief in Counter-Reformation Europe*. London: Routledge, 102–131.

De Renzi, Silvia. 2000. "Writing and Talking of Exotic Animals." In Marina Frasca-Spada and Nick Jardine, eds., *Books and the Sciences in History*. Cambridge: Cambridge University Press, 151–167.

De Renzi, Silvia. 2001. "La natura in tribunale. Conoscenze e pratiche medico-legali a Roma nel XVII secolo." In Cerutti and Pomata 2001a, 799–822.

Dictionary of Scientific Biography. 1970–80. Ed. Charles C. Gillespie. New York: Scribner.

Diehl, Houston. 1986. "Graven Images: Protestant Emblem Books in England." *Renaissance Quarterly* 39:49–66.

Diemer, A., ed. 1968. *System und Klassifikation in Wissenschaft und Dokumentation.* Meisenheim: Anton Hein.

Dilg, Peter. 1969. *Das Botanologicon des Euricius Cordus: ein Beitrag zur botanischen Literatur der Renaissance.* Marburg: Erich Mauersberger.

Dilg, Peter. 1971. "Studia humanitatis et res herbaria: Euricus Cordus als Humanist und Botaniker." *Rete* 1:71–85.

Di Martino, V., and M. Belati. 1986. *Qui arrivò il Tevere: le inondazioni del Tevere nelle testimonianze e nei ricordi storici.* Rome: Multigrafica Editrice, 1986.

Dionisotti, Carlotta. 1997. "Les chapitres entre l'historiographie et le roman." In Jean-Claude Fredouille, Marie-Odile Goulet-Cazé, Philippe Hoffmann, and Pierre Petitmengin, with Simone Deléani, eds., *Titres et articulations du texte dans les oeuvres antiques. Actes du Colloque International de Chantilly, 13–15 décembre 1994.* Collection des Etudes Augustiniennes, Série Antiquité, 152. Paris: Institut d'Etudes Augustiniennes, 529–547.

Dionisotti, Carlo. 1999. "Tradizione classica e volgarizzamenti." In Dionisotti, *Geografia e storia della letteratura italiana.* Turin: Einaudi, 125–178.

Ditchfield, Simon. 1995. *Liturgy, Sanctity and History in Tridentine Italy: Pietro Maria Campi and the Preservation of the Particular.* Cambridge: Cambridge University Press.

Ditchfield, Simon. 1998. "'In Search of Local Knowledge': Rewriting Early Modern Italian Religious History." *Cristianesimo nella storia* 19:255–296.

Ditchfield, Simon. 2000. "Leggere e vedere Roma come icona culturale (1500–1800 circa)." In *Storia d'Italia.* Annali 16. *Roma, la città del papa,* ed. Luigi Fiorani e Adriano Prosperi. Turin: Giulio Einaudi editore, 33–72.

Dizionario biografico degli Italiani. 1960–. Rome: Istituto della Enciclopedia italiana.

D'Onofrio, Cesare. 1980. *Il Tevere.* Rome: Romana Società Editrice.

D'Onofrio, Cesare. 1986. *Le fontane di Roma.* Rome: Romana Società Editrice.

Dubois, Claude-Gilbert. 1977. *La conception de l'histoire en France au XVIe siècle.* Paris: Nizet.

Dufournier, Bernard. 1936. "Th. Zwinger de Bâle et la scholastique de l'histoire au XVIe siècle." *Revue d'histoire moderne* 9:323–331.

Dunstan, Gordon R., ed. 1990. *The Human Embryo: Aristotle and Arabic and European Traditions.* Exeter: University of Exeter Press.

Dupront, Alphonse. 1930. *Pierre-Daniel Huet et l'exegèse comparatiste au XVIIIe siècle.* Paris: Leroux.

Durel, Henri. 1998. "Francis Bacon lecteur d'Aristote à Cambridge." *Nouvelles de la république des letters* 18:29–60.

Durling, Richard J. 1961. "A Chronological Census of Renaissance Editions and Translations of Galen." *Journal of the Warburg and Courtauld Institutes* 24:230–305.

Eamon, William. 1994. *Science and the Secrets of Nature: Book of Secrets in Medieval and Early Modern Culture*. Princeton: Princeton University Press.

Eamon, William, and Gundolf Keil. 1987. " 'Plebs amat empirica': Nicholas of Poland and His Critique of the Medieval Medical Establishment." *Sudhoffs Archiv* 71:180–196.

Elukin, Jonathan. 2002. "Maimonides and the Rise and Fall of the Sabians: Explaining Mosaic Laws and the Limits of Scholarship." *Journal of the History of Ideas* 63:619–637.

Erasmus, H. J. 1962. *The Origins of Rome in Historiography from Petrarch to Perizonius*. Assen: Van Gorcum.

Farmer, Stephen A. 1998. *Syncretism in the West: Pico's 900 Theses (1486); the Evolution of Traditional Religious and Philosophical Systems*. Tempe, Arizona: Medieval and Renaissance Texts and Studies [Arizona State University].

Fattori, Marta. 1980. *Lessico del Novum Organum di Francesco Bacone*, Rome: Edizioni dell'Ateneo & Bizzarri.

Fattori, Marta. 2002. "Experientia-experimentum: un confronto tra il corpus latino e inglese di Francis Bacon." In Marco Veneziani, ed., *Experientia. X Colloquio Internazionale del Lessico Intellettuale Europeo, Roma, 4–6 gennaio 2001*. Florence: Olschki, 243–259.

Fauth, Wolfgang. 1995. *Helios Megistos. Zur synkretistischen Theologie der Spätantike*. Leiden: Brill.

Febvre, Lucien. 1936. "Le mot histoire." *Annales d'histoire économique et sociale* 8:301–302.

Febvre, Lucien. 1968. *Le problème de l'incroyance au XVI siècle: la réligion de Rabelais*. 2d ed. Paris: Michel.

Feingold, Mordechai, Joseph Freedman, and Wolfgang Rother, eds. 2001. *The Influence of Petrus Ramus: Studies in Sixteenth and Seventeenth Century Philosophy and Sciences*. Basel: Schwabe.

Ferrari, Giovanna. 1996. *L'esperienza del passato: Alessandro Benedetti filologo e medico umanista*. Florence: Olschki.

Ferrary, Jean-Louis. 1996. *Onofrio Panvinio et les antiquités romaines*. Rome: Ecole Française de Rome.

Ferri, Sara, ed. 1997. *Pietro Andrea Mattioli, Siena 1501–Trento 1578, la vita e le opere, con l'identificazione delle piante*. Ponte San Giovanni, Perugia: Quattroemme.

Feyerabend, Paul K. 1970. "Problems of Empiricism." Part II. In R. G. Colodny, ed., *The Nature and Function of Scientific Theories*. Pittsburgh: University of Pittsburgh Press, 275–353.

Findlen, Paula. 1994. *Possessing Nature: Museums, Collecting, and Scientific Culture in Early Modern Italy*. Berkeley: University of California Press.

Findlen, Paula. 2000. "The Formation of a Scientific Community: Natural History in Sixteenth-Century Italy." In Grafton and Siraisi 1999, 369–400.

Findlen, Paula, ed. 2004. *Athanasius Kircher: The Last Man Who Knew Everything*. London: Routledge.

Fiorentino Grassi, Silvia. 1982. "Note sull'antiquaria romana nella seconda metà del secolo XVI." In Romeo de Maio et al., eds., *Baronio storico e la Controriforma*. Sora: Centro di Studi Sorani "Vincenzo Patriarca," 199–211.

Fischer, Henry G. 1980. "Hunde." In *Lexikon der Ägyptologie*, vol. 3. Wiesbaden: Harrassowitz, 77–81.

Foucault, Michel. 1966. *Les mots et les choses*. Paris: Gallimard.

Fowden, Garth. 1993. *Empire to Commonwealth: Consequences of Monotheism in Late Antiquity*. Princeton: Princeton University Press.

Fragnito, Gigliola. 1993. "Cardinals' Courts in Sixteenth-Century Rome." *Journal of Modern History* 65:26–56.

Franklin, Julian. 1963. *Jean Bodin and the Sixteenth-Century Revolution in the Methodology of Law and History*. New York: Columbia University Press.

Frede, Michael. 1981. "On Galen's Epistemology." In Vivian Nutton, ed., *Galen: Problems and Prospects*. London: Wellcome Institute, 1981, 65–86.

Frede, Michael. 1985. "Introduction." In Galen, *Three Treatises on the Nature of Science*. Trans. Richard Walzer and Michael Frede. Indianapolis: Hackett Publishing, ix–xxxiv.

Frede, Michael. 1987. "The Ancient Empiricists." In Frede, *Essays in Ancient Philosophy*. Minneapolis: University of Minnesota Press.

Fredouille, Jean-Claude. 1981. "Götzendienst." In *Reallexikon für Antike und Christentum*. Stuttgart: Hiersemann, 11:825–895.

Freedberg, David. 1989. *The Power of Images: Studies in the History and Theory of Response*. Chicago: University of Chicago Press.

Freedberg, David. 2002. *The Eye of the Lynx: Galileo, His Friends and the Beginnings of Modern Natural History*. Chicago: University of Chicago Press.

Freedberg, David, and Enrico Baldini, eds. 1997. *The Paper Museum of Cassiano dal Pozzo: Series B: Natural History*, pt. 1. London: Harvey Miller.

Freedman, Joseph S. 1988. *European Academic Philosophy in the Late Sixteenth and Seventeenth Centuries: The Life, Significance and Philosophy of Clemens Timpler (1563/4–1624)*. Hildesheim: G. Olms.

French, Roger. 1994. *William Harvey's Natural Philosophy*. Cambridge: Cambridge University Press.

French, Roger, and Frank Greenaway, eds. 1986. *Science in the Early Roman Empire: Pliny the Elder, His Sources and Influence*. Totowa, N.J.: Barnes & Noble Books.

Friedenwald, Harry. 1944. *The Jews and Medicine*. 2 vols. Baltimore: Johns Hopkins University Press.

Friedrich, Udo. 1995. *Naturgeschichte zwischen artes liberales und frühneuzeitlicher Wissenschaft; Conrad Gessners Historia animalium und ihre volkssprachige Rezeption*. Tübingen: Max Niemeyer.

Funkenstein, Amos. 1986. *Theology and the Scientific Imagination from the Middle Ages to the Seventeenth Century*. Princeton: Princeton University Press.

Garavini, Fausta. 1992. "Montaigne et le *Theatrum vitae humanae*." In *Montaigne et l'Europe. Actes du colloque international de Bordeaux*. Mont-de-Marsan: Editions Interuniversitaires, 31–45.

Garcia Ballester, Luis. 1993. "On the Origin of the 'Six Non-Natural Things' in Galen." In Jutta Kollesch and Diethart Nikel, eds., *Galen und das hellenistische Erbe*. Stuttgart: F. Steiner, 105–115.

Garfagnini, Manuela Doni. 2002. *Il teatro della storia fra rappresentazione e realtà: storiografia e trattatistica fra Quattrocento e Seicento*. Rome: Edizioni di Storia e Letteratura.

Garin, Eugenio. 1961. "Motivi della cultura filosofica ferrarese." In Garin, *La cultura filosofica del Rinascimento italiano. Ricerche e documenti*. Florence: Sansoni, 402–431.

Garofalo, Fausto. 1950. *Quattro secoli di vita del Protomedicato e del Collegio dei Medici di Roma (Regesto dei documenti dal 1471 al 1870)*. Rome: Istituto di Storia della Medicina dell'Università di Roma.

Gaston, Robert W., ed. 1988. *Pirro Ligorio: Artist and Antiquarian*. Florence: Silvana.

Gazich, Roberto. 1986. "Tecnica di inserzione e struttura dell'*exemplum* narrativo nella *Naturalis historia*." In Pier Vincenzo Cova, Roberto Gazich, Gian Enrico Manzoni, and Graziano Melzani, *Studi sulla lingua di Plinio il Vecchio*. Pubblicazioni della Università Cattolica del Sacro Cuore. Milan: Vita e pensiero, 143–169.

Gentilcore, David. 1994. "'All That Pertains to Medicine': Protomedici and Protomedicati in Early Modern Italy." *Medical History* 38:121–142.

George, Wilma, and Brunsdon Yapp. 1991. *The Naming of the Beasts: Natural History in the Medieval Bestiary*. London: Duckworth.

Gerbi, Antonello. 1985. *Nature in the New World: From Christopher Columbus to Gonzalo Fernández Oviedo*. Trans. Jeremy Moyle. Pittsburgh: University of Pittsburgh Press.

Gilbert, Felix. 1965. *Machiavelli and Guicciardini: Politics and History in Sixteenth-Century Florence*. Princeton: Princeton University Press.

Gilbert, Neal Ward. 1956. "Concepts of Method in the Renaissance and Their Ancient and Medieval Antecedents." Thesis, Columbia University.

Gilbert, Neal Ward. 1960. *Renaissance Concepts of Method*. New York: Columbia University Press.

Gillet, J. F. A. 1860–61. *Crato von Crafftheim und seine Freunde.* Frankfurt: Brönner.

Gilly, Carlos. 1977–79. "Zwischen Erfahrung und Spekulation: Theodor Zwinger und die religiöse und kulturelle Krise seiner Zeit." *Basler Zeitschrift für Geschichte und Altertumskunde* 77:57–137; 79:125–223.

Gilly, Carlos. 1985. *Spanien und der Basler Buchdruck bis 1600. Ein Querschnitt durch die spanische Geistesgeschichte aus der Sicht einer europäischen Buchdruckerstadt.* Basler Beiträge zur Geschichtswissenschaft, 151. Basel and Frankfurt: Helbing & Lichtenhahn.

Gilly, Carlos. 2001. *Die Manuskripte in der Bibliothek des Johannes Oporinus.* Basel: Schwabe.

Gilly, Carlos. 2002. "Il 'Theatrum humanae vitae' di Theodor Zwinger: da una 'historia naturalis' dell'uomo ad un 'Novum Organum delle scienze / Theodor Zwinger's 'Theatrum humanae vitae': From Natural Anthropology to the 'Novum Organum' of Sciences." In Carlos Gilly and C. van Heertum, eds., *Magia, alchimia, scienza dal '400 al '700. L'influsso di Ermete Trismegisto.* 2 vols. Florence: Centro Di, 1:253–274.

Gil Sotres, Pedro. 1993. "Le regole della salute." In Mirko Grmek, ed., *Storia del pensiero medico occidentale. 1. Antichità e Medioevo.* Rome and Bari: Laterza, 399–438.

Gil Sotres, Pedro. 1996. Introduction to Arnald of Villanova, *Regimen sanitatis ad regem Aragonum.* Ed. Luis Garcia Ballester, Juan Antonio Paniagua, and Michael McVaugh. (*Arnaldi de Villanova Opera medica omnia*, X.1.) Barcelona: Publicacions Universitat de Barcelona, 17–110.

Ginzburg, Carlo. 1988. "Ekphrasis and Quotation." *Tijdschrift voor Filosofie* 50:3–19.

Ginzburg, Carlo. 1989. *Clues, Myths, and the Historical Method.* Baltimore: Johns Hopkins University Press.

Ginzburg, Carlo. 1998a. "Idoli e immagini. Un passo di Origine e la sua fortuna." In Ginzburg, *Occhiacci di legno. Nove riflessioni sulla distanza.* Milan: Feltrinelli, 118–135.

Ginzburg, Carlo. 1998b. "Mito. Distanza e menzogna." In Ginzburg, *Occhiacci di legno. Nove riflessioni sulla distanza.* Milan: Feltrinelli, 40–81.

Ginzburg, Carlo. 2000. *No Island Is an Island.* New York: Columbia University Press.

Ginzburg, Carlo, and Marco Ferrari. 1978. "La colombara ha aperto gli occhi." *Quaderni storici* 13.38:631–639.

Glacken, Clarence J. 1967. *Traces on the Rhodian Shore: Nature and Culture in Western Thought from Ancient Times to the End of the Eighteenth Century.* Berkeley: University of California Press.

Gliozzi, Giuliano. 1977. *Adamo e il nuovo mondo.* Florence: Nuova Italia.

Gmelig-Nijboer, Caroline Aleid. 1977. *Conrad Gesner's "Historia Animalium": An Inventory of Renaissance Zoology*. Communicationes biohistoricae Ultrajectinae, 72. Meppel: Krips Repro.

Godman, Peter. 1998. *From Poliziano to Machiavelli: Florentine Humanism in the High Renaissance*. Princeton: Princeton University Press.

Goez, Werner. 1974. "Die Anfänge der historischen Methoden-Reflexion in der italienischen Renaissance und ihre Aufnahme in der Geschichtsschreibung des deutschen Humanismus." *Archiv für Kulturgeschichte* 56:25–48.

Goldgar, Anne. 1995. *Impolite Learning: Conduct and Community in the Republic of Letters*. New Haven: Yale University Press.

Goltz, Dietlinde. 1986. "Der leere Uterus. Zum Einfluss von Harveys *De generatione animalium* auf die Lehren von der Konzeption." *Medizinhistorisches Journal* 21:242–268.

Goodman, L. E. 1990. "The Fetus as a Natural Miracle: The Maimonidean View." In Gordon R. Dunstan, ed., *The Human Embryo: Aristotle and Arabic and European Traditions*. Exeter: Exeter University Press, 79–94.

Goyet, Francis. 1986–87. "A propos de 'ces pastissages de lieux communs' (le rôle des notes de lecture dans la genèse des Essais)." *Bulletin de la Société des Amis de Montaigne* 5–6:11–26 and 7–8:9–30.

Graevenitz, Gerhart von. 1987. *Mythos. Zur Geschichte einer Denkgewohnheit*. Stuttgart: Metzler.

Grafton, Anthony. 1985. "The World of the Polyhistors: Humanism and Encyclopedism." *Central European History* 18:31–47.

Grafton, Anthony. 1990a. *Forgers and Critics*. Princeton: Princeton University Press.

Grafton, Anthony. 1990b. "Humanism, Magic, and Science." In Anthony Goodman and Angus McKay, eds., *The Impact of Humanism on Western Europe*. London: Longman, 99–117.

Grafton, Anthony. 1991a. "Humanism and Science in Rudolphine Prague: Kepler in Context." In Grafton, *Defenders of the Text: The Traditions of Scholarship in an Age of Science, 1450–1800*. Cambridge, Mass.: Harvard University Press, 178–203.

Grafton, Anthony. 1991b. "Traditions of Invention and Invention of Traditions in Renaissance Italy: Annius of Viterbo." In Grafton, *Defenders of the Text: The Traditions of Scholarship in an Age of Science, 1450–1800*. Cambridge, Mass.: Harvard University Press, 76–103.

Grafton, Anthony. 1995a. "The Ancient City Restored: Archaeology, Ecclesiastical History, and Egyptology." In Grafton 1995b, 87–124.

Grafton, Anthony, ed. 1995b. *Rome Reborn: The Vatican Library and Renaissance Culture*. Washington: Library of Congress.

Grafton, Anthony. 1997. *Commerce with the Classics*. Ann Arbor: University of Michigan Press.

Grafton, Anthony. 1999. "Historia and Istoria: Alberti's Terminology in Context." *I Tatti Studies* 8:37–68.

Grafton, Anthony. 2000. *Leon Battista Alberti: Master Builder of the Italian Renaissance*. New York: Hill and Wang.

Grafton, Anthony. 2001a. *Bring Out Your Dead: The Past as Revelation*. Cambridge, Mass.: Harvard University Press.

Grafton, Anthony. 2001b. "Where was Salomon's House? Ecclesiastical History and the Intellectual Origins of Bacon's New Atlantis." In Herbert Jaumann, ed., *Die europäische Gelehrtenrepublik im Zeitalter des Konfessionalismus*. Wiesbaden: Harrassowitz Verlag, 21–38.

Grafton, Anthony, with April E. Shelford and Nancy G. Siraisi. 1992. *New Worlds, Ancient Texts: The Power of Tradition and the Shock of Discovery*. Cambridge, Mass.: Harvard University Press and the New York Public Library.

Grafton, Anthony, and Nancy Siraisi, eds. 1999. *Natural Particulars: Nature and the Disciplines in Renaissance Europe*. Cambridge, Mass.: MIT Press.

Grégoire, Réginald. 1979. "'Servizio dell'anima quanto del corpo' nell'ospedale romano di Santo Spirito (1623)." In *Ricerche per la storia religiosa di Roma: Studi, documenti, inventari*. Rome: Edizioni di Storia e Letteratura, 221–254, with edition of Domenico Borgarucci, "Relazione del modo che si tiene da religiosi di Santo Spirito in Sassia di Roma," at 233–254.

Grell, Chantal. 1983. "Les origines de Rome: mythe et critique. Essai sur l'histoire aux XVIIe et XVIIIe siècles." *HES* 255–280.

Grell, Chantal. 1993. *L'histoire entre érudition et philosophie: étude sur la connaissance historique à l'âge des Lumières*. Paris: Presses universitaires de France.

Grell, Chantal. 1995. *Le dix-huitième siècle et l'antiquité en France*. 2 vols. Oxford: Voltaire Foundation.

Grell, Chantal, and Catherine Volpilhac-Auger, eds. 1994. *Nicolas Fréret: légende et vérité. Colloque des 18 et 19 octobre 1991*. Oxford: Voltaire Foundation.

Grendler, Paul F. 1989. *Schooling in Renaissance Italy: Literacy and Learning, 1300–1600*. Baltimore: Johns Hopkins University Press.

Gualdo, Riccardo. 1996. *Il lessico medico del 'De regimine pregnantium' di Michele Savonarola*. Florence: presso l'Accademia della Crusca.

Gualdo, Riccardo. 1999. "Sul lessico medico di Michele Savonarola: derivazione, sinonimia, gerarchia di parole." *Studi di lessicografia italiana* 16:163–125.

Gualdo, Riccardo, ed. 2001. *Le parole della scienza. Scritture tecniche e scientifiche in volgare (secoli XIII–XV)*. Galatina: Congedo Ed.

Gualdo, Riccardo (forthcoming). "Le cure e i bagni del principe nelle opere volgari di Michele Savonarola." In Paolo Viti, ed., *Gli umanisti e le terme*. Galatina: Congedo Ed.

Guenée, Bernard. 1973. "Histoire, annales, chroniques: essai sur le genres historiques au Moyen Age." *Annales. Economies, Sociétés, Civilisations* 28:997–1016.

Gundersheimer, Werner L. 1988. *Ferrara estense. Lo stile del potere*. Modena: Panini. (Trans. of *Ferrara: The Style of a Renaissance Despotism*, Princeton: Princeton University Press, 1973.)

Habermas Jürgen. 1988. *On the Logic of the Social Sciences*. Cambridge, Mass.: MIT Press.

Hadot, Pierre. 1981. *Exercices spirituels et philosophie antique*. Paris: Etudes augustiniennes.

Häfner, Ralph. 2003. *Götter im Exil. Poesie im Spannungsfeld prophetischen Wissens und philologischer Kritik am Ende des christlichen Humanismus (ca. 1600–1736)*. Tübingen: Niemeyer.

Häfner, Ralph (forthcoming). "Zur Rezeption von Aulus Gellius in der Frühen Neuzeit." In Häfner and Markus Völkel, eds., *Der Kommentar in der Frühen Neuzeit*.

Hamesse, Jacqueline. 1999. "The Scholastic Model of Reading." In Guglielmo Cavallo and Roger Chartier, eds., *A History of Reading in the West*. Oxford: Polity Press, 103–119.

Hamesse, Jacqueline. 2002. "*Experientia/experimentum* dans les lexiques médiévaux et dans les textes philosophiques antérieurs au 14e siècle." In Marco Veneziani, ed., *Experientia. X Colloquio Internazionale del Lessico Intellettuale Europeo, Roma, 4–6 gennaio 2001*. Florence: Olschki Editore, 77–90.

Hammerstein, Notker. 1972. *Jus und Historie. Ein Beitrag zur Geschichte, des historischen Denkens an deutschen Universitäten im späten 17. und im 18. Jahrhundert*. Göttingen: Vandenhoeck und Ruprecht.

Hampton, Timothy. 1990. *Writing from History: The Rhetoric of Exemplarity in Renaissance Literature*. Ithaca: Cornell University Press.

Hankinson, R. J. 1989. "Galen and the Best of All Possible Worlds." *Classical Quarterly*, n.s. 39:206–227.

Hankinson, R. J. 1995. "The Growth of Medical Empiricism." In Don Bates, ed., *Knowledge and the Scholarly Medical Traditions*. Cambridge: Cambridge University Press, 60–83.

Harms, Wolfgang. 1985. "On Natural History and Emblematics in the Sixteenth Century." In *The Natural Sciences and the Arts: Aspects of Interaction from the Renaissance to the Twentieth Century: An International Symposium*. Uppsala: Almqvist & Wiksell.

Hartmann, Martina. 2001. *Humanismus und Kirchenkritik: Matthias Flacius Illyricus als Erforscher des Mittelalters*. Stuttgart: Jan Thorbecke.

Haskell, Francis, et al., eds. 1989. *Il museo cartaceo di Cassiano dal Pozzo. Cassiano naturalista*. Ivrea: Olivetti.

Haskell, Francis. 1993. *History and Its Images: Art and the Interpretation of the Past*. New Haven: Yale University Press.

Hassinger, Erich. 1978. *Empirisch-rationaler Historismus*. Bern and Munich: Francke.

Hazard, Paul. 1935. *La crise de la conscience européenne (1680–1715).* 3 vols. Paris: Boivin.

Hecht, Christian. 1997. *Katholische Bildertheologie im Zeitalter von Gegenreformation und Barock: Studien zu Traktaten von Johannes Molanus, Gabriele Paleotti und anderen Autoren.* Berlin: Mann.

Henkel, Arthur, and Albrecht Schone, eds. 1967. *Emblemata: Handbuch zur Sinnbildkunst des XVI. und XVII. Jahrhunderts.* Stuttgart: J. B. Metzler.

Herklotz, Ingo. 1999. *Cassiano Dal Pozzo und die Archäeologie des 17. Jahrhunderts.* Munich: Hirmer Verlag.

Heusser, Martin. 1987. "Emblems in the Anatomy of Melancholy." *Notes and Queries,* n.s. 34:298–301.

Hirsch, August. 1884–1889. *Biographisches Lexicon der hervorragenden Ärtze aller Zeiten und Völker.* 5 vols. Vienna and Leipzig: Urban and Schwarzenberg.

Hirschberg, Julius. 1899–1918. *Geschichte der Augenheilkunde.* 4 vols. in 9. Leipzig and Berlin: Engelmann.

Holtz, Sabine, and Dieter Mertens, eds. 1999. *Nicodemus Frischlin: poetische und prosaische Praxis unter den Bedingungen des konfessionellen Zeitalters.* Stuttgart: Frommann-Holzboog.

Hoppe, Brigitte. 1969. *Das Kräuterbuch des Hieronymus Bock, wissenschaftshistorische Untersuchung: mit einem Verzeichnis sämtlicher Pflanzen des Werkes, der literarischen Quellen der Heilanzeigen und der Anwendungen der Pflanzen.* Stuttgart: Hiersemann.

Houtzager, H. L., ed. 1989. *Pieter van Foreest: een Hollands medicus in de zestiende eeuw.* Amsterdam: Rodopi.

Hulliung, Mark. 1994. *The Autocritique of Enlightenment: Rousseau and the Philosophes.* Cambridge, Mass.: Harvard University Press.

Humbert, Pierre. 1931. "La première carte de la lune." *Revue des questions scientifiques* 20:193–204.

Humbert, Pierre. 1934. "Un manuscrit inédit de Gassendi." *Revue des questions scientifiques* 53:5–11.

Humbert, Pierre. 1936. *L'oeuvre astronomique de Gassendi.* Paris: Hermann.

Humbert, Pierre. 1944. "Sur l'éclipse lunaire du 8 novembre 1631." *Ciel et terre* 42:42–44.

Humbert, Pierre. 1948. "Joseph Gaultier de La Valette astronome provençal (1564–1647)." *Revue d'histoire des sciences et de leurs applications* 1:314–322.

Humbert, Pierre. 1950. "A propos du passage de Mercure en 1631." *Revue d'histoire des sciences et de leurs applications* 3:27–31.

Humbert, Pierre. 1951a. "Les études de Peiresc sur la vision." *Archives internationales d'histoire des sciences* 30:654–659.

Humbert, Pierre. 1951b. "Peiresc et le microscope." *Revue d'histoire des sciences et de leurs applications* 4:154–158.

Huppert, George. 1970. *The Idea of Perfect History: Historical Erudition and Historical Philosophy in Renaissance France*. Urbana: University of Illinois Press.

Ianziti, Gary. 1988. *Humanistic Historiography under the Sforzas: Politics and Propaganda in Fifteenth-Century Milan*. Oxford: Clarendon Press.

Israel, Jonathan. 2001. *Radical Enlightenment: Philosophy and the Making of Modernity, 1650–1750*. Oxford: Oxford University Press.

Jacquart, Danielle. 1987. "Medical Explanations of Sexual Behavior in the Middle Ages." *Homo carnalis* 14:1–21.

Jacquart, Danielle. 1990. "Theory, Everyday Practice and Three Fifteenth-Century Physicians." *Osiris* 6:140–160.

Jacquart, Danielle. 1998. *La médecine médiévale dans le cadre parisien, XIVe–XVe siècle*. Paris: Fayard.

Jacquart, Danielle. 2001. "Le latin des sciences: quelques réflexions." In Monique Goullet and Michel Parisse, eds., *Les historiens et le latin médiéval*. Paris: Publications de la Sorbonne, 237–244.

Jaffé, David. 1993. "Aspects of Gem Collecting in the Early Seventeenth Century: Nicolas-Claude Peiresc and Lelio Pasqualini." *Burlington Magazine* 135:103–120.

Jaffé, David. 1990. "Mellan and Peiresc." *Print Quarterly* 7:168–175.

Jardine, Lisa. 1990. "Experientia literata or Novum Organum? The Dilemma of Bacon's Scientific Method." In William A. Sessions, ed., *Francis Bacon's Legacy of Texts*. New York: AMS Press, 47–67.

Jardine, Nicholas. 1984. *The Birth of History and Philosophy of Science. Kepler's "A Defence of Tycho against Ursus" with Essays on Its Provenance and Significance*. Cambridge: Cambridge University Press.

Jardine, Nicholas. 1988. "Epistemology of the Sciences." In Charles B. Schmitt and Quentin Skinner, eds., *The Cambridge History of Renaissance Philosophy*. Cambridge: Cambridge University Press, 685–693.

Jardine, Nicholas, J. A. Secord, and Emma C. Spary, eds. 1996. *Cultures of Natural History*. Cambridge: Cambridge University Press.

Jauss, Hans Robert. 1985. "I generi minori del discorso esemplare come sistema di comunicazione letteraria." Italian translation in Michelangelo Picone, ed., *Il racconto*. Bologna: Il Mulino Ed., 53–72.

Jensen, Kristian. 2000. "Description, Division, Definition: Caesalpinus and the Study of Plants as an Independent Discipline." In Marianne Pade, ed., *Renaissance Readings of the Corpus Aristotelicum*. Copenhagen: Museum Tusculanum Press, 185–206.

Jonsson, Ritva. 1968. *Historia. Etudes sur la genèse des offices versifiés*. Stockholm: Almqvist & Wiksell.

Joret, Charles. 1893–94. "Listes des noms de plantes envoyées par Peiresc a Clusius (Charles de l'Écluse)." *Revue des Langues Romanes*, 4th ser. 7:437–442.

Joutsivuo, Timo. 1999. *Scholastic Tradition and Humanist Innovation: The Concept of Neutrum in Renaissance Medicine.* Helsinki: Academia Scientiarum Fennica.

Joy, Lynn Sumida. 1987. *Gassendi the Atomist: Advocate of History in an Age of Science.* Cambridge: Cambridge University Press.

Kablitz, Andreas. 2001. "Lorenzo Vallas Konzept der Geschichte und der Fall der Konstantinischen Schenkung. Zur 'Modernität' von *De falso credita et ementita Constantini donatione.*" In Most 2001, 45–67.

Kany, Roland. 1987. *Mnemosyne als Programm: Geschichte, Erinnerung und die Andacht zum Unbedeutenden im Werk von Usener, Warburg und Benjamin.* Tübingen: Niemeyer.

Karcher, Johannes. 1956. *Theodor Zwinger und seine Zeitgenossen: Episode aus dem Ringen der Basler Ärzte um die Grundlehren der Medizin im Zeitalter des Barocks.* Basel: Helbing und Lichtenhahn.

Katchen, Aaron. 1984. *Christian Hebraists and Dutch Rabbis: Seventeenth Century Apologetics and the Study of Maimonides' Mishneh Torah.* Cambridge, Mass.: Harvard University Press.

Keil, Gundolf, and Peter Assion, eds. 1974. *Fachprosaforschung: acht Vorträge zur mittelalterlichen Artesliteratur.* Berlin: E. Schmidt.

Kelley, Donald R. 1964. "Historia Integra: François Baudouin and His Conception of History." *Journal of the History of Ideas* 35:35–57.

Kelley, Donald. 1970. *Foundations of Modern Historical Scholarship: Language, Law and History in the French Renaissance.* New York: Columbia University Press.

Kelley, Donald R. 1973. "Development and Context of Bodin's Method." In Horst Denzer, ed., *Jean Bodin: Verhandlungen der Internationalen Bodin Tagung in München.* Munich: C. H. Beck, 1123–1150.

Kelley, Donald R. 1984. "Giambattista Vico." In George Stade, ed., *European Writers: The Age of Reason and the Enlightenment.* New York: Charles Scribner's Sons, 293–316.

Kelley, Donald R. 1988a. "Humanism and History." In Albert Rabil, ed., *Renaissance Humanism: Foundations, Forms, and Legacy.* Philadelpha: University of Pennsylvania Press, 3:236–270.

Kelley, Donald R. 1988b. "The Theory of History." In Charles B. Schmitt and Quentin Skinner, eds., *Cambridge History of Renaissance Philosophy.* Cambridge: Cambridge University Press, 746–762.

Kelley, Donald R., ed. 1990. *Versions of History: From Antiquity to the Enlightenment.* New Haven: Yale University Press.

Kelley, Donald R., ed. 1997. *History and the Disciplines: The Reclassification of Knowledge in Early Modern Europe.* Rochester: University of Rochester Press.

Kelley, Donald R. 1998. *Faces of History: Historical Inquiry from Herodotus to Herder.* New Haven: Yale University Press.

Kelley, Donald. 1999. "Writing Cultural History in Early Modern France: Christophe Milieu and His Project." *Renaissance Quarterly* 52:342–365.

Kelley, Donald R. 2003a. *Fortunes of History: Historical Inquiry from Herder to Huizinga.* New Haven: Yale University Press.

Kelley, Donald R. 2003b. "The Rise of Prehistory." *Journal of World History* 14:17–36.

Kemp, Martin. 1993. " 'The Mark of Truth': Looking and Learning in Some Anatomical Illustrations from the Renaissance and the Eighteenth Century." In Bynum and Porter 1993, 85–121.

Kessler, Eckhard. 1982. "Das rhetorische Modell der Historiographie." In Reinhart Koselleck et al., eds., *Formen der Geschichtsschreibung.* Munich: Deutscher Taschenbuch Verlag, 37–85.

Kessler, Eckhard. 1999. "Introducing Aristotle to the Sixteenth Century: The Lefevre Enterprise." In Constance Blackwell and Sachiko Kusukawa, eds., *Philosophy in the Sixteenth and Seventeenth Centuries.* Aldershot: Ashgate, 1–21.

Kessler, Eckhard, and Ian Maclean, eds. 2002. *Res et verba in der Renaissance.* Wiesbaden: Harrassowitz.

Keuck, Karl. 1934. *Historia. Geschichte des Wortes und seiner Bedeutungen in der Antike und in den romanischen Sprachen.* Emsdetten: Heinr. and J. Lechte.

Kibre, Pearl. 1985. *Hippocrates latinus.* New York: Fordham University Press.

Kibre, Pearl, and Richard J. Durling. 1991. "A List of Latin Manuscripts Containing Medieval Versions of the *Methodus medendi.*" In Fridolf Kudlien and Richard J. Durling, eds., *Galen's Method of Healing.* Leiden: Brill, 1991, 117–122.

Klapisch-Zuber, Christiane. 1985. "An Ethnology of Marriage in the Age of Humanism." In Klapisch-Zuber, *Women, Family, and Ritual in Renaissance Italy,* trans. Lydia Cochrane. Chicago: University of Chicago Press, 247–260.

Klauser, Theodor. 1954. "Bohne." In *Reallexikon für Antike und Christentum.* Stuttgart: Hiersemann, 2:489–502.

Klempt, Adalbert. 1960. *Die Säkularisierung der universalhistorischen Auffassung: zum Wandel des Geschichtsdenkens im 16. und 17. Jahrhundert.* Göttingen: Musterschmidt.

Knape, Joachim. 1984. *"Historie" im Mittelalter und früher Neuzeit: begriffs- und gattungsgeschichtliche Untersuchungen in interdisziplinären Kontext.* Baden-Baden: Valentin Koerner.

Koelbing, Huldrych M. 1967. *Der Renaissance der Augenheilkunde, 1540–1630.* Bern and Stuttgart: Huber.

Koelbing, Huldrych M. 1968. "Ocular Physiology in the Seventeenth Century and Its Acceptance by the Medical Profession." In *Analecta medico-historica,* 3: *Steno and Brain Research in the Seventeenth Century.* Oxford: Pergamon, 219–224.

Kolsky, Stephen. 1987. "Culture and Politics in Renaissance Rome: Marco Antonio Altieri's Roman Weddings." *Renaissance Quarterly* 40:49–90.

Kopal, Zdenek, and Robert W. Carder. 1974. *Mapping of the Moon: Past and Present*. Dordrecht: Reidel.

Koselleck, Reinhart. 1984. "*Historia Magistra Vitae*: über die Auflösung des Topos im Horizont neuzeitlich bewegter Geschichte." In *Vergangene Zukunft*. Frankfurt: Suhrkamp, 38–66.

Krauss, Rolf. 1980. "Isis." In *Lexikon der Ägyptologie*. Wiesbaden: Harrassowitz, 3:186–204.

Kraye, Jill. 1986. "Moral Philosophy." In Charles B. Schmitt and Quentin Skinner, eds., *Cambridge History of Renaissance Philosophy*. Cambridge: Cambridge University Press, 303–386.

Kusukawa, Sachiko. 1997. "Leonhart Fuchs on the Importance of Pictures." *Journal of the History of Ideas* 58:403–427.

Ladner, Gerhart. 1983. "The Concept of Image in the Greek Fathers and the Byzantine Iconoclastic Controversy." In Ladner, *Images and Ideas in the Middle Ages*, vol. 1. Rome: Edizioni di Storia e Letteratura.

Laín Entralgo, Pedro. 1950. *La historia clínica: historia y teoría del relato patográfico*. Madrid: Consejo Superior de Investigaciones Científicas.

Lamberton, Robert. 1986. *Homer the Theologian: Neoplatonist Allegorical Reading and the Growth of the Epic Tradition*. Berkeley: University of California Press.

Lamprecht, Franz. 1950. *Zur Theorie der humanistischen Geschichtsschreibung: Mensch und Geschichte bei Francesco Patrizi*. Winterthur: Ziegler.

Lanciani, Rodolfo. 1989. *Storia degli scavi di Roma*. Rome: Quasar.

Landfester, Rüdiger. 1972. *Historia magistra vitae: Untersuchungen zur humanistischen Geschichtstheorie des 14. bis 16. Jahrhunderts*. Geneva: Librairie Droz.

Lateiner, Donald. 1986. "The Empirical Element in the Methods of the Early Greek Medical Writers and Herodotus: A Shared Epistemological Response." *Antichthon* 20:1–20.

Lazzaro, Claudia. 1990. *The Italian Renaissance Garden*. New Haven: Yale University Press.

Lechner, Joan Marie. 1962. *Renaissance Concepts of the Commonplaces: An Historical Investigation of the General and Universal Ideas Used in All Argumentation and Persuasion, with Special Emphasis on the Educational and Literary Tradition of the Sixteenth and Seventeenth Centuries*. New York: Pageant Press.

Leclerq, Henri. 1934. "Monuments de la monarchie française." In Fernand Cabrol and Henri Leclerq, eds., *Dictionnaire d'archéologie chrétienne et de liturgie*. 15 vols. Paris: Letouzey et Ané, 9:2708–2747.

Lefevre, Wolfgang, Jürgen Renn, and Urs Schoepflin, eds. 2003. *The Power of Images in Early Modern Science*. Basel: Birkhäuser.

Legré, Ludovic. 1899–1904. *La botanique en Provence au XVIe siècle*. Marseille: [n.p].

Leijenhorst, Cees, and Christoph Lüthy. 2002. "The Erosion of Aristotelianism: Confessional Physics in Early Modern Germany and the Dutch Republic." In Cees Leijenhorst, Christoph Lüthy, and Johannes Thijssen, eds., *The Dynamics of Aristotelian Natural Philosophy*. Leiden: Brill, 375–411.

Lennox, J. G. 1985. "Theophrastus and the Limits of Teleology." In William Fortenbaugh et al., eds., *Theophrastus of Eresus*. New Brunswick: Transaction Books, 143–164.

Le Paige, C. 1891–92. "Un astronome belge du XVIIe siècle: Godefroy Wendelin." *Ciel et terre* 12:57–66, 81–90.

Leu, Urs B. 1990. *Conrad Gesner als Theologe: ein Beitrag zur Zürcher Geistesgeschichte des 16. Jahrhunderts*. Zürcher Beiträge zur Reformationsgeschichte, 14. Bern: P. Lang.

Leventhal, Robert. 1994. *The Disciplines of Interpretation: Lessing, Herder, Schlegel and Hermeneutics in Germany*. Berlin: De Gruyter.

Lhote, Jean-François, and Danielle Joyal, eds. 1995. *Correspondance de Peiresc et Aleandro*. 2 vols. Clermont-Ferrand: Adosa.

Licoppe, Christian. 1994. "The Crystallization of a New Narrative Form in Experimental Reports (1660–1690)." *Science in Context* 7:206–244.

Liebeschütz, Wolf. 1999. "The Significance of the Speech of Praetextatus." In Polymnia Athanassiadi and Michael Frede, eds., *Pagan Monotheism in Late Antiquity*. Oxford: Clarendon Press, 185–205.

Lindberg, David G. 1976. *Theories of Vision from al-Kindi to Kepler*. Chicago: University of Chicago Press.

Loemker, Leroy E. 1972. *Struggle for Synthesis: The Seventeenth-Century Background of Leibniz's Synthesis of Order and Freedom*. Cambridge, Mass.: Harvard University Press.

Lohr, Charles. 1988. "Metaphysics." In Charles B. Schmitt and Quentin Skinner, eds., *The Cambridge History of Renaissance Philosophy*. Cambridge: Cambridge University Press, 537–638.

Lonie, Iain. 1985. "The 'Paris Hippocratics': Teaching and Research in Paris in the Second Half of the Sixteenth Century." In A. Wear, R. K. French, and Iain M. Lonie, eds., *The Medical Renaissance of the Sixteenth Century*. Cambridge: Cambridge University Press, 155–174, 318–326.

López Piñero, José, and Francisco Calero. 1988. *Los temas polémicos de la medicina renacentista: las Controversias (1556) de Francisco Valles*. Madrid: Consejo Superior de Investigaciones Científicas.

Louis, Pierre. 1955. "Le mot ἱστορία chez Aristote." *Revue de philologie, de littérature et d'histoire anciennes*, ser. 3, 29:39–44.

Louthan, Howard. 1994. *Johann Crato and the Austrian Habsburgs: Reforming a Counter-reform Court*. Princeton: Princeton Theological Seminar.

Lovejoy, Arthur O. 1964 [1936]. *The Great Chain of Being: A Study of the History of an Idea*. Cambridge, Mass.: Harvard University Press.

Lowood, Henry. 1995. "The New World and the European Catalog of Nature." In Karen Ordahl Kupperman, ed., *America in European Consciousness, 1493–1750*. Chapel Hill: University of North Carolina Press, for the Institute of Early American History and Culture, Williamsburg, Va., 295–323.

Lugli, Adalgisa. 1983. *Naturalia et mirabilia: il collezionismo enciclopedico nelle Wunderkammern d'Europa*. Milan: Mazzotta.

Lusignan, Serge. 1987. *Parler vulgairement: les intellectuels et la langue française aux XIIIe et XIVe siècles*. Paris: Vrin; Montréal: Presses de l'Université de Montréal.

Lux, David S. 1989. *Patronage and Royal Science in Seventeenth-Century France: The Académie de Physique in Caen*. Ithaca: Cornell University Press.

Lyon, Gregory. 2003. "Baudouin, Flacius and the Plan for the Magdeburg Centuries." *Journal of the History of Ideas* 64:253–272.

Lyons, John D. 1989. *Exemplum: The Rhetoric of Example in Early Modern France and Italy*. Princeton: Princeton University Press.

MacCormack, Sabine. 2001. "Approaches to Historicization: Romans and Incas in the Light of Early Modern Spanish Scholarship." In Most 2001, 69–101.

Maclean, Ian. 1992. *Interpretation and Meaning in the Renaissance: The Case of Law*. Cambridge: Cambridge University Press.

Maclean, Ian. 2001. "Logical Division and Visual Dichotomies: Ramus in the Context of Renaissance Legal and Medical Writing." In Feingold, Freedman, and Rother 2001, 229–249.

Maclean, Ian. 2002. *Logic, Signs, and Nature in the Renaissance: The Case of Learned Medicine*. Cambridge: Cambridge University Press.

Malcolm, Noel. 2002. *Aspects of Hobbes*. Oxford: Clarendon Press.

Maloney, G., and R. Savoie. 1982. *Cinq cent ans de bibliographie hippocratique, 1473–1982*. St.-Jean-Chrysostome, Quebec: Sphinx.

Mandosio, Jean-Marc. 1995. "L'histoire dans les classifications des sciences et des arts à la Renaissance." *Corpus* 28:43–72.

Mandosio, Jean-Marc. 2002. "La bibliographie de l'histoire chez Conrad Gesner." In *L'histoire en marge de l'histoire à la Renaissance*. Paris: Editions rue d'Ulm, 13–47.

Mandowsky, Erna, and Charles Mitchell. 1963. *Pirro Ligorio's Roman Antiquities*. London: Warburg Institute.

Manuel, Frank. 1959. *The Eighteenth Century Confronts the Gods*. Cambridge, Mass.: Harvard University Press.

Manuel, Frank. 1963. *Isaac Newton, Historian*. Cambridge, Mass.: Harvard University Press.

Marino, Luigi. 1975. *I maestri della Germania: Göttingen 1770–1820*. Turin: Einaudi. Trans. as Marino 1995.

Marino, Luigi. 1995. *Praeceptores Germaniae: Göttingen 1770–1820*. Trans. Brigitte Szabó-Bechstein. Göttingen: Vandenhoeck & Ruprecht.

Marti, Hanspeter. 1982. *Philosophische Dissertationen deutscher Universitäten, 1660–1750: eine Auswahlbibliographie*. Munich: Saur.

Martorelli Vico, Romana. 2002. *Medicina e filosofia. Per una storia dell'embriologia medievale nel XIII e XIV secolo*. Milan: Guerini.

Marx, Jacques. 1974. "L'art d'observer au XVIIIe siècle: Jean Senebier et Charles Bonnet." *Janus* 61:201–220.

Mastronardi, Maria Aurelia. 1993–94. "Retorica e ideologia alla corte di Borso d'Este." *Annali della facoltà di lettere e filosofia* (Università degli Studi della Basilicata) 7:313–349.

Mayr, Otto. 1986. *Authority, Liberty, and Automatic Machinery in Early Modern Europe*. Baltimore: Johns Hopkins University Press.

Mazzarino, Santo. 1960. *Il pensiero storico classico*. 3 vols. Rome and Bari: Laterza.

Mazzetti, Serafino. 1988. *Repertorio di tutti i professori, antichi e moderni, della famosa Università e del celebre Istituto delle scienze di Bologna*. (Facsimile of the edition of Bologna, 1848.) Bologna: Arnaldo Forni.

McCuaig, William. 1991. "The Fasti Capitolini and the Study of Roman Chronology in the Sixteenth Century." *Athenaeum* 79:141–159.

McCulloch, Florence. 1960. *Medieval Latin and French Bestiaries*. Chapel Hill: University of North Carolina Press.

McKnight, Stephen A. 1991. *The Modern Age and the Recovery of Ancient Wisdom: A Reconsideration of Historical Consciousness, 1450–1650*. Columbia: University of Missouri Press.

McRae, Kenneth. 1955. "Ramist Tendencies in the Thought of Jean Bodin." *Journal of the History of Ideas* 16:306–323.

McVaugh, Michael. 1971. "The Experimenta of Arnald of Villanova." *Journal of Medieval and Renaissance Studies* 1:107–118.

McVaugh, Michael. 1975. Introduction to Arnald of Villanova, *Aphorismi de gradibus*, ed. M. McVaugh. *(Arnaldi de Villanova opera medica omnia*, 2.) Granada and Barcelona: Seminarium Historiae Medicae Granadensis.

Meek, Ronald. 1976. *Social Science and the Ignoble Savage*. Cambridge: Cambridge University Press.

Meek, Ronald. 1977. *Smith, Marx, and After*. London: Chapman & Hall.

Meerhoff, Kees. 1986. *Rhétorique et poétique au XVIe siècle en France: du Bellay, Ramus et les autres*. Leiden: Brill.

Meijer, Th. J. 1971. *Kritiek als Herwaardering: het levenswerk van Jacob Perizonius (1651–1715)*. Leiden: Leiden University Press.

Meinecke, Friedrich. 1959. *Die Entstehung des Historismus*. Munich: R. Oldenbourg Verlag. (Originally published 1936.)

Meinecke, Friedrich. 1972. *Historism: The Rise of a New Historical Outlook.* Translation of Meinecke 1959. London. Routledge & Kegan Paul.

Menzel, Michael. 1998. *Predigt und Geschichte. Historische Exempel in der geistlichen Rhetorik des Mittelalters.* Cologne: Böhlau.

Mercati, Giovanni. 1917. "Su Francesco Calvo da Menaggio primo stampatore e Marco Fabio Calvo da Ravenna primo traduttore latino del corpo ippocratico." *Notizie varie di antica letteratura medica e di bibliografia.* Studi e testi 31. Vatican City: Biblioteca Apostolica Vaticana, 47–71.

Meulen, Marjon van der. 1997. "Nicolas-Claude Fabri de Peiresc and Antique Glyptic." In Clifford Malcolm Brown, ed., *Engraved Gems: Survivals and Revivals.* Washington: National Gallery of Art, 195–227.

Mieder, Wolfgang, ed. 1994. *Wise Words: Essays on the Proverb.* New York: Garland.

Mieder, Wolfgang, and Alan Dundes, eds. 1994. *The Wisdom of Many: Essays on the Proverb.* Madison: University of Wisconsin Press.

Mikkeli, Heikki. 1999. *Hygiene in Early Modern Medical Tradition.* Helsinki: Finnish Academy of Science and Letters.

Miller, Peter N. 2000. *Peiresc's Europe: Learning and Virtue in the Seventeenth Century.* New Haven: Yale University Press.

Miller, Peter N. 2001a. "The Antiquary's Art of Comparison: Peiresc and Abraxas." In Ralph Häfner, ed., *Philologie und Erkenntnis. Beiträge zu Begriff und Problem frühneuzeitlicher 'Philologie.'* Tübingen: Max Niemeyer Verlag, 57–94.

Miller, Peter N. 2001b. "Taking Paganism Seriously: Anthropology and Antiquarianism in Early Seventeenth-Century Histories of Religion." *Archiv für Religionsgeschichte* 3:183–209.

Minelli, Alessandro, ed. 1995. *The Botanical Garden of Padua, 1545–1995.* Venice: Marsilio.

Momigliano, Arnaldo. 1950. "Ancient History and the Antiquarian." *Journal of the Warburg and Courtauld Institutes* 13:285–315. (Reprinted in Momigliano, *Studies in Historiography.* London: Weidenfeld and Nicolson, 1966, 1–39.)

Momigliano, Arnaldo. 1955. "Genesi e funzione del concetto di Ellenismo." In Momigliano, *Contributo alla storia degli studi classici.* Rome: Edizioni di Storia e Letteratura, 165–193.

Momigliano, Arnaldo. 1958. "L'eredità della filologia antica e il metodo storico." *Rivista storica italiana* 70:442–458.

Momigliano, Arnaldo. 1985. "La storia tra medicina e retorica." In Momigliano, *Tra storia e storicismo.* Pisa: Nistri-Lischi, 1–24.

Momigliano, Arnaldo. 1987. "History between Medicine and Rhetoric." In Momigliano, *Ottavo contributo alla storia degli studi classici e del mondo antico.* Rome: Edizioni di Storia e Letteratura, 13–25.

Momigliano, Arnaldo. 1990. *The Classical Foundations of Modern Historiography.* Berkeley: University of California Press.

Montesquiou-Fezensac, Blaise de, and Danièle Gaborit-Chopin. 1973–77. *Le Thresor de Saint-Denis.* 3 vols. Paris: Picard.

Moraux, Paul. 1985. "Galen and Aristotle's *De partibus animalium*." In *Aristotle on Nature and Living Things: Philosophical and Historical Studies Presented to David M. Balme on His Seventieth Birthday.* Pittsburgh: Mathesis, 327–344.

Moreau-Reibel, Jean. 1933. *Jean Bodin et le droit public comparé dans ses rapports avec la philosophie de l'histoire.* Paris: Vrin.

Moss, Ann. 1996. *Printed Commonplace-books and the Structuring of Renaissance Thought.* Oxford: Oxford University Press.

Most, Glenn W. 1984. "Rhetorik und Hermeneutik: Zur Konstitution der Neuzeitlichkeit." *Antike und Abendland* 30:62–79.

Most, Glenn W., ed. 2001. *Historicization—Historisierung.* Göttingen: Vandenhoeck & Ruprecht.

Muhlack, Ulrich. 1991. *Geschichtswissenschaft im Humanismus und in der Aufklärung: die Vorgeschichte des Historismus.* Munich: Beck.

Müller, Ingo Wilhelm. 1991. *Iatromechanische Theorie und ärztliche Praxis im Vergleich zur galenistischen Medizin.* Stuttgart: Steiner.

Müller-Wille, Staffan. 2001. "History Redoubled: The Synthesis of Facts in Linnaean Natural History." In *A History of Facts*, Preprint 174. Berlin: Max-Planck-Institut für Wissenschaftsgeschichte, 23–36. (Also as "La storia raddoppiata: La sintesi dei fatti nella storia naturale di Linneo," in Cerutti and Pomata 2001a, 823–842.)

Mulsow, Martin. 1998. *Frühneuzeitliche Selbsterhaltung. Telesio und die Naturphilosophie der Renaissance.* Tübingen: Niemeyer.

Mulsow, Martin. 2001. "John Seldens *De diis Syris*. Idolatriekritik und vergleichende Religionsgeschichte im 17. Jahrhundert." *Archiv für Religionsgeschichte* 3:1–24.

Mulsow, Martin, ed. 2002. *Das Ende des Hermetismus. Historische Kritik und neue Naturphilosophie in der Spätrenaissance.* Tübingen: Niemeyer.

Mulsow, Martin. 2003. "Polyhistorie." In Gert Ueding, ed., *Historisches Wörterbuch der Rhetorik.* Darmstadt: Wissenschaftliche Buchgesellschaft, 6:1522–1526.

Mulsow, Martin (forthcoming). "Ein unbekanntes Gespräch Telesios. Sensualismus, Aristoteleskritik und die Theorie des Lichtes im 16. Jahrhundert."

Münster, Ladislao. 1963. "La cultura e le scienze nell'ambiente medico umanistico rinascimentale di Ferrara." In *Atti del Convegno Internazionale per la celebrazione del V centenario della nascita di Giovanni Manardi (1462–1536).* Ferrara: Università degli Studi di Ferrara, 57–93.

Murdoch, John. 1982. "The Analytic Character of Late Medieval Learning: Natural Philosophy without Nature." In Lawrence D. Roberts, ed., *Approaches to Nature in the Middle Ages.* Binghamton, N.Y.: Center for Medieval and Renaissance Studies, 171–213.

Murdoch, John. 1989. "The Involvement of Logic in Late Medieval Natural Philosophy." In Stefano Caroti, ed., *Studies in Medieval Natural Philosophy*. Florence: Olschki, 3–28.

Murray Jones, Peter. 1991. "Consilium, narratio, memorandum: Types of Medieval Case History." Paper presented at "The History of the Case History," conference at Robert Bosch Institut für die Geschichte der Medizin, Stuttgart, June 1991.

Nadel, George. 1964. "Philosophy of History before Historicism." *History & Theory* 3:291–315.

Nagel, Silvia. 1983. "*Puer* e *pueritia* nella letteratura medica del XIII secolo." *Annali della Fondazione Giangiacomo Feltrinelli* (Milan) 23:87–107.

Nance, Brian. 2001. *Turquet de Mayerne as Baroque Physician: The Art of Medical Portraiture*. Amsterdam: Rodopi.

Nardi, Raymond. 1980. "Le naturaliste entre Pline et Fabre." In Jacques Ferrier, ed., *Fioretti du Quadricentenaire de Fabri de Peiresc*. Avignon: Aubanel.

Naso, Irma. 2000. *Università e sapere medico nel Quattrocento. Pantaleone da Confienza e le sue opere*. Cuneo: Società per gli studi storici di Cuneo.

Nauert, Charles G., Jr. 1979. "Humanists, Scientists, and Pliny: Changing Approaches to a Classical Author." *American Historical Review* 84:72–85.

Neveu, Bruno. 1994. *Erudition et religion aux XVIIe et XVIIIe siècles*. Paris: Albin Michel.

Nicoud, Marilyn. 1994. "'Che manza fichi, semina rogna': problèmes d'identification d'une dermatose au Moyen Age." *Médiévales* 26:85–101.

Nicoud, Marilyn. 1998. "Aux origines d'une médécine préventive: les traitées de diététique en Italie et en France (XIIIe–XVe siècle)." Thèse de doctorat, Ecole Pratique des Hautes Etudes, IVe Section, Paris. Lille: Atelier national de Reproduction des Thèses, 2000.

Nicoud, Marilyn. 2000. "Expérience de la maladie et échange epistolaire: les derniers moments de Bianca Maria Visconti (mai-octobre 1468)." *Mélanges de l'Ecole française de Rome. Moyen Age* 112:311–458.

Nicoud, Marilyn. 2002. "Les médecins et le bain thermal à la fin du Moyen Age." In Didier Boisseuil, ed., *Le bain: espaces et pratiques*. Médiévales, 43. Saint-Denis: Presses Universitaires de Vincennes.

Nutton, Vivian. 1987. *John Caius and the Manuscripts of Galen*. Cambridge: Cambridge Philological Society.

Nutton, Vivian. 1989a. "Hippocrates in the Renaissance." In Gerhard Baader and Rolf Winau, eds., *Die hippokratischen Epidemien, Theorie-Praxis-Tradition. Sudhoffs Archiv*, Beiheft 27. Stuttgart: Steiner, 420–439.

Nutton, Vivian. 1989b. "Pieter van Foreest and the Plagues of Europe: Some Observations on the *Observationes*." In Houtzager 1989, 25–39.

Nutton, Vivian. 1991a. "Case Histories in the Early Renaissance." Paper presented at "The History of the Case History," conference at Robert Bosch Institut für die Geschichte der Medizin, Stuttgart, June 1991.

Nutton, Vivian. 1991b. "Style and Context in the Method of Healing." In Fridolf Kudlien, ed., *Galen's Method of Healing*. Leiden: Brill, 1–25.

Nutton, Vivian. 1993a. "Galen at the Bedside: The Methods of a Medical Detective." In Bynum and Porter 1993.

Nutton, Vivian. 1993b. "Greek Science in the Sixteenth-Century Renaissance." In J. V. Field and Frank A. J. L. James, eds., *Renaissance and Revolution: Humanists, Scholars, Craftsmen and Natural Philosophers in Early Modern Europe*. Cambridge: Cambridge University Press, 15–28.

Nutton, Vivian. 1997. "The Rise of Medical Humanism: Ferrara 1464–1555." *Renaissance Studies* 11:2–19.

Ogilvie, Brian. 1997. "Observation and Experience in Early Modern Natural History." Ph.D. dissertation, University of Chicago.

Ogilvie, Brian W. (forthcoming 2006). *The Science of Describing: Natural History in Renaissance Europe, 1490–1620*. Chicago: University of Chicago Press.

Olmi, Giuseppe. 1991. "Molti amici in vari luoghi: studio della natura e rapporti epistolari nel secolo XVI." *Nuncius* 6:3–31.

Olmi, Giuseppe. 1992. *L'inventario del mondo. Catalogazione della natura e luoghi del sapere nella prima età moderna*. Bologna: Il Mulino.

Olmi, Giuseppe, Lucia Tongiorgi Tomasi, and Attilio Zanca, eds. 2000. *Natura-cultura: l'interpretazione del mondo fisico nei testi e nelle immagini*. Florence: Olschki.

O'Neill, Ynez V. 1974. "The 'fera' or Blighted Twin Phenomena." *Medical History* 18:222–239.

O'Neill, Ynez V. 1975. "Giovanni Michele Savonarola: An Atypical Renaissance Practitioner." *Clio Medica* 10:77–93.

Ong, Walter. 1958. *Ramus, Method and the Decay of Dialogue*. Cambridge, Mass.: Harvard University Press.

Ong, Walter. 1976. "Commonplace Rhapsody: Ravisius Textor, Zwinger and Shakespeare." In R. R. Bolgar, ed., *Classical Influences on European Culture AD 1500–1700*. Cambridge: Cambridge University Press, 91–126.

Ong, Walter. 1977. *Interfaces of the Word: Studies in the Evolution of Consciousness and Culture*. Ithaca: Cornell University Press.

Osborne, John, et al., eds. 1997. *The Paper Museum of Cassiano dal Pozzo: Early Christian and Medieval Antiquities*. London: Harvey Miller.

Osler, Margaret J. 2001. "Whose Ends? Teleology in Early Modern Natural Philosophy." *Osiris* 16:151–168.

Otto, Stephan. 1979. "Die mögliche Wahrheit der Geschichte: die *Dieci Dialoghi della historia* des Francesco Patrizi in ihrer geistes-geschichtlichen Bedeutung." In Otto, *Materialien zur Theorie der Geistesgeschichte*. Munich: Fink.

Palmer, Richard. 1981. "Physicians and the State in Post-medieval Italy." In Andrew W. Russell, ed., *The Town and State Physician in Europe from the Middle Ages to the Enlightenment*. Wolfenbüttel: Herzog August Bibliothek, 47–60.

Palmer, Richard. 1985. "Medical Botany in Northern Italy in the Renaissance." *Journal of the Royal Society of Medicine* 78:149–159.

Pandel, Hans-Jürgen. 1990. *Historik und Didaktik: das Problem der Distribution historiographisch erzeugten Wissens in der deutschen Geschichtswissenschaft von der Spätaufklärung zum Frühhistorismus (1765–1830)*. Baden-Baden: Frommann-Holzboog.

Panofsky, Erwin. 1960. *Renaissance and Renascences in Western Art*. Uppsala: Almquist & Wiksell.

Papagno, Giuseppe, and Amedeo Quondam, eds. 1982. *La corte e lo spazio: Ferrara estense*. 3 vols. Rome: Bulzoni.

Papy, Jean. 1999. "The Attitude towards Aristotelian Biological Thought in the Louvain Medical Treatises during the Sixteenth and Early Seventeenth Century: The Case of Embryology." In Carlos Steel, Guy Guldentops, and Pieter Beullens, eds., *Aristotle's Animals in the Middle Ages and Renaissance*. Leuven: Leuven University Press, 317–337.

Park, Katharine. 1985. *Doctors and Medicine in Early Renaissance Florence*. Princeton: Princeton University Press.

Park, Katharine. 1999. "Natural Particulars: Medical Epistemology, Practice, and the Literature of Healing Springs." In Grafton and Siraisi 1999, 347–367.

Parodi, Massimo. 1987. "Congettura e metafisica." In Mariateresa Fumagalli Beonio Brocchieri, *Le bugie di Isotta: immagini della mente medievale*. Bari: Laterza, 109–150.

Pastine, Dino. 1978. *La nascita dell'idolatria. L'Oriente religioso di Athanasius Kircher*. Florence: Nuova Italia.

Pastoureau, Michel. 1982a. "Le bestiaire héraldique au Moyen Age." In *L'Hermine et le Sinople. Études d'héraldique médiévale*. Paris: Léopard d'or, 105–116.

Pastoureau, Michel. 1982b. "Du Cange héraldiste." In *L'Hermine et le Sinople. Études d'Héraldique Médiévale*. Paris: Léopard d'or, 335–342.

Pastoureau, Michel. 1982c. "Le sanglier dans les sceaux du Moyen Age." In *L'Hermine et le Sinople. Études d'Héraldique Médiévale*. Paris: Léopard d'or, 117–126.

Pazzini, A. 1957. "William Harvey, Disciple of Girolamo Fabrizi d'Acquapendente and the Paduan School." *Journal of the History of Medicine and Allied Sciences* 12:197–201.

Pecchiai, Pio. 1944. *Acquedotti e fontane di Roma nel Cinquecento (con documenti inediti)*. Rome: Staderini Editore.

Pélissier, Leon G. 1888. "Les amis d'Holstenius, III: Aleandro le Jeune." *Mélanges d'archéologie et d'histoire* 8:323–402, 521–608.

Pellegrin, Pierre. 1982. *La classification des animaux chez Aristote: statut de la biologie et unité de l'aristotélisme*. Paris: Belles Lettres.

Perarnau, Josep, ed. 1995. *Actes de la I Trobada internacional d'estudis sobre Arnau de Vilanova.* 2 vols. Barcelona: Institut d'Estudis Catalans i Facultat de Teologia de Catalunya.

Pereira, Michela. 1996. "Medioevo, alchimia, 'coscienza femminile'. Da una ricerca storica a una proposta epistemologica." In *Seminario del Dipartimento di Filosofia e Scienze sociali dell'Università di Siena*, Pontignano (typescript).

Perfetti, Stefano. 1999a. "Docebo vos dubitare. Il commento inedito di Pietro Pomponazzi al *De partibus animalium* (Bologna 1521-1524)." *Documenti e studi sulla tradizione filosofica medievale* 10:439–466.

Perfetti, Stefano. 1999b. "Three Different Ways of Interpreting Aristotle's *De partibus animalium*: Pietro Pomponazzi, Niccolò Leonico Tomeo and Agostino Nifo." In Carlos Steel, Guy Guldentops, and Pieter Beullens, eds., *Aristotle's Animals in the Middle Ages and Renaissance.* Leuven: Leuven University Press, 299–316.

Perfetti, Stefano. 2000. *Aristotle's Zoology and Its Renaissance Commentators.* Leuven: Leuven University Press.

Perfetti, Stefano. 2002. "How and When the Medieval Commentary Died Out: The Case of Aristotle's Zoological Writings." In Gianfranco Fioravanti, Claudio Leonardi, and Stefano Perfetti, eds., *Il commento filosofico nell'Occidente latino (secoli XIII–XV).* Turnhout: Brepols, 429–443.

Perilli, Lorenzo. 2004. *Menodoto di Nicomedia. Contributo a una storia galeniana della medicina empirica.* Munich and Leipzig: K. G. Saur.

Pertile, Antonio. 1902. *Storia del diritto italiano dalla caduta dell'impero romano alla codificazione.* 2d ed. (Facsimile reprint 1966.) Bologna: Arnaldo Forni Editore.

Pesenti, Tiziana. 1977. "Michele Savonarola a Padova: l'ambiente, le opere, la cultura medica." *Quaderni per la storia dell'Università di Padova* 9–10:45–102.

Pesenti, Tiziana. 1978. "Professores chirurgie, medici ciroici e barbitonsores a Padova nell'età di Leonardo Buffi da Bertapaglia (+dopo il 1448)." *Quaderni per la storia dell'Università di Padova* 11:1–38.

Pesenti, Tiziana. 1983. "Generi e pubblico della letteratura medica padovana nel Tre e Quattrocento." In *Università e società nei secoli, Atti del Nono Convegno Internazionale del Centro Italiano di Studi di Storia e d'Arte, Pistoia, 1982.* Bologna: Centro Italiano di Studi di Storia e d'Arte, 523–545.

Pesenti, Tiziana. 1984. *Professori e promotori di medicina nello studio di Padova dal 1405 al 1509. Repertorio biobibliografico.* Padua and Trieste: Lindt.

Pesenti, Tiziana. 1997. "Medici di corte e università." *Medicina nei secoli* 9:391–401.

Pesenti, Tiziana. 2003. *Marsilio Santasofia tra corti e università. La carriera di un "Monarcha medicine" del Trecento.* Treviso: Edizioni Antilia.

Petrucci, A. 1979. "Ceccarelli, Alfonso." In *Dizionario biografico degli italiani.* Rome: Istituto dell'Enciclopedia Italiana, 23:199–202.

Petrucci, F. 1982. "Colonna, Ascanio." In *Dizionario biografico degli italiani*. Rome: Istituto dell'Enciclopedia Italiana, 27:275–278.

Phillips, Mark. 1979. "Machiavelli, Guicciardini, and the Tradition of Vernacular Historiography in Florence." *American Historical Review* 84:86–105.

Phillips, Mark. 1996. "Reconsiderations on History and Antiquarianism: Arnaldo Momigliano and the Historiography of Eighteenth-Century Britain." *Journal of the History of Ideas* 57:297–316.

Pinon, Laurent. 1995. *Livres de zoologie de la Renaissance: une anthologie*. Paris: Kincksieck.

Pinon, Laurent. 2002. "Clématite bleue contre poissons séchés: sept lettres inédites d'Ippolito Salviani à Ulisse Aldrovandi." *Mélanges de l'Ecole Française de Rome. Italie et Méditerranée* 114:477–492.

Pinon, Laurent. 2003. "Entre compilation et observation: l'écriture de l'ornithologie d'Ulisse Aldrovandi." *Genesis* 20 (special issue, "L'écriture scientifique"), 53–70.

Pintard, René. 1943. *Le libertinage érudit dans la première moitié du 17. siècle*. Paris: Boivin. (Rev. ed., Geneva: Slatkine, 2000).

Pitassi, Maria Cristina. 1987. *Entre croire et savoir: le problème de la méthode chez Jean Le Clerc*. Leiden: Brill.

Pittion, J.-P. 1987. "Scepticism and Medicine in the Renaissance." In R. H. Popkin and Charles B. Schmitt, eds., *Scepticism from the Renaissance to the Enlightenment*. Wiesbaden: In Kommission bei O. Harrassowitz, 103–132.

Pocock, J. G. A. 1957. *The Ancient Constitution and the Feudal Law*. Cambridge: Cambridge University Press.

Pomata, Gianna. 1996. "'Observatio' ovvero 'Historia.' Note su empirismo e storia in età moderna." *Quaderni storici* 31.91:173–198.

Pomata, Gianna. 1998. *Contracting a Cure: Patients, Healers and the Law in Early Modern Bologna*. Baltimore: Johns Hopkins University Press. (Originally published as *La promessa di guarigione: malati e curatori in antico regime: Bologna XVI–XVIII secolo*, Rome: Laterza, 1994.)

Pomata, Gianna. 2002. "Lecturing on Discovery. Pavia 1625: Gaspare Aselli Lectures on His Discovery of the Lacteals." Paper presented at the 8th International Summer School in History of Science: "Rethinking Scientific Knowledge in the 16th and 17th Centuries." Centre Alexandre Koyré, Paris, September 16–20.

Pomian, Krzysztof. 1987. *Collectionneurs, amateurs et curieux: Paris-Venise 16.-18 siècle*. Paris: Gallimard.

Poovey, Mary. 1998. *A History of the Modern Fact: Problems of Knowledge in the Sciences of Wealth and Society*. Chicago: University of Chicago Press.

Popkin, Richard H. 1979. *The History of Scepticism from Erasmus to Spinoza*. Berkeley: University of California Press.

Poppi, Antonino, ed. 1993. *Scienza e filosofia all'Università di Padova nel Quattrocento*. Padua and Trieste: Lindt.

Potter, Paul. 1989. "Epidemien I/III: Form und Absicht der zweiundzwanzig Fallbeschreibungen." In Gerhard Baader and Rolf Winau, eds., *Die hippokratischen Epidemien. Theorie-Praxis-Tradition. Sudhoffs Archiv*, Beiheft 27, pp. 9–19.

Praz, Mario. 1964. *Studies in Seventeenth-Century Imagery*. 2d ed. Rome: Edizioni di Storia e Letteratura.

Premuda, Loris. 1993. *Storia dell'iconografia anatomica*. Milan: Ciba Edizioni.

Premuda, Loris. 1974. "Mercati, Michele." In *Dictionary of Scientific Biography* 9:308–309.

Quaglioni, Diego. 1987. "Il modello del principe cristiano: gli *specula principum* tra medioevo e età moderna." In Vittor Ivo Comparato, ed., *Modelli nella storia del pensiero politico*. Florence: Olschki, 1:103–122.

Quondam, Amedeo. 1978. "Le biblioteche della corte estense." In Quondam, ed., *Il libro a corte*. Rome: Bulzoni, 7–38.

Rademaker, Cornelis S. M. 1981. *Life and Work of Gerardus Joannes Vossius*. Assen: Van Gorcum.

Raggio, Osvaldo. 2000. *Storia di una passione. Cultura aristocratica e collezionismo alla fine dell'ancien régime*. Venice: Marsilio.

Rapisarda, Stefano. 2001. "Appunti sulla circolazione del 'Secretum secretorum' in Italia." In Riccardo Gualdo, ed., *Le parole della scienza: scritture tecniche e scientifiche in volgare (secoli XIII–XV): atti del convegno, Lecce, 16–18 aprile 1999*. Galatina: Congedo, 77–97.

Raskolnikova, Muza. 1992. *Histoire romaine et critique historique dans l'Europe des Lumières: la naissance de l'hypercritique dans l'historiographie de la Rome antique*. Rome: Ecole Française de Rome.

Rath, Gernot. 1950–51. "Die Briefe Konrad Gessners aus der Trewschen Sammlung." *Gesnerus* 7:140–170; 8:195–215.

Rechenauer, Georg. 1991. *Thukydides und die hippokratischen Medizin. Naturwissenschaftliche Methodik als Modell für Geschichtsdeutung*. Hildesheim: G. Olms.

Redondi, Pietro. 1983. *Galileo eretico*. Turin: Einaudi.

Reeds, Karen M. 1976. "Renaissance Humanism and Botany." *Annals of Science* 33:519–542.

Reeds, Karen M. 1991. *Botany in Medieval and Renaissance Universities*. New York: Garland.

Regoliosi, Mariangela. 1991. "Riflessioni umanistiche sullo 'scrivere storia.'" *Rinascimento*, ser. II, 31:3–37.

Regoliosi, Mariangela. 1994. "Lorenzo Valla e la concezione della storia." *La storiografia umanistica. Convegno internazionale dell'Associazione per il Medioevo e l'Umanesimo latini, Messina, 22–25 ottobre 1987*. Messina: Sicania, vol. I, pt. 2, 549–571.

Regoliosi, Mariangela. 1995a. "'Res gestae patriae' e 'res gestae ex universa Italia': la lettera di Lapo da Castiglionchio a Biondo Flavio." In Claudia Bastia and Maria Bolognani, eds., *La memoria e la città*. Bologna (Istituto per i beni artistici, culturali e naturali della Regione Emilia-Romagna): Il Nove, 273–305.

Regoliosi, Mariangela. 1995b. "Tradizione contro verità: Cortesi, Sandei, Mansi e l'orazione del Valla sulla 'Donazione di Costantino.'" *Momus* 3–4:47–57.

Reina, Maria Elena. 2002. *"Hoc hic et nunc." Buridano, Marsilio di Inghen e la conoscenza del singolare*. Florence: Olschki.

Reinhardt, Volker. 1991. *Überleben in der frühneuzeitlichen Stadt. Annona und Getreideversorgung in Rom 1563–1797*. Tübingen: Max Niemeyer Verlag.

Renard, Marcel. 1953. "Iuno Istoria." *Latomus* 12:137–154.

Renazzi, Filippo Maria. 1804. *Storia dell'Università degli studi di Roma*. Rome, 2:188–191.

Reynolds, Beatrice. 1953. "Shifting Currents in Historical Criticism." *Journal of the History of Ideas* 14:471–492.

Reynolds, L. D. 1983. *Texts and Transmission: A Survey of the Latin Classics*. Oxford: Clarendon Press.

Rickert, Heinrich. 1902. *Die Grenzen der naturwissenschaftlichen Begriffsbildung*. Tübingen: Mohr. (Trans. as *The Limits of Concept Formation in Natural Science*. Cambridge: Cambridge University Press, 1986.)

Ricklin, Thomas, ed. (forthcoming). *"Exempla docent." Les exempla philosophiques de l'Antiquité à la Renaissance, Colloque international, Université de Neuchâtel*. Paris: Vrin.

Riedl-Dorn, Christa. 1989. *Wissenschaft und Fabelwesen, ein kritischer Versuch über Conrad Gessner und Ulisse Aldrovandi*. Vienna: Böhlau.

Rinne, Katherine W. 2000. "Fluid Precision: Giacomo della Porta and the Acqua Vergine Fountains of Rome." In Jan Birksted, ed., *Landscapes of Memory and Experience*. London and New York: Spon Press.

Rinne, Katherine W. 2001–02. "The Landscape of Laundry in Late Cinquecento Rome." *Studies in the Decorative Arts* 9:34–60.

Riondato, Ezio. 1961. *Storia e metafisica nel pensiero di Aristotele*. Padua: Editrice Antenore.

Ritschl, Otto. 1906. *System und systematische Methode: Geschichte des wissenschaftlichen Sprachgebrauchs und der philosophischen Methodologie*. Bonn: A. Marcus and E. Weber.

Rizza, Cecilia. 1961. "Galileo nella corrispondenza di Peiresc." *Studi francesi* 15:433–451.

Roberts, K. B., and J. D. W. Tomlinson. 1992. *The Fabric of the Body: European Traditions of Anatomical Illustration*. Oxford: Clarendon Press.

Robertson, Clare. 1992. *Il Gran Cardinale: Alessandro Farnese, Patron of the Arts*. New Haven: Yale University Press.

Robin, Léon. 1944. *Pyrrhon et le scepticisme grec*. Paris: Presses Universitaires de France.

Roger, Jacques. 1973. "L'humanisme médical de Symphorien Champier." In *L'humanisme français au début de la Renaissance, Colloque international d'études humanistes*. Paris: J. Vrin, 261–272.

Roger, Jacques. 1980. "The Living World." In G. S. Rousseau and Roy Porter, eds., *The Ferment of Knowledge*. Cambridge: Cambridge University Press.

Ross, David J. A. 1988. *Alexander Historiatus: A Guide to Medieval Illustrated Alexander Literature*. 2d ed. Beiträge zur Klassischen Philologie, 186. Frankfurt am Main: Athenäum.

Rossi, Mario Manlio. 1947. *La vita, le opere, i tempi di Edoardo Herbert di Chirbury*. Florence: Sansoni.

Rotondò, Antonio. 1974. "Pietro Perna e la vita culturale e religiosa di Basilea fra il 1570 e il 1580." In *Studi e ricerche di storia ereticale italiana del Cinquecento*. Turin: Edizioni Giappichelli, 1:273–391.

Rotzoll, Maike. 2000. *Pierleone da Spoleto. Vita e opere di un medico del Rinascimento*. Florence: Olschki.

Rouse, Richard H. 1981. "L'évolution des attitudes envers l'autorité écrite: le développement des instruments de travail au XIIIe siècle." In Geneviève Hasenohr and Jean Longère, eds., *Culture et travail intellectuel dans l'Occident médiéval*. Paris: Editions du Centre National de la Recherche Scientifique, 115–144.

Rowland, Ingrid. 1998. *The Culture of the High Renaissance: Ancients and Moderns in Sixteenth-Century Rome*. Cambridge: Cambridge University Press.

Rubiés, Joan-Pau. 1996. "Teaching the Eye to See." *History and Anthropology* 9:139–190.

Rubiés, Joan-Pau. 2000. *Travel and Ethnology in the Renaissance: South India Through European Eyes, 1250–1625*. Cambridge: Cambridge University Press.

Ruiz, Elisa. 1976. "Los años romanos de Pedro Chacon: vida y obras." *Cuadernos de filologia clásica* 10:189–247.

Sabra, A. I. 1981. *Theories of Light, from Descartes to Newton*, 2d ed. Cambridge: Cambridge University Press.

Saffrey, H. D. 1994. "L'homme-microcosme dans une estampe médico-philosophique du seizième siècle." *Journal of the Warburg and Courtauld Institutes* 5:89–122.

Salmon, John. 1997. "Precept, Example, and Truth: Degory Wheare and the ars historica." In Donald Kelley and David Sacks, eds., *The Historical Imagination in Early Modern Britain: History, Rhetoric, and Fiction, 1500–1800*. Washington: Woodrow Wilson Center Press; Cambridge: Cambridge University Press, 11–36.

Samaritani, Antonio. 1976. "Michele Savonarola riformatore cattolico nella corte estense a metà del secolo XV." *Atti e memorie della Deputazione provinciale ferrarese di Storia Patria*, ser. III, 22:44–85.

San Juan, Rose Marie. 2001. *Rome: A City Out of Print*. Minneapolis: University of Minnesota Press.

Sansa, Renato. 2002. "L'odore del contagio. Ambiente urbano e prevenzione delle epidemie nella prima età moderna." *Medicina e storia* 2:83–108.

Sapegno, Maria Serena. 1984. "Il trattato politico e utopico." In Alberto Asor Rosa, ed., *La letteratura italiana*. III, *Le forme del testo*. II, *La prosa*. Turin: Einaudi, 949–977.

Sartori, Marco. 1982. "L'incertitudine dei primi secoli di Roma e il metodo storico nella prima metà del Settecento." *Clio* 18:7–35.

Sartori, Marco. 1985. "Voltaire, Newton, Fréret: la cronologia e la storia delle antiche nazioni." *Studi settecenteschi* 7–8:167–189.

Saxl, Fritz. 1957. *Lectures*. London: Warburg Institute.

Scheele, Meta. 1930. *Wissen und Glaube in der Geschichtswissenschaft: Studien zum historischen Pyrrhonismus in Frankreich und Deutschland*. Heidelberg: Winter.

Scheible, Heinz. 1966. *Die Entstehung der Magdeburger Zenturien*. Gütersloh: G. Mohr.

Schlamm, Carl C. 1978. "Graduation Speches of Gentile da Foligno." *Mediaeval Studies* 40:96–119.

Schmid, Francis. 1990. "La discussion sur l'origine de l'idolatrie aux XVIIe et XVIIIe siècles." In *L'idolatrie. Rencontres de l'Ecole du Louvre*. Paris: Documentation française, 53–68.

Schmidt-Biggemann, Wilhelm. 1983. *Topica universalis: eine Modellgeschichte humanistischer und barocker Wissenschaft*. Hamburg: Meiner.

Schmidt-Biggemann, Wilhelm. 1998. *Philosophia perennis*. Frankfurt: Suhrkamp.

Schmitt, Charles B. 1969. "Experience and Experiment: A Comparison of Zabarella's View with Galileo's in 'De Motu.'" *Studies in the Renaissance* 16:80–138.

Schmitt, Charles B. 1971. "Theophrastus." In Paul Oskar Kristeller and F. Edward Kranz, eds., *Catalogus translationum et commentariorum: Mediaeval and Renaissance Latin Translations and Commentaries*. Washington: Catholic University of America Press, 2:239–322.

Schnapp, Alain. 1993. *La conquête du passé. Aux origines de l'archéologie*. Paris: Ed. Carré.

Schneewind, J. B. 1990. "The Misfortunes of Virtue." *Ethics* 101:42–63.

Schreurs, Anna. 2000. *Antikenbild und Kunstanschauungen des neapolitanischen Malers, Architekten und antiquars Pirro Ligorio (1513–1583)*. Cologne: Walther Konig.

Scott, John B. 1991. *Images of Nepotism: The Painted Ceilings of Palazzo Barberini*. Princeton: Princeton University Press.

Segarizzi, Arnaldo. 1900. *Della vita e delle opere di Michele Savonarola, medico padovano del secolo XV*. Padua: Tipografia Fratelli Gallina.

Seifert, Arno. 1976. *Cognitio historica: die Geschichte als Namengeberin der frühneuzeitlichen Empirie*. Berlin: Duncker & Humblot.

Seifert, Arno. 1977. "Historia im Mittelalter." *Archiv für Begriffsgeschichte* 21:226–284.

Serrai, Alfredo. 1990. *Conrad Gesner*. Ed. Maria Cochetti, with bibliography by Marco Menato. Rome: Bulzoni.

Seznec, Jean. 1931. "Un essai de mythologie comparée au début du XVIIe siècle." *Ecole française de Rome. Mélanges d'archéologie et d'histoire* 48:268–281.

Seznec, Jean. 1953. *The Survival of the Pagan Gods: The Mythological Tradition and Its Place in Renaissance Humanism and Art*. New York: Harper & Row.

Shapin, Steven. 1984. "Pump and Circumstance: Robert Boyle's Literary Technology." *Social Studies of Science* 14:481–520.

Shapin, Steven, and Simon Schaffer. 1985. *Leviathan and the Air Pump*. Princeton: Princeton University Press.

Shapiro, Barbara. 1979. "History and Natural History in Sixteenth- and Seventeenth-Century England: An Essay on the Relationship between Humanism and Science." In Barbara Shapiro and Robert G. Frank, Jr., *English Scientific Virtuosi in the Sixteenth and Seventeenth Centuries*. Los Angeles: William Andrews Clark Memorial Library, University of California, 3–55.

Shapiro, Barbara. 1983. *Probability and Certainty in Seventeenth Century England*. Princeton: Princeton University Press.

Shapiro, Barbara. 1991. "Early Modern Intellectual Life: Humanism, Religion and Science in Seventeenth Century England." *History of Science* 29:45–71.

Shapiro, Barbara. 1994. "The Concept of 'Fact': Legal Origins and Cultural Diffusion." *Albion* 26:227–252.

Shapiro, Barbara J. 2000. *A Culture of Fact: England, 1550–1720*. Ithaca: Cornell University Press.

Sharpe, Kevin. 1982. "The Foundation of the Chairs of History at Oxford and Cambridge: An Episode in Jacobean Politics." *History of Universities* 2:127–152.

Simili, A. 1970. "Alcune lettere inedite di Andrea Bacci a Ulisse Aldrovandi." In *Atti del XXIV Congresso Nazionale di Storia della Medicina, Taranto-Bari, 1969*. Rome, 428–437.

Siraisi, Nancy G. 1987. *Avicenna in Renaissance Italy: The Canon and Medical Teaching in Italian Universities after 1500*. Princeton: Princeton University Press.

Siraisi, Nancy G. 1993. "Life Sciences and Medicine in the Renaissance World." In Anthony Grafton, ed., *Rome Reborn: The Vatican Library and Renaissance Culture*. Washington: Library of Congress, 169–197.

Siraisi, Nancy G. 1994a. "How to Write a Latin Book on Surgery: Organizing Principles and Authorial Devices in Guglielmo da Saliceto and Dino del Garbo." In Luis Garcia Ballester et al., eds., *Practical Medicine from Salerno to the Black Death*. Cambridge: Cambridge University Press, 88–109.

Siraisi, Nancy G. 1994b. "Vesalius and Human Diversity in *De humani corporis fabrica.*" *Journal of the Warburg and Courtauld Institutes* 57:60–88.

Siraisi, Nancy G. 1996. "L'individuale' nella medicina tra Medioevo e Umanesimo: i casi clinici." In Cardini and Regoliosi 1996, 33–62.

Siraisi, Nancy G. 1997. *The Clock and the Mirror: Girolamo Cardano and Renaissance Medicine.* Princeton: Princeton University Press.

Siraisi, Nancy G. 1998. "La comunicazione del sapere anatomico ai confini tra diritto e agiografia: due casi del secolo XVI." In Massimo Galuzzi et al., eds., *Le forme della comunicazione scientifica.* Milan: Franco Angeli, 419–438. (English version, "Signs and Evidence: Autopsy and Sanctity in Late Sixteenth-Century Italy" in Siraisi 2001, 356–380.)

Siraisi, Nancy. 2000. "Anatomizing the Past: Physicians and History in Renaissance Culture." *Renaissance Quarterly* 53:1–30.

Siraisi, Nancy G. 2001. *Medicine and the Italian Universities, 1250–1600.* Leiden: Brill.

Siraisi, Nancy G. 2002a. "Disease and Symptom as Problematic Concepts in Renaissance Medicine." In Eckhard Kessler and Ian Maclean, eds., *Res et verba in the Renaissance.* Wiesbaden: Harrassowitz, 217–239.

Siraisi, Nancy G. 2002b. "Hermes among the Physicians." In Martin Mulsow, ed., *Das Ende des Hermetismus. Historische Kritik und neue Naturphilosophie in der Spätrenaissance.* Tübingen: Niemeyer, 189–212.

Siraisi, Nancy G. 2003. "History, Antiquarianism and Medicine: The Case of Girolamo Mercuriale." *Journal of the History of Ideas* 64:231–250.

Skalnik, James Veazie. 2002. *Ramus and Reform: University and Church at the End of the Renaissance.* Sixteenth Century Essays and Studies, vol. 59. Kirksville, Mo.: Truman State University Press.

Slack, Paul. 1979. "Mirrors of Health and Treasures of Poor Men: The Use of the Vernacular Medical Literature of Tudor England." In Charles Webster, ed., *Health, Medicine and Mortality in the Sixteenth Century.* Cambridge: Cambridge University Press, 237–273.

Smalley, Beryl. 1974. *Historians in the Middle Ages.* London: Thames and Hudson.

Smith, Rowland. 1995. *Julian's Gods: Religion and Philosophy in the Thought and Action of Julian the Apostate.* London: Routledge.

Smith, Wesley D. 1979. *The Hippocratic Tradition.* Ithaca: Cornell University Press.

Soleil à la Renaissance. 1965. *Le soleil à la Renaissance: sciences et mythes. Colloque international tenu en avril 1963.* Brussels: Presses Universitaires de Bruxelles.

Solinas, Francesco, ed. 1989. *Cassiano dal Pozzo: atti del seminario internazionale di studi.* Rome: De Luca.

Sparn, Walter. 1976. *Wiederkehr der Metaphysik: die ontologische Frage in der lutherischen Theologie des frühen 17. Jahrhunderts.* Stuttgart: Calwer Verlag.

Sparti, Donatella. 1992. *Le collezioni dal Pozzo: storia di una famiglia e del suo museo nella Roma seicentesca*. Modena: Franco Cosimo Panini.

Spencer, Diana. 2002. *The Roman Alexander: Reading a Cultural Myth*. Exeter: University of Exeter Press.

Spini, Giorgio. 1948. "I trattatisti dell'arte storica della Controriforma italiana." *Quaderni di Belfagor* 1:109–136.

Spini, Giorgio. 1970. "Historiography: The Art of History in the Italian Counter Reformation." In Eric Cochrane, ed., *The Late Italian Renaissance, 1525–1630*. New York: Harper, 91–133.

Spinosa, Giacinta. 2002. "'Empeiria/experientia': modelli di 'prova' tra antichità, medioevo e età cartesiana." In Marco Veneziani, ed., *Experientia. X Colloquio Internazionale del Lessico Intellettuale Europeo, Roma, 4–6 gennaio 2001*. Florence: Leo S. Olsckhi Editore, 169–198.

Spolsky, Ellen. 2001. *Satisfying Skepticism: Embodied Knowledge in the Early Modern World*. Aldershot: Ashgate.

Stabile, Giorgio. 1973. "Cagnati, Marsilio." In *Dizionario biografico degli italiani*. Rome: Istituto dell'Enciclopedia Italiana, 16:301–303.

Stannard, Jerry. 1978. "Natural History." In David C. Lindberg, ed., *Science in the Middle Ages*. Chicago: University of Chicago Press, 429–460.

Stebbins, Sara. 1980. *Maxima in minimis: zum Empirie- und Autoritätsverständnis in der physikotheologischen Literatur der Frühaufklärung*. Frankfurt am Main: Peter D. Lang.

Stefanutti, Ugo. 1979. "Andrea Bacci e la sua opera sui vini." In *Atti del XXIX Congresso Nazionale di Storia della Medicina (Casale Monferrato 22–23–24 Settembre 1978)*. Casale Monferrato: Operaia Artigiana, 269–274.

Stephany, Erich. 1957. "Die Bilder aus Aachen für Monsieur Peiresc, 1607/08." *Zeitschrift des Aachener Geschichtsvereins* 69:67–70.

Stephens, Walter. 1989. *Giants in Those Days*. Lincoln: University of Nebraska Press.

Stern, Henri. 1956. "Un sarcophage de la Gayole découvert par Peiresc." *Académie des Inscriptions & Belles-Lettres. Comptes Rendus*, 250–256.

Storiografia umanistica. 1994. *La storiografia umanistica. Atti del Convegno AMUL, Messina 1987*. Messina: Sicania.

Strauss, Leo. 1936. *The Political Philosophy of Hobbes*. Trans. E. M. Sinclair. Oxford: Oxford University Press.

Stroumsa, Guy G. 2001. "John Spencer and the Roots of Idolatry." *History of Religions* 40:1–23.

Struever, Nancy S. 1970. *The Language of History in the Renaissance: Rhetoric and Historical Consciousness in Florentine Humanism*. Princeton: Princeton University Press.

Sudhoff, Karl. 1907. "Augenanatomiebilder im 15. und 16. Jahrhundert." In *Tradition und Naturbeobachtung in den Illustrationen medizinischer Handschriften*. Leipzig: J. A. Barth, 19–26.

Sudhoff, Karl. 1910. "Abermals eine neue Handschrift der anatomischen Fünfbilderserie. Versuch einer Wiederherstellung des lateinischen Textes dieses illustrierten Leitsadens der Anatomie." *Archiv für Geschichte der Medizin* 3:353–368.

Taavitsainen, Irma. 1994. "On the Evolution of Scientific Writings from 1375 to 1675: Repertoire of Emotive Features." In F. Fernandez, M. Fuster, and J. Calvo, eds., *English Historical Linguistics*. Amsterdam: J. Benjamins, 329–342.

Tachau, Katherine. 1988. *Vision and Certitude in the Age of Ockham*. Leiden: Brill.

Temkin, Owsei. 1929. "Krankengeschichte und Sinnsphäre der Medizin." *Kyklos* 2:42–66.

Temkin, Owsei. 1973. *Galenism: Rise and Decline of a Medical Philosophy*. Ithaca: Cornell University Press.

Thomann, Johannes. 1997. *Studien zum Speculum physiognomie des Michele Savonarola*. Zurich: Copy Quick.

Thomas, Rosalind. 2000. *Herodotus in Context: Ethnography, Science and the Art of Persuasion*. Cambridge: Cambridge University Press.

Thorndike, Lynn. 1959. *A History of Magic and Experimental Science*, vol. 4. New York: Columbia University Press. (One of 8 vols. originally published 1923–1958.)

Tissoni Benvenuti, Antonia. 1987a. "Le armi e le lettere nell'educazione del signore nelle corti padane del Quattrocento." *Mélanges de l'Ecole française de Rome. Moyen Age* 99:435–446.

Tissoni Benvenuti, Antonia. 1987b. "Il mondo cavalleresco e la corte estense." In *I libri di Orlando Innamorato*. Modena: Panini, 13–33.

Tissoni Benvenuti, Antonia. 1991. "Guarino, i suoi libri, e le letture della corte estense." In *Le muse e il principe. Arte di corte del primo Rinascimento padano*, 2, *Saggi*. Modena: Panini, 63–79.

Tissoni Benvenuti, Antonia. 1994. "L'antico a corte: da Guarino a Boiardo." In Marco Bertozzi, ed., *Alla corte degli Estensi. Filosofia, arte e cultura a Ferrara nei secoli XV e XVI*. Ferrara: Università degli Studi, 389–404.

Tobias Stimmer. 1984. *Tobias Stimmer, 1539–1584: Spätrenaissance am Oberrhein. Ausstellung im Kunstmuseum Basel*. Basel: Kunstmuseum.

Tooley, Marian. 1953. "Bodin and the Medieval Theory of Climate." *Speculum* 28:64–83.

Tracconaglia, Giovanni. 1922. *Contributo alla storia dell'Italianismo in Francia. IV. Femminismo e platonismo in un libro raro del 1503: la Nef des Dames di Symphorien Champier*. Lodi: Tip. C. Dell'Avo.

Turner, G. L'E. 1969. "The History of Optical Instruments: A Brief Survey of Sources and Modern Studies." *History of Science* 8:53–93.

The Uses of Historical Evidence in Early Modern Europe. 2003. Special issue of *Journal of the History of Ideas* 64.2.

Vallerani, Massimo. 2001. "I fatti nella logica del processo medievale. Note introduttive." In Cerutti and Pomata 2001a, 665–693.

van den Broek, Gerard. 1991. "Sensible, Logical, Godly and Sexual Order in Botanical Practice." *Semiotica* 84:43–99.

Van de Vyver, O. 1971. "Lunar Maps of the Seventeenth Century." *Vatican Observatory Publications* 1:69–83.

Van Helden, Albert. 1976. "The Importance of the Transit of Mercury of 1631." *Journal for the History of Astronomy* 7:1–10.

Vasoli, Cesare. 1960. "Temi e fonti della tradizione ermetica in uno scritto di Symphorien Champier." *Archivio di filosofia* 30:235–247.

Vasoli, Cesare. 1979. "The Contribution of Humanism to the Birth of Modern Science." *Renaissance and Reformation* 3:1–15.

Vasoli, Cesare. 1989. *Francesco Patrizi da Cherso*. Rome: Bulzoni.

Vegetti, Mario. 1994. "L'immagine del medico e lo statuto epistemologico della medicina in Galeno." In Wolfgang Haase, ed., *Aufstieg und Niedergang der römischen Welt*. Vol. 2, part 2. Berlin: De Gruyter, 1672–1717.

Veneziani, Marco, ed. 2002. *Experientia. X Colloquio Internazionale del Lessico Intellettuale Europeo, Roma, 4–6 gennaio 2001*. Florence: Olschki.

Vernacularization of Science. 1998. The Vernacularization of Science: Medicine and Technology in Late Medieval Europe. Special issue of *Early Science and Medicine*, 3.2.

Villey, Pierre. 1908. *Les sources et l'évolution des Essais de Montaigne*. 2 vols. Paris, Hachette.

Vitale Brovarone, Alessandro. 1980. "L'exemplum tra due retoriche (VI–XII sec.)." *Mélanges de l'Ecole française de Rome. Moyen Age* 92:87–112.

Voigts, Linda. 1990. "The 'Sloane Group': Related Scientific and Medical Manuscripts from the Fifteenth Century in the Sloane Collection." *British Library Journal* 16:26–57.

Voigts, Linda. 1996. "What's the Word? Bilingualism in the Late Medieval England." *Speculum* 71:813–826.

Völkel, Markus. 1987. *"Pyrrhonismus historicus" und "fides historica": die Entwicklung der deutschen historischen Methodologie unter dem Gesichtspunkt der historischen Skepsis*. Frankfurt am Main: Peter Lang

Volpi, Caterina. 1992. "Lorenzo Pignoria e i suoi corrispondenti." *Nouvelles de la république des lettres* 12:71–127.

Volpi, Caterina. 1996. *Le immagini degli dei di Vincenzo Cartari*. Rome: De Luca.

Von Moos, Peter. 1988. *Geschichte als Topik: das rhetorische Exemplum von der Antike zur Neuzeit und die historiae im "Policraticus" Johanns von Salisbury*. Hildesheim: G. Olms.

Wack, Mary France. 1990. *Lovesickness in the Middle Ages: The "Viaticum" and Its Commentaries*. Philadelpia: University of Pennsylvania Press.

Walker, D. P. 1972. "Edward Lord Herbert of Cherbury and Christian Apologetic." In Walker, *The Ancient Theology: Studies in Christian Platonism from the Fifteenth to the Eighteenth Century*. Ithaca: Cornell University Press.

Wallis, Faith. 1995. "The Experience of the Book: Manuscripts, Texts, and the Role of Epistemology in Early Medieval Medicine." In Don Bates, ed., *Knowledge and the Scholarly Medical Tradition*. Cambridge: Cambridge University Press, 101–126.

Walther, Gerrit. 1983. *Niebuhrs Forschung*. Stuttgart: Steiner.

Waquet, Françoise, ed. 2000. *Mapping the World of Learning: The "Polyhistor" of Daniel Georg Morhof*. Wiesbaden: Harrassowitz.

Watanabe-O'Kelly, Helen, and Anne Simon. 2000. *Festivals and Ceremonies: A Bibliography of Works Relating to Court, Civic, and Religious Festivals in Europe 1500–1800*. London: Mansell.

Wear, Andrew. 1985. "Explorations in Renaissance Writings on the Practice of Medicine." In A. Wear, R. K. French, and Iain M. Lonie, eds., *The Medical Renaissance of the Sixteenth Century*. Cambridge: Cambridge University Press, 118–145.

Wear, Andrew. 1995. "Epistemology and Learned Medicine in Early Modern England." In Don Bates, ed., *Knowledge and the Scholarly Medical Traditions*. Cambridge: Cambridge University Press, 151–173.

Webb, Ruth. 1999. "Ekphrasis Ancient and Modern: The Invention of a Genre." *Word and Image* (special issue on "Ekphrasis") 15:7–18.

Weiss, Roberto. 1973. *The Renaissance Discovery of Classical Antiquity*. Oxford: Blackwell.

Wellisch, Hans H. 1984. *Conrad Gessner, a Bio-bibliography*. Zurich: IDC.

Welter, Jean-Thiébaut. 1973. *L'exemplum dans la littérature religieuse et didactique du Moyen Age*. Geneva: Slatkine Reprints. (Originally published 1927.)

West, Martin L. 1999. "Towards Monotheism." In Polymnia Athanassiadi and Michael Frede, eds., *Pagan Monotheism in Late Antiquity*. Oxford: Clarendon Press, 21–40.

West, William. 2002. *Theatres and Encyclopedias in Early Modern Europe*. Cambridge: Cambridge University Press.

Whitaker, E. A. 1989. "Selenography in the Seventeenth Century." In R. Taton and C. Wilson, eds., *Planetary Astronomy from the Renaissance to the Rise of Astrophysics. Part A. Tycho Brahe to Newton*. Cambridge: Cambridge University Press, 119–143.

Whitaker, E. A. 1999. *Mapping and Naming the Moon: A History of Lunar Cartography and Nomenclature*. Cambridge: Cambridge University Press.

Wickenden, Nicholas. 1993. *G. J. Vossius and the Humanist Concept of History*. Assen: Van Gorcum.

Wilde, Emil. 1838. *Geschichte der Optik vom Ursprunge dieser Wissenschaft bis auf die gegenwärtige Zeit*. 2 vols. Berlin: Rücker und Püchler.

Williams, Steven. 2003. *The "Secret of Secrets": The Scholarly Career of a Pseudo-Aristotelian Text in the Latin Middle Ages.* Ann Arbor: University of Michigan Press.

Williams, Steven. 2004. "Giving Advice and Taking It: The Reception by Rulers of the Pseudo-Aristotelian 'Secretum secretorum' as a 'speculum principis.'" In Carla Casagrande et al., eds., *"Consilium." Teorie e pratiche del consigliare nella cultura medievale.* Florence: Edizioni del Galluzzo-SISMEL, 139–180.

Wind, Edgar. 1958. *Pagan Mysteries in the Renaissance.* London: Faber & Faber.

Winkler, Mary G., and Albert Van Helden. 1992. "Representing the Heavens: Galileo and Visual Astronomy." *Isis* 83:195–217.

Winkler, Mary G., and Albert Van Helden. 1993. "Johannes Hevelius and the Visual Language of Astronomy." In J. V. Field and Frank A. J. L. James, eds., *Renaissance and Revolution: Humanists, Scholars, Craftsmen and Natural Philosophers in Early Modern Europe.* Cambridge: Cambridge University Press, 97–116.

Winterbottom, Michael. 1983. "Curtius Rufus." In L. D. Reynolds, ed., *Texts and Transmission.* Oxford: Clarendon Press, 148–149.

Wisch, Barbara, and Susan Scott Munshower. 1990. *Art and Pageantry in the Renaissance and Baroque.* University Park: Pennsylvania State University Press.

Witschi-Bernz, Astrid. 1972a. "Bibliography of Works in the Philosophy of History, 1500–1800." *History & Theory* Beiheft 12:3–50.

Witschi-Bernz, Astrid. 1972b. "Main Trends in Historical-Method Literature, Sixteenth to Eighteenth Centuries." *History & Theory* Beiheft 12:51–90.

Witt, Ronald G. 2000. *In the Footsteps of the Ancients: The Origins of Humanism from Lovato to Bruni.* Leiden: Brill.

Wolff, Emil. 1977 [1910]. *Francis Bacon und seine Quellen.* Nendeln/Liechtenstein: Kraus-Thomson.

Yeo, Richard. 2001. *Encyclopedic Visions: Scientific Dictionaries and Enlightenment Culture.* Cambridge: Cambridge University Press.

Zagorin, Perez. 1990. *Ways of Lying: Dissimulation, Persecution, and Conformity in Early Modern Europe.* Cambridge, Mass.: Harvard University Press.

Zapperi, Roberto. 1994. *Der Neid und die Macht. Die Farnese und Aldobrandini im barocken Rom.* Munich: Beck.

Zedelmaier, Helmut. 1992. *Bibliotheca universalis und Bibliotheca selecta. Das Problem der Ordnung des gelehrten Wissens in der frühen Neuzeit.* Cologne: Böhlau.

Zen, Stefano. 1994. *Baronio storico: Controriforma e crisi del metodo umanistico.* Naples: Vivarium.

Ziegler, Joseph. 1998. *Medicine and Religion c. 1300: The Case of Arnau de Vilanova.* York: York Medieval Press.

Zimmerman, T. C. Price. 1995. *Paolo Giovio: The Historian and the Crisis of Sixteenth Century Italy*. Princeton: Princeton University Press.

Zorzetti, Nevio. 1980. "Dimostrare e convincere: l'exemplum nel ragionamento induttivo e nella dimostrazione." In *Rhétorique et histoire. L'exemplum et le modèle de comportement dans le discours antique et médiéval, Mélanges de l'Ecole française de Rome. Moyen Age* 92:33–65.

Zuccolin, Gabriella. 2003. "Michele Savonarola medico tra addestramento pratico e divulgazione. Stratificazione di generi, destinatari e scopi nel 'De regimine pregnantium'." Tesi di laurea, Facoltà di Lettere e Filosofia, Università di Pavia.

Contributors

Ann Blair is professor of history at Harvard University. She is the author of *The Theater of Nature: Jean Bodin and Renaissance Science* (Princeton: Princeton University Press, 1997).

Chiara Crisciani is professor of history of medieval philosophy at the University of Pavia. Her publications include *Il papa e l'alchimia: Felice V, Guglielmo Fabri e l'elixir* (Rome: Viella, 2002), and, with Jole Agrimi, *Edocere medicos: medicina scolastica nei secoli XIII–XV* (Milan: Guerini e associati, 1988) and *Les "consilia" médicaux* (Turnhout: Brepols, 1994).

Anthony Grafton teaches European history and history of science at Princeton University. His publications include *Joseph Scaliger: A Study in the History of Classical Scholarship* (2 vols., Oxford: Clarendon Press, 1983, 1993), *Cardano's Cosmos* (Cambridge, Mass.: Havard University Press, 1999), and *Leon Battista Alberti, Master Builder of the Italian Renaissance* (New York: Hill and Wang, 2000). In the series Transformations: Studies in the History of Science and Technology he is the co-editor, with William R. Newman, of *Secrets of Nature: Astrology and Alchemy in Early Modern Europe* (Cambridge, Mass.: MIT Press, 2001).

Donald R. Kelley is professor of history at Rutgers University, New Brunswick. His recent publications include *Faces of History: Historical Inquiry from Herodotus to Herder* (New Haven: Yale University Press, 1998) and *Fortunes of History: Historical Inquiry from Herder to Huizinga* (New Haven: Yale University Press, 2003) and the edited collection *History and the Disciplines: The Reclassification of Knowledge in Early Modern Europe* (Rochester: University of Rochester Press, 1997).

Ian Maclean is Professor of Renaissance Studies at the University of Oxford and a Senior Research Fellow of All Souls College. Among his publications are *The Renaissance Notion of Woman* (1980), *Meaning and Interpretation in the Renaissance: The Case of Law* (1992), *Montaigne philosophe* (1996), *Logic, Signs and Nature in the Renaissance: The Case of Learned Medicine* (2001), and an edition of Cardano's *De libris propriis* (2004).

Peter N. Miller is professor of history at the Bard Graduate Center, New York. He is the author, most recently, of *Peiresc's Europe: Learning and Virtue in the Seventeenth Century* (New Haven: Yale University Press, 2000).

Martin Mulsow is professor of history at Rutgers University. Among his publications are *Frühneuzeitliche Selbsterhaltung: Telesio und die Naturphilosophie der Renaissance* (Tübingen: Niemeyer, 1998), *Moderne aus dem Untergrund. Radikale Frühaufklärung in Deutschland 1680–1720* (Hamburg: Felix Meiner Verlag, 2002), and the edited collection *Das Ende des Hermetismus: historische Kritik und neue Naturphilosophie in der Spätrenaissance; Dokumentation und Analyse der Debatte um die Datierung der hermetischen Schriften von Genebrard bis Casaubon (1567–1614)* (Tübingen: Mohr Siebeck, 2002).

Brian W. Ogilvie is associate professor of history at the University of Massachusetts Amherst. His book *The Science of Describing: Natural History in Renaissance Europe* is forthcoming from the University of Chicago Press in 2006.

Laurent Pinon teaches early modern history and history of science at the École normale supérieure, Paris. His publications include *Livres de zoologie de la Renaissance, une anthologie (1450–1700)* (Paris: Klincksieck, 1995).

Gianna Pomata is associate professor of history at the University of Bologna. Her publications include *La promessa di guarigione: malati e curatori in antico regime* (1994), translated into English as *Contracting a Cure: Patients, Healers, and the Law in Early Modern Bologna* (Baltimore: Johns Hopkins University Press, 1998) and " 'Observatio' ovvero 'historia.' Note su empirismo e storia in età moderna" (*Quaderni storici* 31.91 [1996], 173–198). With Simona Cerutti she is the co-editor of *Fatti: storie dell'evidenza empirica*, a special issue of *Quaderni storici* (36.108, 2001) and with Lorraine Daston of *The Faces of Nature in Enlightenment Europe* (Berlin: Berliner Wissenschafts-Verlag, 2003).

Nancy G. Siraisi recently retired from teaching history at Hunter College and the Graduate School of the City University of New York. Among her publications are *Medieval and Early Renaissance Medicine* (Chicago: University of Chicago Press, 1990) and *The Clock and the Mirror: Girolamo Cardano and Renaissance Medicine* (Princeton: Princeton University Press, 1997). With Anthony Grafton she is co-editor of *Natural Particulars: Nature and the Disciplines in Renaissance Europe* (Cambridge, Mass.: MIT Press, 1999).

Index

Abano, Pietro d', 10, 303–304
Accademia dei Lincei, 184
Achillini, Alessandro, 140 n. 53
Aconcio, Jacopo, 56, 70 n. 50
Acta eruditorum, 135
Acta medica et philosophica Hafniensia, 135
Advancement of Learning (Bacon), 148, 166–167
Aelian, 190, 252, 263
Aesop, 89
Ago, Renata, 35 n. 52
Agricola, Rudolph, 163
Agrimi, Jole, 122
Aicardi, Paolo, 189
Airs, Waters, Places (Hippocrates), 165, 330
Alardus Amstelredamus, 177 n. 84
Alberti, Leon Battista, 327
Alberti, Salomon, 216
Albertus Magnus, 78–79, 256
Albricus, 183
Alciati, Andrea, 33 n. 35, 249
Aldrovandi, Ulisse, 18, 19, 88, 98, 164, 179 n. 98, 263, 326, 329, 368
Aleandro, Girolamo, 184–185, 187, 189, 195, 196, 197–198, 199, 201, 202
Alembert, Jean Le Rond d', 226
Alexander, the Great, 42, 45, 66, 67, 299
Alfonso of Aragon, 42–43, 66
Alleaume, Jacques, 363
Alpers, Svetlana, 356–357
Alpino, Prospero, 189, 368
Alsted, Johann, 222, 223
Altieri, Marco Antonio, 327
Amatus Lusitanus, 125–126, 127, 130–131, 133, 134, 146 n. 138
Amerbach, Basil, 272, 286
Anatomica corporis . . . historia (Bauhin), 105
anatomy, 2–3, 13–14, 17, 19, 26, 30, 105–106, 111, 114–122, 133, 215, 356, 357, 358–359. *See also* medicine
Anatomy of Melancholy (Burton), 88
Ancient Medicine (Hippocrates), 8
animals, *historia* of, 18, 19, 108–110, 116, 212. *See also* zoology
Annales ecclesiastici, 194
Annius of Viterbo. *See* Nanni, Giovanni
anthropology, early modern, 380–390
Antiquae tabulae marmoreae (Aleandro), 184
antiquarianism, 3, 4, 6–7, 11, 15–16, 17, 18, 23, 24–29, 30, 52–53, 64–68, 186, 187, 189–190, 195–196, 198–199, 203, 325–327, 336–337, 346, 355, 357, 360, 362–363
Antiquario, Jacopo, 115
Antwerp, 81
Aphorisms (Hippocrates), 342
Aquatilium animalium historia (Salviani), 261, 262

Arbax, 279
Archimatthaeus, 122
Arcos, S. d', 381
Argyropulos, Johannes, 108
Aristotelianism, 4, 13–15, 30, 108, 111, 113, 147–159, 161–162, 164, 167, 189, 190, 192, 200, 205 n. 28, 215, 222, 223, 283, 297–299, 326, 356
Aristotle, 8–9, 10–11, 14, 18, 53–54, 54, 58, 75, 77, 78, 80, 81, 94, 96, 107, 116, 118, 130, 148–149, 151–160, 163–164, 167, 182, 212–213, 221, 243–244, 245, 263, 272, 299. *See also works by title*
Arnald of Villanova, 115, 126, 318 n. 14
Arnold, Gottfried, 228
Ars critica (Le Clerc), 12, 42–43, 47–48
ars excerpendi. See excerpting
Ars historica (Vossius), 51
artes historicae, 4, 10, 11, 51–58, 63–64, 215, 217, 219, 221, 227. *See also* rhetoric
Artis historicae penus (Wolf), 51, 106, 272
Aselli, Gaspare, 14, 19, 25, 118–121, 355, 356, 359
Ashworth, William, Jr., 251
astronomy, 26, 199, 202, 226, 373–380
Athenaeus, 263
Auch eine Philosophie (Herder), 230
Augustine, Saint, 76, 213, 220, 221, 234
Augustus, emperor, 332
Aurelian, 198
Averroes, 154–155
Avicenna, 308, 322 n. 71

Bacci, Andrea, 23–24, 325, 326, 328–340, 342–343, 346–347
Bacon, Francis, 5, 15, 91, 148, 166–168, 213, 215, 220–221, 231, 284, 285, 289, 358–359, 374, 376

Baillet, Adrien, 279–280
Bailly, Jean-Sylvain, 226
Barbaro, Ermolao, 79, 221
Barberini court, 184–185, 195, 198, 201
Barkan, Leonard, 190
Baron, Eguinaire, 220
Baronio, Cesare, 52, 55, 194
Bartholin, Thomas, 145 n. 134
Bartholomaeus Anglicus, 78
Barzizza, Cristoforo, 321 n. 67
Basel, 272
Basilides, 194
baths, public. *See* medicine: waters
Baudouin, François, 55–57, 217–219, 223
Bauhin, Caspar, 81, 105, 163, 165–166, 168 n. 2
Bauhin, Jean, 134, 260
Baumgarten, Siegmund Jakob, 227, 229–230
Baxandall, Michael, 358
Becichemo, Marino, 109
Bellarmino, Roberto, 161, 194
Belmissero, Paolo, 171 n. 30
Belon, Pierre, 18–19, 241, 247–248, 254, 257, 260–261, 263
Benedetti, Alessandro, 33 n. 20, 105, 109, 114–116, 132
Benedetti, Natalino, 207 n. 55
Benedictines, 41
Benivieni, Antonio, 132–134
Bernard de Gordon, 318 n. 14
Berosus, 54, 57
Berr, Henri, 234
Berval, Richer de, 368
bestiaries, medieval, 82, 86–87
Beurer, Johann Jakob, 185–186
Beyerlinck, Laurentius, 271
Bibliotheca anatomica (Le Clerc and Manget), 121
Bibliotheca universalis (Gessner), 241–242
Bierling, F. W., 228
Bignon, Jérôme, 366
Biondi, Albano, 57
Biondo, Flavio, 327

Bisticci, Vespasiano da, 65
Black, Robert, 54
Blair, Ann, 19, 21–22, 115, 133, 298
Boccaccio, Giovanni, 183
Boccadiferro, Lodovico, 108
Bock, Hans, 292 n. 19
Bock, Hieronymus, 80
Bodin, Jean, 51–52, 56–57, 62–64, 82, 91–93, 99, 185, 217, 219–222, 229, 230, 232, 243, 276
Boethius, Anicius Manlius Severinus, 150–151, 163
Bolingbroke, Henry St. John, viscount, 230
Bolognetti, Giorgio, 370
Bonifacio, Baldassarre, 181, 182, 184
Borgia family, 55
Bossuet, Jacques-Bénigne, 229
botany, 153, 182, 215, 243, 247, 368. *See also* plants
Bouilliau, Ismael, 371
Boyle, Robert, 27, 97
Bracelli, Giacomo, 33 n. 35
Bramante, Donato, 333
Brethren of the Common Life, 83
Bridgewater Treatises, 98
Browne, Thomas, 88–89
Brucker, J. J., 223, 231
Brunfels, Otto, 80
Bruni, Leonardo, 76, 303, 320 n. 41
Budé, Guillaume, 63
Buongiovanni, Quinzio, 190, 192
Burton, Robert, 88

cabinets of curiosities, 28, 285
Cadamusto, Alvise, 338
Caesar, Julius, 18, 251–252, 305, 339
Cagnati, Marsilio, 23–24, 328–331, 333–335, 340–343, 346–347
Cairo, 381
Caius (Key), John, 258
Calco, Tristano, 54
Calepino, Ambrogio, 252
Caligula, 62
Camden, William, 365, 387

Camerarius, Joachim, 85–89, 92, 98, 134, 164, 260
Campanella, Tommaso, 52, 82, 91, 201
Campofulgosus. *See* Fregoso
Canada, 381
Cano, Melchior, 56–57
Canon (Avicenna), 308, 322 n. 71
Carcano, Giambattista, 141 n. 69
Cardano, Girolamo, 34 n. 37, 125, 127, 134, 329
Carne, Edward, 339
Carpocrates, 194
Cartari, Vincenzo, 192, 202
Cayre (anatomist), 372, 373
Cellarius, Conrad, 229
Celsus, 112, 331
Centuriae curationum (Amatus Lusitanus), 125
Centuriae symbolorum (Camerarius), 85–86
Centuriators of Magdeburg. *See* Magdeburg Centuriators
Ceremonial de France (Godefroy), 382
Cermisone, Antonio, 301
Cesalpino, Andrea, 81, 147, 152, 155, 158, 159, 160, 163, 165–166, 168 n. 2, 326
Cesarius, Johannes, 116
Cesi, Federico, 184
Chacon, Pedro, 329
Chaldaean Oracles, 199
Champier, Symphorien, 33 n. 20, 105–106, 123–124, 130, 133
Charlemagne, 26, 366–367
Charles I, king of England, 382
Charles V, emperor, 334
Charles VIII, king of France, 116
Chasseneux, Barthélemy de, 338
Chladenius, J. M., 233
Chrysoloras, Manuel, 357–358
Chytraeus, David, 57
Cicero, Marcus Tullius, 46, 47, 53, 54, 60, 82, 93–94, 96, 163, 211, 217, 220, 333
City of God (Augustine), 76

478 Index

Clement VIII, pope, 334
Clusius (L'Ecluse), Carolus, 81, 90, 368
Cobelluccio, Scipione, 194
Coccejus, Henricus, 221
Colerus, Christopher, 43
collectionism, 28
Collenuccio, Pandolfo, 79
Colmar, Johann, 229
Colonna, Ascanio, 328, 335, 337
Comenius, Jan Amos, 284
Condillac, Etienne Bonnot de, 232
Connan, François, 220
consilia. See medicine
Constantine the Great, emperor, 55
Consultationes (Da Monte), 128–129
Consultationes (Vittori), 329, 344–345
Convivium religiosum (Erasmus), 83–84
Copernicus, Nicolaus, 187, 224
Cornarius, Diomedes, 131–132
Cosmographia (Munster), 252
Costeo, Giovanni, 119
Cotroneo, Girolamo, 110
Cotton, Robert, 387
Council of Trent, 161
Counter-Reformation, 327
Cousin, Victor, 232
Crato von Crafftheim, Johann, 53, 124, 129, 142 n. 94
Crippa, Bernardino, 161, 171 n. 30
Crisciani, Chiara, 12, 22, 122–123
Crivelli, Lodrisio, 320 n. 48
Croce, Benedetto, 230, 234
Cruydeboeck (Dodoens), 81
Cunningham, Andrew, 163, 356
curationes. See medicine: case histories
Curationes empiricae et historicae (Ruland), 127
Curationes et praedictiones (Cardano), 125
Curtius Rufus, Quintus, 12, 42–50, 64, 66–68

Dalla Riva, Bonvesin, 320 n. 41
Dal Pozzo, Cassiano, 187, 379
Da Monte, Giovanni Battista, 128, 142 n. 94
Daston, Lorraine, 21–22, 284, 285, 369
De abditis ac mirandis . . . causis (Benivieni), 132–133
De anima (Aristotle), 107–109, 153
De animalibus (Albertus Magnus), 78–79
Dear, Peter, 27, 284, 360
De arte gymnastica (Mercuriale), 337, 339–340
De balneis (Savonarola), 302
De bello Gallico (Caesar), 251–252
De causis plantarum (Theophrastus), 77, 153, 244
Decembrio, Angelo, 12, 64–67, 303
Decembrio, Pier Candido, 320 n. 41
Declamatiuncula (Lorenzo Valla), 59
De coelo (Aristotle), 167
De deis gentium varia . . . historia (Giraldi), 183
De demonstratione (Galen), 154
De diis Syris (Selden), 196
De experientia medica (Galen), 111
De generatione animalium (Aristotle), 110, 117, 149, 244
De gotta (Savonarola), 302, 311, 313, 315
De historiae institutione (Morcillo), 110
De historia stirpium (Fuchs), 80
De ingenio sanitatis (Galen), 123
De institutione historiae (Baudouin), 55, 217
De inventione dialectica (Agricola), 163
De inventoribus rerum (P. Vergil), 78
De la decorante Ferrara (Savonarola), 304
Del felice progresso di Borso d'Este (Savonarola), 302, 304–307, 316
Delft, 130
Della historia (Patrizi), 60, 105
Della Porta, Giovanni Vincenzo, 368

De locis affectis (Galen), 123
De medica historia (Donati), 132
De medicina (Celsus), 112
De medicina Aegyptiorum (Alpino), 189
De medicorum principuum historia (Zacutus Lusitanus), 106, 135
De methodis (Zabarella), 158
Democritus, 167
De motu cordis (Harvey), 117–118, 356
De natura animalium (Aelian), 252
De natura deorum (Cicero), 93
De naturali vinorum historia (Bacci), 335–336
De natura rerum (Cantimpré), 78
De optima secta, 111–112
De partibus animalium (Aristotle), 79, 110, 117, 148, 149, 151–153, 154, 155, 156, 158–160, 162, 164, 167, 244
De placitis Hippocratis (Galen), 154
De plantis (Cesalpino), 81
De plantis (Theophrastus), 153
De politia litteraria (Decembrio), 64
De providentia numinis (Lessius), 94
De rebus expetendis (Giorgio Valla), 79
De regimine pregnantium (Savonarola), 302, 311–315
De rerum inventoribus (Vergil), 214
Descartes, René, 49, 96, 225, 231, 232, 360, 372
De scribenda universitatis rerum historia (Milieu), 70 n. 50, 217–218
Description of Greece (Pausanias), 31
De sui ipsius . . . ignorantia (Petrarch), 75–76
De theologia gentili (Vossius), 181, 188, 197, 201
De thermis (Bacci), 331, 335, 337–340, 347
De usu partium corporis (Galen), 154
De vera et digna seculari militia (Savonarola), 302

De veritate (Herbert), 201
De vermibus (Savonarola), 301
Diaccetto, Francesco Cattani da, 272
Diaria de bello Carolino (Benedetti), 116
Diego of Alcalá, 345
Dilthey, Wilhelm, 3
Dio Cassius, 333
Diodorus Siculus, 52, 62
Diogenes Laertius, 223, 225, 277
Dionysius of Halicarnassus, 50, 212, 222, 230
Dioptrique (Descartes), 372
Dioscorides, 79–81, 153
Directorium ad actum (Savonarola), 301
Discalcio, Ottonello, 205 n. 28
Discourse on Method (Descartes), 225
Dissertatio peripatetico-theologica (Tanner), 203
Dodoens, Rembert, 81, 130, 132–133, 134, 146 n. 138
Donati, Marcello, 132, 134
Doni, Anton Francesco, 20
Drebbel, Cornelius, 360
Du Boardigne, Jan, 362
Du Cange, Charles du Fresne, 212
Du Chesne, André, 366
Du Laurens, André, 105, 116
Dullaert, Joannes, 161
Dupuy, Pierre and Jacques, 369, 372
Du Vair, Guillaume, 382, 388

ecclesiastical history, 32 n. 9, 55, 345–346, 363, 383–387
Egnazio (Egnatius), Giovanni Battista, 87, 279
Egyptian mythology, 189–190, 192, 194, 198, 200
ekphrasis, 26, 357
Eliot, George, 114
Emanuele Filiberto, duke of Turin, 129
emblematics, 85–89, 192
Empedocles, 167

empiricism, early modern, 5, 7–8, 11–12, 13–14, 17, 24–25, 28, 29–30, 29–31, 117
empiricism, late medieval, 29, 298
Empiricists, ancient, 111–112
encyclopedism, 1, 5, 83, 105, 182, 222, 226–227, 229–230
English Parliament, 388, 389
Enlightenment, 16, 103 n. 114, 233
Ephemerides medico-physicae, 135
Epidemics (Hippocrates), 3, 13, 24, 112, 124–125, 128–130, 132, 135, 341–343, 347
Episcopius, Nicolaus, 286
Erasmus, Desiderius, 42–43, 83–84, 286
Estates-General (France), 388, 390
Este, Borso d', 304–306, 311
Este, Leonello d', 64–67, 301, 305, 311
Este, Nicolò d', 301, 305
Este court, 300
Estienne, Charles, 107, 116
ethics and natural history, 82–89
Etymologies (Isidore), 9
Euripides, 8
Eusebius, 196, 229
Eustachius, Bartholomaeus (Bartolomeo Eustachi), 166
excerpting, 19–21, 105–106, 115, 123–124, 133, 181, 269–270. *See also* exemplar history
exemplar history, 19–20, 269–270, 273–280, 289, 306–307, 316
exemplar morality, 273–275, 278
Exempla virtutum (Herold), 270, 277
Exercitationes de generatione (Harvey), 121

fable, 181, 183, 185, 195, 198, 201, 214, 361. *See also* emblematics; mythography
Fabricius, Georg, 53, 111, 254
Fabricius, Hieronymus (Girolamo Fabrizi), of Acquapendente, 2, 116–118, 163, 355–356
Fabricius, Johann Andreas, 222

Facio, Bartolomeo, 42
"fact," 30, 284–289, 298
Facta ac dicta memorabilia (Valerius Maximus), 269
Falloppia, Gabriele, 116, 141 n. 69
Farnese, Alessandro, 327
Farnese, Odoardo, 184
Fernández de Oviedo y Valdés, Gonzalo, 80, 215
Fernel, Jean, 134
Ferrara, 22, 64–65, 189, 301, 302, 306
Ferrari da Grado, Giovanni Matteo, 300, 321 n. 67
Fiamma, Gaetano, 107–108
Fichte, Johann Gottlieb, 233
Flacius Illyricus, Matthias, 55, 58
Florence, 300
Foglietta (Foletus), Uberto, 61, 62, 221
Foligno, Gentile da, 303
Fontenay, Guy de, 291 n. 6
Foreest (Forestus), Pieter van, 24, 130–131, 134, 137, 146 n. 138
forgeries, 54, 56, 57
Fort, Pierre, 373
Foucault, Michel, 284
Fox Morcillo, Sebastián, 110, 221
Fracastoro, Girolamo, 340
François I, king of France, 254
Franeker, 41
Frankfurt an der Oder, 53
Frédeau, Claude, 379
Freedberg, David, 369
Fregoso (Campofulgosus, Fulgos(i)us), Battista, 87, 277, 291 n. 6, 293 n. 39
French, Roger, 118
French Academy, 48
Frischlin, Nicodemus, 43
Froben, Johannes, 286
Frontinus, Sextus Julius, 291 n. 5, 331, 339
Fuchs, Leonhart, 80–81, 89–90, 164, 215, 247
Fulgentius, Fabius Planciades, 183
Fulgos(i)us. *See* Fregoso

Fumanelli, Antonio, 340
Funkenstein, Amos, 95
Furlanus, Daniel, 155, 158, 159

Gaillard, Pierre Droit de, 220
Galen, 13, 35 n. 55, 36 n. 56, 42, 97–98, 111–112, 123–127, 130, 133, 135, 136, 154–155, 162, 212, 272, 341, 355
Galeni historiae medicinales (Selvatico), 106
Galeni historiales campi (Champier), 105, 123, 133
Galileo Galilei, 26, 52, 82, 94, 184, 224–225, 368, 374–376, 378
Garavini, Fausta, 288
Gassendi, Pierre, 25–26, 97, 355–356, 363, 370–371, 372, 378–380, 382
Gatterer, Johann Christoph, 230
Gaultier, Joseph, 373, 379
Gaza, Theodore, 11–12, 94, 155–156, 212
Gellius, Aulus, 64, 76, 181, 277
Genealogiae deorum (Boccaccio), 183
Gérando, Joseph-Marie, baron de, 232
Gerard, John, 90
Gessner, Conrad, 18–19, 20, 22, 80–81, 84–85, 90, 92–93, 98, 164, 215, 241–264, 286
Gesta Ferdinandi Regis Aragonum (Lorenzo Valla), 53
Giacomo della Torre da Forlì, 301
Gilbert, Felix, 51
Ginzburg, Carlo, 200, 357–358
Giovanni Conversini da Ravenna, 300
Giovanni da Ferrara, 320 n. 48
Giphanius, Obertus, 53
Giraldi, Lilio Gregorio, 183, 196
Glisson, Francis, 113
Goclenius, Rodolphus, 111, 161, 212, 213
Godefroy, Théodore, 382
Goethe, Johann Wolfgang von, 233
Gondi, Henri de, 383

Grafton, Anthony, 16
Grange, Balthasar, 369
Greek culture, classical, 8, 31, 41, 44
Gregory XIII, pope, 330
Grolier, Jean, 260
Grotius, Hugo, 232
Guainerio, Antonio, 321 n. 68, 322 n. 70, 338
Gualdo, Paolo, 374
Gualenghi, Giovanni, 65
Guarino, Battista, 76
Guarinoni, Cristoforo, 155–156, 158
Guarino of Verona, 53, 64, 76
Guicciardini, Francesco, 54, 57, 168
Guilandini, Melchior, 189
Guillemin, Denis, 362, 367
Guizot, François, 233

Hanappus, Nicolas, 290 n. 4
Hannibal, 43, 65–66
Harriot, Thomas, 147
Harvey, William, 2, 14, 25, 116–118, 121–122, 355, 356, 371, 374
Hausen, Carl, 229
Hazard, Paul, 50
Heeren, Arnold Hermann Ludwig, 230
Hegel, Georg Wilhelm Friedrich, 233
Heineccius, Johann Gottlieb, 223
Heliodorus, 190
Helmstedt, 53
Henri de Mondeville, 318 n. 14
Henrietta Maria, queen of England, 382
heraldry, 365
Herbal (Gerard), 90
Herbarum vivae eicones (Bruntels and Weiditz), 80
Herberstein, Sigmund von, 252–253
Herbert (archpriest of St. Marie Madeleine), 386–387
Herbert, Edward, of Cherbury, 201–203
Herbster, Christine, 272
Herbster, Johannes, 272, 277
Herder, Johann Gottfried von, 16, 229, 230

hermeneutics, 42
Hermeticism, 187
Herodotus, 8, 67, 211, 213, 234, 357
Herold, Johannes, 270, 277, 291 n. 5
Hevelius, Johannes, 380
Heyne, G. B., 357
Hippocrates, 8, 13, 24, 42, 129, 130, 136, 165, 272, 331, 341, 345, 355. *See also* Epidemics (Hippocrates)
Histoire de la nature des Oyseaux (Belon), 248
Histoire des histoires (La Popelinière), 222–223
Histoire des plantes (Dodoens), 81
Histoire naturelle, générelle et particulière (Buffon), 98
historia. *See also* animals, *historia* of; Aristotelianism; Aristotle; medicine; plants; Pliny the Elder; Scholasticism
  as epistemic category, 3, 4, 5, 13, 109–110, 113
  versus history, 1–2, 4, 11–12, 16, 106–108, 306–307
  versus *philosophia*, 10, 107–108
  pre-Aristotelian, 4, 8, 108–110, 113
  semantic transformations, 8–10, 183, 211–215
Historia anatomica (Du Laurens), 105
Historia animalium (Aristotle), 18, 77, 78–79, 108, 110, 116, 117, 152, 153, 156, 158, 164–165, 212, 221, 244, 260
Historia animalium (Gessner), 18, 81, 85, 241, 243–245, 246, 247–249, 250, 251, 253–254, 255, 256, 258–259, 263–264
Historia aquatilium (Marschalk), 80
Historia corporis (Benedetti), 105, 116
Historia deorum (Pomey), 184
Historiae admirandae (Cornarius), 132
Historiae anatomicae rariores (Bartholin), 145 n. 134

Historiae navigationum (Peter Martyr), 256
Historia insectorum (Ray), 82
Historia ludicra (Bonifacio), 181
Historia naturale, dell' (Imperato), 82
Historia piscium (Ray), 81
Historia plantarum (Gaza), 110
Historia plantarum (Gessner), 80
Historia plantarum (Ray), 82
Historia plantarum (Theophrastus), 221, 244
Historia plerarunque partium humani corporis (Alberti), 216
historicism, 3–5, 233
Historie der Gelehrsamkeit (Fabricius), 222
Historie der Weisheit und Torheit (Thomasius), 224
historiography, 44–47, 50, 51, 53–57, 63–64, 68 n. 7, 213, 226. *See also* rhetoric
  medieval, 75
history, distinguished from *historia*, 1–2, 4, 11–12, 16, 106–108, 306–307
history, philosophy of, 16, 222
History of Foure-Footed Beastes (Topsell), 92
History of Plants (Theophrastus), 77
Hodgson, Margaret, 380
Hohenburg, Herwart von, 189
Holbach, Paul Henri Thiry, baron d', 232
Holstenius, Lucas, 196
Homer, 197, 211
Horace (Quintus Horatius Flaccus), 221
Hortus sanitatis, 87
Huet, Pierre-Daniel, 195, 225
humanism, 4–5, 10–13, 17, 46–47, 75–76, 82–83, 105–106, 211
Humbert, Pierre, 372, 373
Huyghens, Constantijn, 360

Icones animalium (Gessner), 179 n. 98, 244, 247, 253
idolatry. *See* mythography; religion

Imagini delli dei de gl'antichi (Cartari), 192
Imitatio Christi, 83
Imperato, Ferrante, 82
In librum de constitutione (Zwinger), 272
Instauratio magna (Bacon), 358–359
Isagoge (Porphyry), 149–151
Isidore of Seville, 9, 78, 214, 258
Israel, Jonathan, 50

Jacchaeus, Gilbert, 171 n. 30
Jansenists, 41
Jensen, Kristian, 158
Jesuits, 41
John VIII Palaeologus, emperor, 357
Johnson, Thomas, 90
Johnston (Jonstonus), John, 98
Journal des savants, 135
Julian, emperor, 196, 198
Julius, count of Salma and Neoburg, 281
jurisprudence, 4, 218–220
Justin (Marcus Junianus Justinus), 60, 62
Justin Martyr, 197

Kant, Immanuel, 4, 5, 230, 233
Keckermann, Bartholomäus, 51, 56, 161, 224, 226–227
Kelley, Donald, 16
Kemp, Martin, 357
Kepler, Johannes, 147, 376, 378, 382
Kessler, Eckhard, 54, 57
Keuck, Karl, 213
Kircher, Athanasius, 194
Knape, Joachim, 213
Koselleck, Reinhart, 57
Koyré, Alexandre, 29
Kreütterbuch (Bock), 80
Kuhn, Thomas, 29
Kusukawa, Sachiko, 164

Lactantius, 194
La Fontaine, Jean de, 89
Laín Entralgo, Pedro, 122, 124–125
Lanzoni, Giuseppe, 33 n. 20

La Popelinière, Henri Lancelot Voisin, sieur de, 222–223
law, 29
Lazius, Wolfgang, 33 n. 20
Leblanc, Vincent, 361
Le Clerc, Jean, 12, 42, 43–51, 58, 61, 63–64
Lectiones antiquae (Rhodiginus), 181
Le Febre (acquaintance of Peiresc), 395 n. 99
Leibniz, Gottfried Wilhelm, 225–226, 233
Leiden, 41, 145 n. 124
Lembus, Heraclides, 291 n. 5
Leo III, pope, 367
Leonardo da Vinci, 357
Leoniceno, Niccolò, 77, 79
Le Roy, Louis, 232
Lessius, Leonard, 94, 96
Leto, Pomponio, 201, 327
Lexicon philosophicum (Goclenius), 111
Libellus (Savonarola), 303–304
liberal arts, 211, 223, 242
Liber anathomiae (Zerbi), 114
Liber creaturarum (Raymond), 94
Liber nonus Almansoris, 308
Libraria, La (Doni), 20
Libretto de tutte le cose che se manzano (Savonarola), 302, 311
Libri quinque de bello Hispaniensi (Bracelli), 33 n. 35
Ligorio, Pirro, 187
Linnaeus, Carolus, 35 n. 46, 264
Lipsius, Justus, 43, 52–53
Livy (Titus Livius), 20, 42–43, 62, 65, 332
Lobkowitz, J. Caramuel, 396 n. 115
Locke, John, 41
logic. *See* Aristotelianism
Lombard, Mr., 369
Lomenie (correspondent of Peiresc), 386
Lorraine, 123
Loyola, Ignatius, 328, 345
Lublinus, Valentinus, 143 n. 110
Lucian of Samosata, 53, 54, 109

Luther, Martin, 142 n. 94
Lutheran histories, 57–58. *See also* Magdeburg Centuriators
Lycosthenes, Conrad, 272
Lyon, 123

Mabillon, Jean, 363
Macer, Floridus Odo, 78
Machiavelli, Niccolò, 168
Maclean, Ian, 14–15, 16, 35 n. 55
Macrobius, Ambrosius Aurelius Theodosius, 76, 195, 197, 198, 199, 202, 277
Maestlin, Michael, 378
Magdeburg Centuriators, 32 n. 9, 55
Magnum theatrum vitae humanae (Beyerlinck), 269, 271
Maimonides, Moses, 196–197, 202
Malatesta, Battista, 76
Malpighi, Marcello, 26
Mandosio, Jean-Marc, 243
Manetho, 54, 57
Mannucci, Vincenzo, 343
Marchesa del Monferrato, 310, 311
Marguerite of Navarre, 254
Marschalk, Nicolaus, 80
Marulus, Marcus, 291 n. 4
Mastronardi, Maria Amelia, 304
Matteacio, Angelo, 205 n. 28
Mattioli, Pietro Andrea, 85
Maximilian II, emperor, 131, 281
Maximus of Tyre, 197
Max Planck Institut für Wissenschaftsgeschichte, Berlin, 1
Medica disputatio (Cagnati), 333, 343
Medici, Cosimo de', 332
Medicinalium observationum exempla (Dodoens), 130
medicine, 2, 6–7, 13–14, 22–24, 29–30, 35 n. 55, 78, 79, 83, 189, 215, 223, 250, 276, 277, 297–299, 301, 308–309. *See also* anatomy; Galen; Hippocrates; public health
  autopsy narratives, 2, 132, 343, 345–346
  case histories, 2–3, 6, 14, 24, 112, 122, 124–135, 310, 311–312, 341–347, 356
  court physicians, 299–300
  disease, metaphorical explanantions, 313–315
  disease, theory of, 161–163
  Empiricists, ancient, 111–112, 135, 136–137
  exercise, 340
  folk healing, 126–127
  health rules, 302, 308–309, 311–313, 317, 330, 336
  medieval, 112, 114, 122–123, 125–126, 298, 299–300, 314, 315
  plants, 79–81, 90
  proverbs, 312–313
  publishing, 134–135
  waters, 337–340
Meinecke, Friedrich, 3, 233
Melanchthon, Philipp, 142 n. 94
Meliara, Biagio da, 310
Mellan, Claude, 379–380
Mémoires pour servir à l'historie naturelle (Réaumur), 98
Memorabilia Socratis (Xenophon), 167
Mensa Isiaca (Pignoria), 191
Mercati, Michele, 326, 329, 346
Mercenario, Arcangelo, 171 n. 30
Mercuriale, Girolamo, 33 n. 20, 327, 337, 339–340
Mersenne, Marin, 372
Merula, Giorgio, 54
Metallotheca (Mercati), 329
Metaphysics (Aristotle), 130, 148, 152, 155, 156, 158–159, 161
Metasthenes, 54
Meteors (Descartes), 360
Methodus ad facilem historiarum cognitionem (Bodin), 91, 217, 219–220, 229, 230
Methodus apodemica (Zwinger), 272, 280
Methodus legendi cognoscendique historiam (Reineck), 53

Methodus medendi (Galen), 154, 162, 341
Methodus rustica (Zwinger), 272
Mexía, Pedro, 269
Michael of Ephesus, 155
Michael Scot, 78
Michelangelo Buonarroti, 333
Michelet, Jules, 233, 234
microscopy, 369–370
Middle Platonism, 197
Milieu, Christoph, 70 n. 50, 217–219, 222, 232
Miller, Peter, 25, 26
Minucius Felix, Marcus, 201
Minuti, Théophile, 369
Mirabelli, Domenico Nani, 271
Mishne Torah, 197
Mocenigo, Giovanni, 190
Modio, Giovanni Battista, 325, 331
Momigliano, Arnaldo, 2, 4, 25, 325, 356–357
Monde, Le (Descartes), 96
Mondella, Luigi, 124
Mont, M. de, 381
Montagnana, Bartolomeo, 301, 338
Montaigne, Michel de, 94, 95, 276, 288–289
Montano, Benito Arias, 194
Montepulciano (cardinal), 332
Monteux, Sébastien de, 164
Montfaucon, Bernard de, 362, 363
Montmaur, M. de (correspondent of Peiresc), 387
Morhof, Daniel Georg, 214, 222
Muhlack, Ulrich, 57
Müller-Wille, Staffan, 35 n. 46, 264
Mulsow, Martin, 15–16
Münster, Sebastian, 252
Muratori, Lodovico Antonio, 363
music, 9
myth, allegorical interpretation of, 187, 189–190, 195–199, 204
Mythobius, Hector, 254
mythography, 15, 182–186, 189–190
mythology, classical, 41, 192, 197–198

Nadel, George, 57
Nanni, Giovanni, 54–57
Naples, 42, 53
natural history, 12–13, 14–15, 18–19, 30, 75–78, 79–91, 109–110, 182, 215. *See also* Bacon, Francis; botany; Gessner, Conrad; plants; zoology
medieval, 78, 86–87
Natural History (Pliny), 76–78, 115
Natural hystoria de las Indias (Fernández de Oviedo), 80, 215
natural philosophy, 2, 5, 6, 97, 118, 297
natural theology, 89–99, 278
Naudé, Gabriel, 33 n. 20, 214
Neile, Richard, 92
Neoplatonism, 110, 187, 196, 197
Neri, Filippo, 329, 343, 345–346
Newton, Isaac, 41
Nicholas V, pope, 155, 326, 327
Nicholas of Poland, 317 n. 4
Nicomachean Ethics (Aristotle), 148, 272
Niebuhr, Barthold Georg, 230
Nifo, Agostino, 155–156, 159
Noël (acquaintance of Peiresc), 395 n. 99
Novalis (Friedrich von Hardenberg), 233
novel, 114
Novum organum (Bacon), 166
numismatics, 190, 366–368
Nuremberg, 85
Nutton, Vivian, 17, 125, 127

Obel, Mathieu de l', 82
observation, theories of, 111–112, 275
observationes. *See* medicine: case histories
Observationes (Dodoens), 132
Observationes anatomicae (Falloppia), 142 n. 94
Observationes et curationes (van Foreest), 130

Observationes medicinales (Cornarius), 132
Observationes medicinales (Valleriola), 129
Observationes variae (Cagnati), 330
Odonus, Caesar, 164
Ogilvie, Brian, 12–13, 15, 16
Olmi, Giuseppe, 6
Ong, Walter, 283–284
Ophusius, Johannes Francus, 24
Oppian, 258, 263
optics, 369–373
Orphic hymns, 199
Orsini, Fulvio, 329
Orta, Garcia da, 81
Ortega y Gasset, José, 234
Ottelio, Marcantonio, 206 n. 28
Ovid (Publius Ovidius Naso), 192
Oviedo. *See* Fernández de Oviedo y Valdés, Gonzalo
Oxford University, 43

Pace, Guido, 395 n. 99
Padua, 13, 22, 128, 272, 301, 302, 303–304, 309, 311, 355
Palaephatus, 183
Paley, William, 98
Pancirolo, Guido, 205 n. 28
Panepistemon (Poliziano), 10, 242–243
Pantaleone da Confienza, 300
Panzani, Gregorio, 201
Paolo Veneto, 301
Paracelsianism, 272
Paratereseon (Schenck), 133, 136
Parato, Guido, 300
Paris, 272, 363, 365, 384–386, 388
Parrhasiana (Le Clerc), 47
Pasqualini, Lelio, 366–367, 368
Pasti, Matteo de', 65
Patin, Charles, 33 n. 20
Patrizi, Francesco, 11, 52–54, 54, 60–64, 105, 107, 109–110, 113–114, 221, 330
Paul, St. (apostle), 182
Pausanias, 31, 190
Pavia, 118

Pecquet, Jean, 359
Peiresc, Nicolas Claude Fabri de, 25–28, 189, 195, 224, 355–390
  affairs of state, documentation of, 382–383, 387–390
  antiquarian interests, 362–363, 365, 378
  archive, 356–358, 360, 362, 382
  astronomical investigations, 373–380
  biography, 355–356
  Charlemagne, interest in, 366–368
  Church affairs, documentation of, 383–387
  cultural studies, 380–381
  drawings, 361–363, 365–366, 375–376, 377, 389
  as naturalist, 368–369
  optical investigations, 369–373
Pelacani, Biagio, 301
Pelagianism, 94
Perfetti, Stefano, 155–156
Perizonius, Jacob, 41–42, 47–51, 63
Perlini, Gerolamo, 135
Persio, Antonio, 190, 192
Peter Martyr (Pietro Martire d'Anghiera), 256
Petrarch (Petrarca), Francesco, 20–21, 73 n. 110, 75–76, 83, 87, 93–94, 96, 97
Petroni, Alessandro, 23–24, 134, 325, 328–329, 331, 334–335, 343–344
Phaedo (Plato), 8, 110
philology, 7, 15–16, 17, 18–19, 42, 85, 186, 196, 203, 228, 231
Philo of Alexandria, 292 n. 29
Philoponus, 108
Philosophical Transactions (Royal Society), 135
Philosophie de l'histoire (Voltaire), 230
philosophy, history of, 223–224
Phornutus, 183
physico-theology, 13, 95–97
Physics (Aristotle), 148, 156
Physiologus (Theobald), 87

Piccolomini, Aeneas Sylvius, 76, 303, 320 n. 41
Piccolomini, Francesco, 206 n. 28
Piedmont, 300
Pierleone, da Spoleto, 300
Pignoria, Lorenzo, 187, 189–190, 192, 194–195, 196, 197, 198, 205 n. 28
Pinax (Bauhin), 165–166
Pinelli, Gianvincenzo, 189
Pinon, Laurent, 18, 164
Pisanello (Antonio Pisano), 65
Pius V, pope, 332
Plantin, Christophe, 81–82
plants
  historia, 80–82, 109, 110
  medicinal, 79–81, 90
  natural theology, 89–90
Plato, 8, 86, 159
Platonism, 167, 187, 189, 192, 195, 272, 297
Platter, Felix, 272
Pliny the Elder (Gaius Plinius Secundus), 11, 52, 75, 76–78, 79, 80, 109–110, 115–116, 147, 154, 182, 201, 221, 248, 254, 263, 332, 333, 336
Pluche, Noël Antoine, 98
Plutarch, 189, 276, 333
Poetics (Aristotle), 53–54
Polcastro, Sigismondo, 310
Poliziano, Angelo, 10–11, 77, 263
Polyanthea (Mirabelli), 271
Polybius, 43, 50, 57, 218
Polyhistor (Morhof), 214
Pomata, Gianna, 13–15, 16, 299, 356
Pomey, François Antoine, 184
Pomponazzi, Pietro, 148, 155–156
Ponce de Leon, Gonzalo, 87
Pontano, Giovanni, 54, 61, 221
Poovey, Mary, 284, 285
Porphyry, 149–151, 159, 195
Possevino, Antonio, 55
Posterior Analytics (Aristotle), 149
Practica (Velasco de Tarenta), 132

Practica canonica (Savonarola), 301, 302, 323 n. 95
Practica major (Savonarola), 301, 302, 310–311, 323 n. 95
Principles and Duties of Natural Religion (Wilkins), 98
Prior Analytics (Aristotle), 149
Proclus, 196
Properties of Things (Bartholomaeus Anglicus), 78
public health, 23, 329, 330–335, 346–347. *See also* medicine
Pufendorf, Samuel, 232
Pythagoras, 181

Q. Curtius in integrum restitutus (Perizonius), 48
Quadripartitum de quercu, 78
Querenghi, Antonio, 189
Quintilian (Marcus Fabius Quintilianus), 64, 163, 214
Quod animi mores (Galen), 154

Rabelais, François, 83
Ramus, Petrus, 272, 294 n. 55
Ranke, Leopold von, 215, 234
Rariorum plantarum historia (Clusius), 90
Ravisius Textor, Johannes, 87
Ray, John, 81, 95–96
Raymond of Sabunde, 94–95
Réaumur, René-Antoine Fourchauld de, 98
Reeds, Karen, 165
Reformation, 55, 161
regimina sanitatis. *See* medicine: health rules
Regiomontanus, Joannes, 24
Reguardati, Benedetto, 300
Reineck, Reiner, 53
religion. *See also* Counter-Reformation; ecclesiastical history; mythography; Reformation; theology
  anthropological interpretation, 192, 202
  comparative study, 184, 203

religion (cont.)
  history *(historia)* of, 15–16,
    182–183, 185–186, 195, 200,
    202–204, 203
  Judaism, 186, 195
  origin theories, 192, 194–197, 199
  sun worship, 184, 187, 196–198,
    200–203
Republic of Letters, 41, 47, 224
Reychersdorff, Georg, 338
Reyes Franco, Gaspar de los,
  146 n. 134
Rhaedus, Thomas, 161
rhetoric, 11–12, 41, 44–49, 51, 53,
  58–63, 211, 217, 278–280, 286,
  317
Rhetz, Cardinal de, 386
Rhodiginus, Coelius, 181
Richelieu, Armand-Jean du Plessis,
  cardinal, 383
Rickert, Heinrich, 3
Robin, Jean, 368
Robortello, Francesco, 53, 54, 221
Roger, Jacques, 1
Roman Academy, 327
Rome, 184–185, 189, 195, 198,
  320 n. 41, 325, 326–328,
  330–335, 339–340, 382
Rondelet, Guillaume, 83, 165, 254,
  372
Rostock, 161
Royal Society of London, 284, 287,
  356
Rüdin, Valeria, 272
Rudolph II, emperor, 281
Ruland, Martin, the Elder, 127, 133,
  134, 137, 146 n. 138

Sabellicus, Marcus Antonius, 87,
  277, 291 n. 6, 293 n. 39
Sabolini (acquaintance of Peiresc),
  373
Saccardini, Costantino, 127
Saliceto, Guglielmo da, 310
Sallustius, 198
Salviani, Ippolito, 18, 241, 261–263,
  326

Sambucus, Joannes, 53
Santasofia, Galeazzo, 301
Santasofia, Marsilio di, 310
Sardanapalus, 279
Sardi, Alessandro, 53, 62
Sarpi, Paolo, 356
Saturnalia (Macrobius), 197
Saulvat, Claude, 379
Savonarola, Michele, 22–23, 122,
  297–317
  biography, 299–301, 312
  civic encomiums, 303–304
  medical writings, 301–302,
    308–310, 316–317, 338 (*see also
    works by title*)
  political-historical encomiums,
    304–307
Savoy, 123
Scala, Bartolomeo, 54
Scaliger, Julius Caesar, 155, 160,
  168 n. 2, 182
Schaffer, Simon, 284
Schedel, Hartmann, 229
Schegk, Jakob, 155, 159
Schenck, Johann, von Grafenberg,
  133–134, 146 n. 138, 344
Schikard, Wilhelm, 379
Schlegel, Friedrich, 232
Schlözer, August Ludwig von, 230
Schneewind, J. B., 96
Scholasticism, 4, 9–10, 106,
  107–108, 110, 111, 113, 122, 128,
  131, 147, 214, 219, 297
Scholzius, Laurentius, 143 n. 94
Schonerus, Joannes, 24
Schrader, Justus, 121–122
Scienza nuova (Vico), 231, 232
Scripture. *See* theology
Secretum secretorum (pseudo-
  Aristotle), 301
Seifert, Arno, 4–5, 31, 107, 108, 113,
  213, 358
Selden, John, 196, 197, 199, 363,
  388
Selvatico, Giovan Battista, 106, 135
Settala, Luigi, 119
Sforza, Francesco, 320 n. 48

Sforza (cardinal), 332
Sforza court, 300
Shapin, Steven, 284
Shapiro, Barbara, 6, 284
Sidereus nuncius (Galilei), 82, 374–375
Siena, 328
Silva de varia lección (Mexia), 269
Simiand, François, 234
simples, medical. *See* medicine
Simplicius of Cilicia, 108
Siraisi, Nancy, 22, 125, 162, 166
Sixtus of Siena, 55
Sixtus V, pope, 328
skepticism, 94
Sleidan, Johann, 229
Smith, Adam, 230
Socrates, 8
Spectacle de la nature (Pluche), 98
specula, 86
Speculum historiale (Vincent of Beauvais), 214
Speculum phisionomie (Savonarola), 302
Spelman, Henry, 387
Speroni, Sperone, 54, 109
Spini, Giorgio, 70 n. 50
Spinoza, Baruch, 49, 96
Spon, Jacob, 33 n. 20
Stewart, Dugald, 230
Stirpium adversaria nova (Obel), 82
Stoicism, 94–96, 197, 200
Strozzi, Tito Vespasiano, 65
Subfiguratio empirica (Galen), 111
Suetonius, 23, 332
Sumario (Fernández de Oviedo), 80
Summa curationis (Saliceto), 310
Summa de pulsibus (Savonarola), 301
Sylva sylvarum (Bacon), 285, 289
Symbolorum et emblematum . . . centuria (Camerarius), 193
systems, historical, 224, 229–233

"Tabula Bembina," 187
Tacitus, Cornelius, 23, 43, 63, 332, 333
Tadino, Alessandro, 119

Tanner, Adam, 203
Taurellus, Nicolaus, 158, 159, 171 n. 30
Telesio, Bernardino, 190, 192
Tertullian (Quintus Septimius Florens Tertullianus), 198
Tevere, del (Bacci), 331, 333, 335–336
Textor, Benedictus, 165
Theatrum Galeni (Mondella), 124
Theatrum humanae vitae (Zwinger), 19–20, 21, 212, 248, 269, 270–276, 278, 280–283, 285–289
Theobald, bishop, 87
Theologia naturalis (Raymond), 94, 95
theology, 82, 84–86, 160–161, 168, 185, 196–197, 201, 213–214, 314. *See also* natural theology; physico-theology
Theophrastus, 75, 77, 80, 81, 109, 110, 148, 155, 168 n. 2, 221, 244, 245, 326
Thomas Aquinas, 107, 160
Thomasius, Christian, 224
Thomasius, Jakob, 225
Thomas of Cantimpré, 78
Thucydides, 11, 46, 52, 62
Tiberius, emperor, 332
Tillotson, John, 96
Timpler, Clemens, 161
Toland, John, 95–97
Tomeo, Niccolò Leonico, 155–156, 157, 159
Tommasini, Giacomo Filippo, 187
Topics (Aristotle), 149, 151, 159, 163
Topsell, Edward, 92–93
Tossignano, Pietro da, 310
Traité des animauls (J. Bauhin), 260
Troeltsch, Ernst, 223
Trogus, Pompeius, 60
Tunis, 381
Turgot, Anne-Robert-Jacques, 231
Turin, 129

Ugolino da Montecatini, 300, 338
Urban VIII, pope, 201

Valerius Maximus, 87, 269, 277, 279, 280, 285, 286, 291 n. 6
Valla, Gianpietro, 83
Valla, Giorgio, 79, 83, 105, 108, 109, 115
Valla, Lorenzo, 42, 53–53, 58–59, 110, 217
Vallavez, Palamède de Fabri, sieur de, 366, 382–383, 388, 390, 393 n. 61
Valleriola, François, 129–131, 134, 146 n. 138
Vallés, Francisco, 130
Varro, Marcus Terentius, 221
Varthema, Ludovico, 338
Velasco de Tarenta, 132, 321 n. 68
Vergerio, Pier Paolo, 76
Vergil, Polydore, 77–78, 82, 214, 217, 232, 339
Vesalius, Andreas, 17, 116, 117, 179 n. 106
Vico, Giambattista, 16, 230, 231, 233
Vienna Medical College, 132
Vincent of Beauvais, 9, 10, 214
Viperano, Giovanni Antonio, 61, 138 n. 21, 221
Virgil (Publius Vergilius Maro), 332
Vita Peireskii (Gassendi), 380
Vittori, Angelo, 23–24, 328–329, 343–347
Vittorino da Feltre, 300
Vives, Juan Luis, 232
Voltaire (François-Marie Arouet), 230
Vossius, Dionysius, 197
Vossius, Gerhard Johannes, 50–51, 58, 62, 181–183, 188, 197, 201–203, 221–222

Walker, D. P., 203
Weiditz, Hans, 80
Welser, Markus, 187, 189
Werdmüller, Otto, 164
Werner, Georg, 338
Wheare, Degory, 43, 56
Wilkins, John, 96, 98
Willughby, Francis, 82
Windelband, Wilhelm, 3

Winghe, Jérôme, 368
Wisdom of God (Ray), 95, 96
Wittenberg, 53
Wolf, Johannes, 51, 217, 219
Wolffhart, Conrad, 272
World in a Nutshell (Colmar), 229
Wotton, Edward, 171 n. 30

Xavier, Francis, 345
Xenophon, 167

Zabarella, Jacopo, 110–111, 158, 161, 165, 206 n. 28, 328
Zacchia, Paolo, 354 n. 97
Zacutus Lusitanus, Abraham, 106, 135
Zerbi, Gabriele, 114
zoology, 13, 14–15, 26, 182, 215, 243–245, 250–251, 259–261, 263–264. *See also* animals, *historia* of; bestiaries, medieval
  Aristotelian commentaries, 154–156, 163–164
  dissection, 370, 371–373
  monsters, 254, 256, 372
  taxonomy, 151–154, 157–166
Zurich, 241
Zwinger, Jakob, 294 n. 55
Zwinger, Theodor, 20–21, 22, 133, 145 n. 132, 185, 212, 248, 269–289
  biography, 272–273
  correspondence, 291 n. 14
  and "factual" reasoning, 283–289
  publications, 272, 289–290